Applied Industrial Catalysis

Applied Industrial Catalysis

Editor:

Juan Carlos Serrano Ruiz, Ph.D.

ArclerPress

www.arclerpress.com

Applied Industrial Catalysis

Editor: Juan Carlos Serrano Ruiz, Ph.D.

708 3rd Avenue, 6th Floor
New York
NY 10017
United States of America
www.arclerpress.com

ISBN: 978-1-68094-429-7

Library of Congress Control Number: 2016951139

About the Editor

Juan Carlos Serrano Ruiz, Ph.D.

Dr. Juan Carlos Serrano-Ruiz, graduated in Chemistry (University of Granada, Spain, 2001) and PhD in Catalysis and Materials Science (University of Alicante, Spain, 2006). In 2007 he moved to Madrid to for a post-doc position at the Spanish National Council of Research (CSIC) and in 2008 he was awarded a 2 years Fulbright-MEC fellowship to conduct studies on catalytic biomass conversion to fuels and chemicals in Professor James Dumesic's research group at the University of Wisconsin-Madison (USA). Dr. Serrano-Ruiz is co-author of about 60 papers in SCI journals including one in Science (considered one of the 30 biggest chemical breakthroughs of the year according to Chemistry World), 5 book chapters, and a good number of high-impact reviews (Energy and Environmental Science, Chemical Society Reviews, Annual Reviews) on catalytic biomass conversion. Dr. Juan Carlos Serrano has an h index of 25 and accumulates nearly 3000 citations (90 % of them within the last 5 years). Since 2012, he works as a Senior Researcher at Abengoa Research, the R&D branch of the multinational company Abengoa developing technologies for advanced biofuels production via catalytic conversion of biomass, among other projects.

Contents

List of Contributors

Morris D. Argyle
Chemical Engineering Department, Brigham Young University, Provo, UT 84602, USA

Calvin H. Bartholomew
Chemical Engineering Department, Brigham Young University, Provo, UT 84602, USA

Erling Rytter
Department of Chemical Engineering, Norwegian University of Science and Technology (NTNU), N-7491 Trondheim, Norway
SINTEF Materials and Chemistry, N-7465 Trondheim, Norway

Anders Holmen
Department of Chemical Engineering, Norwegian University of Science and Technology (NTNU), N-7491 Trondheim, Norway

Sylvain Keav
Empa, Swiss Federal Laboratories for Materials Science and Technology, Laboratory for Solid State Chemistry and Catalysis, Ueberlandstrasse 129, Dübendorf CH-8600, Switzerland

Santhosh Kumar Matam
Empa, Swiss Federal Laboratories for Materials Science and Technology, Laboratory for Solid State Chemistry and Catalysis, Ueberlandstrasse 129, Dübendorf CH-8600, Switzerland

Davide Ferri
Paul Scherrer Institut, Villigen CH-5232, Switzerland

Anke Weidenkaff
Empa, Swiss Federal Laboratories for Materials Science and Technology, Laboratory for Solid State Chemistry and Catalysis, Ueberlandstrasse 129, Dübendorf CH-8600, Switzerland
Materials Chemistry, Institute for Materials Science, University of Stuttgart, Heisenbergstrasse 3, Stuttgart DE-70569, Germany

Arturo Martínez-Arias
Institute of Catalysis and Petroleumchemistry, National Council of Scientific Research (CSIC), C/Marie Curie 2, Campus Cantoblanco, 28049 Madrid, Spain

Daniel Gamarra
Institute of Catalysis and Petroleumchemistry, National Council of Scientific Research (CSIC), C/Marie Curie 2, Campus Cantoblanco, 28049 Madrid, Spain

Ana B. Hungría
Department of Materials Science, Metallurgical Engineering and Inorganic Chemistry, Faculty of Sciences, University of Cádiz, 11510 Puerto Real, Cádiz, Spain

Marcos Fernández-García
Institute of Catalysis and Petroleumchemistry, National Council of Scientific Research (CSIC), C/Marie Curie 2, Campus Cantoblanco, 28049 Madrid, Spain

Guillermo Munuera
Department of Inorganic Chemistry, University of Sevilla, 41092 Sevilla, Spain

Aitor Hornés
Institute of Catalysis and Petroleumchemistry, National Council of Scientific Research (CSIC), C/Marie Curie 2, Campus Cantoblanco, 28049 Madrid, Spain

Parthasarathi Bera
Institute of Catalysis and Petroleumchemistry, National Council of Scientific Research (CSIC), C/Marie Curie 2, Campus Cantoblanco, 28049 Madrid, Spain

José C. Conesa
Institute of Catalysis and Petroleumchemistry, National Council of Scientific Research (CSIC), C/Marie Curie 2, Campus Cantoblanco, 28049 Madrid, Spain

Antonio López Cámara
Institute of Catalysis and Petroleumchemistry, National Council of Scientific Research (CSIC), C/Marie Curie 2, Campus Cantoblanco, 28049 Madrid, Spain

Georgiy B. Shul'pin
Department of Kinetics and Catalysis, Semenov Institute of Chemical Physics, Russian Academy of Sciences, ulitsa Kosygina, dom 4, Moscow 119991, Russia; Tel.: +7-495-939-7317
Chair of Chemistry and Physics, Plekhanov Russian University of Economics, Stremyannyi pereulok, dom 36, Moscow 117997, Russia

P. Schacht
Instituto Mexicano del Petróleo, México, D.F., México

L. Díaz-García
Instituto Mexicano del Petróleo, México, D.F., México

J. Aguilar
Universidad Autónoma Metropolitana,0División de Ciencias Básicas e Ingeniería, México, D.F., México

S. Ramírez
Instituto Mexicano del Petróleo, México, D.F., México

P. Pérez-Romo
Instituto Mexicano del Petróleo, México, D.F., México

M. Espinosa
Instituto Mexicano del Petróleo, México, D.F., México

Hongjing Wu
Department of Applied Physics, Northwestern Polytechnical University (NPU), 127 Youyi Xilu, 710072 Xi'an, China
Institute for The Study of Nanostructured Materials (ISMN)-CNR, Via Ugo La Malfa 153, 90146 Palermo, Italy

Valeria La Parola
Institute for The Study of Nanostructured Materials (ISMN)-CNR, Via Ugo La Malfa 153, 90146 Palermo, Italy

Giuseppe Pantaleo
Institute for The Study of Nanostructured Materials (ISMN)-CNR, Via Ugo La Malfa 153, 90146 Palermo, Italy

Fabrizio Puleo
Institute for The Study of Nanostructured Materials (ISMN)-CNR, Via Ugo La Malfa 153, 90146 Palermo, Italy
Department of Physics and Chemistry, University of Palermo, Viale delle Scienze Ed.17, 90128 Palermo, Italy

Anna M. Venezia
Institute for The Study of Nanostructured Materials (ISMN)-CNR, Via Ugo La Malfa 153, 90146 Palermo, Italy

Leonarda F. Liotta
Institute for The Study of Nanostructured Materials (ISMN)-CNR, Via Ugo La Malfa 153, 90146 Palermo, Italy

Hamed Gharehbaghi
Department of Chemical Engineering, University of Sistan and Baluchestan, P.O. Box 98164-161, Zahedan, Iran

Jafar Sadeghi
Department of Chemical Engineering, University of Sistan and Baluchestan, P.O. Box 98164-161, Zahedan, Iran

Andreas Sundermann
hte GmbH, Kurpfalzring 104, 69123 Heidelberg, Germany

Olga Gerlach
hte GmbH, Kurpfalzring 104, 69123 Heidelberg, Germany

Toshiaki Hanaoka
Biomass Treatment Group, Research Institute for Sustainable Chemistry, Department of Materials and Chemistry, National Institute of Advanced Industrial Science and Technology (AIST), 3-11-32 Kagamiyama, Higashihiroshima, Hiroshima 739-0046, Japan

Tomohisa Miyazawa
Planning Headquarters General Planning Office, National Institute of Advanced Industrial Science and Technology (AIST), 1-3-1 Kasumi-gaseki, Chiyoda-ku, Tokyo 100-8921, Japan

Katsuya Shimura
Advanced Heterogeneous Catalysis Team, Interdisciplinary Research Center for Catalytic Chemistry, Department of Materials and Chemistry, National Institute of Advanced Industrial Science and Technology (AIST), 1-1-1 Umezono, Tsukuba, Ibaraki 305-8568, Japan

Satoshi Hirata
National Institute of Advanced Industrial Science and Technology (AIST), Central 1, 1-1-1 Umezono, Tsukuba, Ibaraki 305-8560, Japan

Gcina Mamba
Nanotechnology and Water Sustainability Research Unit, College of Science, Engineering and Technology, University of South Africa, Florida 1709, Johannesburg, South Africa

Ajay Mishra
Nanotechnology and Water Sustainability Research Unit, College of Science, Engineering and Technology, University of South Africa, Florida 1709, Johannesburg, South Africa

Rubén de Regil
Unidad de Biotecnología Industrial, CIATEJ, A.C. Av. Normalistas 800, Col. Colinas de la Normal, Guadalajara, Jal, C.P. 44270, Mexico

Georgina Sandoval
Unidad de Biotecnología Industrial, CIATEJ, A.C. Av. Normalistas 800, Col. Colinas de la Normal, Guadalajara, Jal, C.P. 44270, Mexico

Preface

Catalysis is the science that allows facilitating the chemical processes by employing a substance (catalyst) that actively participates in the reaction mechanism while not being consumed during the process. The practical totality of the most relevant chemical processes (by volume of product) up to date involves the utilization of catalysts as a crucial part of the technology. Thus, some of the most important bulk chemicals (nitric, sulfuric, and terephtalic acids) are produced via catalytic oxidation while others such as ammonia involves catalytic hydrogenation processes. Bulk polymers such as polyethylene and polypropylene and many high added-value fine chemicals produced at lower scale are generated with the key participation of a catalyst. This book intends to provide an overview of some of the most relevant industrial catalytic technologies available today. Thus, important heterogeneous catalytic processes such as Fischer-Tropsch, three-way conversion of gaseous pollutants, and preferential oxidation of CO in H2 streams have dedicated chapters in this compendium. Deactivation of catalysts, a central issue in any catalytic technology, is adequately tackled by excellent reviews from relevant authors. The upgrading of petroleum to fuels and chemicals is also covered by including chapters on hydrocracking of heavy oils and Fischer-Tropsch-derived streams. Finally, enzymes, which can be considered as natural catalysts, are described in a review chapter as essential components involved in the production of pharmaceutical, cosmetic, food, and even bulk chemicals.

Editor
Juan Carlos Serrano Ruiz, Ph.D.

INTRODUCTION

The first time a catalyst was used in the industry was in 1746 by J. Roebuck in the manufacture of lead chamber sulfuric acid. Since then catalysts have been in use in a large portion of the chemical industry. In the start only pure components were used as catalysts, but after the year 1900 multicomponent catalysts were studied and are now commonly used in the industry.

In the chemical industry and industrial research, catalysis play an important role. Different catalysts are in constant development to fulfill economic, political and environmental demands. When using a catalyst it is possible to replace a polluting chemical reaction with a more environmentally friendly alternative. Today, and in the future, this can be vital for the chemical industry. In addition it's important for a company/researcher to pay attention to market development. If a company's catalyst is not continually improved, another company can make progress in research on that particular catalyst and gain market share. For a company, a new and improved catalyst can be a huge advantage for a competitive manufacturing cost. It's extremely expensive for a company to shut down the plant because of an error in the catalyst, so the correct selection of a catalyst or a new improvement can be key to industrial success.

To achieve the best understanding and development of a catalyst it is important that different special fields work together. These fields can be: organic chemistry, analytic chemistry, inorganic chemistry, chemical engineers and surface chemistry. The economics must also be taken into account. One of the issues that must be considered is if the company should use money on doing the catalyst research themselves or buy the technology from someone else. As the analytical tools are becoming more advanced, the catalysts used in the industry are improving. One example of an improvement can be to develop a catalyst with a longer lifetime than the previous version. Some of the advantages an improved catalyst gives, that affects people's lives, are: cheaper and more effective fuel, new drugs and medications and new polymers.

Some of the large chemical processes that use catalysis today are the production of methanol and ammonia. Both methanol and ammonia synthesis take advantage of the water-gas shift reaction and heterogeneous catalysis, while other chemical industries use homogenous catalysis. If the catalyst exists in the same phase as the reactants it is said to be homogenous; otherwise it is heterogeneous.

The water gas shift reaction was first used industrially at the beginning of the 20th century. Today the WGS reaction is used primarily to produce hydrogen that can be used for further production of methanol and ammonia.[3]

WGS reaction:

$$CO + H_2O \leftrightarrow H_2 + CO_2 \tag{1}$$

The reaction refers to carbon monoxide (CO) that reacts with water (H_2O) to form carbon dioxide (CO_2) and hydrogen (H_2). The reaction is exothermic with ΔH= -41.1 kJ/mol and have an adiabatic temperature rise of 8–10 °C per percent CO converted to CO_2 and H_2.

The most common catalysts used in the water-gas shift reaction are the high temperature shift (HTS) catalyst and the low temperature shift (LTS) catalyst. The HTS catalyst consists of iron oxide stabilized by chromium oxide, while the LTS catalyst is based on copper. The main purpose of the LTS catalyst is to reduce CO content in the reformate which is especially important in the ammonia production for high yield of H_2. Both catalysts are necessary for thermal stability, since using the LTS reactor alone increases exit-stream temperatures to unacceptable levels.

The equilibrium constant for the reaction is given as

$$K_p=(p_{H_2} \times p_{CO_2})/(p_{CO} \times p_{H_2O}) \tag{2}$$

$$K_p=e^{((4577.8K/T-4.22))} \tag{3}$$

Low temperatures will therefore shift the reaction to the right, and more products will be produced. The equilibrium constant is extremely dependent on the reaction temperature, for example is the Kp equal to 228 at 200 °C, but only 11.8 at 400 °C. The WGS reaction can be performed both homogenously and heterogeneously, but only the heterogeneously way is used commercially.

High temperature shift (HTS) catalyst

The first step in the WGS reaction is the high temperature shift which is carried out at temperatures between 320 °C and 450 °C. As mentioned before, the catalyst is a composition of iron-oxide, Fe_2O_3(90-95%), and chromium oxides Cr_2O_3 (5-10%) which have an ideal activity and selectivity at these temperatures. When preparing this catalyst, one of the most important step is washing to remove sulfate that can turn into hydrogen sulfide and poison the LTS catalyst later in the process. Chromium is added to the catalyst to stabilize the catalyst activity over time and to delay sintering of iron oxide. Sintering will decrease the active catalyst area, so by decreasing the sintering rate the lifetime of the catalyst will be extended. The catalyst is usually used in pellets form, and the size play an important role. Large pellets will be strong, but the reaction rate will be limited.

In the end, the dominate phase in the catalyst consist of Cr_3+ in α-Fe_2O_3 but the catalyst is still not active. To be active α-Fe_2O_3 must be reduced to Fe and CrO_3 must be reduced to Cr in presence of H_2. This usually happens in the reactor start-up phase and because the reduction reactions are exothermic the reduction should happen under controlled circumstances. The lifetime of the iron-chrome catalyst is approximately 3–5 years, depending on how the catalyst is handled.

Even though the mechanism for the HTS catalyst has been done a lot of research on, there is no final agreement on the kinetics/mechanism. Research has narrowed it down to two possible mechanisms: a regenerative redox mechanism and an adsorptive (associative) mechanism.

The redox mechanism is given below:

First a CO molecule reduces an O molecule, yielding CO2 and a vacant surface center:

$$CO+(O) \rightarrow CO_2 + (^*) \tag{4}$$

The vacant side is then reoxidized by water, and the oxide center is regenerated:

$$H_2O+(^*) \rightarrow H_2 + (O) \tag{5}$$

The adsorptive mechanism assumes that format species is produced when an adsorbed CO molecule reacts with a surface hydroxyl group:

$$H_2O \rightarrow OH(ads) + H(ads) \tag{6}$$

$$CO(ads) + OH(ads) \rightarrow COOH\ (ads) \tag{7}$$

The format decomposes then in the presence of steam:

$$COOH(ads) \rightarrow CO_2 + H(ads) \tag{8}$$

$$2H(ads) \rightarrow H_2 \tag{9}$$

Low temperature shift (LTS) catalyst

The low temperature process is the second stage in the process, and is designed to take advantage of higher hydrogen equilibrium at low temperatures. The reaction is carried out between 200 °C and 250 °C, and the most commonly used catalyst is based on copper. While the HTS reactor used an iron-chrome based catalyst, the copper-catalyst is more active at lower temperatures thereby yielding a lower equilibrium concentration of CO and a higher equilibrium concentration of H_2. The disadvantage with a copper catalysts is that it is very sensitive when it comes to sulfide poisoning, a future use of for example a cobalt- molybdenum catalyst could solve this problem. The catalyst mainly used in the industry today is a copper-zinc-alumina ($Cu/ZnO/Al_2O_3$) based catalyst.

Catalysts are substances that speed up reactions by providing an alternative pathway for the breaking and making of bonds. Key to this alternative pathway is a lower activation energy than that required for the unanalyzed reaction. Catalysts are often specific for one particular reaction and this is particularly so for enzymes which catalyze biological reactions, for example in the fermentation of carbohydrates to produce biofuels.

Much fundamental and applied research is done by industrial companies and university research laboratories to find out how catalysts work and to improve their effectiveness. If catalytic activity can be improved, it may be possible to lower the temperature and/or the pressure at which the process operates and thus save fuel which is one of the major costs in a large-scale chemical process. Further, it may be possible to reduce the amount of reactants that are wasted forming unwanted by-products.

If the catalyst is in the same phase as the reactants, it is referred to as a homogeneous catalyst. A heterogeneous catalyst on the other hand is in a different phase to the reactants and products, and is often favored in industry, being easily separated from the products, although it is often less specific and allows side reactions to occur.

Heterogeneous catalysis

The most common examples of heterogeneous catalysis in industry involve the reactions of gases being passed over the surface of a solid, often a metal, a metal oxide or a zeolite

Table 1: Examples of industrial processes using heterogeneous catalysis.

Process	Catalyst	Equation
Making ammonia	Iron	$N_2(g) + 3H_2(g) \rightleftharpoons 2NH_3(g)$
Making synthesis gas (carbon monoxide and hydrogen)	Nickel	$CH_4(g) + H_2O(g) \rightleftharpoons CO(g) + 3H_2(g)$
Catalytic cracking of gas oil	Zeolite	Produces: a gas (e.g. ethene, propene) a liquid (e.g.petrol) a residue (e.g. fuel oil)
Reforming of naphtha	Platinum and rhenium on alumina	$CH_3CH_2CH_2CH_2CH_2CH_3(g) \rightarrow$ 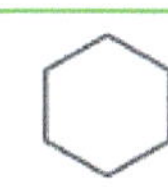 $(g) + H_2(g)$
Making epoxyethane	Silver on alumina	$C_2H_4(g) + ½O_2(g) \rightarrow$ 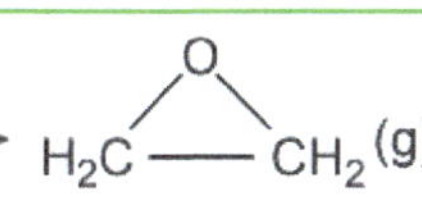
Making sulfuric acid	Vanadium(V) oxide on silica	$SO_2(g) + ½O_2(g) \rightarrow SO_3(g)$
Making nitric acid	Platinum and rhodium	$4NH_3(g) + 5O_2(g) \rightarrow 4NO(g) + 6H_2O(g)$

The gas molecules interact with atoms or ions on the surface of the solid. The first process usually involves the formation of very weak intermolecular bonds, a process known as physisorption, followed by chemical bonds being formed, a process known as chemisorption.

Physisorption can be likened to a physical process such as liquefaction. Indeed the enthalpy changes that occur in physisorption are ca -20 to -50 kJ mol^{-1}, similar to those of enthalpy changes when a gas condenses to form a liquid. The enthalpies of chemisorption are similar to the values found for enthalpies of reaction. They have a very wide range, just like the range for non-catalytic chemical reactions.

An example of the stepwise processes that occur in heterogeneous catalysis is the oxidation of carbon monoxide to carbon dioxide over palladium. This is a very important process in everyday life. Motor vehicles are fitted with catalytic converters. These consist of a metal casing in which there are two metals, palladium and rhodium, dispersed very finely on the surface of a ceramic support that resembles a honeycomb of holes. The converter is placed between the engine and the outlet of the exhaust pipe.

The exhaust gases contain carbon monoxide and unburned hydrocarbons that react with the excess oxygen to form carbon dioxide and water vapor, the reaction being catalyzed principally by the palladium:

$$2CO(g) + O_2(g) \rightarrow 2CO_2(g)$$

$$C_xH_y(g) + (x+y/4)O_2(g) \rightarrow xCO_2(g) + y/2 H_2O(g)$$

The exhaust gases also contain nitrogen (II) oxide (nitric oxide, NO), and this is removed by reactions catalyzed principally by the rhodium:

$$2CO(g) + 2NO(g) \rightarrow 2CO_2(g) + N_2(g)$$

The accepted mechanism for the oxidation of carbon monoxide to carbon dioxide involves the chemisorption of both carbon monoxide molecules and oxygen molecules on the surface of the metals. The adsorbed oxygen molecules dissociate into separate atoms of oxygen. Each of these oxygen atoms can combine with a chemisorbed carbon monoxide molecule to form a carbon dioxide molecule. The carbon dioxide molecules are then desorbed from the surface of the catalyst.

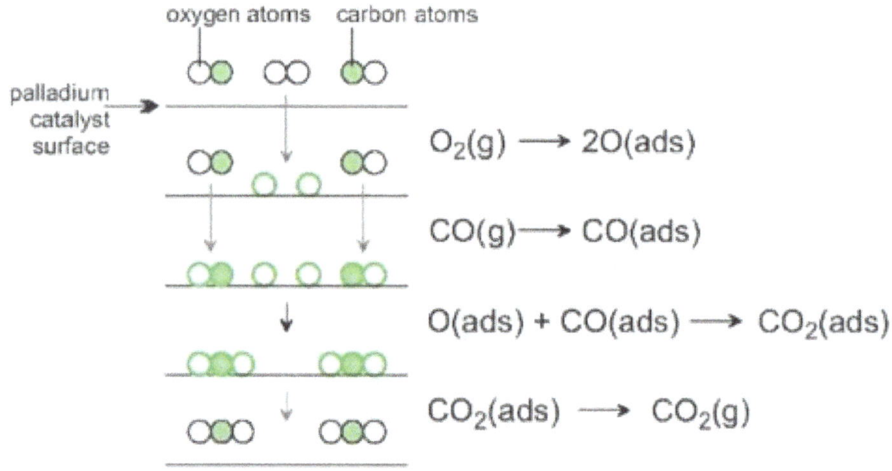

Figure 1 A mechanism for the oxidation of carbon monoxide.

Each of these steps has a much lower activation energy than the homogeneous reaction between the carbon monoxide and oxygen.

The removal of carbon monoxide, unburned hydrocarbons and nitrogen (II) oxide from car and lorry exhausts is very important for this mixture leads to photochemical smogs which aggravate respiratory diseases such as asthma. Platinum, palladium and rhodium are all used but are very expensive metals and indeed each is more expensive than gold. Recently, much work has been devoted to making catalysts with very tiny particles of the metals, an example of the advances being made by nanotechnology. It is not simply the ability of the heterogeneous catalyst's surface to interact with the reactant molecules, chemisorption, that makes it a good catalyst. If the adsorption is too exothermic, i.e. the enthalpy of chemisorption is too high, further reaction is likely to be too endothermic to proceed. The enthalpy of chemisorption has to be sufficiently exothermic for chemisorption to take place, but not so high that it does not allow further reaction to proceed. For example, in the oxidation of carbon monoxide, molybdenum might at first sight be favored as a choice, as oxygen is readily chemisorbed by the metal. However, the resulting oxygen atoms do not react further as they are too strongly adsorbed on the surface. Platinum and palladium, on the other hand, have lower enthalpies of chemisorption with oxygen, and the oxygen atoms can then react further with adsorbed carbon monoxide. Another point to consider in choosing a catalyst is that the product must not be able to adsorb too strongly to its surface. Carbon dioxide does not adsorb strongly on platinum and palladium and so it is rapidly desorbed into the gas phase.

$$N_2(g) + 3H_2(g) \rightleftharpoons 2NH_3(g)$$

Ertl obtained crucial evidence on how iron catalysis the dissociation of the nitrogen molecules and hydrogen molecules leading to the formation of ammonia.

$$H_2(g) \longrightarrow 2H(ads)$$

$$N_2(g) \longrightarrow N_2(ads)$$

$$N_2(ads) \longrightarrow 2N(ads)$$

$$N(ads) \xrightarrow{+\,H(ads)} NH(ads) \xrightarrow{+\,H(ads)} NH_2(ads) \xrightarrow{+\,H(ads)} NH_3(ads)$$

$$NH_3(ads) \longrightarrow NH_3(g)$$

Figure 2: A mechanism for the catalytic synthesis of ammonia.

Figure 3 shows how the activation energy barriers are much lower than the estimated activation energy barrier (which would be at least 200 kJ mol1) for the unanalyzed synthesis of ammonia.

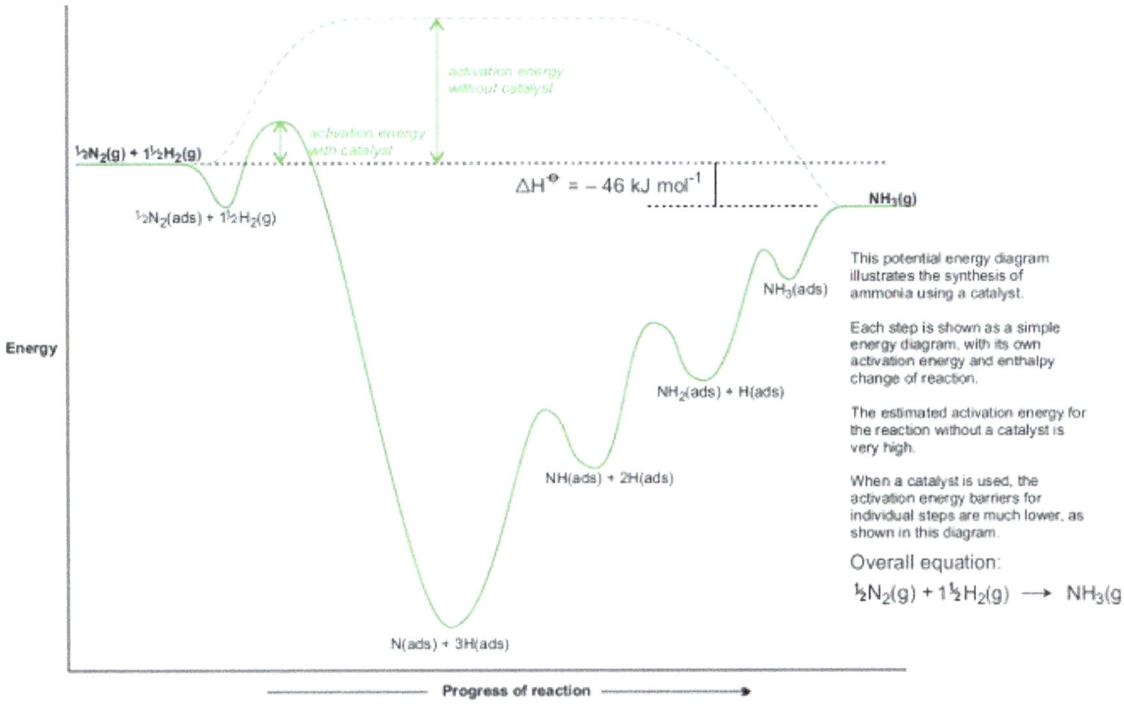

Figure 3: The activation energy barriers for the reactions occurring during the catalytic synthesis of ammonia.

General requirements for a heterogeneous catalyst

To be successful the catalyst must allow the reaction to proceed at a suitable rate under conditions that are economically desirable, at as low a temperature and pressure as possible. It must also be long lasting. Some reactions lead to undesirable side products. For example in the cracking of gas oil, carbon is formed which is deposited on the surface of the catalyst, a zeolite, and leads to a rapid deterioration of its effectiveness. Many catalysts are prone to poisoning which occurs when an impurity attaches itself to the surface of the catalyst and prevents adsorption of the reactants. Minute traces of such a substance can ruin the process, One example is sulfur dioxide, which poisons the surface of platinum and palladium. Thus all traces of sulfur compounds must be removed from the petrol used in cars fitted with catalytic converters.

Further, solid catalysts are much more effective if they are finely divided as this increases the surface area.

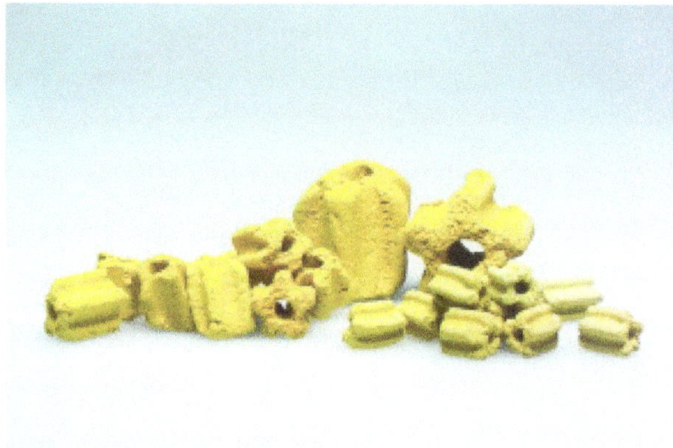

At high temperatures, the particles of a finely divided catalyst tend to fuse together and the powder may 'cake', a process known as sintering. This reduces the activity of the catalyst and steps must be taken to avoid this. One way is to add another substance, known as a promoter. When iron is used as the catalyst in the Haber process, aluminum oxide is added and acts as a barrier to the fusion of the metal particles. A second promoter is added, potassium oxide that appears to cause the nitrogen atoms to be chemisorbed more readily, thus accelerating the slowest step in the reaction scheme.

Figure 4: A platinum-rhodium gauze is used as a catalyst in the reaction between ammonia and methane to produce hydrogen cyanide, an intermediate in the production of methyl 2-methylpropenoate. The gauze operates at 1270 K and is thus glowing. The photograph was taken though a sight glass located on the reactor.

Aluminum oxide, silicon dioxide, aluminosilicates and zeolites

One of the most important reactions in which aluminium oxide, Al_2O_3, (often referred to as alumina) takes part in an industrial reaction is in platforming, in which naphtha is reformed over aluminina impregnated with platinum or rhenium. Both the oxide and the metals have catalytic roles to play, an example of bifunctional catalysis. There are hydroxyl groups on the surface of alumina which are, in effect, sites which are negatively charged to which a hydrogen ion is attached that can act as an acid catalyst.

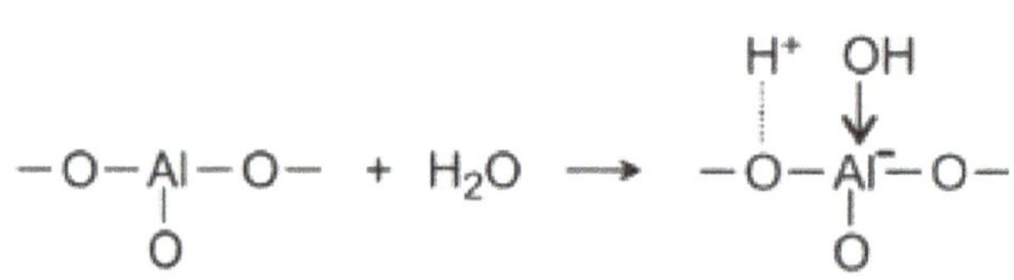

Silicon dioxide (silica) is another acidic oxide used in industry. It becomes particularly active if it has been coated with an acid (such as phosphoric acid), thereby increasing the number of active acidic sites. For example, the manufacture of ethanol is achieved by the hydration of ethane using silica, coated with phosphoric acid:

$$C_2H_4(g) + H_2O(g) \longrightarrow C_2H_5OH(g)$$

The mechanism involves the formation of a carbocation

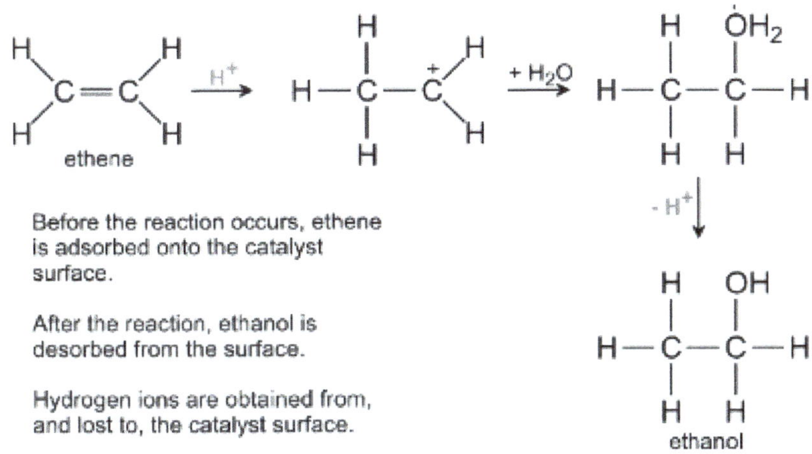

Figure 5 A mechanism for the hydration of ethane to ethanol.

Aluminosilicates are also used as catalysts when an acid site is required. These are made from silicon dioxide (silica) and aluminium oxide. They contain silicate ions, SiO44- that have a tetrahedral structure which can be linked together in several ways. When some of the Si atoms are replaced with Al atoms, the result is an aluminosilicate. Hydrogen ions are again associated with the aluminium atoms:

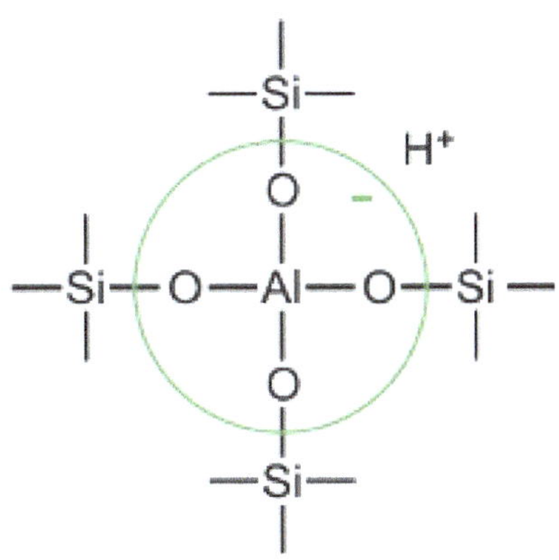

Zeolite catalysts

A particular class of aluminosilicates that has excited huge interest in recent years is the zeolites. There are many different zeolites because of the different ways in which the atoms can be arranged. Their structure of silicate and aluminate ions can have large vacant spaces in three dimensional structures that give room for cations such as sodium and calcium and molecules such as water. The spaces are interconnected and form long channels and pores which are of different sizes in different zeolites.

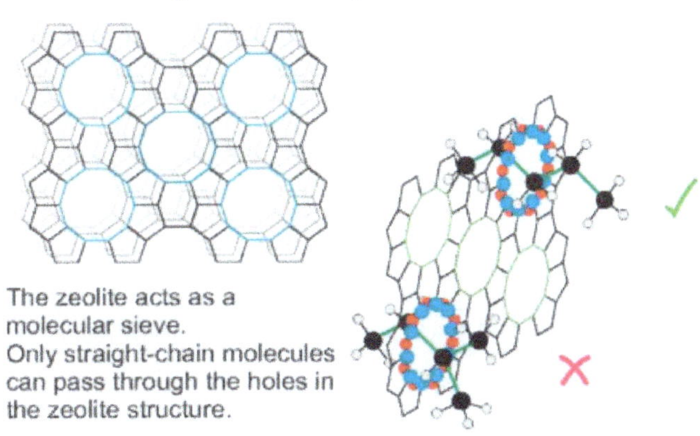

Figure 6 The structure of a zeolite.

A zeolite which is commonly used in many catalytic reactions is ZSM-5 which is prepared from sodium aluminate (a solution of aluminum oxide in aqueous sodium hydroxide) and a colloidal solution of silica, sodium hydroxide, sulfuric acid and tetrapropylammonium bromide.

It is, for example, a very effective catalyst for the conversion of methylbenzene (toluene) to the three dimethyl benzenes (xylenes). Alas, the mixture produced only contains about 25% 1,4-dimethylbenzene, (p-xylene) the isomer needed for the manufacture of the polyesters and the rest, 1,2- (o-xylene) and 1,3-dimethylbenzenes (m-xylene), is not wanted in such large quantities.

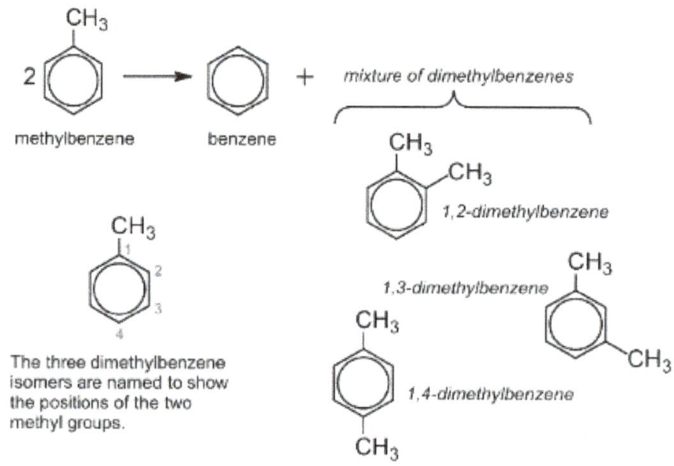

However, if the zeolite is washed with phosphoric acid and heated strongly, minute particles of phosphorus (V) oxide are deposited on the surface making the pores slightly smaller. This restricts the diffusion of the 1,2- and 1,3-isomers and they are held in the pores until they are converted into the 1,4-isomer and can escape (Figure). This remarkable selectivity enables the yield of the 1,4-isomer to be increased from 25% to 97%.

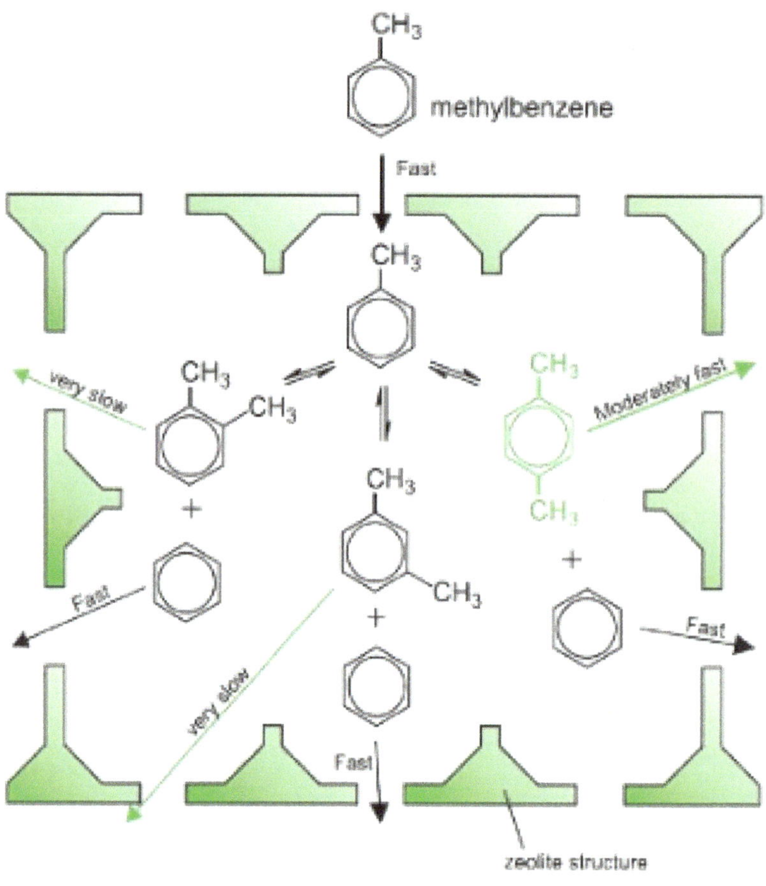

Figure 7: A zeolite acting on a molecular sieve and a catalyst during the formation of 1,4-dimethylbenzene from methylbenzene.

The ability of the zeolite to adsorb some molecules and to reject others gives it the ability to act as a molecular sieve. For example, in the manufacture of ethanol from ethane or from biomass, an aqueous solution of ethanol is produced, in which there is 4% water still present even after repeated distillations. Further purification of ethanol requires the use of a zeolite which absorbs the water preferentially.

CHAPTER 1

HETEROGENEOUS CATALYST

Morris D. Argyle and Calvin H. Bartholomew

Keywords: Heterogeneous catalysis; deactivation; regeneration

INTRODUCTION

Catalyst deactivation, the loss over time of catalytic activity and/or selectivity, is a problem of great and continuing concern in the practice of industrial catalytic processes. Costs to industry for catalyst replacement and process shutdown total billions of dollars per year. Time scales for catalyst deactivation vary considerably; for example, in the case of cracking catalysts, catalyst mortality may be on the order of seconds, while in ammonia synthesis the iron catalyst may last for 5–10 years. However, it is inevitable that all catalysts will decay.

Typically, the loss of activity in a well-controlled process occurs slowly. However, process upsets or poorly designed hardware can bring about catastrophic failure. For example, in steam reforming of methane or naphtha, great care must be taken to avoid reactor operation at excessively high temperatures or at steam-to-hydrocarbon ratios below a critical value. Indeed, these conditions can cause formation of large quantities of carbon filaments that plug catalyst pores and voids, pulverize catalyst pellets, and bring about process shutdown, all within a few hours.

While catalyst deactivation is inevitable for most processes, some of its immediate, drastic consequences may be avoided, postponed, or even reversed. Thus, deactivation issues (*i.e.*, extent, rate, and reactivation) greatly impact research, development, design, and operation of commercial processes. Accordingly, there is considerable motivation to understand and treat catalyst decay. Over the past three decades, the science of catalyst deactivation has been steadily developing, while literature addressing this topic has expanded considerably to include books [1,2,3,4], comprehensive reviews [5,6,7,8], proceedings of international symposia [9,10,11,12,13,14], topical journal issues (e.g., [15]), and more than 20,000 U.S. patents for the period of 1976–2013. (In a U.S. patent search conducted in November 2013 for the keywords catalyst and deactivation, catalyst and life, and catalyst and regeneration, 14,712, 62,945, and 22,520 patents were found respectively.) This area of research provides a critical understanding that is the foundation for modeling deactivation processes, designing stable catalysts, and optimizing processes to prevent or slow catalyst deactivation.

The purpose of this article is to provide the reader with a comprehensive overview of the scientific and practical aspects of catalyst deactivation with a focus on mechanisms of catalyst decay, prevention of deactivation, and regeneration of catalysts. Case studies of deactivation and regeneration of Co Fischer-Tropsch catalysts and of commercial catalysts for selective catalytic reduction of nitrogen oxides in stationary sources have been included.

MECHANISMS OF DEACTIVATION

There are many paths for heterogeneous catalyst decay. For example, a catalyst solid may be poisoned by any one of a dozen contaminants present in the feed. Its surface, pores, and voids may be fouled by carbon or coke produced by cracking/condensation reactions of hydrocarbon reactants, intermediates, and/or products. In the treatment of a power plant flue gas, the catalyst can be dusted or eroded by and/or plugged with fly ash. Catalytic converters used to reduce emissions from gasoline or diesel engines may be poisoned or fouled by fuel or lubricant additives and/or engine corrosion products. If the catalytic reaction is conducted at high temperatures, thermal

Citation: Morris D. Argyle and Calvin H. Bartholomew, Heterogeneous Catalyst, doi:10.3390/catal5010145.

degradation may occur in the form of active phase crystallite growth, collapse of the carrier (support) pore structure, and/or solid-state reactions of the active phase with the carrier or promoters. In addition, the presence of oxygen or chlorine in the feed gas can lead to formation of volatile oxides or chlorides of the active phase, followed by gas-phase transport from the reactor. Similarly, changes in the oxidation state of the active catalytic phase can be induced by the presence of reactive gases in the feed.

Thus, the mechanisms of solid catalyst deactivation are many; nevertheless, they can be grouped into six intrinsic mechanisms of catalyst decay: (1) poisoning, (2) fouling, (3) thermal degradation, (4) vapor compound formation and/or leaching accompanied by transport from the catalyst surface or particle, (5) vapor–solid and/or solid–solid reactions, and (6) attrition/crushing. As mechanisms 1, 4, and 5 are chemical in nature while 2 and 6 are mechanical, the causes of deactivation are basically threefold: chemical, mechanical, and thermal. Each of the six basic mechanisms is defined briefly in Table 1 and treated in some detail in the subsections that follow, with an emphasis on the first three. Mechanisms 4 and 5 are treated together, since 4 is a subset of 5.

Table 1: Mechanisms of catalyst deactivation.

Mechanism	Type	Brief definition/description
Poisoning	Chemical	Strong chemisorption of species on catalytic sites which block sites for catalytic reaction
Fouling	Mechanical	Physical deposition of species from fluid phase onto the catalytic surface and in catalyst pores
Thermal degradation and sintering	Thermal Thermal/chemical	Thermally induced loss of catalytic surface area, support area, and active phase-support reactions
Vapor formation	Chemical	Reaction of gas with catalyst phase to produce volatile compound
Vapor–solid and solid–solid reactions	Chemical	Reaction of vapor, support, or promoter with catalytic phase to produce inactive phase
Attrition/crushing	Mechanical	Loss of catalytic material due to abrasion; loss of internal surface area due to mechanical-induced crushing of the catalyst particle

Poisoning

Poisoning [3,16,17,18,19,20,21,22] is the strong chemisorption of reactants, products, or impurities on sites otherwise available for catalysis. Thus, poisoning has operational meaning; that is, whether a species acts as a poison depends upon its adsorption strength relative to the other species competing for catalytic sites. For example, oxygen can be a reactant in partial oxidation of ethylene to ethylene oxide on a silver catalyst and a poison in hydrogenation of ethylene on nickel. In addition to physically blocking of adsorption sites, adsorbed poisons may induce changes in the electronic or geometric structure of the surface [17,21]. Finally, poisoning may be reversible or irreversible. An example of reversible poisoning is the deactivation of acid sites in fluid catalytic cracking catalysts by nitrogen compounds in the feed. Although the effects can be severe, they are temporary and are generally eliminated within a few hours to days after the nitrogen source is removed from the feed. Similar effects have been observed for nitrogen compound (e.g., ammonia and cyanide) addition to the syngas of cobalt Fischer-Tropsch catalysts, although these surface species require weeks to months before the lost activity is regained [23]. However, most poisons are irreversibly chemisorbed to the catalytic surface sites, as is the case for sulfur on most metals, as discussed in detail below. Regardless of whether the poisoning is reversible or irreversible, the deactivation effects while the poison is adsorbed on the surface are the same.

Many poisons occur naturally in feed streams that are treated in catalytic processes. For example, crude oil contains sulfur and metals, such as vanadium and nickel, that act as catalyst poisons for many petroleum refinery processes, especially those that use precious metal catalysts, like catalytic reforming, and those that treat heavier hydrocarbon fractions in which the sulfur concentrates and metals are almost exclusively found, such as fluid catalytic cracking and residuum hydroprocessing. Coal contains numerous potential poisons, again including sulfur and others like arsenic, phosphorous, and selenium, often concentrated in the ash, that can poison selective catalytic reduction catalysts as discussed later inSection 4.3.3.1. As a final example, some poisons may be added purposefully, either to moderate the activity and/or to alter the selectivity of fresh catalysts, as discussed as the end of this section, or to improve the performance of a product that is later reprocessed catalytically. An example of this latter case is lubricating oils that contain additives like zinc and phosphorous to improve their lubricating properties and stability, which become poisons when the lubricants are reprocessed in a hydrotreater or a fluid catalytic cracking unit.

Mechanisms by which a poison may affect catalytic activity are multifold, as illustrated by a conceptual two-dimensional model of sulfur poisoning of ethylene hydrogenation on a metal surface shown in Figure 1. To begin with, a strongly adsorbed atom of sulfur physically blocks at least one three- or fourfold adsorption/reaction site (projecting into three dimensions) and three or four topside sites on the metal surface. Second, by virtue of its strong chemical bond, it electronically modifies its nearest neighbor metal atoms and possibly its next-nearest neighbor atoms, thereby modifying their abilities to adsorb and/or dissociate reactant molecules (in this case H_2 and ethylene molecules), although these effects do not extend beyond about 5 atomic units [21]. A third effect may be the restructuring of the surface by the strongly adsorbed poison, possibly causing dramatic changes in catalytic properties, especially for reactions sensitive to surface structure. In addition, the adsorbed poison blocks access of adsorbed reactants to each other (a fourth effect) and finally prevents or slows the surface diffusion of adsorbed reactants (effect number five).

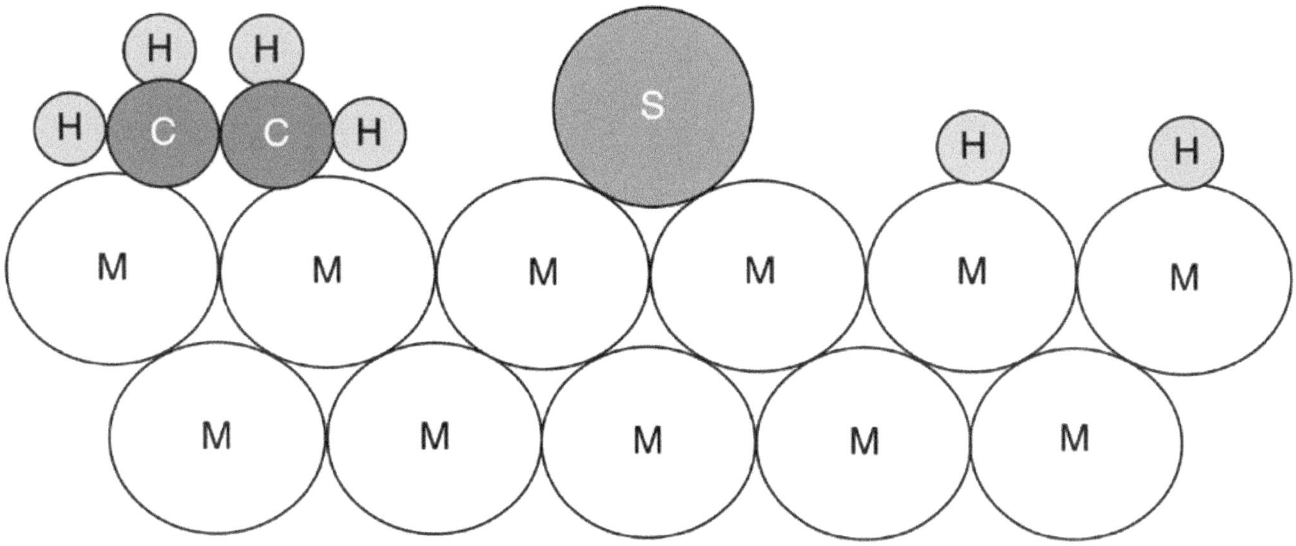

Figure 1: Conceptual model of poisoning by sulfur atoms of a metal surface during ethylene hydrogenation. Reproduced from [8]. Copyright 2006, Wiley-Interscience.

Catalyst poisons can be classified according to their chemical makeup, selectivity for active sites, and the types of reactions poisoned. Table 2 lists four groups of catalyst poisons classified according to chemical origin and their type of interaction with metals. It should be emphasized that interactions of Group VA–VIIA elements with catalytic metal phases depend on the oxidation state of the former, e.g., how many electron pairs are available for bonding and the degree of shielding of the sulfur ion by ligands [16]. Thus, the order of decreasing toxicity for poisoning of a given metal by different sulfur species is H_2S, SO_2, SO_4^{2-}, *i.e.*, in the order of increased shielding by oxygen. Toxicity also increases with increasing atomic or molecular size and electronegativity, but decreases if the poison can be gasified by O_2, H_2O, or H_2 present in the reactant stream [21]; for example, adsorbed carbon can be gasified by O_2 to CO or CO_2 or by H_2 to CH_4.

Table 2: Common poisons classified according to chemical structure.

Chemical type	Examples	Type of interaction with metals
Groups VA and VIA	N, P, As, Sb, O, S, Se, Te	Through *s* and *p* orbitals; shielded structures are less toxic
Group VIIA	F, Cl, Br, I	Through *s* and *p* orbitals; formation of volatile halides
Toxic heavy metals and ions	As, Pb, Hg, Bi, Sn, Cd, Cu, Fe	Occupy *d* orbitals; may form alloys
Molecules that adsorb with multiple bonds	CO, NO, HCN, benzene, acetylene, other unsaturated hydrocarbons	Chemisorption through multiple bonds and back bonding

Table 3 lists a number of common poisons for selected catalysts in important representative reactions. It is apparent that organic bases (e.g., amines) and ammonia are common poisons for acidic solids, such as silica–aluminas and zeolites in cracking and hydrocracking reactions, while sulfur- and arsenic-containing compounds are typical poisons for metals in hydrogenation, dehydrogenation, and steam reforming reactions. Metal compounds (e.g., of Ni, Pb, V, and Zn) are poisons in automotive emissions control, catalytic cracking, and hydrotreating. Acetylene is a poison for ethylene oxidation, while asphaltenes are poisons in hydrotreating of petroleum residuum.

Table 3: Poisons for selected catalysts in important representative reactions.

Catalyst	Reaction	Poisons
Silica–alumina, zeolites	Cracking	Organic bases, hydrocarbons, heavy metals
Nickel, platinum, palladium	Hydrogenation/dehydrogenation	Compounds of S, P, As, Zn, Hg, halides, Pb, NH_3, C_2H_2
Nickel	Steam reforming of methane, naphtha	H_2S, As
Iron, ruthenium	Ammonia synthesis	O_2, H_2O, CO, S, C_2H_2, H_2O
Cobalt, iron	Fischer–Tropsch synthesis	H_2S, COS, As, NH_3, metal carbonyls
Noble metals on zeolites	Hydrocracking	NH_3, S, Se, Te, P
Silver	Ethylene oxidation to ethylene oxide	C_2H_2
Vanadium oxide	Oxidation/selective catalytic reduction	As/Fe, K, Na from fly ash
Platinum, palladium	Oxidation of CO and hydrocarbons	Pb, P, Zn, SO_2, Fe
Cobalt and molybdenum sulfides	Hydrotreating of residuum	Asphaltenes, N compounds, Ni, V

Poisoning selectivity is illustrated in Figure 2, a plot of activity (the reaction rate normalized to initial rate) *versus*normalized poison concentration. "Selective" poisoning involves preferential adsorption of the poison on the most active sites at low concentrations. If sites of lesser activity are blocked initially, the poisoning is "antiselective". If the activity loss is proportional to the concentration of adsorbed poison, the poisoning is "nonselective." An example of selective poisoning is the deactivation of platinum by CO for the para-H_2 conversion (Figure 3a) [24] while Pb poisoning of CO oxidation on platinum is apparently antiselective (Figure 3b) [25], and arsenic poisoning of cyclopropane hydrogenation on Pt is nonselective (Figure 3c) [26]. For nonselective poisoning, the linear decrease in activity with poison concentration or susceptibility (σ) is defined by the slope of the activity *versus* poison concentration curve. Several other important terms associated with poisoning are defined in Table 4. Poison tolerance, the activity at saturation coverage of the poison, and resistance (the inverse of deactivation rate) are important concepts that are often encountered in discussions of poisoning including those below.

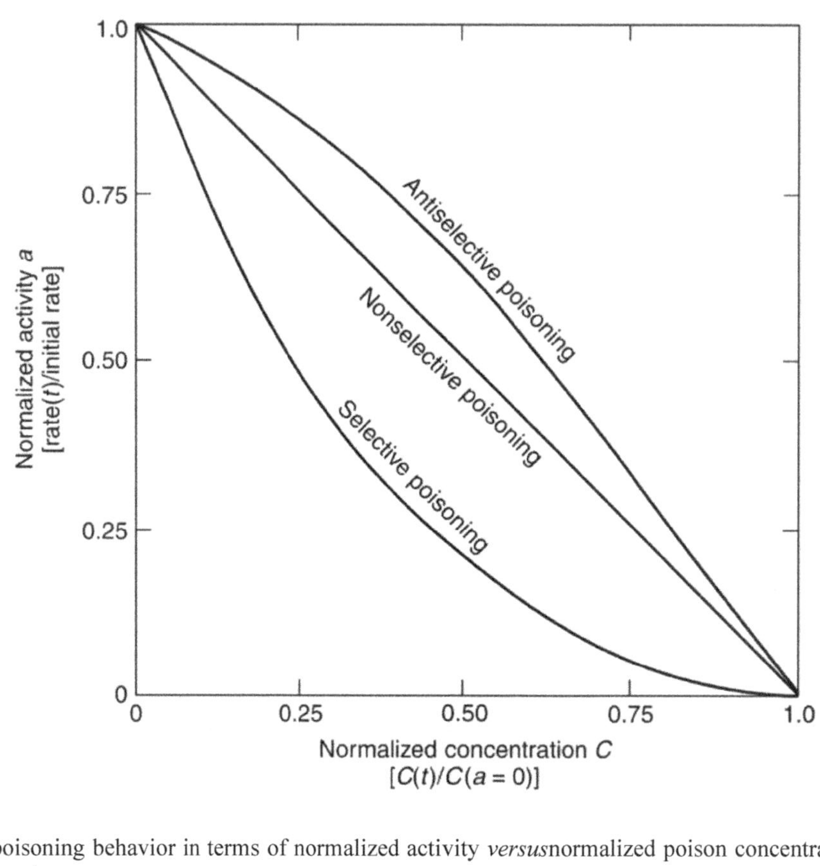

Figure 2: Three kinds of poisoning behavior in terms of normalized activity *versus*normalized poison concentration. Reproduced from [8]. Copyright 2006, Wiley-Interscience.

Table 4: Important Poisoning Parameters.

Parameter	Definition
Activity (a)	Reaction rate at time t relative to that at $t = 0$
Susceptibility (σ)	Negative slope of the activity *versus* poison concentration curve [$\sigma = (a - 1)/C(t)$]. Measure of a catalyst's sensitivity to a given poison
Toxicity	Susceptibility of a given catalyst for a poison relative to that for another poison
Resistance	Inverse of the deactivation rate. Property that determines how rapidly a catalyst deactivates
Tolerance ($a(C_{sat})$)	Activity of the catalyst at saturation coverage (some catalysts may have negligible activity at saturation coverage)

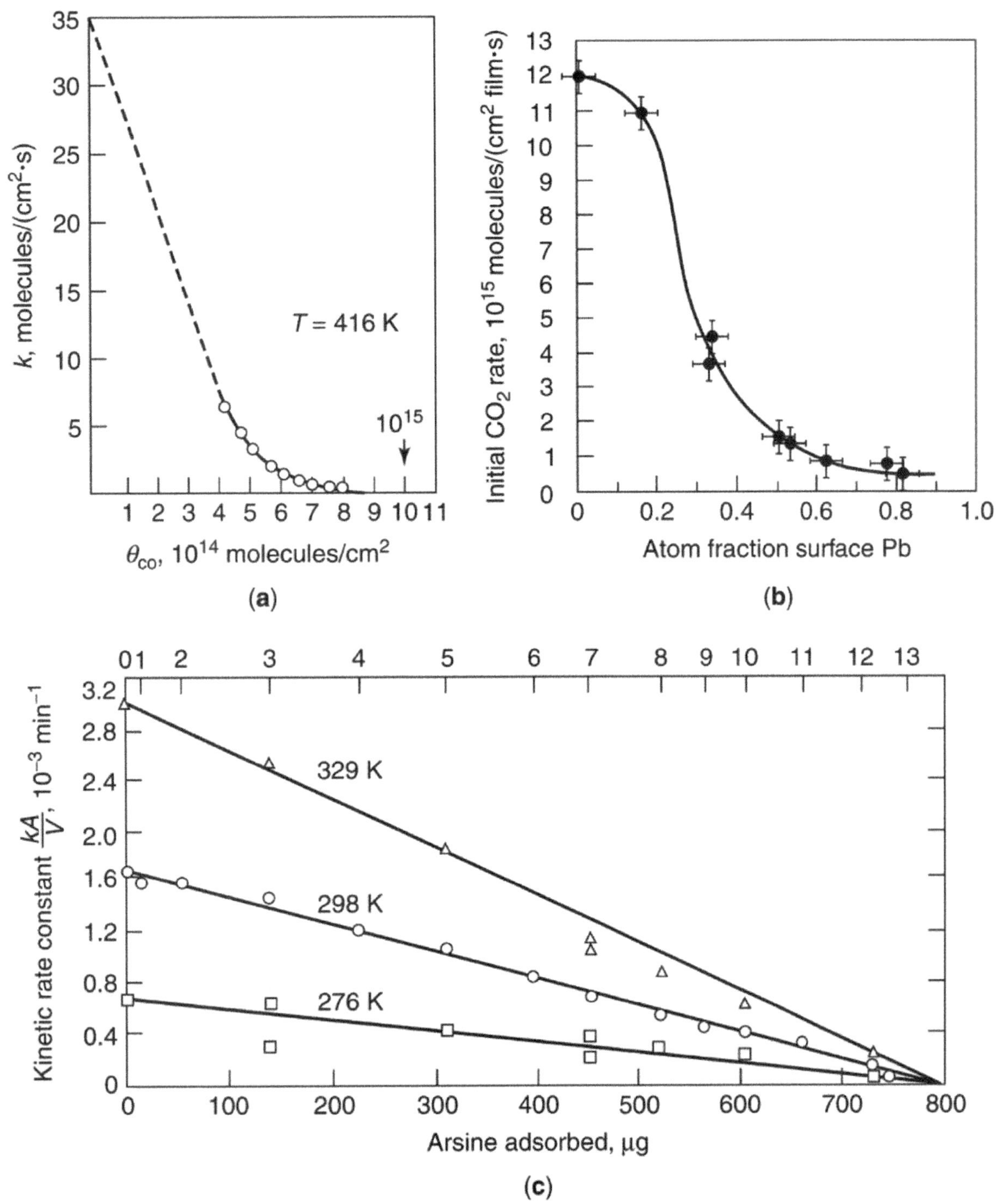

Figure 3: (**a**) CO poisoning of para-H_2 conversion over a Pt foil, reproduced from [24], copyright 1974, Wiley-VHC; (**b**) effect of lead coverage on the rate of CO oxidation of Pt film, reproduced from [25], copyright 1978, Elsevier; (**c**) rate constants of cyclopropane hydrogenolysis over a Pt film as a function of the amount of AsH_3adsorbed, reproduced from [26], copyright 1970, Elsevier.

The activity *versus* poison concentration patterns illustrated in Figure 2 are based on the assumption of uniform poisoning of the catalyst surface and surface reaction rate controlling, *i.e.*, negligible pore-diffusional resistance. These assumptions, however, are rarely met in typical industrial processes because the severe reaction conditions of high temperature and high pressure bring about a high pore-diffusional resistance for either the main or poisoning reaction or both. In physical terms, this means that the reaction may occur preferentially in the outer shell of the catalysts particle, or that poison is preferentially adsorbed in the outer shell of the catalyst particle, or both. The nonuniformly distributed reaction and/or poison leads to nonlinear activity *versus* poison concentration curves that mimic the patterns in Figure 2 but do not represent truly selective or antiselective poisoning. For example, if the main reaction is limited to an outer shell in a pellet where poison is concentrated, the drop in activity with concentration will be precipitous. Pore diffusional effects in poisoning (nonuniform poison) are treated later in this review.

As sulfur poisoning is a difficult problem in many important catalytic processes (e.g., hydrogenation, methanation, Fischer–Tropsch synthesis, steam reforming, and fuel cell power production), it merits separate discussion as an example of catalyst poisoning phenomena. Studies of sulfur poisoning in hydrogenation and CO hydrogenation reactions have been thoroughly reviewed [8,21,27,28,29,30,31]. Much of the previous work focused on poisoning of nickel metal catalysts by H_2S, the primary sulfur poison in many important catalytic processes, and thus provides some useful case studies of poisoning.

Previous adsorption studies [28,29,30] indicate that H_2S adsorbs strongly and dissociatively on nickel metal surfaces. The high stability and low reversibility of adsorbed sulfur is illustrated by the data in Figure 4 [28], in which most of the previous equilibrium data for nickel are represented on a single plot of log ($P_{H2}S/P_{H2}$) *versus* reciprocal temperature. The solid line corresponds to the equilibrium data for formation of bulk Ni_3S_2. Based on the equation $\Delta G = RT \ln(P_{H2}S/P_{H2}) = \Delta H - T\Delta S$, the slope of this line is $\Delta H/R$, where $\Delta H = -75$ kJ/mol and the intercept is $-\Delta S/R$. Most of the adsorption data lie between the dashed lines corresponding to $\Delta H = -125$ and -165 kJ/mol for coverages ranging from 0.5 to 0.9, indicating that adsorbed sulfur is more stable than the bulk sulfide. Indeed, extrapolation of high temperature data to zero coverage using a Tempkin isotherm [29] yields an enthalpy of adsorption of −250 kJ/mol; in other words, at low sulfur coverages, surface nickel–sulfur bonds are a factor of 3 more stable than bulk nickel–sulfur bonds. It is apparent from Figure 4 that the absolute heat of adsorption increases with decreasing coverage and that the equilibrium partial pressure of H_2S increases with increasing temperature and increasing coverage. For instance, at 725 K (450 °C) and $\theta = 0.5$, the values of $P_{H2}S/P_{H2}$ range from about 10^{-8} to 10^{-9}. In other words, half coverage occurs at 1–10 ppb H_2S, a concentration range at the lower limit of our present analytical capability. At the same temperature (450 °C), almost complete coverage ($\theta > 0.9$) occurs at values of$P_{H2}S/P_{H2}$ of 10^{-7}–10^{-6} (0.1–1 ppm) or at H_2S concentrations encountered in many catalytic processes after the gas has been processed to remove sulfur compounds. These data are typical of sulfur adsorption on most catalytic metals. Thus, we can expect that H_2S (and other sulfur impurities) will adsorb essentially irreversibly to high coverage in most catalytic processes involving metal catalysts.

Two important keys to reaching a deeper understanding of poisoning phenomena include (1) determining surface structures of poisons adsorbed on metal surfaces and (2) understanding how surface structure and hence adsorption stoichiometry change with increasing coverage of the poison. Studies of structures of adsorbed sulfur on single crystal metals (especially Ni) [3,28,32,33,34,35,36,37,38] provide such information. They reveal, for example, that sulfur adsorbs on Ni(100) in an ordered p(2 × 2) overlayer, bonded to four Ni atoms at S/Ni_s <0.25 and in a c(2 × 2) overlayer to two Ni atoms for S/Ni_s = 0.25–0.50 (see Figure 5; Ni_s denotes a surface atom of Ni); saturation coverage of sulfur on Ni(100) occurs at S/Ni_s= 0.5. Adsorption of sulfur on Ni(110), Ni(111), and higher index planes of Ni is more complicated; while the same p(2 × 2) structure is observed at low coverage, complex overlayers appear at higher coverages—for example, at S/Ni_s > 0.3 on Ni(111) a $(5\sqrt{3} \times 2)$S overlayer is formed [32,33,34]. In more open surface structures, such as Ni(110) and Ni(210), saturation coverage occurs at S/Ni_s = 0.74 and 1.09 respectively; indeed, there is a trend of increasing S/Ni_s with decreasing planar density and increasing surface roughness for Ni, while the saturation sulfur concentration remains constant at 44 ng/cm² Ni (see Table 5).

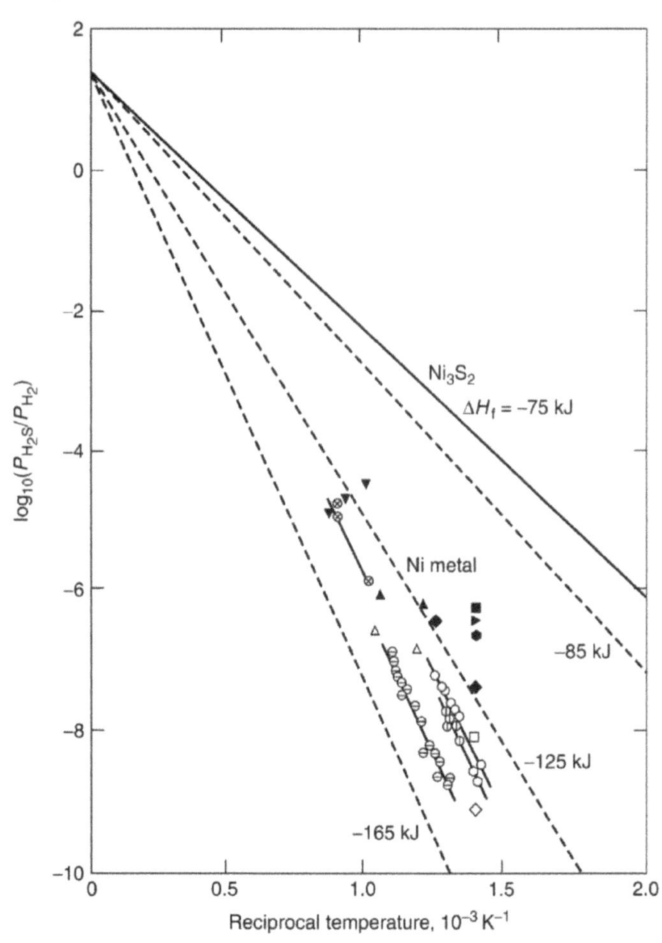

Figure 4: Equilibrium partial pressure of H_2S *versus* reciprocal temperature (values of ΔH_f based on 1 mole of H_2S); open symbols: θ = 0.5–0.6; closed symbols: θ = 0.8–0.9. Reproduced from [28]. Copyright 1982, Academic Press.

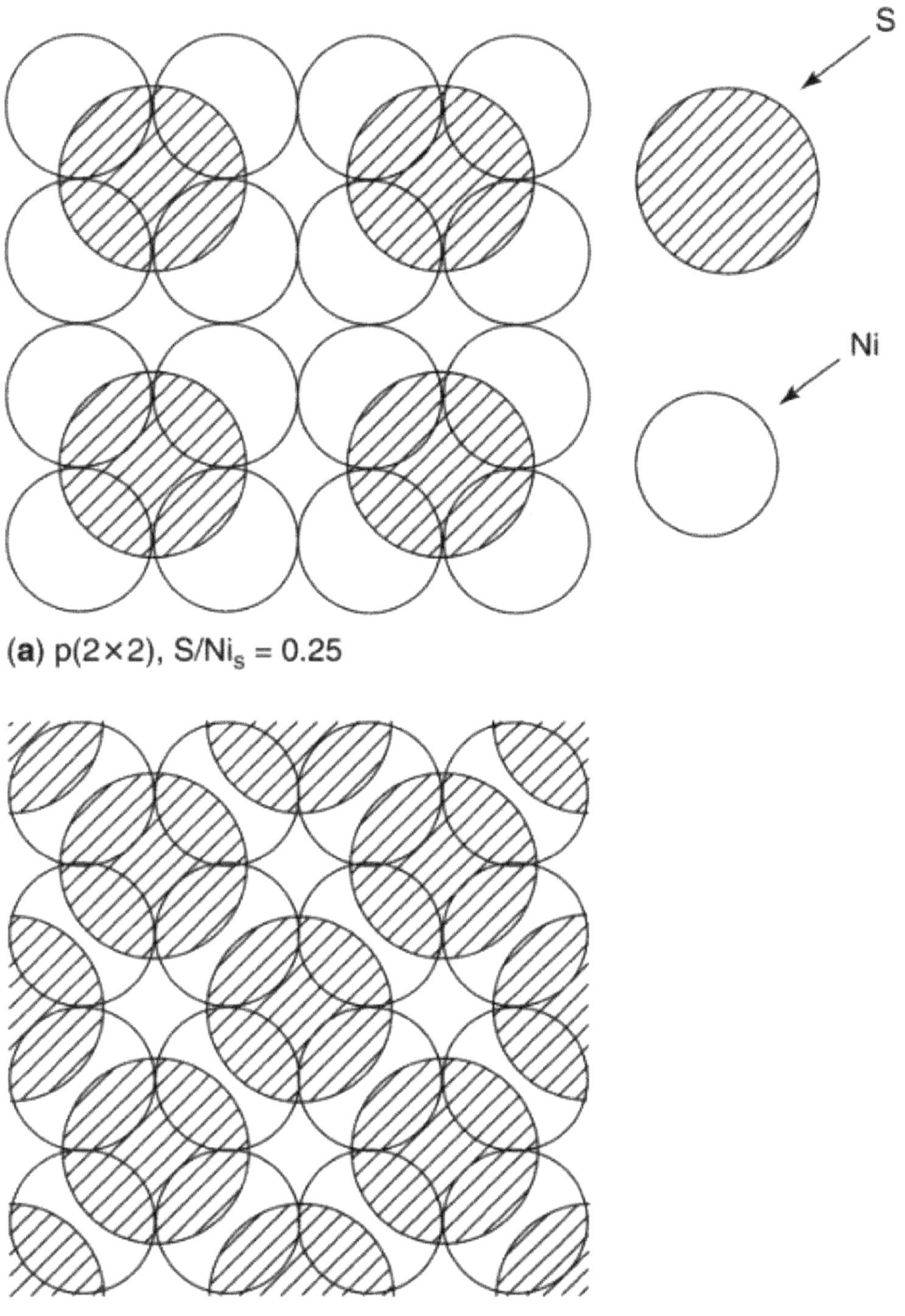

Figure 5: Schematic view of sulfur adsorbed on a Ni(100) surface at a (**a**) S/Ni$_s$ = 0.25 in a p(2 × 2) structure and (**b**) S/Ni$_s$ = 0.50 in a c(2 × 2) structure. Reproduced from [39]. Copyright 2001, Elsevier.

Table 5: Sulfur Adsorption Densities on Various Crystal Faces of Nickel [a].

Crystal face	Sulfur conc. at saturation, ng·S/cm^2	Number of S atoms/cm^2 ($\times 10^{15}$)	Number of Ni atoms/cm^2 ($\times 10^{15}$)	S atoms per surface Ni atom
(111)	47 ± 1	0.86	1.8	0.48
(100)	43 ± 1	0.80	1.6	0.50
(110)	44.5 ± 1	0.82	1.1	0.74
(210)	42 ± 1	0.78	0.72	1.09
Polycrystalline	44.5 ± 1	0.82	—	—

[a] Data from [35].

Reported saturation stoichiometries for sulfur adsorption on polycrystalline and supported Ni catalysts (S/Ni_s) vary from 0.25 to 1.3 [28]. The values of saturation coverage greater than $S/Ni_s = 0.5$ may be explained by (1) a higher fractional coverage of sites of lower coordination number, *i.e.*, atoms located on edges or corners of rough, high-index planes (Table 5); (2) enhanced adsorption capacity at higher gas phase concentrations of H_2S in line with the observed trend of increasing saturation coverage with increasing H_2S concentration in Figure 4; and/or (3) reconstruction of planar surfaces to rougher planes by adsorbed sulfur at moderately high coverages and adsorption temperatures.

The first effect would be favored, and in fact is observed, for supported metals of higher dispersion [28]. The second effect may explain the typically lower observed values of S/Ni_s for single crystal Ni, which are measured at extremely low pressures (high vacuum) relative to the higher values of S/Ni_s for polycrystalline and supported Ni, typically measured at orders of magnitude higher pressure; thus, in the case of the single crystal studies, the surface is not in equilibrium with gas phase H_2S/H_2.

The third effect, reconstruction of nickel surfaces by adsorbed sulfur, has been reported by a number of workers [28,32,33,36,37,38]; for example, McCarroll and co-workers [37,38] found that sulfur adsorbed at near saturation coverage on a Ni(111) face was initially in a hexagonal pattern, but upon heating above 700 K reoriented to a distorted c(2 × 2) (100) overlayer. Oudar [36] reported that sulfur adsorbed on a Ni(810) surface caused decomposition to (100) and (410) facets. During adsorption of H2S at RT, Ruan *et al.* [33] observed surface restructuring of Ni(111) from a p(2 × 2) at low coverage to a missing-row $(53\surd \times 2)$S terrace structure (0.4 monolayer) sparsely covered with small, irregular islands composed of sulfur adsorbed on disordered nickel; upon annealing to 460 K for 5 min, the islands ordered to the $(53\surd \times 2)$S phase and their size increased, suggesting further diffusion of Ni atoms from the terraces. The reconstruction of Ni (111) involving ejection and migration of Ni atoms was attributed to compressive surface stresses induced by sulfur adsorption; the role of compressive surface stress due to sulfur coverages exceeding 0.3 was confirmed by Grossmann *et al.* [32]. From these and similar studies, it is concluded that at moderately high temperatures (300–600 K) and coverages greater than 0.3, restructuring by sulfur of different facets of Ni to rougher, more open, stable structures is probably a general phenomenon. Thus, reconstruction probably accounts at least in part for observed increases in saturation S coverage with decreasing Ni site density.

The nature of reconstruction of a surface by a poison may depend on its pretreatment. For example, in a scanning tunneling microscopy (STM) study of room temperature H_2S adsorption on Ni(110), Ruan and co-workers [40] found that the S/Ni structure at saturation varied with the initial state of the surface, *i.e.*, whether clean or oxygen covered. Beginning with a clean Ni(110) surface, oxygen adsorbs dissociatively to form a (2 × 1)O overlayer at 1/2 monolayer coverage (Figure 6a); this is accompanied by a homogeneous nucleation of low-coordinated -Ni-O- rows along the [001] direction. As the oxygen-covered surface is exposed stepwise to 3 and then 8 Langmuirs (L) of H_2S, oxygen atoms are removed by reaction with hydrogen to water; the surface is first roughened, after which white islands and black troughs having a c(2 × 2) structure are formed as sulfur atoms replace oxygen atoms (Figure 6b). Upon exposure to 25 L of H_2S, the c(2 × 2) islands dissolve, while low-coordinated rows (periodicity of 1) form in the [001] direction, developing into ordered regions with a periodicity of 4 in the $[1\ \bar{1}\ 0]$ direction (Figure 6c). After exposure to 50 L of H_2S (Figure 6d), a stable, well-ordered (4 × 1) S structure appears, a surface clearly reconstructed relative to the original Ni(110). Moreover, the reconstructed surface inFigure 6d is very different from that observed upon direct exposure of the Ni(110) to H_2S at room temperature, *i.e.*, a c(2 × 2)S overlying the original Ni(110) (similar to Figure 5b); in other words, it appears that no reconstruction occurs by direct exposure to H_2S at room temperature, rather only in the presence of O_2 (or air). This emphasizes the complexities inherent in predicting the structure and stability of a given poison adsorbed on a given catalyst during a specified reaction as a function of different pretreatments or process disruptions, e.g., exposure to air.

In the previous discussion of Figure 4, $-\Delta H_{ads}$ was observed to decrease with increasing sulfur coverage; data in Figure 7[41] show that $-\Delta H_{ads}$ decreases with increasing gas-phase H_2S concentration and coverage. However, in contrast to the data in Figure 4, those in Figure 7 [41] show that at very high H_2S concentrations and high adsorption temperatures, $-\Delta H_{ads}$ falls well below the $-\Delta H_{formation}$ of bulk Ni_3S_2; at the same time, the S/Ni_s ratio approaches that of Ni_2S_3. This is a unique result, since all of the data obtained at lower temperatures and H_2S concentrations [28] show $-\Delta H_{ads}$ to be greater than $-\Delta H_{formation}$ of Ni_3S_2.

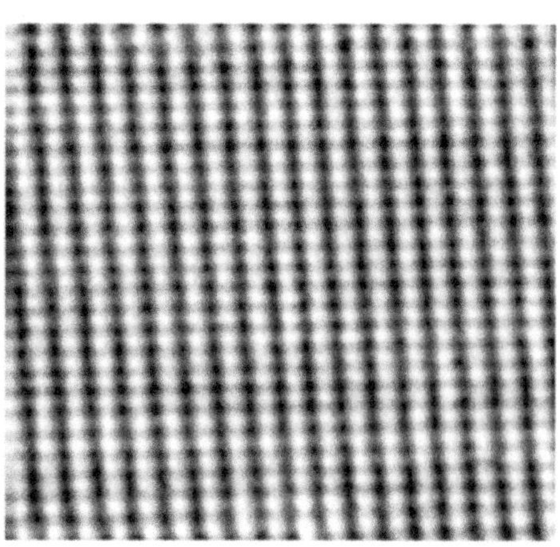

(a)

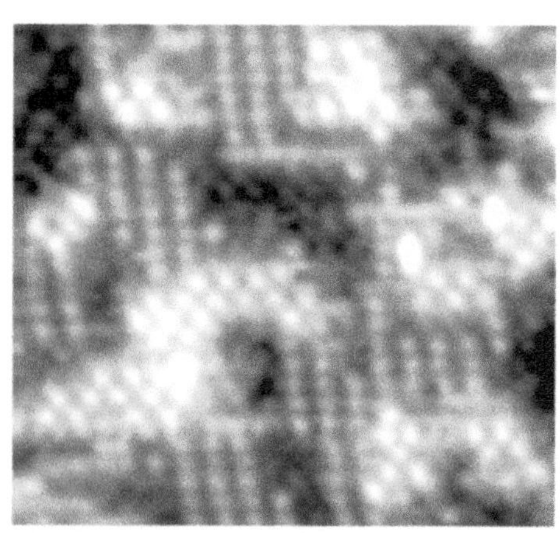

(b)

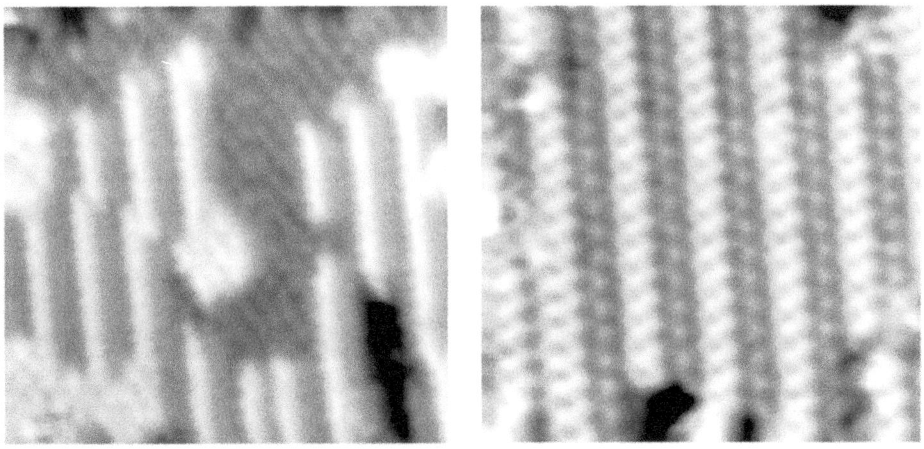

Figure 6: A series of *in situ* scanning tunneling microscope (STM) images recorded after exposure of Ni(110) to oxygen and then progressively higher exposures of H_2S: (**a**) (2 × 1)O overlayer; (**b**) white islands and black troughs with a c(2 × 2)S structure after exposure to 3 and 8 L of H_2S; (**c**) 25 L, islands transform to low-coordinated rows in the [001] direction; and (**d**) 50 L, stable, well-ordered (4 × 1)S. Reproduced from [40]. Copyright 1992, American Physical Society.

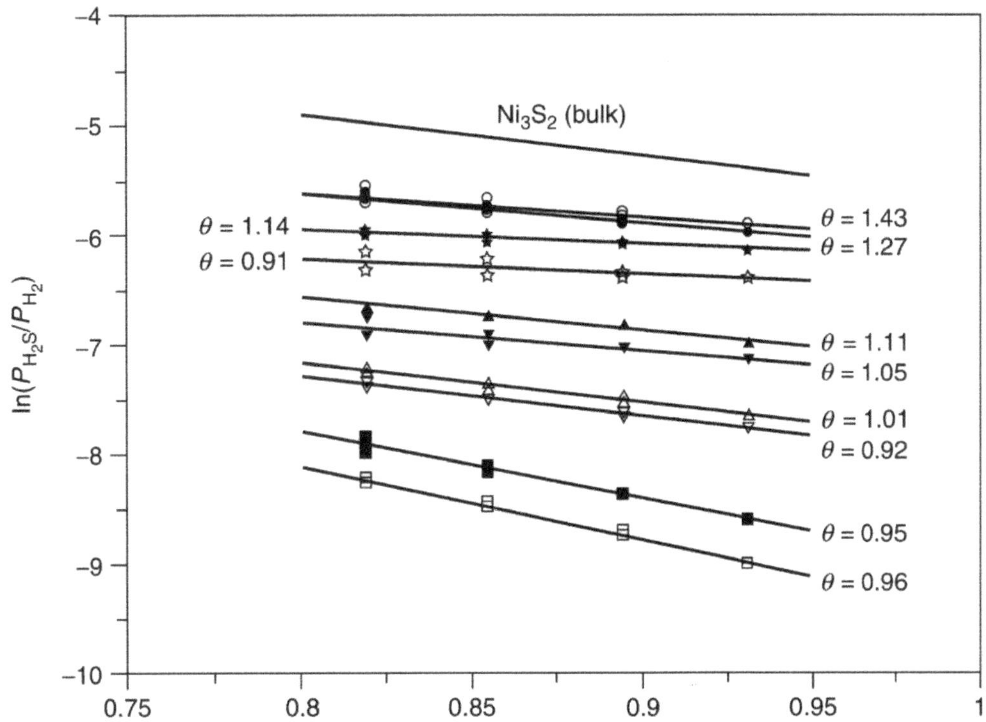

Figure 7: Sulfur chemisorption isosteres on a Ni/α-Al_2O_3 catalyst at high temperatures and high H_2S concentrations. Reproduced from [41]. Copyright 1999,Elsevier.

From the above discussion, the structure and stoichiometry of sulfur adsorbed on nickel evidently are complex functions of temperature, H_2S concentration, sulfur coverage, and pretreatment, phenomena that account at least in part for the complex nature of nickel poisoning by sulfur. Could one expect similar complexities in the poisoning of other metals? Probably, since poisoning of nickel is prototypical, *i.e.*, similar principles operate and similar poisoning behaviors are observed in other poison/metal systems, although none have been studied to the same depth as sulfur/nickel.

Since one of the necessary steps in a catalytic reaction is the adsorption of one or more reactants, investigation of the effects of adsorbed sulfur on the adsorption of other molecules, can provide useful insights into the poisoning process [21,28]. Previous investigations [28,42,43,44,45,46,47,48] indicate that both H_2 and CO adsorptions on nickel are poisoned by adsorbed sulfur. For example, thermal desorption studies of CO from presulfided Ni(100) [44] reveal a weakening of the CO adsorption bond and a rapid, nonlinear decline in the most strongly bound β_2 state (bridged CO) with increasing sulfur coverage, corresponding to a poisoning of about 8–10 Ni atoms for bridged CO adsorption per adsorbed sulfur atom at low sulfur coverage (see Figure 8); moreover, the β_2 CO species is completely poisoned at about 0.2–0.4 mL of sulfur relative to a saturation coverage of 0.5 mL. Hydrogen adsorption is poisoned in a similar nonlinear fashion. On the other hand, the coverage of the β_1 state (linear CO) is constant with increasing sulfur coverage. The sharp nonlinear drop in CO and hydrogen adsorptions at low sulfur coverages has been interpreted in terms of a combination of short-range electronic and steric effects operating over a range of less than 5 atomic units [13]. The different effects of sulfur on β_1 and β_2 states of CO have important implications for sulfur poisoning in reactions involving CO; that is, sulfur poisoning can affect reaction selectivity as well as activity [28].

Because sulfur adsorbs so strongly on metals and prevents or modifies the further adsorption of reactant molecules, its presence on a catalyst surface usually effects substantial or complete loss of activity in many important reactions. This is illustrated by the data in Figure 9 showing the steady-state methanation activities of Ni, Co, Fe, and Ru relative to the fresh, unpoisoned surface activity as a function of gas phase H_2S concentration. These data indicate that Ni, Co, Fe, and Ru all suffer 3–4 orders of magnitude loss in activity at 15–100 ppb of H_2S, *i.e.*, their sulfur tolerances are extremely low. Moreover, the sharp drop in activity with increasing H_2S concentration suggests highly selective poisoning. Nevertheless, the rate of sulfur poisoning and hence sulfur resistance varies from catalyst to catalyst and is apparently a function of catalyst composition [28] and reaction conditions [49]. Indeed, it is possible to significantly improve sulfur resistance of Ni, Co, and Fe with catalyst additives such as Mo and B that selectively adsorb sulfur. Because the adsorption of sulfur compounds is generally rapid and irreversible, surface sulfur concentrations in catalyst particles and beds are nonuniform, e.g., H_2S adsorbs selectively at the entrance to a packed bed and on the outer surface of catalyst particles, making the experimental study and modeling of sulfur poisoning extremely difficult.

There are other complications in the study of sulfur poisoning. For example, the adsorption stoichiometry of sulfur in CO hydrogenation on Ni is apparently a function of the temperature, H_2/CO ratio, and water partial pressure [49]. Moreover, at high CO partial pressures sulfur may be removed from the surface as COS, which is not as strongly adsorbed as H_2S. At low temperature conditions, e.g., those representative of Fischer–Tropsch synthesis or liquid phase hydrogenations, the gas phase concentration of H_2S in poisoning studies must be kept very low, *i.e.*, below 0.1–5 ppm, to avoid formation of bulk metal sulfides—a phenomenon that seriously compromises the validity of the results. Thus, the importance of studying poisoning phenomena *in situ* under realistic reaction conditions, at low process-relevant poison concentrations, and over a process-representative range of temperature and concentration conditions is emphasized.

As mentioned earlier, there are a number of industrial processes in which one intentionally poisons the catalyst in order to improve its selectivity. For example, Pt-containing naphtha reforming catalysts are often pre-sulfided to minimize unwanted cracking reactions. On basic Pt/KL zeolite catalysts, these short term, low concentration exposures are beneficial to produce Pt ensemble sizes that promote aromatization, while longer term or higher concentration exposures poison the catalyst both by forming Pt-S bonds and producing large crystallites that block pores, as shown by transmission electron microscopy (TEM) and X-ray absorption fine structure spectroscopy (EXAFS), and favor only dehydrogenation [50,51,52,53]. Other examples are sulfur added to Fischer-Tropsch catalysts that have been reported to have either beneficial or negligibly harmful effects, which are important considerations in setting the minimum gas clean-up requirements [27,30,54,55,56]. S and P are added to Ni catalysts to improve isomerization selectivity in the fats and oils hydrogenation industry, while S and Cu are added to Ni catalysts in steam reforming to minimize coking. In catalytic reforming, sulfided Re or Sn is added to Pt to enhance the dehydrogenation of paraffins to olefins while poisoning hydrogenolysis/coking reactions. V_2O_5 is added to Pt to suppress SO_2 oxidation to SO_3 in diesel emissions control catalysts.

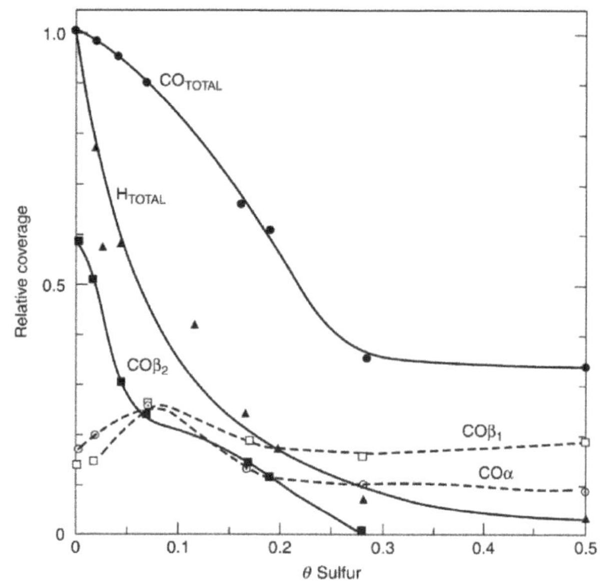

Figure 8: Area under thermal programmed desorption spectra for H_2 and the α, β_1, β_2, and total CO adsorption curves, as a function of sulfur precoverage. Reproduced from [44]. Copyright 1981, Elsevier.

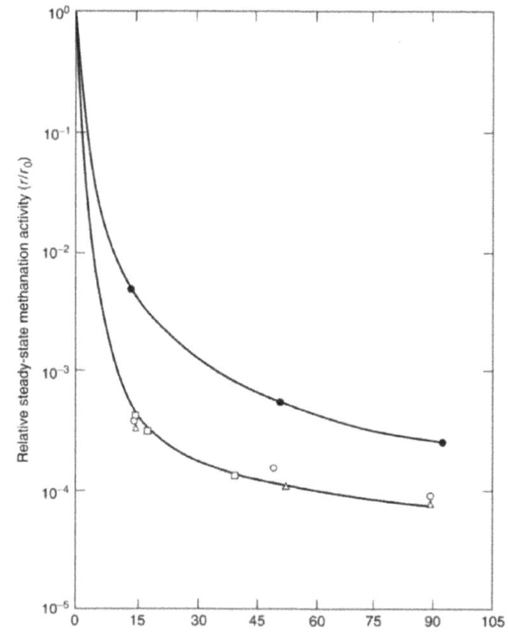

Figure 9: Relative steady-state methanation activity profiles for Ni (●), Co (Δ), Fe (○), and Ru (□) as a function of gas-phase H_2S concentration. Reaction conditions: 100 kPa, 400 °C, 1% CO/99% H_2 for Co, Fe, and Ru, 4% CO/96% H_2 for Ni. Reproduced from [28]. Copyright 1982, Academic Press.

Fouling, Coking, and Carbon Deposition

Fouling

Fouling is the physical (mechanical) deposition of species from the fluid phase onto the catalyst surface, which results in activity loss due to blockage of sites and/or pores. In its advanced stages, it may result in disintegration of catalyst particles and plugging of the reactor voids. Important examples include mechanical deposits of carbon and coke in porous catalysts, although carbon- and coke-forming processes also involve chemisorption of different kinds of carbons or condensed hydrocarbons that may act as catalyst poisons. The definitions of carbon and coke are somewhat arbitrary and by convention related to their origin. Carbon is typically a product of CO disproportionation while coke is produced by decomposition or condensation of hydrocarbons on catalyst surfaces and typically consists of polymerized heavy hydrocarbons. Nevertheless, coke forms may vary from high molecular weight hydrocarbons to primarily carbons such as graphite, depending upon the conditions under which the coke was formed and aged. A number of books and reviews treat the formation of carbons and coke on catalysts and the attendant deactivation of the catalysts [1,4,57,58,59,60,61,62].

The chemical structures of cokes or carbons formed in catalytic processes vary with reaction type, catalyst type, and reaction conditions. Menon [62] suggested that catalytic reactions accompanied by carbon or coke formation can be broadly classified as either coke-sensitive or coke-insensitive, analogous to Boudart's more general classification of structure-sensitive and structure-insensitive catalytic reactions. In coke-sensitive reactions, unreactive coke is deposited on active sites, leading to activity decline, while in coke-insensitive reactions, relatively reactive coke precursors formed on active sites are readily removed by hydrogen (or other gasifying agents). Examples of coke-sensitive reactions include catalytic cracking and hydrogenolysis; on the other hand, Fischer–Tropsch synthesis, catalytic reforming, and methanol synthesis are examples of coke-insensitive reactions. On the basis of this classification, Menon [62] reasoned that the structure and location of a coke are more important than its quantity in affecting catalytic activity.

Consistent with Menon's classification, it is also generally observed that not only structure and location of coke vary but also its mechanism of formation varies with catalyst type, e.g., whether it is a metal or metal oxide (or sulfide, sulfides being similar to oxides). Because of these significant differences in mechanism, formation of carbon and coke is discussed below separately for supported metals and for metal oxides and sulfides.

Carbon and Coke Formation on Supported Metal Catalysts

Possible effects of fouling by carbon (or coke) on the functioning of a supported metal catalyst are illustrated in Figure 10. Carbon may (1) chemisorb strongly as a monolayer or physically adsorb in multilayers and in either case block access of reactants to metal surface sites, (2) totally encapsulate a metal particle and thereby completely deactivate that particle, and (3) plug micro- and mesopores such that access of reactants is denied to many crystallites inside these pores. Finally, in extreme cases, strong carbon filaments may build up in pores to the extent that they stress and fracture the support material, ultimately causing the disintegration of catalyst pellets and plugging of reactor voids. For example, in steam methane reforming (SMR) catalysts, which are typically nickel supported on alumina with alkaline earth oxides, the carbon can diffuse through and begin to grow filaments from the back side of the nickel particles (structural type 3 in Table 6) especially at high reaction temperatures and low steam to methane ratios, which push the nickel particles off the support surface. Thermal or mechanical shock can then cause the carbon filaments to fall off the support, thus permanently deactivating the catalyst [8,60]. However, the behavior is complex because for other reaction conditions and other metals, the filaments may grow from the top surface of the metal particles or the carbon may diffuse into the metal and form bulk carbides [8].

An example of recent interest for biomass reactions that points to the complex interaction between the active metal and the support during carbon deposition is the steam reforming of light alcohols and other oxygenates, in which deactivation occurs primarily through coking. For traditional SMR catalysts (e.g., $Ni/MgAl_2O_4$) the coke is believed to originate primarily from alkene formation [63,64]. However, for the case of Ni/La_2O_3 catalysts, carbon appears to form at the interface between the active metal and the support to block the active phase [65].

Mechanisms of carbon deposition and coke formation on metal catalysts from carbon monoxide and hydrocarbons, including methane during SMR for hydrogen production [4,57,58,59,60,61], are illustrated in Figure 11 and Figure 12. Different kinds of carbon and coke that vary in morphology and reactivity are formed in these reactions (see Table 6 andTable 7). For example, CO dissociates on metals to form C_α, an adsorbed atomic carbon; C_α can react to C_β, a polymeric carbon film. The more reactive, amorphous forms of carbon formed at low temperatures (e.g., C_α and C_β) are converted at high temperatures over a period of time to less reactive, graphitic forms [60]

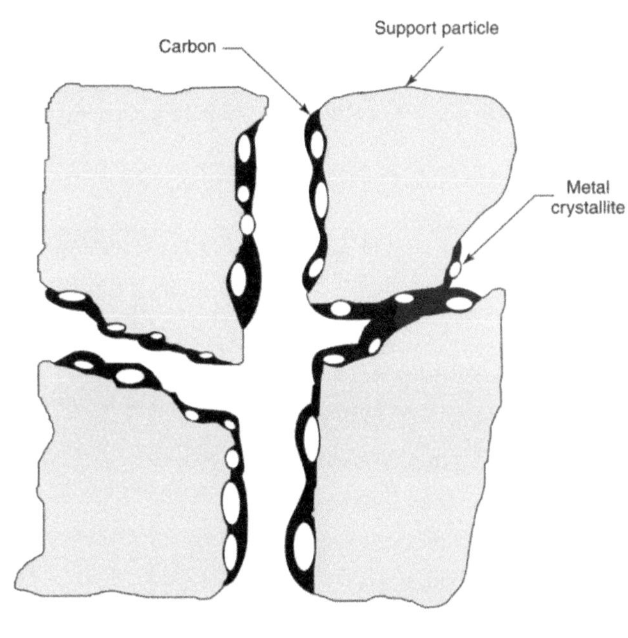

Figure 10: Conceptual model of fouling, crystallite encapsulation, and pore plugging of a supported metal catalyst owing to carbon deposition. Reproduced from [8]. Copyright 2006, Wiley-Interscience.

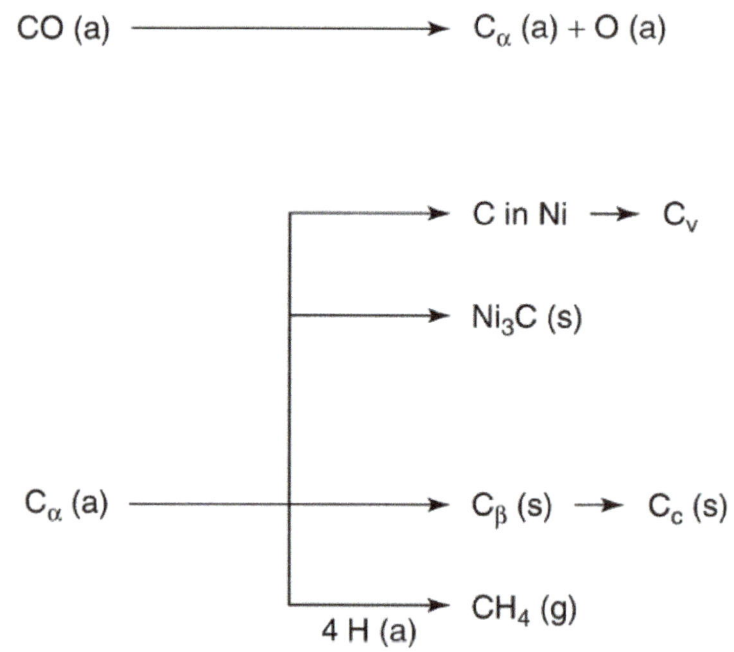

Figure 11: Formation, transformation, and gasification of carbon on nickel (a, g, s refer to adsorbed, gaseous, and solid states respectively). Reproduced from [60]. Copyright 1983, Elsevier.

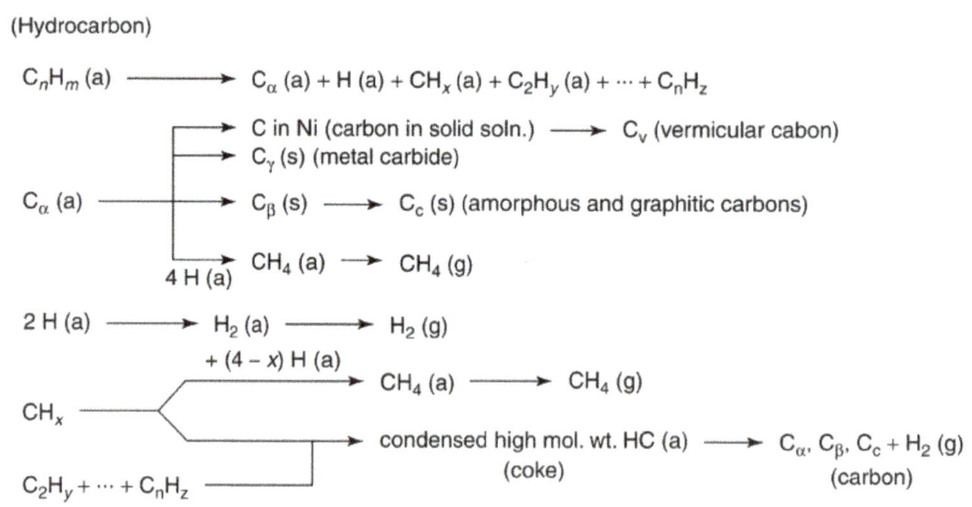

Figure 12: Formation and transformation of coke on metal surfaces (a, g, s refer to adsorbed, gaseous, and solid states respectively); gas phase reactions are not considered. Reproduced from [60]. Copyright 1983, Elsevier.

Table 6: Forms and Reactivities of Carbon Species Formed by Decomposition of CO on Nickel [a].

Structural type	Designation	Temp. formed, °C	Peak temp. for reaction with H_2, °C
1. Adsorbed, atomic (surface carbide)	C_α	200–400	200
2. Polymeric, amorphous films or filaments	C_β	250–500	400
3. Vermicular filaments, fibers, and/or whiskers	C_v	300–1000	400–600
4. Nickel carbide (bulk)	C_γ	150–250	275
5. Graphitic (crystalline) platelets or films	C_c	500–550	550–850

[a] Ref. [60].

Table 7: Carbon Species Formed in Steam Reforming of Hydrocarbons on Nickel Catalysts [a].

Attribute	Encapsulating film	Whisker-like	Pyrolytic carbon
Formation	Slow polymerization of C_nH_m radicals on Ni surface, into encapsulating film	Diffusion of C through Ni crystal, nucleation and whisker growth with Ni crystal at top	Thermal cracking of hydrocarbon; deposition of C precursors on catalyst
Effects	Progressive deactivation	No deactivation of Ni surface. Breakdown of catalyst and increasing ΔP	Encapsulation of catalyst particle; deactivation and increasing ΔP
Temp. range, °C	<500	>450	>600
Critical parameters	Low temperature, low H_2O/C_nH_m, low H_2/C_nH_m, aromatic feed	High temperature, low H_2O/C_nH_m, no enhanced H_2O adsorption, low activity, aromatic feed	High temperature, high void fraction, low H_2O/C_nH_m, high pressure, acidic catalyst

[a] Ref. [60].

It should also be emphasized that some forms of carbon result in loss of catalytic activity and some do not. For example, at low reaction temperatures (<300–375 °C) condensed polymer or β-carbon films and at high temperatures (>650 °C) graphitic carbon films encapsulate the metal surfaces of methanation and steam reforming catalysts [60]. Deactivation of steam reforming catalysts at high reaction temperatures (500–900 °C) may be caused by precipitation of atomic (carbidic) carbon dissolved in the Ni surface layers to a depth of more than 50–70 nm [62,66]. If it accumulates on the metal surface (at high or low temperatures), adsorbed atomic carbon can deactivate metal sites for adsorption and/or reaction. For example, Durer and co-workers [67] demonstrated that carbon atoms residing in the fourfold hollow sites of Rh(100) block the adsorption of hydrogen (and hence could block sites for hydrogenation). In the intermediate temperature range of 375–650 °C, carbon filaments (Figure 13) are formed by precipitation of dissolved carbon at the rear side of metal crystallites, causing the metal particles to grow away from the support [57]. Filament growth ceases when sufficient carbon accumulates on the free surface to cause encapsulation by a carbon layer; however, encapsulation of the metal particles does not occur if H_2/CO or H_2O/hydrocarbon ratios are sufficiently high. Thus, carbon filaments sometimes formed in CO hydrogenation or steam reforming of hydrocarbons would not necessarily cause a loss of intrinsic catalyst activity unless they are formed in sufficient quantities to cause plugging of the pores [60] or loss of metal occurs as the carbon fibers are removed during regeneration [68,69]. However, in practice, regions of carbon forming potential in steam reforming must be carefully avoided, since once initiated, the rates of filamentous carbon formation are sufficiently high to cause catastrophic pore plugging and catalyst failure within a few hours to days.

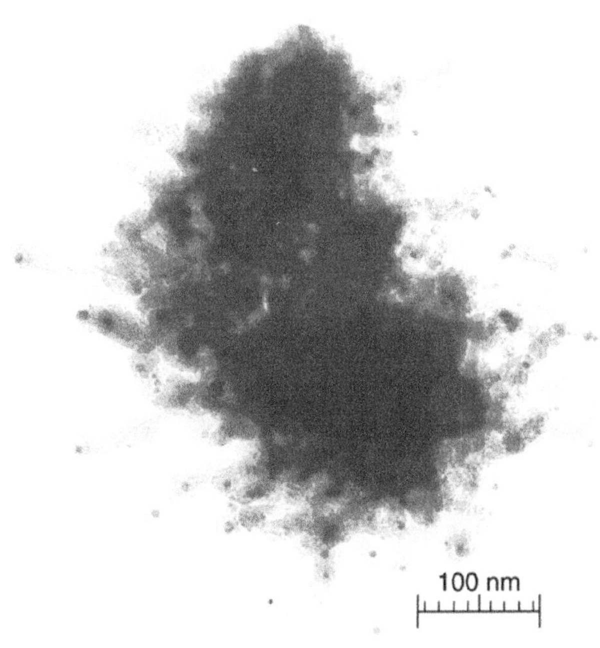

Figure 13: Electron micrograph of 14% Ni/Al_2O_3 having undergone extensive carbon deposition during CO disproportionation at 673 K, P_{CO} = 4.55 kPa (magnification of 200,000). Courtesy of the BYU Catalysis Laboratory.

The rate at which deactivation occurs for a given catalyst and reaction depends greatly on reaction conditions—especially temperature and reactant composition. A fundamental principle for coke-insensitive reactions on metals (e.g., methanation, Fischer–Tropsch synthesis, steam reforming, catalytic reforming, and methanol synthesis) is that deactivation rate depends greatly on the difference in rates of formation and gasification of carbon/coke precursors, *i.e.*, $r_d = r_f - r_g$. If the rate of gasification, r_g, is equal to or greater than that of formation, r_f, carbon/coke is not deposited. Rates of carbon/coke precursor formation and gasification both increase exponentially with temperature, although the difference between them varies a great deal with temperature because of differences in preexponential factors and activation energies. Thus, carbon/coke formation is avoided in regions of temperature in which precursor gasification rate exceeds deposition rate. This is illustrated in Figure 14, an Arrhenius plot for rates of formation and hydrogenation of alpha and beta carbons on nickel during CO methanation. Since at temperatures below 600 K ($1/T > 1.66 \times 10^{-3}$ K^{-1}) the rate of C_α gasification exceeds that of C_α formation, no carbon is deposited. However above 600 K, C_α accumulates on the surface since the rate of C_α formation exceeds that of C_α gasification. As C_α accumulates (at 600–700 K), it is converted to a C_β polymeric chain or film that deactivates the nickel catalyst; however, above 700 K ($1/T < 1.43 \times 10^{-3}$ K^{-1}), the rate of C_β hydrogenation exceeds that of formation and no deactivation occurs. Thus, the "safe" regions of methanation for avoiding deactivation by carbon are below 600 K and above 700 K; of course, these regions will vary somewhat with reactant concentrations and catalyst activity. A similar principle operates in steam reforming, *i.e.*, at a sufficiently low reaction temperature, the rate of hydrocarbon adsorption exceeds the rate of hydrocracking and a deactivating polymer film is formed [70]; accordingly, it is necessary to operate above this temperature to avoid deactivation.

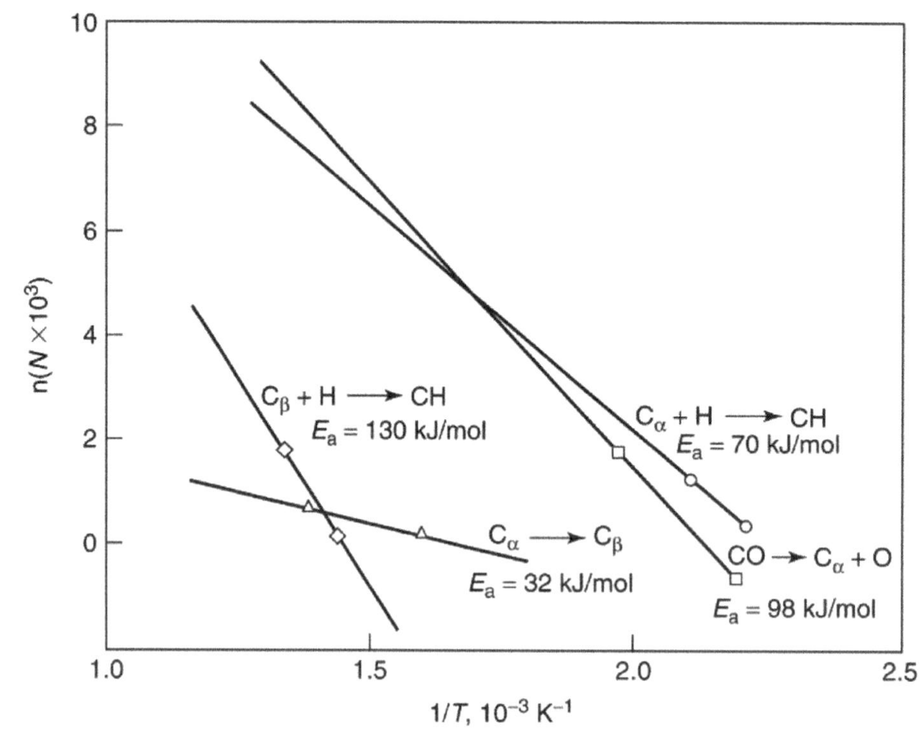

Figure 14: Rates of formation (log scale) and hydrogenation of C_α and C_β *versus*reciprocal temperature. Reproduced from [60]. Copyright 1983, Elsevier.

In steam reforming, filamentous carbon formation rate is a strong function of reactant hydrocarbon structure; for example, it decreases in the order acetylenes, olefins, paraffins, *i.e.*, in order of decreasing reactivity, although activation energies for nickel are in the same range (125–139 kJ) independent of hydrocarbon structure and about the same as those observed for formation of filamentous carbon from decomposition of CO [60]. This latter observation suggests that the reactions of CO and different hydrocarbons to filamentous carbon proceed by a common mechanism and rate-determining step—probably the diffusion of carbon through the metal crystallites [60].

The rate at which a carbon or coke is accumulated in a given reaction under given conditions can vary significantly with catalyst structure, including metal type, metal crystallite size, promoter, and catalyst support. For example, supported Co, Fe, and Ni are active above 350–400 °C for filamentous carbon formation from CO and hydrocarbons; the order of decreasing activity is reportedly Fe > Co > Ni [60]. Pt, Ru, and Rh catalysts, on the other hand, while equally or more active than Ni, Co, or Fe in steam reforming, produce little or no coke or carbon. This is attributed to reduced mobility and/or solubility of carbon in the noble metals, thus retarding the nucleation process. Thus, it is not surprising that addition of noble metals to base metals retards carbon formation; for example, addition of Pt in Ni lowers carbon deposition rate during methanation, while addition of Cu or Au to Ni substantially lowers carbon formation in steam reforming [60,71]. In contrast to the moderating effects of noble metal additives, addition of 0.5% Sn to cobalt substantially increases the rate of carbon filament formation from ethylene [72], an effect desirable in the commercial production of carbon filament fibers.

Since carbon formation and gasification rates are influenced differently by modifications in metal crystallite surface chemistry, which are in turn a function of catalyst structure, oxide additives or oxide supports may be used to moderate the rate of undesirable carbon or coke accumulation. For example, Bartholomew and Strasburg [73] found the specific rate (turnover frequency) of filamentous carbon deposition on nickel during methanation at 350 °C to decrease in the order $Ni/TiO_2 > NiAl_2O_3 > Ni/SiO_2$, while Vance and Bartholomew [74] observed C_α hydrogenation rates at 170 °C to decrease in this same order (the same as for methanation at 225 °C). This behavior was explained in terms of promotional or inhibiting effects due to decoration of metal crystallites by the support, for example silica, inhibiting both CO dissociation and carbon hydrogenation. This hypothesis is consistent with observations [75,76] that silica evaporated on metal surfaces and supported metals inhibits formation of filamentous carbon. Similarly Bitter and co-workers [77] observed rates of carbon formation in CO_2/CH_4 reforming to decrease in the order $Pt/\gamma\text{-}Al_2O_3 \rightarrow Pt/TiO_2 > Pt/ZrO_2$; while 90% of the carbon deposited on the support, the authors linked deactivation to carbon accumulated on the metal owing to an imbalance between carbon formed by methane dissociation and oxidation by chemisorbed CO_2. The rate of formation of coke in steam reforming is delayed and occurs at lower rates in nickel catalysts promoted with alkali or supported on basic MgO [78].

Since formation of coke, graphite, or filamentous carbon involves the formation of C-C bonds on multiple atoms sites, one might expect that coke or carbon formation on metals is structure-sensitive, *i.e.*, sensitive to surface structure and metal crystallite size. Indeed, Bitter and co-workers [77] found that catalysts containing larger Pt crystallites deactivate more rapidly than those containing small crystallites. Moreover, a crystallite size effect, observed in steam reforming of methane on nickel [60,78], appears to operate in the same direction, *i.e.*, formation of filamentous carbon occurs at lower rates in catalysts containing smaller metal crystallites.

In summary, deactivation of supported metals by carbon or coke may occur chemically, owing to chemisorption or carbide formation, or physically and mechanically, owing to blocking of surface sites, metal crystallite encapsulation, plugging of pores, and destruction of catalyst pellets by carbon filaments. Blocking of catalytic sites by chemisorbed hydrocarbons, surface carbides, or relatively reactive films is generally reversible in hydrogen, steam, CO_2, or oxygen. Further details of the thermodynamics, kinetics, and mechanisms of carbon and coke formation in methanation and steam reforming reactions are available in reviews by Bartholomew [60] and Rostrup-Nielsen [70,78]. In recent reviews addressing deactivation of Co catalysts by carbon during Fischer-Tropsch synthesis [79,80], the same or similar carbon species, e.g., α, β, polymeric, and graphitic carbons, are observed on Co surfaces as on Ni; moreover, poisoning or fouling of the Co surfaces with β, polymeric, and graphitic carbon layers are found to be major causes of deactivation.

Coke Formation on Metal Oxide and Sulfide Catalysts

In reactions involving hydrocarbons, coke may be formed in the gas phase and on both noncatalytic and catalytic surfaces. Nevertheless, formation of coke on oxides and sulfides is principally a result of cracking reactions involving coke precursors (typically olefins or aromatics) catalyzed by acid sites [81,82]. Dehydrogenation and cyclization reactions of carbocation intermediates formed on acid sites lead to aromatics, which react further to higher molecular weight polynuclear aromatics that condense as coke (see Figure 15). Reactions 1–3 in Figure 15 illustrate the polymerization of olefins, reactions 4–8 illustrate cyclization from olefins, and reactions 9–14 illustrate chain reaction formation of polynuclear aromatics that condense as coke on the catalyst surface. Because of the high stability of the polynuclear carbocations (formed in reactions 10–13), they can continue to grow on the surface for a relatively long time before a termination reaction occurs through the back donation of a proton.

From this mechanistic scheme (Figure 15), it is clear that olefins, benzene and benzene derivatives, and polynuclear aromatics are precursors to coke formation. However, the order of reactivity for coke formation is clearly structure dependent, *i.e.*, decreases in the order polynuclear aromatics > aromatics > olefins > branched alkanes > normal alkanes. For example, the weight percent coke formed on silica–alumina at 500 °C is 0.06, 3.8, 12.5, and 23% for benzene, naphthalene, fluoranthene, and anthracene respectively [83].

Coking reactions in processes involving heavy hydrocarbons are very complex; different kinds of coke may be formed and they may range in composition from CH to C and have a wide range of reactivities with oxygen and hydrogen, depending upon the time on stream and temperature to which they are exposed. For example, coke deposits occurring in hydrodesulfurization of residuum have been classified into three types [84]:

(1) Type I deposits are reversibly adsorbed normal aromatics deposited during the first part of the cycle at low temperature.

(2) Type II deposits are reversibly adsorbed asphaltenes deposited early in the coking process.

(3) Type III deposits result from condensation of aromatic concentrates into clusters and then crystals that constitute a "mesophase." This crystalline phase is formed after long reaction times at high temperature. This hardened coke causes severe deactivation of the catalyst [84].

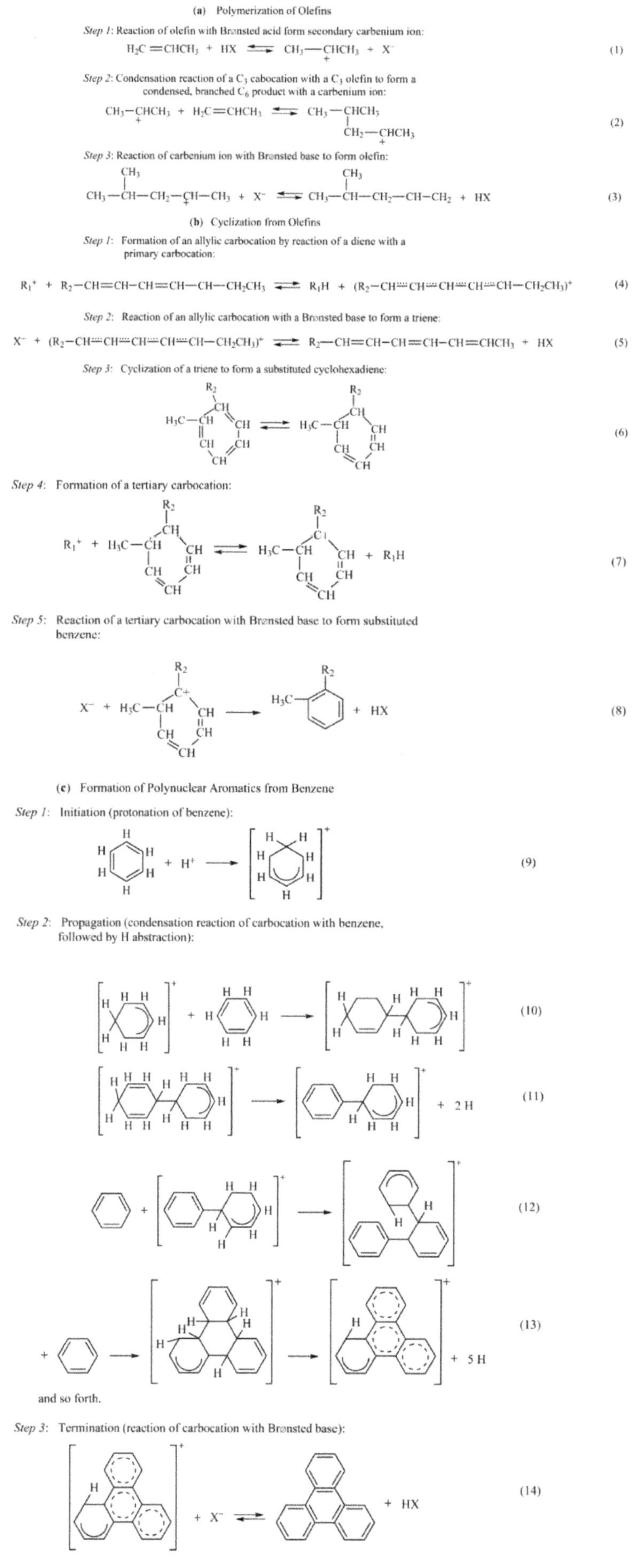

Figure 15: Coke-forming reactions of alkenes and aromatics on oxide and sulfide catalysts: (**a**) polymerization of alkenes, (**b**) cyclization from alkenes, and (**c**) formation of polynuclear aromatics from benzene. Reproduced from [8], Copyright 2006, Wiley-Interscience.

In addition to hydrocarbon structure and reaction conditions, extent and rate of coke formation are also a function of the acidity and pore structure of the catalyst. Generally, the rate and extent of coke formation increase with increasing acid strength and concentration. Coke yield decreases with decreasing pore size (for a fixed acid strength and concentration); this is especially true in zeolites where shape selectivity plays an important role in coke formation. For example, coke yield in fluid catalytic cracking is only 0.4% for ZSM-5 (pore diameters of 0.54 × 0.56 nm) compared to 2.2% for Y-faujasite (aperture diameter of 0.72 nm) [82]. However, in pores of molecular diameter, a relatively small quantity of coke can cause substantial loss of activity. It should be emphasized that coke yield can vary considerably into the interior pores of a catalyst particle or along a catalyst bed, depending upon the extent to which the main and deactivation reactions are affected by film mass transport and pore diffusional resistance.

The mechanisms by which coke deactivates oxide and sulfide catalysts are, as in the case of supported metals, both chemical and physical. However, some aspects of the chemistry are quite different. The principal chemical loss of activity in oxides and sulfides is due to the strong adsorption of coke molecules on acidic sites. However, as discussed earlier, strong acid sites also play an important role in the formation of coke precursors, which subsequently undergo condensation reactions to produce large polynuclear aromatic molecules that physically coat catalytic surfaces. Physical loss of activity also occurs as coke accumulates, ultimately partially or completely blocking catalyst pores as in supported metal catalysts. For example, in isomerization of *cis*-butene on SiO_2/Al_2O_3 [85] catalyst deactivation occurs by rapid, selective poisoning of strong acid sites; coke evolved early in the reaction is soluble in dichloromethane and pyridine and is slightly aromatic. Apparently, the blocking of active sites does not significantly affect porosity or catalyst surface area, as SiO_2/Al_2O_3 contains relatively large mesopores.

In the case of supported bifunctional metal/metal oxide catalysts, different kinds of coke are formed on the metal and the acidic oxide support, e.g., soft coke (high H/C ratio) on Pt or Pt–Re metals and hard coke (low H/C ratio) on the alumina support in catalytic reforming [86]. In this case, coke precursors may be formed on the metal via hydrogenolysis, following which they migrate to the support and undergo polymerization and cyclization reactions, after which the larger molecules are dehydrogenated on the metal and finally accumulate on the support, causing loss of isomerization activity. Mild sulfiding of these catalysts (especially Pt–Re/alumina) substantially reduces the rate of hydrogenolysis and the overall formation of coke on both metal and support; it especially reduces the hard coke, which is mainly responsible for deactivation.

Several recent studies [82,87,88,89,90,91,92,93,94,95,96,97] have focused on coke formation during hydrocarbon reactions in zeolites including (1) the detailed chemistry of coke precursors and coke molecules formed in zeolite pores and pore intersections (or supercages) and (2) the relative importance of adsorption on acid sites *versus* pore blockage. The principal conclusions from these studies can be summarized as follows: (1) the formation of coke and the manner in which it deactivates a zeolite catalyst are shape-selective processes, (2) deactivation is mainly due to the formation and retention of heavy aromatic clusters in pores and pore intersections, and (3) while both acid-site poisoning and pore blockage participate in the deactivation, the former dominates at low coking rates, low coke coverages (e.g., in Y-zeolite below 2 wt%), and high temperatures, while the latter process dominates at high reaction rates, high coke coverages, and low temperatures. Thus, pore size and pore structure are probably more important than acid strength and density under typical commercial process conditions. Indeed, deactivation is typically more rapid in zeolites having small pores or apertures and/or a monodimensional structure [95]. Figure 16 illustrates four possible modes of deactivation of HZSM-5 by carbonaceous deposits with increasing severity of coking [95].

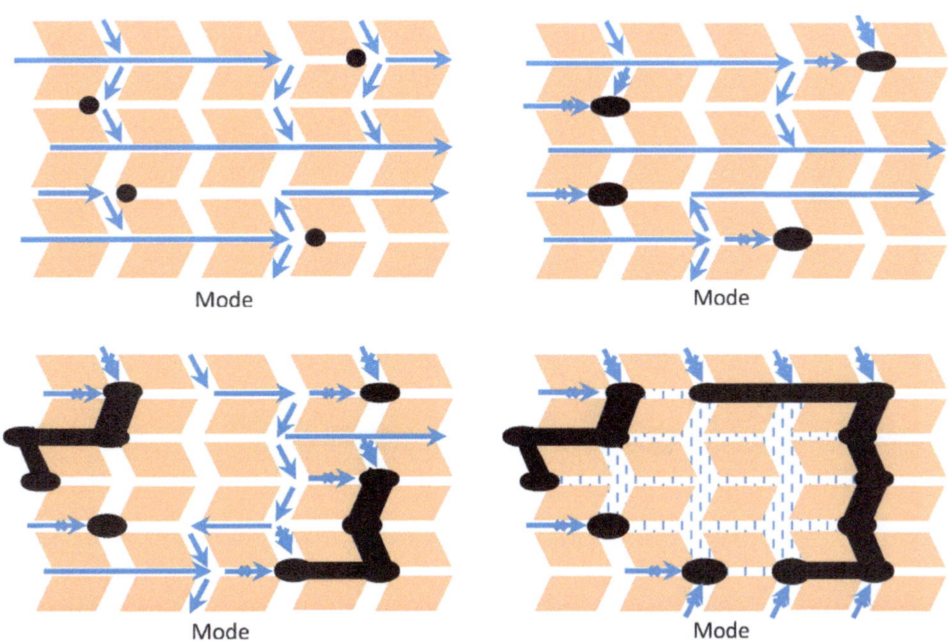

Figure 16: Schematic of the four possible modes of deactivation by carbonaceous deposits in HZSM-5: (1) reversible adsorption on acid sites, (2) irreversible adsorption on sites with partial blocking of pore intersections, (3) partial steric blocking of pores, and (4) extensive steric blocking of pores by exterior deposits. Adapted from [95].

These conclusions (in the previous paragraph) are borne out, for example, in the study by Cerqueira and co-workers [97] of USHY zeolite deactivation during methylcyclohexane transformation at 450 °C, showing the following:

(1) Coke is probably mainly formed by rapid transformation of toluenic C_7 carbenium ions with lesser contributions from reactions of cyclopentadiene, C_3–C_6 olefins, and aromatics.

(2) Soluble coke consists of polynuclear aromatic clusters containing three to seven five- and six-membered rings having a typical compositions of $C_{30}H_{40}$ to $C_{40}H_{44}$ and having dimensions of 0.9 × 1.1 nm to 1.1 × 1.5 nm, *i.e.*, sizes that would cause them to be trapped in the supercages of Y-zeolite.

(3) At short contact times, coking is relatively slow and deactivation is mainly due to acid-site poisoning, while at long contact times, coking is much faster because of the high concentrations of coke precursors; under these latter conditions coke is preferentially deposited at the outer pore openings of zeolite crystallites and deactivation is dominated by pore-mouth blockage.

That coke formed at large contact times not only blocks pores and/or pore intersections inside the zeolite, but also migrates to the outside of zeolite crystallites, where it blocks pore entrances, has been observed in several studies [91,93,94,97]. However, the amount, structure, and location of coke in ZSM-5 depends strongly on the coke precursor, e.g., coke formed from mesitylene is deposited on the external zeolite surface, whereas coking with isobutene leads to largely paraffinic deposits inside pores; coke from toluene, on the other hand, is polyaromatic and is deposited both on external and internal zeolite surfaces [91].

Thermal Degradation and Sintering

Background

Thermally induced deactivation of catalysts results from (1) loss of catalytic surface area due to crystallite growth of the catalytic phase, (2) loss of support area due to support collapse and of catalytic surface area due to pore collapse on crystallites of the active phase, and/or (3) chemical transformations of catalytic phases to noncatalytic phases. The first two processes are typically referred to as "sintering". The third is discussed in the next section under solid–solid reactions. Sintering processes generally take place at high reaction temperatures (e.g., > 500 °C) and are generally accelerated by the presence of water vapor.

Most of the previous sintering and redispersion work has focused on supported metals. Experimental and theoretical studies of sintering and redispersion of supported metals published before 1997 have been reviewed fairly extensively [8,98,99,100,101,102,103,104,105,106,107]. Three principal mechanisms of metal crystallite growth have been advanced: (1) crystallite migration, (2) atomic migration, and (3) (at very high temperatures) vapor transport. The processes of crystallite and atomic migration are illustrated in Figure 17. Crystallite migration involves the migration of entire crystallites over the support surface, followed by collision and coalescence. Atomic migration involves detachment of metal atoms or molecular metal clusters from crystallites, migration of these atoms over the support surface, and ultimately, capture by larger crystallites. Redispersion, the reverse of crystallite growth in the presence of O_2 and/or Cl_2, may involve (1) formation of volatile metal oxide or metal chloride complexes that attach to the support and are subsequently decomposed to small crystallites upon reduction and/or (2) formation of oxide particles or films that break into small crystallites during subsequent reduction.

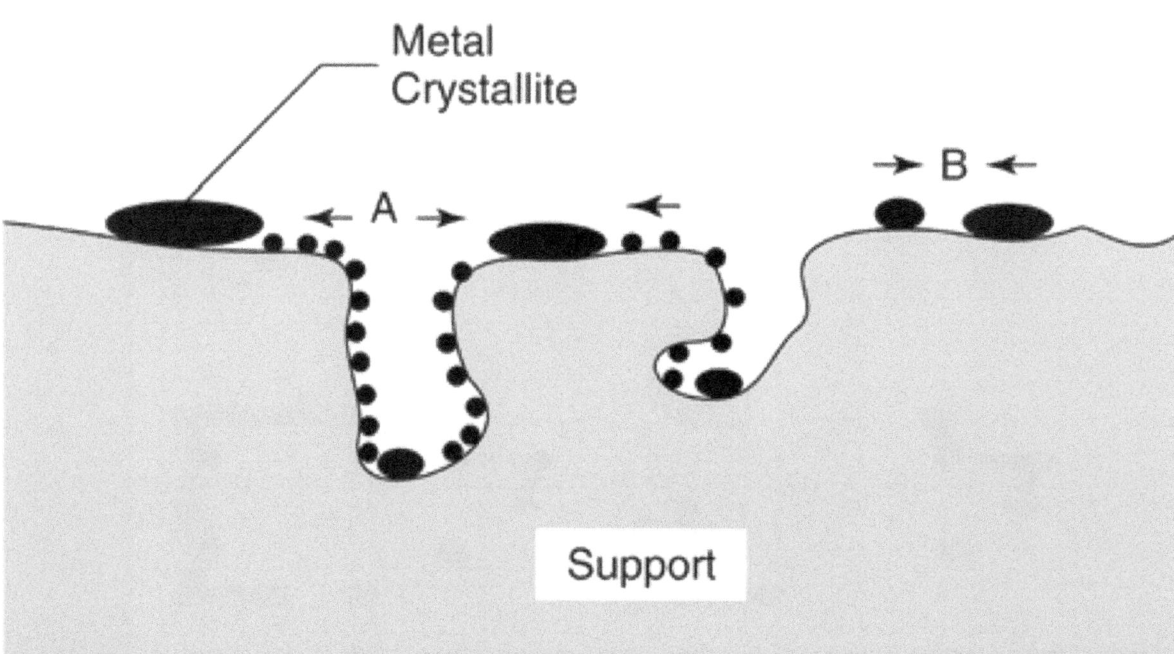

Figure 17: Two conceptual models for crystallite growth due to sintering by (**A**) atomic migration or (**B**) crystallite migration. Reproduced from [8], Copyright 2006, Wiley-Interscience.

There is controversy in the literature regarding which mechanism of sintering (or redispersion) operates at a given set of conditions. Logically, atomic migration would be favored at lower temperatures than crystallite migration, since the higher diffusivities of atoms or small cluster would facilitate their migration, whereas the thermal energy necessary to induce motion of larger crystallites would only be available at higher temperatures. Moreover, migration of small crystallites might be favorable early in the sintering process but unfavorable as crystallites become larger. However, focusing on only one of the three sintering mechanisms (and two dispersion mechanisms) is a simplification that ignores the possibility that all mechanisms may occur simultaneously and may be coupled with each other through complex physicochemical processes, including the following: (1) dissociation and emission of metal atoms or metal-containing molecules from metal crystallites, (2) adsorption and trapping of metal atoms or metal-containing molecules on the support surface, (3) diffusion of metal atoms, metal-containing molecules and/or metal crystallites across support surfaces, (4) metal or metal oxide particle spreading, (5) support surface wetting by metal particles, (6) metal particle nucleation, (7) coalescence of, or bridging between, two metal particles, (8) capture of atoms or molecules by metal particles, (9) liquid formation, (10) metal volatilization through volatile compound formation, (11) splitting of crystallites in O_2 atmosphere owing to formation of oxides of a different specific volume, and (12) metal atom vaporization. Depending upon reaction or redispersion conditions, a few or all of these processes may be important; thus, the complexity of sintering/redispersion processes is emphasized.

In general, thermal sintering processes are kinetically slow (at moderate reaction temperatures) and irreversible or difficult to reverse. Thus, sintering is more easily prevented than cured.

Factors Affecting Metal Particle Growth and Redispersion in Supported Metals

Temperature, atmosphere, metal type, metal dispersion, promoters/impurities, and support surface area, texture, and porosity are the principal parameters affecting rates of sintering and redispersion (see Table 8) [8,103,104,105,106,107]. Sintering rates increase exponentially with temperature. Metals sinter relatively rapidly in oxygen and relatively slowly in hydrogen, although depending upon the support, metal redispersion can be facilitated by exposure at high temperature (e.g., 500–550 °C for Pt/Al_2O_3) to oxygen and chlorine, followed by reduction. Water vapor also increases the sintering rate of supported metals, likely through chemical-assisted sintering effects similar to those described in Section 2.4.3.

Table 8: Effects of Important Reaction and Catalyst Variables on Sintering Rates of Supported Metals Based on General Power-Law Expression (GPLE) Data [a].

Variable	Effect
Temperature	Sintering rates are exponentially dependent on T: E_{act} varies from 30 to 150 kJ/mol. E_{act} decreases with increasing metal loading; it increases in the following order with atmosphere: $NO < O_2 < H_2 < N_2$
Atmosphere	Sintering rates are much higher for noble metals in O_2 than in H_2 and higher for noble and base metals in H_2 relative to N_2. Sintering rate decreases for supported Pt in atmospheres in the following order: $NO > O_2 > H_2 > N_2$
Metal	Observed order of decreasing thermal stability in H_2 is Ru > Ir ≅ Rh > Pt; thermal stability in O_2 is a function of (1) volatility of metal oxide and (2) strength of metal oxide–support interaction
Support	Metal–support interactions are weak (bond strengths of 5–15 kJ/mol); with a few exceptions, thermal stability for a given metal decreases with support in the following order: $Al_2O_3 > SiO_2 >$ carbon
Promoters	Some additives decrease atom mobility, e.g., C, O, CaO, BaO, CeO_2, GeO_2; others increase atom mobility, e.g., Pb, Bi, Cl, F, or S. Oxides of Ba, Ca, or Sr are "trapping agents" that decrease sintering rate
Pore size	Sintering rates are lower for porous *versus* nonporous supports; they decrease as crystallite diameters approach those of the pores

[a] Refs. [8,103–107]. For the definition of a GPLE, see Equation 2 later in this section.

Normalized dispersion (percentage of metal exposed at any time divided by the initial percentage exposed) *versus* time data in Figure 18 show that at temperatures of 650 °C or higher, rates of metal surface area loss (measured by hydrogen chemisorption) due to sintering of Ni/silica in hydrogen atmosphere are significant, causing 70% loss of the original metal surface area within 50 h at 750 °C. In reducing atmosphere, metal crystallite stability generally decreases with decreasing metal melting temperature, *i.e.*, in the order Ru > Ir > Rh > Pt > Pd > Ni > Cu > Ag, although this order may be affected by relatively stronger metal–support interactions, e.g., the observed order of decreasing stability of supported platinum in vacuum is $Pt/Al_2O_3 > Pt/SiO_2 > Pt/C$. In oxidizing atmospheres, metal crystallite stability depends on the volatility of metal oxides and the strength of the metal–oxide–support interaction. For noble metals, metal stability in air decreases in the order Rh > Pt > Ir > Ru; formation of volatile RuO_4 accounts for the relative instability of ruthenium.

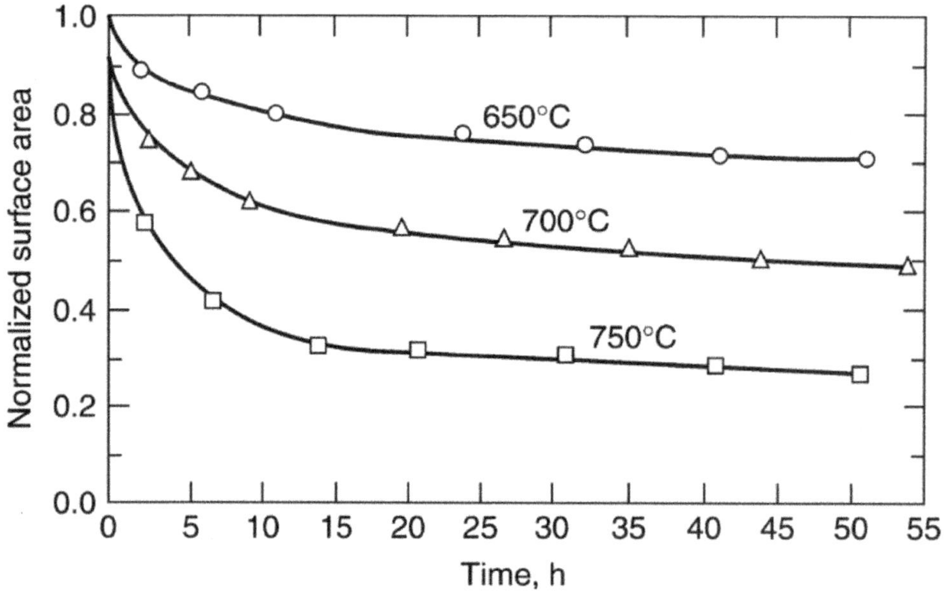

Figure 18: Normalized nickel surface area (based on H_2 adsorption) *versus* time data during sintering of 13.5% Ni/SiO_2 in H_2 at 650, 700, and 750 °C. Reproduced from [108]. Copyright 1983, Elsevier.

The effect of temperature on sintering of metals and oxides can be understood physically in terms of the driving forces for dissociation and diffusion of surface atoms, which are both proportional to the fractional approach to the absolute melting point temperature (T_{mp}). Thus, as temperature increases, the mean lattice vibration of surface atoms increases; when the Hüttig temperature ($0.3T_{mp}$) is reached, less strongly bound surface atoms at defect sites (e.g., edges and corner sites) dissociate and diffuse readily over the surface, while at the Tamman temperature ($0.5T_{mp}$), atoms in the bulk become mobile. Accordingly, sintering rates of a metal or metal oxide are significant above the Hüttig temperature and very high near the Tamman temperature; thus, the relative thermal stability of metals or metal oxides can be correlated in terms of the Hüttig or Tamman temperatures [109]. This can be illustrated from values of the melting and Tamman temperatures for noble and base metals and their compounds listed in Table 9. For example, sintering of copper catalysts for methanol synthesis is promoted by traces of chlorine in the feed, which react at about 225 °C (500 K) with the active metal/metal oxide surface to produce a highly mobile copper chloride phase having a Tamman temperature of only 79–174 °C (352–447 K) relative to 405–527 °C (678–800 K) for copper metal or metal oxides [110].

Table 9: Values of Melting and Tamman Temperatures (°C) for Common Catalytic Metals and Their Compounds [a].

Compound	T_{mp}, K	T_{Tamman}, K	$T_{Hüttig}$, K
Ag	1233	617	370
Au	1336	668	401
Co	1753	877	526
Cu	1356	678	407
CuO	1599	800	480
Cu_2O	1508	754	452
$CuCl_2$	893	447	268
Cu_2Cl_2	703	352	211
Fe	1808	904	542
Mo	2883	1442	865
MoO_3	1068	534	320
MoS_2	1458	729	437
Ni	1725	863	518
NiO	2228	1114	668
$NiCl_2$	1281	641	384
$Ni(CO)_4$	254	127	76
Rh	2258	1129	677
Rh_2O_3	1373	687	412
Ru	2723	1362	817
Pd	1828	914	548
PdO	1023	512	307
Pt	2028	1014	608
PtO	823	412	247
PtO_2	723	362	217
$PtCl_2$	854	427	256
$PtCl_4$	643	322	193
Zn	693	347	208
ZnO	2248	1124	674

[a] Adapted from Ref. [109].

Promoters or impurities affect sintering and redispersion by either increasing (e.g., chlorine and sulfur) or decreasing (e.g., oxygen, calcium, cesium) metal atom mobility on the support; in the latter case, this is due to their high resistance to dissociation and migration due to high melting points, as well as their hindering dissociation and surface diffusion of other atoms. Similarly, support surface defects or pores impede surface migration of metal particles—especially micropores and mesopores with pore diameters about the same size as the metal crystallites.

Historically, sintering rate data were fitted to a simple power-law expression (SPLE) of the form

$$-\frac{d(\frac{D}{D_0})}{dt} = k_s(\frac{D}{D_0})^n \quad (1)$$

where k_s is the sintering rate constant, D_0 the initial dispersion, and n is the sintering order, which for typical catalyst systems may vary from 3 to 15; unfortunately, the SPLE is, in general, not valid for sintering processes because it assumes that surface area or dispersion ultimately reaches zero, given sufficient time, when in fact, for a given temperature and atmosphere, a nonzero or limiting dispersion is observed after long sintering times. Moreover, the use of the SPLE is further questionable because variations in sintering order are observed as a function of time and temperature for a given catalyst in a fixed atmosphere [105,106,107]; thus, data obtained for different samples and different reaction conditions cannot be quantitatively compared. Nevertheless, it has been shown by Fuentes [111,112] and Bartholomew [104,105,106] that the effects of temperature, atmosphere, metal, promoter, and support can be quantitatively determined by fitting sintering kinetic data to the general power-law expression (GPLE)

$$-\frac{d\left(\frac{D}{D_0}\right)}{dt} = k_s\left(\frac{D}{D_0} - \frac{D_{eq}}{D_0}\right)^m \quad (2)$$

which adds a term $-D_{eq}/D_0$ to account for the observed asymptotic approach of the typical dispersion *versus* time curve to a limiting dispersion D_{eq} at infinite time; m, the order of sintering, is found to be either 1 or 2. A recently compiled, comprehensive quantitative treatment of previous sintering rate data based on the GPLE with an order m of 2 [104,105,106] quantitatively addresses the effects of catalyst properties and reaction conditions on sintering rate. Some of these data are summarized in Table 10 [108,113,114,115]. These data show, for example, that the rate constant, and hence the rate of sintering, is less for Ni/Al_2O_3 than for Pt/Al_2O_3, an unexpected result in view of the lower heat of vaporization for Ni. This result is possibly explained by a greater metal support interaction for Ni with alumina.

Table 10: Comparison of Second-Order Sintering Rate Constants and Activation Energies for Pt, Ni, and Ag Catalysts [a].

Catalyst	Atm.	D_0 [b]	k_s [c] (400 °C)	k_s (650 °C)	k_s (700 °C)	k_s (750 °C)	$E_{act.}$ [d] kJ/mol	Ref.
0.6% $Pt/\gamma-Al_2O_3$	H_2	~0.85	0.007	0.310	0.530	1.32	79	[113]
5% $Pt/\gamma-Al_2O_3$	H_2	0.10	0.420	0.76	0.84	0.97	13	[114]
15% $Ni/\gamma-Al_2O_3$	H_2	0.16	0.004	0.083	0.13	0.27	66	[108]
0.6% $Pt/\gamma-Al_2O_3$	Air	~0.85	0.024	0.29	0.41	0.75	52	[113]
5% $Pt/\gamma-Al_2O_3$	Air	0.10	0.014	1.46	2.79	8.51	97	[114]
1.8% $Ag/\eta-Al_2O_3$	Air	0.36	0.69	-	-	-	-	[115]

[a] Refs. [105,106]; [b] Initial metal dispersion or percentage exposed; [c] Second-order sintering rate constant from general power-law expression (GPLE) with units of h^{-1}; [d] Sintering activation energy for GPLE, $-d(D/D_0)/dt = k_s[D/D_0 - D_{eq}/D_0]^m$, where $m = 2$.

Sintering studies of supported metals are generally of two types: (1) studies of commercially relevant supported metal catalysts and (2) studies of model metal–support systems. The former type provides useful rate data that can be used to predict sintering rates, while the latter type provides insights into the mechanisms of metal particle migration and sintering, although the results cannot be quantitatively extrapolated to predict behavior of commercial catalysts. There is direct evidence from the previous studies of model-supported catalysts [104,107] for the occurrence of crystallite migration (mainly in well-dispersed systems early in the sintering process), atomic migration (mainly at longer sintering times), and spreading of metal crystallites (mainly in oxygen atmosphere). There is also evidence that under reaction conditions, the surface is dynamic, *i.e.*, adsorbates and other adatoms rapidly restructure the surface and slowly bring about faceting; moreover, thermal treatments cause gradual changes in the distribution of coordination sites to minimize surface energy. There is a trend in increasing sophistication of spectroscopic tools used to study sintering and redispersion. In the next decade, we might expect additional insights into atomic and molecular processes during reaction at the atomic scale using STM, analytical high resolution transmission electron microscopy (HRTEM), and other such powerful surface science tools.

Sintering of Catalyst Carriers

Sintering of carriers (supports) has been reviewed by Baker and co-workers [103] and Trimm [116]. Single-phase oxide carriers sinter by one or more of the following processes: (1) surface diffusion, (2) solid-state diffusion, (3) evaporation/ condensation of volatile atoms or molecules, (4) grain boundary diffusion, and (5) phase transformations. In oxidizing atmospheres, γ-alumina and silica are the most thermally stable carriers; in reducing atmospheres, carbons are the most thermally stable carriers. Additives and impurities affect the thermal properties of carriers by occupying defect sites or forming new phases. Alkali metals, for example, accelerate sintering; while calcium, barium, nickel, and lanthanum oxides form thermally stable spinel phases with alumina. Steam accelerates support sintering by forming mobile surface hydroxyl groups that are subsequently volatilized at higher temperatures. Chlorine also promotes sintering and grain growth in magnesia and titania during high temperature calcination. This is illustrated in Figure 19 [117]. By contrast, sulfuric acid treatment of hydrated alumina (gibbsite) followed by two-step calcination, results in a very stable transitional alumina with needle-like particle morphology [116]. Dispersed metals in supported metal catalysts can also accelerate support sintering; for example, dispersed nickel accelerates the loss of Al_2O_3 surface area in Ni/Al_2O_3 catalysts.

As an important example of support sintering through phase transformations, Al_2O_3 has a rich phase behavior as a function of temperature and preparation. A few among the many important phases that are stable or metastable, include boehmite, γ-alumina, and α-alumina [8,118,119]. Other phases are possible and the temperatures at which the phase transitions occur depend on crystal size and moisture content of the starting material, but as an example, as temperature is raised, boehmite, which is a hydrated or hydroxyl form of alumina, transforms to γ-alumina between 300 and 450 °C, then to δ-alumina at ~850°C, θ-alumina at ~1000°C, and finally α-alumina at ~1125 °C. The corresponding crystal structures for these five phases are orthorhombic, cubic defective spinel, orthorhombic, deformed monoclinic spinel, and hexagonal close pack (hcp with ABAB stacking) [8,118,119]. The approximate surface areas of these respective phases, as measured by nitrogen physisorption using Brunauer, Emmett, Teller (BET) analysis, are approximately 400, 200, 120, 50, and 1 m^2/g [8]. The dramatic drop in surface area during the transition from θ to α is associated with collapse of the microporous structure and formation of the dense hcp phase.

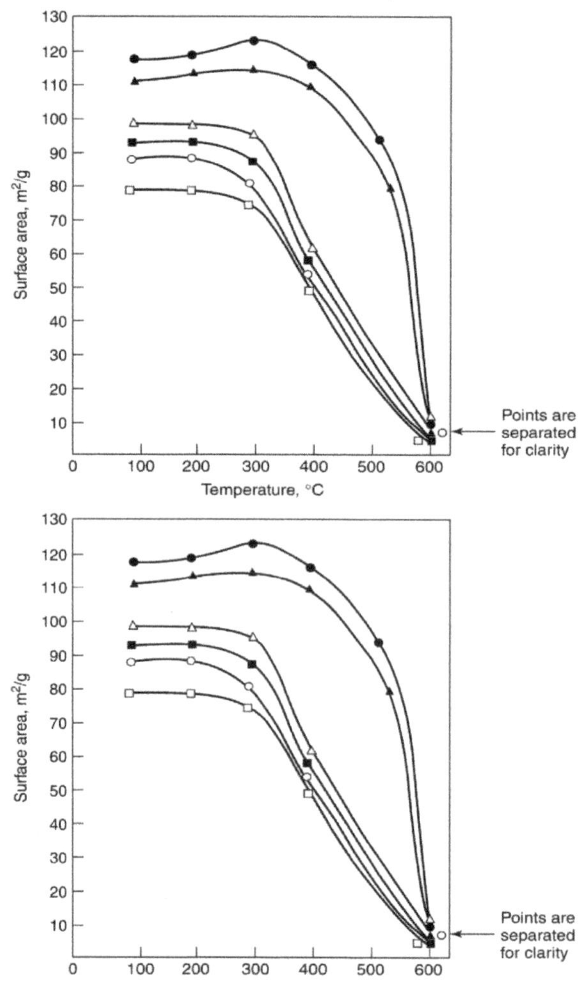

Figure 19: BET surface area of titania as a function of thermal treatment and chlorine content of fresh samples (before pretreatment). Samples were treated at the temperature indicated for 2 h. Reproduced from [117]. Copyright 1985, Elsevier. ● = Blank TiO_2; ▲ = TiO_2 soaked in H_2O; Δ = TiO_2 soaked in HCl/H_2O (2.06 wt% Cl); ■ = TiO_2 soaked in HCl/H_2O (2.40 wt% Cl); ○ = TiO_2 soaked in HCl/H_2O (2.55 wt% Cl); □ = TiO_2 soaked in HCl/H_2O (2.30 wt% Cl).

Effects of Sintering on Catalyst Activity

Baker and co-workers [103] have reviewed the effects of sintering on catalytic activity. Specific activity (based on catalytic surface area) can either increase or decrease with increasing metal crystallite size during sintering if the reaction is structure-sensitive, or it can be independent of changes in metal crystallite size if the reaction is structure-insensitive. Thus, for a structure-sensitive reaction, the impact of sintering may be either magnified or moderated; while for a structure insensitive-reaction, sintering has in principle no effect on specific activity (per unit surface area). In the latter case, the decrease in mass-based activity is proportional to the decrease in metal surface area. Ethane hydrogenolysis and ethane steam reforming are examples of structure-sensitive reactions, while CO hydrogenation on supported cobalt, nickel, iron, and ruthenium is largely structure-insensitive in catalysts of moderate loading and dispersion.

Gas/Vapor–Solid and Solid-State Reactions

In addition to poisoning, there are a number of chemical routes leading to catalyst deactivation: (1) reactions of the vapor phase with the catalyst surface to produce (a) inactive bulk and surface phases (rather than strongly adsorbed species), (b) volatile compounds that exit the catalyst and reactor in the vapor phase, or (c) sintering due to adsorbate interactions, that we call chemical-assisted sintering to distinguish it from thermal sintering previously discussed; (2) catalytic solid-support or catalytic solid-promoter reactions, and (3) solid-state transformations of the catalytic phases during reaction. Each of these routes is discussed in some detail below.

Gas/Vapor–Solid Reactions

Reactions of Gas/Vapor with Solid to Produce Inactive Phases

Dispersed metals, metal oxides, metal sulfides, and metal carbides are typical catalytic phases, the surfaces of which are similar in composition to the bulk phases. For a given reaction, one of these catalyst types is generally substantially more active than the others, e.g., only Fe and Ru metals are active for ammonia synthesis, while the oxides, sulfides, and carbides are inactive. If, therefore, one of these metal catalysts is oxidized, sulfided, or carbided, it will lose essentially all of its activity. While these chemical modifications are closely related to poisoning, the distinction here is that rather than losing activity owing to the presence of an adsorbed species, the loss of activity is due to the formation of a new phase altogether.

Examples of vapor-induced chemical transformations of catalysts to inactive phases are listed in Table 11[8,120,121,122,123,124,125,126,127]. These include the formation of $RhAl_2O_4$ in the three-way Pt–Rh/Al_2O_3 catalyst during high temperature operation in an auto exhaust; oxidation of Fe by low levels of O_2 during ammonia synthesis or by H_2O during regeneration; dealumination (migration of Al from the zeolite framework) of Y-zeolite during high temperature catalytic cracking and regeneration in steam; reaction of SO_3 with the alumina support to form aluminum sulfate leading to support breakdown and catalyst pore plugging in several processes, including CO oxidation in a gas turbine exhaust, conversion of CO and hydrocarbons in a diesel exhaust converter, and selective catalytic reduction (SCR) of NO_x in utility boiler flue gases [8,122,123,124,127]; oxidation of Fe_5C_2 to Fe_3O_4 and of Co metal supported on alumina or silica to Co surface aluminates or silicates during Fischer–Tropsch synthesis at high conversions and hence high $P_{H2}O$; and formation of $NiAl_2O_4$ during reaction and steam regeneration of Ni/Al_2O_3 in a slightly oxidizing atmosphere above about 500 °C, especially if more reactive aluminas, e.g., γ, δ, or θ forms, are used as supports. Because reaction of SO_3 with γ-Al_2O_3 to produce $Al_2(SO_4)_3$ is a serious cause of deactivation of alumina-supported catalysts in several catalytic processes (e.g., diesel exhaust abatement and SCR), TiO_2 or SiO_2 carriers are used rather than Al_2O_3 or in the diesel or automotive exhaust the alumina catalyst is stabilized by addition of BaO, SrO, or ZrO_2 [8,122,123,124,125,126,127].

Table 11: Examples of Reactions of Gases/Vapors with Catalytic Solids to Produce Inactive Phases.

Catalytic process	Gas/vapor composition	Catalytic solid	Deactivating chemical reaction	Ref.
Auto emissions control	N_2, O_2, HCs, CO, NO, H_2O, SO_2	Pt–Rh/Al_2O_3	2 Rh_2O_3 + γ-Al_2O_3→$RhAl_2O_4$ + 0.5 O_2	[120,121]
Ammonia synthesis and regeneration	H_2, N_2	Fe/K/Al_2O_3	Fe→FeO at >50 ppm O_2	[8]
	Traces O_2, H_2O		Fe→FeO at >0.16 ppm H_2O/H_2	
Catalytic cracking	HCs, H_2, H_2O	La-Y-zeolite	H_2O induced Al migration from zeolite framework causing zeolite destruction	[8]
CO oxidation, gas turbine exhaust	N_2, O_2, 400 ppm CO, 100–400 ppm SO_2	Pt/Al_2O_3	2 SO_3 + γ-Al_2O_3→$Al_2(SO_4)_3$ which blocks catalyst pores	[8]
Diesel HC/soot emissions control	N_2, O_2, HCs (gas and liquid), CO, NO, H_2O, soot, SO_2	Pt/Al_2O_3 and β-zeolite; oxides of CaCuFeVK on TiO_2	Formation of $Al_2(SO_4)_3$ or sulfates of Ca, Cu, Fe, or V, which block catalysts pores and lower activity for oxidation; Al_2O_3 stabilized by BaO	[122–124]
Fischer–Tropsch	CO, H_2, H_2O, CO_2, HCs	Fe/K/Cu/SiO_2	Fe_5C_2→Fe_3O_4 due to oxidation at high X_{CO} by product H_2O, CO_2	[125]
Fischer–Tropsch	CO, H_2, H_2O, HCs	Co/SiO_2	Co + SiO_2→CoO·SiO_2 and collapse of SiO_2 by product H_2O	[126]
Selective catalytic reduction (SCR), stationary	N_2, O_2, NO, PM [a], H_2O, SO_2	$V_2O_5/WO_3/TiO_2$	Formation of $Al_2(SO_4)_3$ if Al_2O_3 is used	[127]
Steam reforming and regeneration in H_2O	CH_4, H_2O, CO, H_2, CO_2	Ni/Al_2O_3	Ni + Al_2O_3→$NiAl_2O_4$	[8]

[a] Particulate matter.

Reactions of Gas/Vapor with Solid to Produce Volatile Compounds

Metal loss through direct vaporization is generally an insignificant route to catalyst deactivation. By contrast, metal loss through formation of volatile compounds, e.g., metal carbonyls, oxides, sulfides, and halides in CO, O_2, H_2S, and halogen-containing environments, can be significant over a wide range of conditions, including relatively mild conditions. Classes and examples of volatile compounds are listed in Table 12. Carbonyls are formed at relatively low temperatures but high pressures of CO; halides can be formed at relatively low temperatures and low concentration of the halogens. However, the conditions under which volatile oxides are formed vary considerably with the metal; for example, RuO_3 can be formed at room temperature, while PtO_2 is formed at measurable rates only at temperatures exceeding about 500 °C.

Table 12: Types and Examples of Volatile Compounds Formed in Catalytic Reactions.

Gaseous environment	Compound type	Example of compound
CO, NO	Carbonyls and nitrosyl carbonyls	$Ni(CO)_4$, $Fe(CO)_5$ (0–300 °C) [a]
O_2	Oxides	RuO_3 (25 °C), PbO (>850 °C), PtO_2 (>700 °C)
H_2S	Sulfides	MoS_2 (>550 °C)
Halogens	Halides	$PdBr_2$, $PtCl_4$, PtF_6, $CuCl_2$, Cu_2Cl_2

[a] Temperatures of vapor formation are listed in parentheses.

While the chemical properties of volatile metal carbonyls, oxides, and halides are well known, there is surprisingly little information available on their rates of formation during catalytic reactions. There have been no reviews on this subject and relatively few reported studies to define the effects of metal loss on catalytic activity [28,128,129,130,131,132,133,134,135,136,137,138,139,140,141]. Most of the previous work has focused on volatilization of Ru in automotive converters [128,129,130,131]; nickel carbonyl formation in nickel catalysts during methanation of CO [133,139] or during CO chemisorption at 25 °C [28,135], and formation of Ru carbonyls during Fischer–Tropsch synthesis [136,137]; volatilization of Pt during ammonia oxidation on Pt–Rh gauze catalysts [140,141]; and volatilization of Cu from methanol synthesis and diesel soot oxidation catalysts, leading to sintering in the former and better catalyst–soot contact but also metal loss in the latter case [109].

Results of selected studies are summarized in Table 13. Bartholomew [131] found evidence of significant (50%) Ru loss after testing of a Pd–Ru catalyst in an actual reducing automobile exhaust for 100 h, which he attributed to formation of a volatile ruthenium oxide and which was considered responsible at least in part for a significant loss (20%) of NO reduction activity.

Table 13: Documented Examples of Reactions of Vapor with Solid to Produce Volatile Compounds.

Catalytic process	Catalytic solid	Vapor formed	Comments on deactivation process	Ref.
Automotive converter	$Pd–Ru/Al_2O_3$	RuO_4	50% loss of Ru during 100-h test in reducing automotive exhaust	[131]
Methanation of CO	Ni/Al_2O_3	$Ni(CO)_4$	P_{CO} > 20 kPa and T < 425 °C due to $Ni(CO)_4$ formation, diffusion and decomposition on the support as large crystallites	[133]
CO chemi-sorption	Ni catalysts	$Ni(CO)_4$	P_{CO} > 0.4 kPa and T > 0 °C due to $Ni(CO)_4$ formation; catalyzed by sulfur compounds	[134]
Fischer–Tropsch synthesis (FTS)	Ru/NaY zeolite, Ru/Al_2O_3, Ru/TiO_2	$Ru(CO)_5$, $Ru_3(CO)_{12}$	Loss of Ru during FTS (H_2/CO = 1, 200–250 °C, 1 atm) on Ru/NaY zeolite and Ru/Al_2O_3; up to 40% loss while flowing CO at 175–275 °C over Ru/Al_2O_3 for 24 h. Rate of Ru loss less on titania-supported Ru and for catalysts containing large metal crystallites (3 nm) relative to small metal crystallites (1.3 nm). Surface carbon lowers loss	[136,137]
Ammonia oxidation	Pt–Rh gauze	PtO_2	Loss: 0.05–0.3 g Pt/ton HNO_3; recovered with Pd gauze; loss of Pt leads to surface enrichment with inactive Rh	[8,142]
HCN synthesis	Pt–Rh gauze	PtO_2	Extensive restructuring and loss of mechanical strength	[8,143]
Methanol synthesis	CuZnO	$CuCl_2$, Cu_2Cl_2	Mobile copper chloride phase leads to sintering at reaction temperature (225 °C)	[109]
Diesel soot oxidation	Oxides of K, Cu, Mo, and trace C1	$CuCl_2$, Cu_2Cl_2	Mobile copper chloride improves catalyst–soot contact; catalyst evaporation observed	[109]

Shen and co-workers [133] found that Ni/Al_2O_3 methanation catalysts deactivate rapidly during methanation at high partial pressures of CO (>20 kPa) and temperatures below 425 °C because of $Ni(CO)_4$ formation, diffusion, and decomposition on the support as large crystallites; under severe conditions (very high P_{CO} and relatively low reaction temperatures) loss of nickel metal occurs. Thus, loss of nickel and crystallite growth could be serious problems at the entrance to methanation reactors where the temperature is low enough and P_{CO} high enough for metal carbonyl formation. Agnelli and co-workers [139] investigated kinetics and modeling of sintering due to formation and migration of nickel carbonyl species. They found that the initially sharp crystallite size distribution evolved during several hours of sintering under low temperature (230 °C) reaction conditions to a bimodal system consisting of small spherical crystallites and large faceted crystals favoring (111) planes. The sintering process was modeled in terms of an Ostwald-ripening mechanism coupled with mass transport of mobile subcarbonyl intermediates. Long-term simulations were found to predict reasonably well the ultimate state of the catalyst. On the basis of their work, they proposed two solutions for reducing loss of nickel: (1) increasing reaction temperature and decreasing CO partial pressure in order to lower the rate of carbonyl formation, and (2) changing catalyst composition, e.g., alloying nickel with copper or adding alkali to inhibit carbonyl species migration.

Of note, Kuo and Hwang have shown that the particle morphology itself affects the rate of Ostwald ripening due to different relative chemical potential energies of the surfaces [144]. Using silver nanoparticles, they found that atoms at sharp edges and corners were removed first, resulting in more rounded particles for all starting geometries. Thus, initial particle geometry appears to have an effect in addition to the chemical atmosphere experienced by the particles.

Loss of nickel metal during CO chemisorption on nickel catalysts at temperatures above 0 °C is also a serious problem; moreover, this loss is catalyzed by sulfur poisoning [28]. In view of the toxicity of nickel tetracarbonyl, the rapid loss of nickel metal, and the ill-defined adsorption stoichiometries, researchers are advised to avoid using CO chemisorption for measuring nickel surface areas; instead, hydrogen chemisorption, an accepted ASTM method with a well-defined adsorption stoichiometry, is recommended [145]. Figure 20 illustrates a mechanism for the formation of $Ni(CO)_4$ on a crystallite of nickel in CO atmosphere.

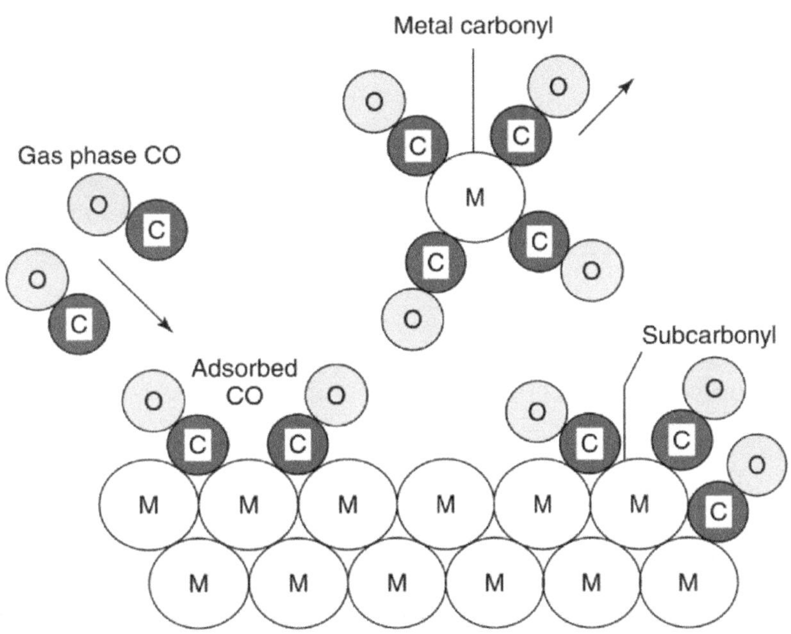

Figure 20: Formation of volatile tetra-nickel carbonyl at the surface of nickel crystallite in CO atmosphere. Reproduced from [8]. Copyright 2006, Wiley-Interscience.

Goodwin and co-workers [136,137] studied the influence of reaction atmosphere, support, and metal particle size on the loss of Ru due to carbonyl formation. They found that the loss of Ru during CO hydrogenation (H_2/CO = 1, 200–250 °C, 1 atm) on Ru/NaY zeolite and Ru/Al_2O_3 for extended periods of time was significant (e.g., up to 40% while flowing CO at 175–275 °C over Ru/Al_2O_3 for 24 h). The loss of Ru was significantly less on titania-supported Ru; moreover, the rate of loss was lower for catalysts containing large metal crystallites (3 nm) relative to those containing small metal crystallites (1.3 nm). Metal loss was inhibited in part at higher reaction temperatures as a result of carbon deposition. Thus, while it is clear that loss of ruthenium could be a serious problem in Fischer–Tropsch synthesis, there are measures in terms of catalyst design and choice of reaction conditions that can be taken to minimize loss.

One of the most dramatic examples of vapor phase loss of the catalyst occurs during NH_3 oxidation on Pt–Rh gauze, an important reaction in the manufacture of nitric acid [8,140,141]. At the high reaction temperature (~900 °C), formation of a volatile platinum oxide (PtO_2) occurs at a very significant rate; in fact, the rate of loss of 0.05–0.3 g Pt/ton of HNO_3 is high enough to provide a substantial economic incentive for Pt recovery [8]. The most effective recovery process involves placing a woven Pd-rich alloy gauze immediately below the Pt–Rh gauze to capture the Pt through formation of a Pd–Pt alloy. Pt loss is also the most significant cause of catalyst deactivation as the gauze surface becomes enriched in nonvolatile but inactive rhodium oxide [142], requiring shutdown and catalyst replacement every 3–12 months [8].

Decomposition of volatile platinum oxide species formed during high temperature reaction may (similar to the previously discussed formation of large crystallites of Ni from $Ni(CO)_4$) lead to formation of large Pt crystallites and/or substantial restructuring of the metal surface. For example, Wu and Phillips [146,147,148] observed surface etching, enhanced sintering, and dramatic surface restructuring of Pt thin films to faceted particles during ethylene oxidation over a relatively narrow temperature range (500–700 °C). The substantially higher rate of sintering and restructuring in O_2/C_2H_4 relative to that in nonreactive atmospheres was attributed to the interaction of free radicals such as HO_2, formed homogeneously in the gas phase, with the metal surface to form metastable mobile intermediates. Etching of Pt–Rh gauze in a H_2/O_2 mixture under the same conditions as Pt surfaces (600 °C, $N_2/O_2/H_2$ = 90/7.5/2.5) has also been reported [143]. A significant weight loss was observed in a laminar flow reactor with little change in surface roughness, while in an impinging jet reactor, there was little weight loss, but substantial restructuring of the surface to particle-like structures, 1–10 μm in diameter; these particles were found to have the same Pt–Rh composition as the original gauze. The nodular structures of about 10-μm diameter formed in these experiments are strikingly similar to those observed on Pt–Rh gauze after use in production of HCN at 1100 °C in 15% NH_3, 13% CH_4, and 72% air (see Figure 21). Moreover, because of the high space velocities during HCN production, turbulent rather than laminar flow would be expected, as in the impinging jet reactor. While little Pt is volatilized from the Pt–Rh gauze catalyst during HCN synthesis, the extensive restructuring leads to mechanical weakening of the gauze [8].

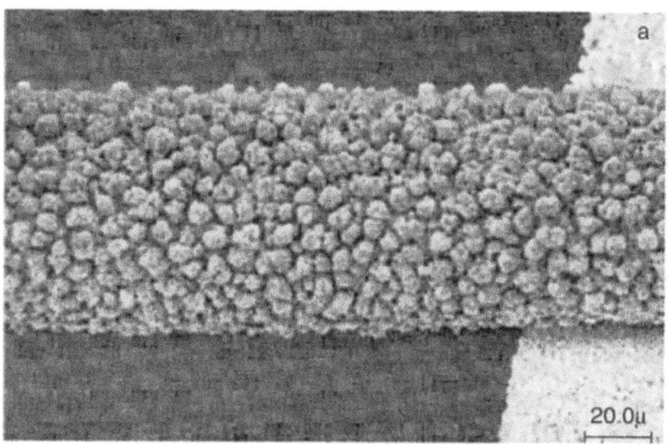

(a)

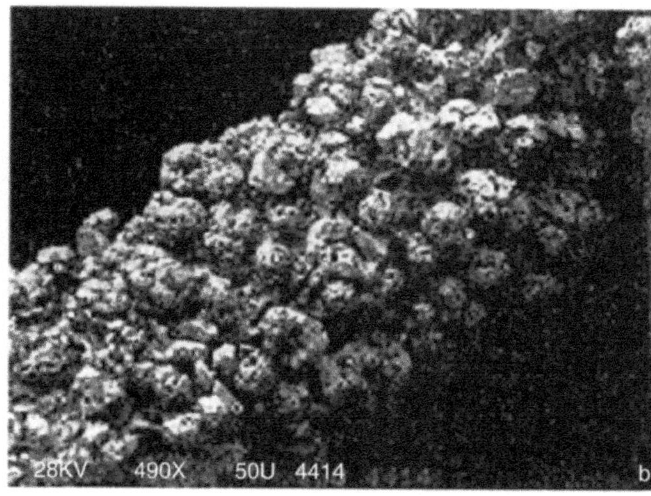

(b)

Figure 21: (**a**) SEM of Pt–Rh gauze after etching in $N_2/O_2/H_2$ = 90/7.5/2.5 at 875 K for 45 h. Reproduced from [143]. Copyright 1992, Elsevier. (**b**) SEM of Pt–Rh gauze after use in production of HCN (magnification 1000×). Photographs courtesy of Ted Koch at DuPont, personal correspondence to the author.

Other examples of catalyst deactivation due to volatile compound formation include (1) loss of the phosphorus promoter from the VPO catalyst used in the fluidized-bed production of maleic anhydride, with an attendant loss of catalyst selectivity [8], (2) vapor-phase loss of the potassium promoter from steam-reforming catalysts in the high temperature, steam-containing environment [8], and (3) loss of Mo from a 12-Mo-V-heteropolyacid due to formation of a volatile Mo species during oxydehydrogenation of isobutyric acid to methacrylic acid [138].

While relatively few definitive studies of deactivation by volatile compound formation have been reported, the previous work does provide the basis for enumerating some general principles. A generalized mechanism of deactivation by formation of volatile metal compounds can be postulated (see Figure 22). In addition, the roles of kinetics and thermodynamics can be stated in general terms:

(1) At low temperatures and partial pressures of the volatilization agent (VA), the overall rate of the process is limited by the rate of volatile compound formation.

(2) At intermediate temperatures and partial pressures of the VA, the rate of formation of the volatile compound exceeds the rate of decomposition. Thus, the rate of vaporization is high, the vapor is stable, and metal loss is high.

(3) At high temperatures and partial pressures of the VA, the rate of formation equals the rate of decomposition, *i.e.*, equilibrium is achieved. However, the volatile compound may be too unstable to form or may decompose before there is an opportunity to be transported from the system. From the previous work, it is also evident that besides temperature and gas phase composition, catalyst properties (crystallite size and support) can play an important role in determining the rate of metal loss.

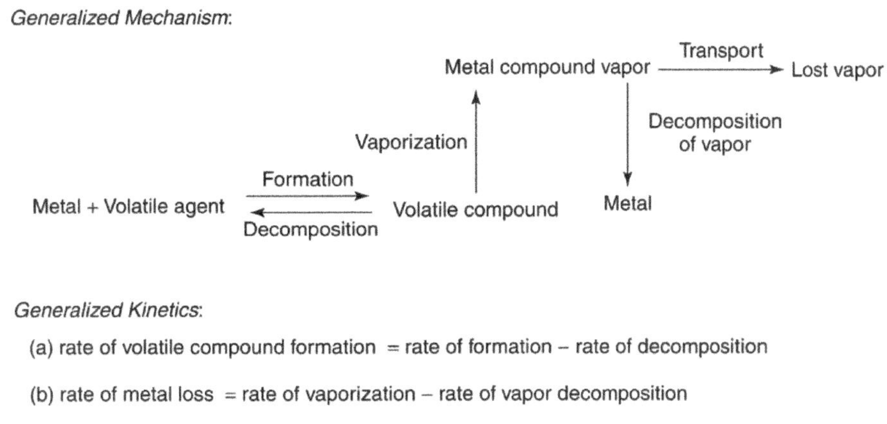

Figure 22: Generalized mechanisms and kinetics for deactivation by metal loss. Reproduced from [8]. Copyright 2006, Wiley-Interscience.

Solid-State Reactions

Catalyst deactivation by solid-state diffusion and reaction appears to be an important mechanism for degradation of complex multicomponent catalysts in dehydrogenation, synthesis, partial oxidation, and total oxidation reactions [8,149,150,151,152,153,154,155,156,157,158,159,160]. However, it is difficult in most of these reactions to know the extent to which the solid-state processes, such as diffusion and solid-state reaction, are affected by surface reactions. For example, the rate of diffusion of Al_2O_3 to the surface to form an aluminate may be enhanced by the presence of gas-phase oxygen or water or the nucleation of a different phase may be induced by either reducing or oxidizing conditions. Recognizing this inherent limitation, the focus here is nevertheless on processes in which formation of a new bulk phase (and presumably the attendant surface phase) leads to substantially lower activity. There is probably some overlap with some of the examples given under Gas/Vapor–Solid Reactions involving reactions of gas/vapor with solid to produce inactive phases.

Examples from the literature of solid-state transformations leading to catalyst deactivation are summarized in Table 14. They include (1) the formation of $KAlO_2$ during ammonia synthesis at the $Fe/K/Al_2O_3$ catalyst surface, (2) decomposition of the active phase PdO to inactive Pd metal during catalytic combustion on PdO/Al_2O_3 and PdO/ZrO_2 catalysts, (3) transformation of active carbides to inactive carbides in Fischer–Tropsch synthesis on Fe/K/Cu catalysts, (4) formation of inactive V(IV) compounds in SO_2 oxidation, and (5) reductive transformation of iron molybdate catalysts during partial oxidation of benzene, methanol, propene, and isobutene.

Table 14: Examples of Solid-State Transformations Leading to Catalyst Deactivation.

Catalytic process	Catalytic solid	Deactivating chemical reaction	Ref.
Ammonia synthesis	$Fe/K/Al_2O_3$	Formation of $KAlO_2$ at catalyst surface	[159]
Catalytic combustion	PdO/Al_2O_3, PdO/ZrO_2	PdO→Pd at $T > 800$ °C	[152]
Catalytic combustion	Co/K on MgO, CeO_2, or La_2O_3	Formation of CoO–MgO solid soln., $LaCoO_3$, or K_2O film on CeO_2	[160]
Dehydrogenation of ethyl benzene to styrene	$Fe_2O_3/Cr_2O_3/K_2O$	K migration to center of pellet caused by thermal gradient	[8]
Fischer–Tropsch	Fe/K, Fe/K/CuO	Transformation of active carbides to inactive carbides	[157,158]
Oxidation of SO_2 to SO_3	$V_2O_5/K_2O/Na_2O/$	Formation of inactive V(IV) compounds at $T < 420$–430 °C	[155]
Partial oxidation of benzene to maleic anhydride	V_2O_5–MoO_3	Decreased selectivity due to loss of MoO_3 and formation of inactive vanadium compounds	[149]
Partial oxidation of methanol to formaldehyde	$Fe_2(MoO_4)_3$ plus MoO_3	Structural reorganization to β-$FeMoO_4$; reduction of MoO_3	[150,156]
Partial oxidation of propene to acrolein	$Fe_2(MoO_4)_3$	Reductive transformation of $Mo_{18}O_{52}$ to Mo_4O_{11}	[153,156]
Partial oxidation of isobutene to methacrolein	$Fe_2(MoO_4)_3$	Reduction to $FeMoO_4$ and MoO_{3-x}	[151,154]

There are basic principles underlying most solid-state reactions in working catalysts that have been enumerated by Delmon [156]: (1) the active catalytic phase is generally a high-surface-area defect structure of high surface energy and as such a precursor to more stable, but less active phases and (2) the basic reaction processes may itself trigger the solid-state conversion of the active phase to an inactive phase; for example, it may involve a redox process, part of which nucleates the inactive phase.

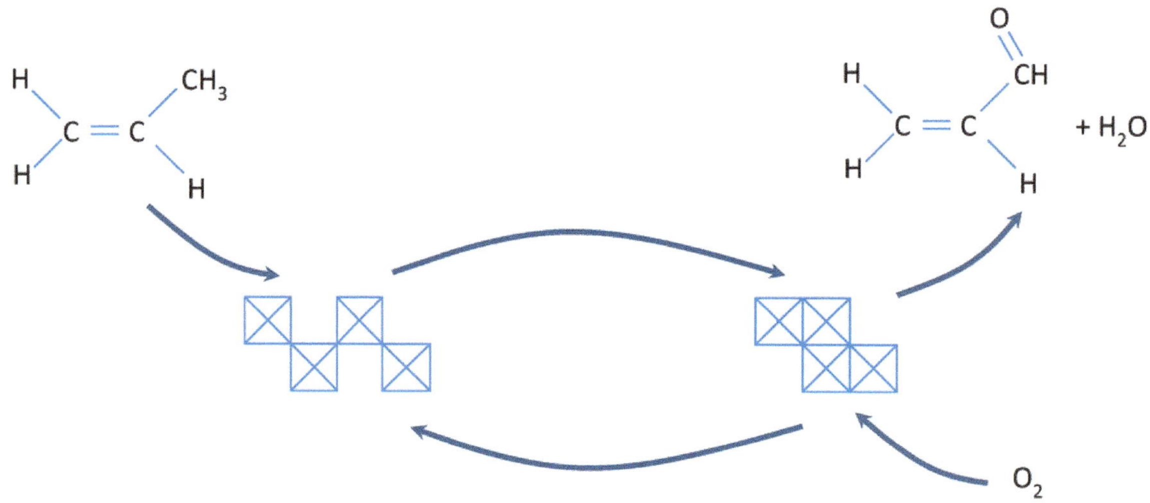

Figure 23: Schematic representation of the cyclic reduction/oxidation of twin pairs of MoO_6 octahedra between the corner and the edge-sharing arrangements (boxes represent MoO_6 octahedra with sharing of oxygen atoms at corners for MoO_3 or edges for MoO_2). The figure is not completely accurate, because it cannot take into account the fact that the arrangements are not perpendicular to the main axes of the lattice. Adapted from [156].

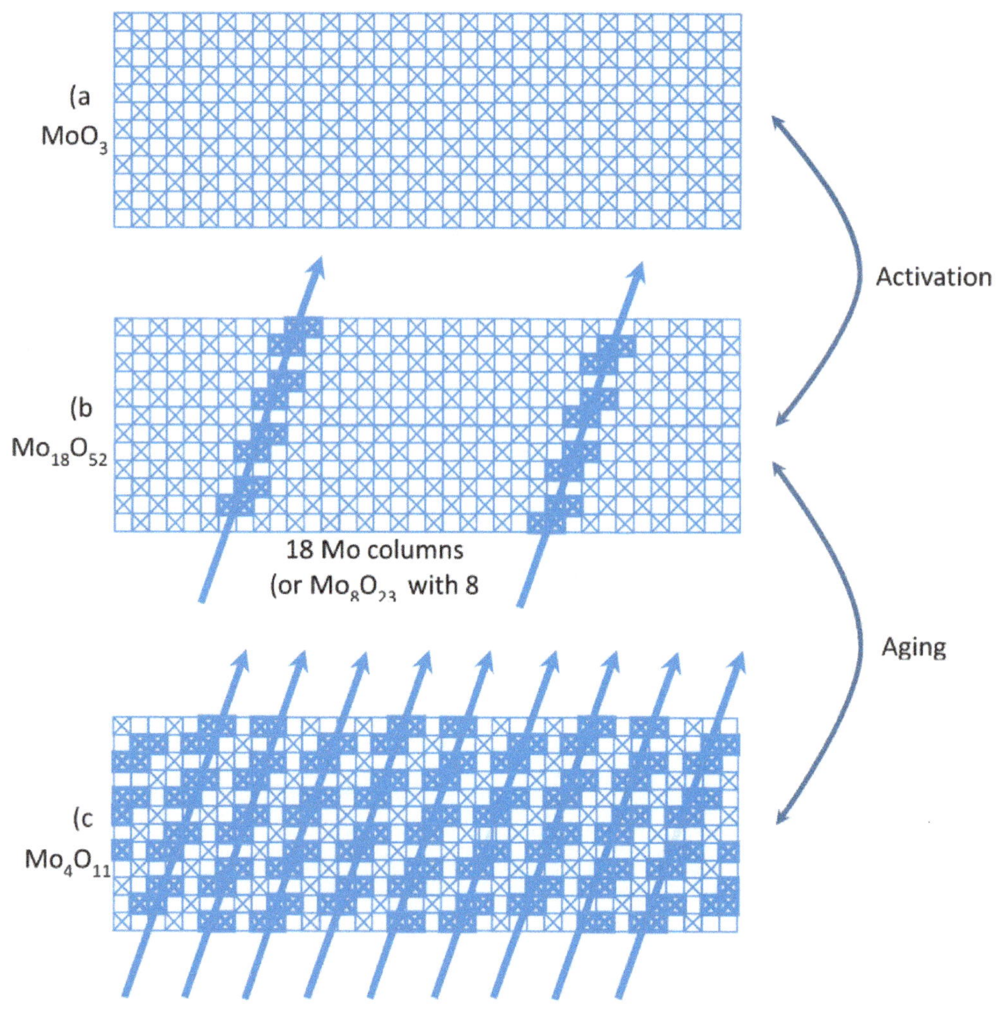

Figure 24: Schematic representation of the structure of MoO_3, $Mo_{18}O_{52}$, and Mo_4O_{11}. The shear planes in $Mo_{18}O_{52}$ and Mo_4O_{11} are represented by the oblique arrows (boxes with an "X" represent MoO_5 octahedra). Adapted from [156].

A well-documented example of these principles occurs in the partial oxidation of propene to acrolein on a $Fe_2(MoO_4)_3$catalyst [153,156]. This oxidation occurs by the "*Mars van Krevelen*" *mechanism*, *i.e.*, a redox mechanism in which lattice oxygen reacts with the adsorbed hydrocarbon to produce the partially oxygenated product; the reduced catalyst is restored to its oxidized state through reaction with gaseous oxygen. In propene oxidation, two atoms of oxygen from the catalyst are used, one for removing two hydrogen atoms from the olefin and the other one in forming the unsaturated aldehyde. The fresh, calcined catalyst MoO_3 consists of corner-sharing MoO_6 octahedra (with Mo at the center and six oxygen atoms at the corners); but upon reduction to MoO_2, octahedra share edges as shown in Figure 23. However, it has been reported [153,156] that only slightly reduced (relative to MoO_3), open structures such as $Mo_{18}O_{52}$ and Mo_8O_{23} are the most active, selective phases; more complete reduction of either of these structures leads to formation of Mo_4O_{11} (see Figure 24) having substantially lower selectivity. Delmon and co-workers [154,156] have shown that addition of an oxygen donor such as Sb_2O_4 facilitates spillover of oxygen and thereby prevents overreduction and deactivation of the catalyst.

Reactions of Gas/Vapor with Solid to Restructure the Surface by Chemical Assisted Sintering

The surfaces of metallic catalysts can be greatly roughened by interactions with the reactants and/or products. However, as opposed to forming volatile species that are transported out of the reactor as discussed in the previous section, these interactions lead to a restructuring of the surface that is similar to that which occurs during thermal sintering, but at temperatures which are below the Tamman or Huttig temperatures, respectively defined as 0.5 and 0.3 of the melting point (T_m) of the material, at which thermal sintering might be expected. Therefore, this surface restructuring must be attributed to the interaction of the gas phase with the solid. The following three examples from the literature highlight the chemical-assisted sintering process caused by adsorbate-surface interactions on Ni, Co, and Pd surfaces.

Chemical sintering of Ni/alumina catalysts in methanation due to formation of volatile $Ni(CO)_4$ followed by its decomposition downstream to large Ni crystallites has been well documented [8,105]. Moreover, deactivation of Ni/alumina by Ni aluminate formation is also observed at the exit of methanators where temperature is moderately high (T = 450 °C) and steam partial pressure is maximum [105].

Wilson and de Groot [161] reported that under high pressure (4 bar, H_2/CO = 2) and moderate temperature (523 K) conditions, single crystal Co (0001) surfaces restructured significantly due to interaction with the CO, which they attributed to an etch-regrowth mechanism. The left hand panel of Figure 25a shows the scanning tunneling microscopy (STM) image of the single crystal surface, while Figure 25b shows the same location after exposure to the H_2/CO atmosphere for 1 h The surface restructuring and roughening is profound, with the peaks approximately four atoms high relative to the previously smooth surface that had only well-defined steps interrupting the (0001) planar surface.

More recently, Parkinson *et al.* [162] have shown that chemical-assisted sintering occurs at room temperature for palladium supported on magnetite under ultra-high vacuum conditions with CO partial pressures of only 5×10^{-10} mbar. Figure 26 shows four STM images from a movie that demonstrates the surface mobility of the Pd at these low CO partial pressures. Figure 26a is the surface prior to CO exposure, while Figure 26b–d show the surface as a function of time up to about an hour of exposure. The authors note that hydroxyl-Pd groups (OH-Pd), identified by the ×'s in the images, serve as anchoring points for the coalescence of larger Pd clusters. The full movie, available with the supplementary material for this article [162], is recommended to fully appreciate the unexpectedly high atomic mobility under these conditions.

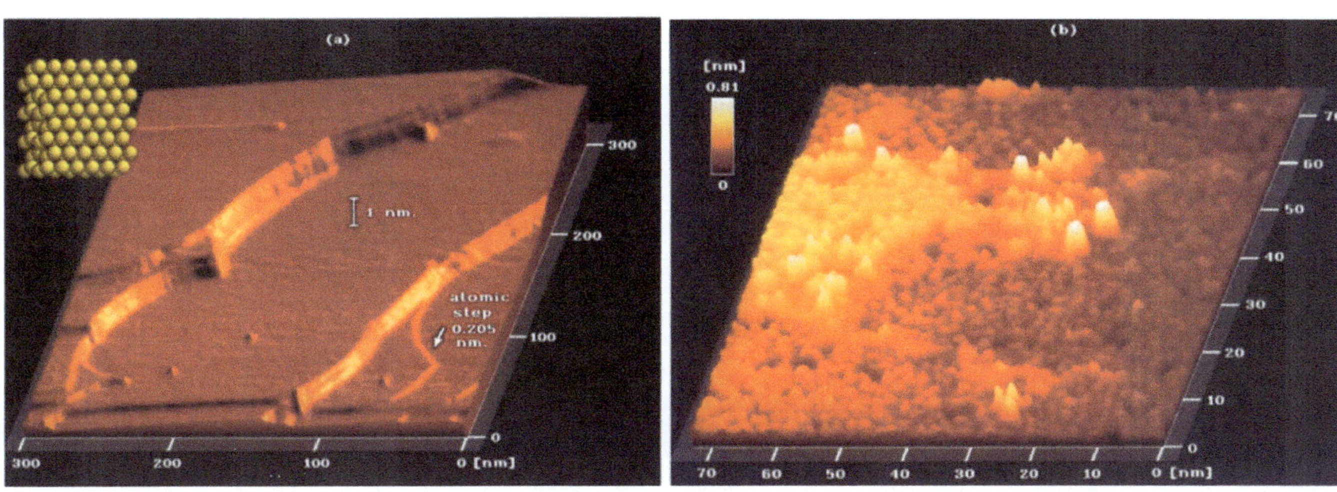

Figure 25: STM images of the Co (0001) surface (**a**) before and (**b**) after 1 h exposure to 4 MPa 2:1 H_2:CO atmosphere at 523 K. Reproduced from [161]. Copyright 1995, American Chemical Society.

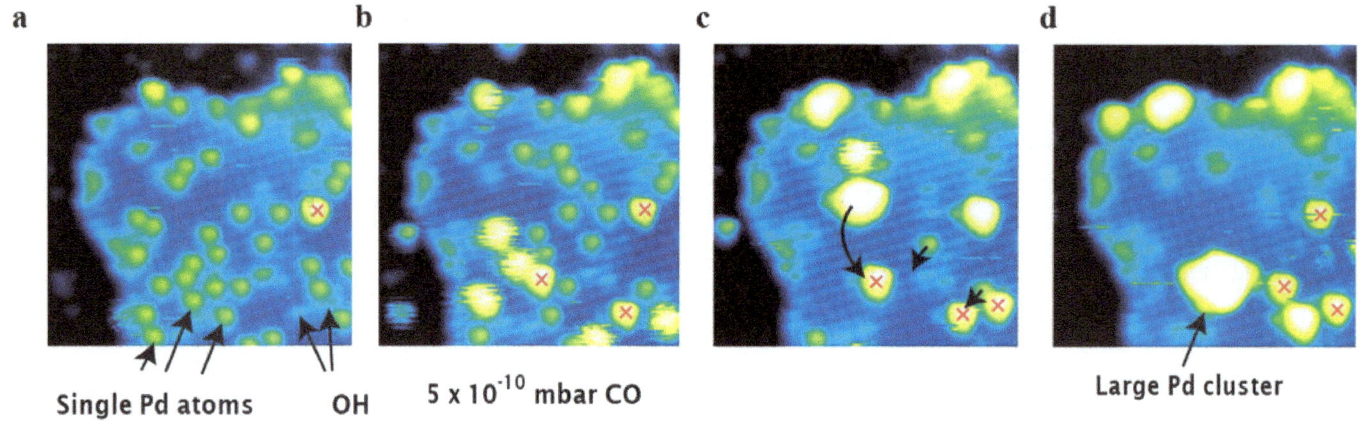

Figure 26: "The CO-induced formation of a large Pd cluster. a – d, Four STM images (14 × 14 nm², +1 V, 0.2 nA) selected from a 36-frame STM movie (duration 1 h 50 min) following the deposition of 0.2 ML Pd [on Fe_3O_4] at RT. Initially (**a**), isolated Pd atoms are present, together with hydroxyl groups and one OH–Pd (red cross). After three frames the background pressure of CO is raised to 5×10^{-10} mbar. Thirty minutes later (frame **b**), several mobile 'fuzzy' Pd carbonyl species, trapped at other Pd atoms, have formed. Shortly afterwards (**c**), three Pd carbonyls and four adatoms have formed a large cluster. Twenty-five minutes later (**d**), the cluster has captured another Pd carbonyl, and diffused to merge with an OH–Pd species.". Reproduced from [162]. Copyright 2013, MacMillan Publishers.

Mechanical Failure of Catalysts

Forms and Mechanisms of Failure

Mechanical failure of catalysts is observed in several different forms that depend on the type of reactor, including (1) crushing of granular, pellet, or monolithic catalyst forms due to a load in fixed beds; (2) attrition, the size reduction, and/or breakup of catalyst granules or pellets to produce fines, especially in fluid or slurry beds; and (3) erosion of catalyst particles or monolith coatings at high fluid velocities in any reactor design. Attrition is evident by a reduction in the particle size or a rounding or smoothing of the catalyst particle easily observed under an optical or electron microscope. Washcoat loss is observed by scanning a cross section of the honeycomb channel with either an optical or an electron microscope. Large increases in pressure drop in a catalytic process are often indicative of fouling, masking, or the fracturing and accumulation of attritted catalyst in the reactor bed.

Commercial catalysts are vulnerable to mechanical failure in large part because of the manner in which they are formed; that is, catalyst granules, spheres, extrudates, and pellets ranging in diameter from 50 μm to several millimeters are in general prepared by agglomeration of 0.02–2 μm aggregates of much smaller primary particles having diameters of 10–100 nm by means of precipitation or gel formation, followed by spray drying, extrusion, or compaction. These agglomerates have, in general, considerably lower strengths than the primary particles and aggregates of particles from which they are formed.

Two principal mechanisms are involved in mechanical failure of catalyst agglomerates: (1) fracture of agglomerates into smaller agglomerates of approximately $0.2d_0$–$0.8d_0$ and (2) erosion (or abrasion) from the surface of the agglomerate of aggregates of primary particles having diameters ranging from 0.1 to 10 μm [163]. While erosion is caused by mechanical stresses, fracture may be due to mechanical, thermal, and/or chemical stresses. Mechanical stresses leading to fracture or erosion in fluidized or slurry beds may result from (1) collisions of particles with each other or with reactor walls or (2) shear forces created by turbulent eddies or collapsing bubbles (cavitation) at high fluid velocities. Thermal stresses occur as catalyst particles are heated and/or cooled rapidly; they are magnified by temperature gradients across particles and by differences in thermal expansion coefficients at the interface of two different materials, e.g., catalyst coating/monolith interfaces; in the latter case the heating or cooling process can lead to fracture and separation of the catalyst coating. Chemical stresses occur as phases of different density are formed within a catalyst particle via chemical reaction; for example, carbiding of primary iron oxide particles increases their specific volume and micromorphology leading to stresses that break up these particles [164]. A further example occurs in supported metal catalysts when large quantities of filamentous carbon (according to reaction mechanisms discussed earlier) overfill catalysts pores, generating enormous stresses that can fracture primary particles and agglomerates.

Role of Physical and Chemical Properties of Ceramic Agglomerates in Determining Strength and Attrition Resistance

Factors Affecting the Magnitude of Stress Required for Agglomerate Breakage and the Mechanisms by Which It Occurs

The extent to which a mechanism, *i.e.*, fracture or erosion, participates in agglomerate size reduction depends upon several factors: (1) the magnitude of a stress, (2) the strength and fracture toughness of the agglomerate, (3) agglomerate size and surface area, and (4) crack size and radius. Erosion (abrasion) occurs when the stress (e.g., force per area due to collision or cavitation pressure) exceeds the agglomerate strength, *i.e.*, the strength of bonding between primary particles. Erosion rate is reportedly [163] proportional to the external surface area of the catalyst; thus, erosion rate increases with decreasing agglomerate size.

Fracture Toughness of Ceramic Agglomerates

Most heterogeneous catalysts are complex, multiphase materials that consist, in large part, of porous ceramic materials,*i.e.*, are typically oxides, sulfides, or metals on an oxide carrier or support. When a tensile stress of a magnitude close to the yield point is applied, ceramics almost always undergo brittle fracture before plastic deformation can occur. Brittle fracture occurs through formation and propagation of cracks through the cross section of a material in a direction perpendicular to the applied stress. Agglomerate fracture due to a tensile stress occurs by propagation of internal and surface flaws; these flaws, created by external stresses or inherent defects, are stress multipliers, *i.e.*, the stress is multiplied by $2(a/r)^{0.5}$, where *a*is the crack length and *r* is the radius of curvature of the crack tip; since *a*/*r* can vary from 2 to 1000, the effective stress at the tip of a crack can be 4–60 times the applied stress. Tensile stress multipliers may be microcracks, internal pores, and grain corners.

The ability of a material to resist fracture is termed *fracture toughness*. The plane strain fracture toughness, K_{Ic}, is defined as

KIc=Yσ(πα)0.5 (3)

where *Y* is a dimensionless parameter (often close to 1.0–2.0), the magnitude of which depends upon both specimen and crack geometries, σ is the applied stress, and *a* is the length of a surface crack or half the length of an internal crack. Crack propagation and fracture are likely if the right hand side of Equation 3 exceeds the experimental value of plane strain fracture toughness (left-hand side of Equation 3). Plane strain fracture toughness values for ceramic materials are significantly smaller than for metals and typically below 10 MPa(m)$^{0.5}$; reported values for nonporous, crystalline alumina (99.9%), fused silica, and zirconia (3 mol% Y_2O_3) are 4–6, 0.8, and 7–12 MPa(m)$^{0.5}$, respectively; flexural strengths (analogous to yield strengths for metals) for the same materials are 280–550, 100, and 800–1500 MPa [165]. Thus, on the basis of both fracture toughness and flexural strength, nonporous, crystalline zirconia is much stronger toward fracture than alumina, which in turn is much stronger than fused silica.

Effects of Porosity on Ceramic Agglomerate Strength

The introduction of porosity to crystalline or polycrystalline ceramic materials will, on the basis of stress amplification, significantly decrease elastic modulus and flexural strength for materials in tension. This is illustrated by data in Figure 27, showing that elastic modulus and flexural strength of a ceramic alumina (probably alpha form) are reduced by 75 and 85% respectively as porosity is increased from 0 to 50% [166]. Thus, according to Figure 27b, the flexural strength of typical porous aluminas used as catalyst supports might lie in the range of 30–40 MPa. However, yield strengths for γ-Al_2O_3(shown in the next section) are factors of 3–50 lower. Nevertheless, the data in Figure 27b suggest that higher strengths may be possible.

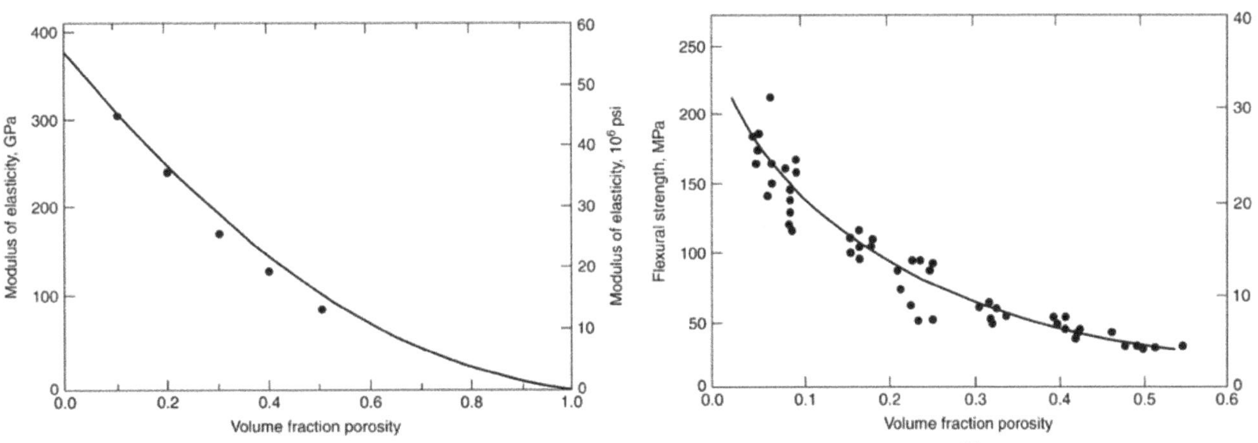

Figure 27: The influence of porosity on (**a**) the modulus of elasticity for aluminum oxide at room temperature and (**b**) the flexural strength for aluminum oxide at room temperature. Reproduced from [166]. Copyright 1956, Wiley.

Compressive Strengths of Ceramic Materials

Thus far, the discussion has focused mainly on tensile strength, the extent of which is greatly reduced by the presence of cracks or pores. However, for ceramic materials in compression, there is no stress amplification due to flaws or pores; thus ceramic materials (including catalytic materials) in compression are much stronger (approximately a factor of 10) than in tension. In addition, the strength of ceramic materials can be dramatically enhanced by imposing a residual compressive stress at the surface through thermal or chemical tempering. Moreover, introduction of binders, such as graphite, enables agglomerates of ceramic powders to undergo significant plastic deformation before fracture.

Tensile Strengths and Attrition Resistance of Catalyst Supports and Catalysts

Tensile Strength Data for Catalyst Support Agglomerates

The strengths cited above for nonporous, annealed crystalline or polycrystalline materials do not necessarily apply to porous catalyst agglomerates, even under compression; rather, agglomerate strength is dependent upon the strengths of chemical and physical bonds, including the cohesive energy, between primary particles. Agglomerate strength would depend greatly on the preparation of the compact. Representative data for catalyst agglomerates (see Table 15) suggest they are generally substantially weaker than polycrystalline ceramic materials prepared by high temperature sintering, such as the alumina cited in Figure 27 [163,165,167,168,169,170,171]. For example, Pham and co-workers [163] found that the breaking strength of a VISTA B alumina agglomerate during uniaxial compaction is in the range of 5–10 MPa—substantially lower than the reported values for heat-treated polycrystalline alumina of 280–550 MPa [165]. A large part of this difference (about 85–95%) can be attributed to porosity; however, the remaining 5–15% must be due to differences in bonding between primary particles. In other words, the bonds between primary particles in catalyst agglomerates (and some ceramic agglomerates prepared by similar methods) are typically physical in nature (e.g., involve van der Waals forces) while those in sintered polycrystalline ceramic agglomerates are principally chemical because of solid bridging of primary particles. Thus, there appears to be considerable potential for strengthening catalyst agglomerates, since their strengths are typically factors of 3–50 lower than for conventional, heat-treated ceramics of similar porosity.

Table 15: Mechanical Strengths and Attrition Rates of Catalyst Supports Compared to Those of Sintered Ceramic Agglomerates.

Catalyst support or ceramic	Preparation/pretreatment/properties	Strength, MPa	Attrition index, wt%/h	Ref.
High surface area catalyst supports				
γ-Al_2O_3, 1.2–4.25-mm spheres	Sol–gel granulation/dried 10 h at 40 °C, calcined 3 h at 450 °C/389 m^2/g, d_{pore} = 3.5 nm	11.6 ± 1.9	0.033	[167]
γ-Al_2O_3, 4.25-mm spheres	Alcoa LD-350	0.7	0.177	[167]
γ-Al_2O_3, 100 μm	VISTA-B-965-500C	6.2 ± 1.3	-	[163]
TiO_2 (anatase), 30 μm	Thermal hydrolysis/dried 110 °C, calcined 2 h at 500 °C/ 92 m^2/g, <10-nm primary crystallites	28[a]	-	[168]
TiO_2 (anatase), 90 μm	Basic precipitation/dried 110 °C, calcined 2 h at 500 °C/81 m^2/g, 10–14-nm primary crystallites	15[a]	-	[168]
TiO_2 (75% anatase, 25% rutile)	Degussa P25, fumed/4-mm extrudates/48 m^2/g, V_{pore} = 0.34 cm^3/g, d_{pore} = 21 nm	0.9	-	[169]
High surface area catalyst supports (cont.)				
TiO_2 (anatase)	Rhone-Poulenc DT51, ppt./4 mm extrudates/92 m^2/g, V_{pore} = 0.40 cm^3/g, d_{pore} = 8, 65 nm	0.9	-	[169]
Low surface area ceramics				
Al_2O_3	Spray dried with organic binder; plastic deformation observed	2.3	-	[170]
Al_2O_3	Heat treated (sintered), 99.9%	282–551	-	[165]
TiO_2 (Rutile)	Partially sintered	194	-	[170]
ZrO_2 (yttria additive)	Commercial samples from three companies, spray-dried	0.035–0.43	-	[171]
ZrO_2 (3% Y_2O_3)	Heat treated (sintered)	800–1500	-	[165]

[a]Rough estimates from break points on relative density *versus* log[applied pressure] curves; data are consistent with mass distribution *versus* pressure curves from ultrasonic tests.

Effects of Preparation and Pretreatment on Catalyst Agglomerate Strength

From the data in Table 15 it is evident that even subtle differences in preparation and pretreatment also affect agglomerate strength. For example, spheres of γ-Al_2O_3 prepared by sol–gel granulation are substantially (17 times) stronger than commercial γ-Al_2O_3 spheres [166]. Moreover, 30- and 90- μm diameter particles of TiO_2 prepared by thermal hydrolysis or basic precipitation are 30 and 15 times stronger than commercially available 4-mm extrudates [169].

Attrition of Catalyst Agglomerates: Mechanisms, Studies, and Test Methods

Catalyst attrition is a difficult problem in the operation of moving-bed, slurry-bed, or fluidized-bed reactors. Generally, stronger materials have greater attrition resistance; this conclusion is supported by representative data in Table 15 for γ-Al_2O_3, showing that the strength of the alumina prepared by sol–gel granulation is 17 times higher, while its attrition rates is 5 times lower.

The mechanism by which attrition occurs (erosion or fracture) can vary with catalyst or support preparation, crush strength, and with reactor environment; it can also vary with the mechanical test method. There is some evidence in the attrition literature, supporting the hypothesis that in the presence of a large stress, weaker oxide materials are prone to failure by fracture, while stronger materials tend to erode. For example, in the fluid catalytic cracking process, as new silica–alumina/zeolite catalyst in the form of 50–150-μm spherical agglomerates is added to replace catalyst lost by attrition, the weaker agglomerates break up fairly rapidly by fracture into smaller subagglomerates, following which the stronger agglomerates are slowly abraded to produce fine particles of 1–10 μm [172]. However, there is also contrary evidence from Thoma and co-workers [168], showing that fracture may be the preferred mechanism for strong TiO_2 agglomerates, while abrasion is favored for weaker agglomerates. That is, when subjected to ultrasonic stress, 30-μm-diameter agglomerates of amorphous anatase (TiO_2) prepared by thermal hydrolysis were observed to undergo fracture to 5–15-μm fragments, while 90-μm agglomerates of polycrystalline anatase prepared by basic precipitation were found to break down by erosion to 0.1–5-μm fragments [168]; in this case, the amorphous anatase was apparently stronger by a factor of 2 (see Table 15). Supporting a third trend, data from Pham and co-workers [163] show that attrition mechanism and rate are independent of agglomerate strength, but depend instead on the type of material. That is, 100-μm-diameter agglomerates of precipitated Fe/Cu/K Fischer–Tropsch catalyst (prepared by United Catalyst Incorporated) and having nearly the same strength shown inTable 15 for Vista-B Al_2O_3 (6.3 *vs.* 6.2 MPa), were found to undergo substantial fracture to 5–30-μm fragments (an increase from 45 to 85%; see Figure 28) as well as substantial erosion to 1 μm or less fragments (increase from 2 to 50%). By comparison, under the same treatment conditions, 90-μm-diameter agglomerates of Vista-B Al_2O_3 underwent much less attrition, mainly by erosion (20% increase in 0.1–5-μm fragments). The very low attrition resistance of the Fe/Cu/K Universal Catalysts, Inc. (UCI) catalyst is further emphasized by the unsatisfactory outcome of a test of this catalyst by the U.S. Department of Energy (DOE) in a pilot-scale slurry-phase bubble-column reactor in LaPorte, TX.; following one day of operation, the filter system was plugged with catalyst fines, preventing catalyst–wax separation and forcing shutdown of the plant [173].

Thus, based on these three representative examples, it follows that which of the two attrition mechanisms predominates depends much more on material composition and type than on agglomerate strength. However, irrespective of mechanism, the rate of attrition is usually greater for the weaker material.

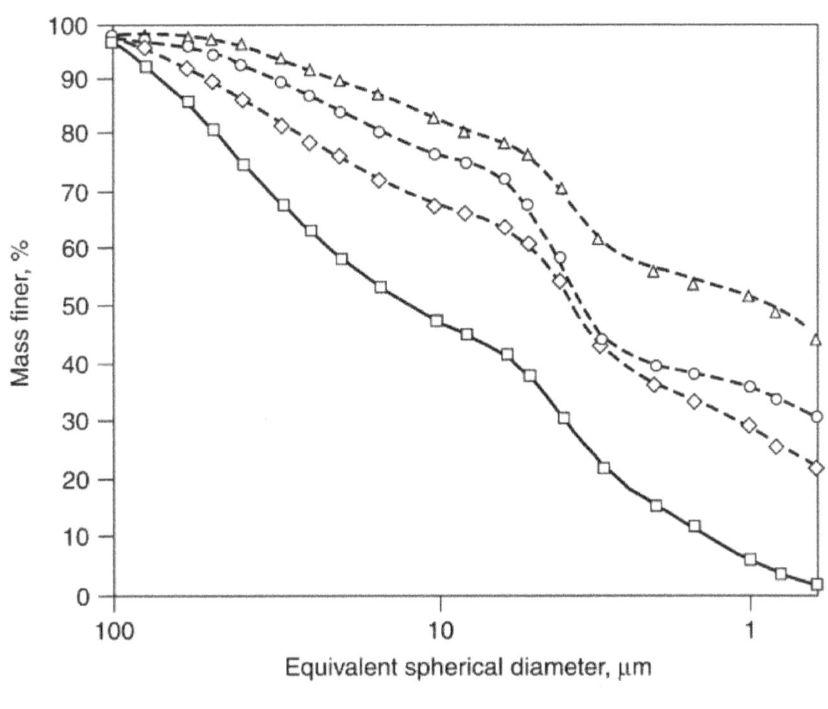

Figure 28: Sedigraph particle size distribution for a United Catalysts, Inc. (UCI) Fischer-Tropsch catalyst (designated as UCI-LAPI-COMP-DRUMC), used previously in Department of Energy (DOE) pilot-plant tests. There is considerable particle breakdown and generation of fine particles after 15 min of ultrasonic irradiation. Reproduced from [163]. Copyright 1999, Elsevier. —□— 0 min; – –◇– –5 min; - -○- - 10 min; – –Δ– –15 min.

Figure 29 illustrates the large effect that catalyst preparation method can have on the attrition resistance of an Fe/Cu Fischer–Tropsch catalyst [174]. This catalyst, prepared by precipitation, undergoes severe attrition during a 25-min treatment with ultrasonic radiation; indeed, the mass fraction finer than 0.1–5 μm increases from 0 to 65%. However, after a spray drying treatment of the same catalyst, an increase of only 0 to 10% in the same fractions is evident.

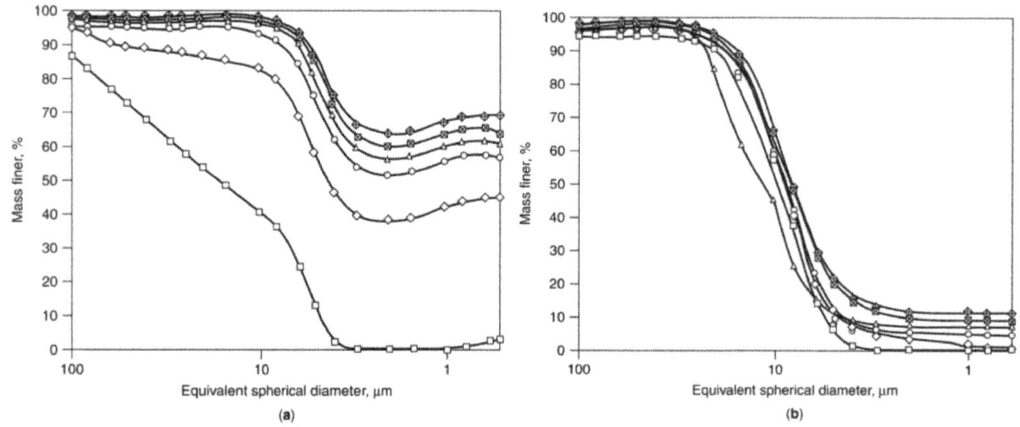

Figure 29: Sedigraph particle size distributions of a precipitated Fe–Cu catalyst, as-prepared and after spray-drying. The as-prepared catalyst (**a**) is weak and breaks down easily after 25 min of ultrasonic irradiation, while spray-drying (**b**) improves its attrition resistance. Reproduced from [174]. Copyright 2000, Elsevier. —□— 0 min; ♦ 5 min; —○— 10 min; —Δ— 15 min; —⊠— 20 min; — ⊕— 25 min.

In their review of attrition and attrition test methods, Bemrose and Bridgewater [175] discuss how attrition varies with reactor type, e.g., involves mainly particle–wall impacts in moving pellet bed reactors and particle–particle impacts in fluidized-bed reactors of high fluid velocity. In fact, jet attrition of catalyst particles in a gas fluidized-bed involving principally abrasion due to collision of high-velocity particles has been modeled in some detail [172,176]. Thus, given such important differences in attrition mechanism, realistic attrition test methods should attempt to model reactor operation as closely as possible. In addition, the ideal test would require only a small catalyst sample, a simple, inexpensive apparatus, and a few minutes to complete the test. Relatively quick, inexpensive single-particle crushing tests have been devised [175]; however, properties of a single particle are rarely representative of those for the bed; moreover, it is difficult to relate the results of this crushing test to the actual abrasion process. Realistic tests have been devised for two reactor types involving a moving catalyst, *i.e.*, an air-jet test for fluidized-bed catalysts [177,178], and a rotating drum apparatus for moving-bed catalysts [179]; however, the air-jet test requires a large quantity (e.g., 50 g) of catalyst, an expensive apparatus, and about 20 h to run. In the past decade, a new jet-cup test has been developed for testing of fluidized-bed catalysts [177,178], which requires only a 5-g sample and about 1 h to complete; comparisons of results for the jet-cup and air-jet tests indicate that the two tests give comparable results [177,178]. Nevertheless, the mechanisms for the two tests are different, *i.e.*, the air-jet (fluid-bed) test is abrasion- (erosion-) dominant, while the jet-cup test includes both abrasion and fracture mechanisms [178]. A 30-min, 10-g ultrasonic attrition test based on cavitation has also been developed in the past decade [168,174,180]; while it likewise involves both abrasion and fracture mechanisms, the results appear to correlate with other methods. For example, particle size distributions for the same Co/silica catalyst after ultrasonic, jet-cup, and laboratory-scale, slurry-bed column reactor (SBCR) tests are very similar (see Figure 30), indicating that both fracture and abrasion mechanisms operate in the small-scale SBCR. Moreover, the good agreement among the three methods suggests that both the jet-cup and ultrasonic tests may provide data representative of the attrition process in laboratory-scale SBCR reactors. It is evident that these two small-scale methods are especially useful for screening of a series of catalysts to determine relative strength.

Nevertheless, the more realistic large-scale tests are probably needed for accurately determining design attrition rates of a commercial catalyst to be used in a full-scale process. The observation that attrition of a fluid catalytic cracking (FCC) catalyst initially involves fracture of weak agglomerates followed by abrasion of strong agglomerates emphasizes the need to collect and analyze the particle size distribution of attrited fines as a function of time in order to define which mechanism (or mechanisms) operates at startup as well as in the steady-state process. Because the mechanism may be time dependent, rapid, small-scale tests may produce misleading results.

While realistic laboratory-scale tests have been developed for simulating attrition in large moving-bed and fluidized-bed reactors, no such laboratory test has been developed and demonstrated yet for simulation of large-scale SBCR reactors, although recent research has focused on the development of such tests. For example, in laboratory-scale, SBCR tests of supported cobalt catalysts over several days [180], the attrition resistance decreases in the order Co/Al_2O_3 > Co/SiO_2 > Co/TiO_2 (especially the anatase form underwent attrition at a high rate); attrition resistance was observed to increase with increasing cobalt loading from 10 to 40 wt%.

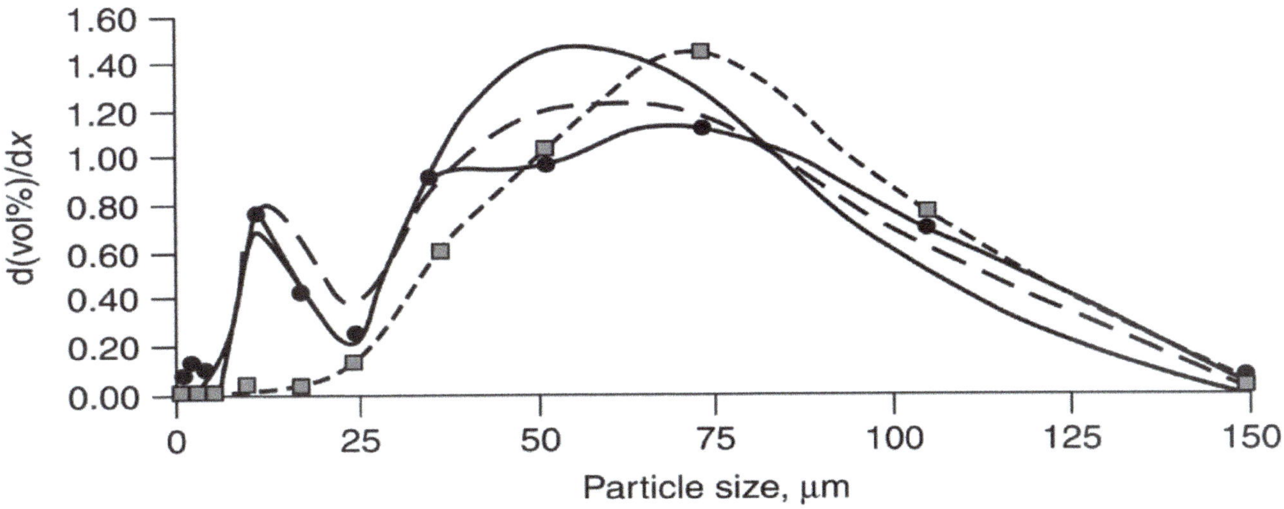

Figure 30: Particle size distributions of Co/SiO_2 catalyst. Adapted from [178]. ——— Ultrasound 250 W (>10 μm);- - - jet cup L/min (>10 μm); —•— Co/SiO_2 after SBCR;—■— Co/SiO_2 fresh.

Implications of Mechanistic Knowledge of Attrition for Catalyst Design

The understanding of mechanisms important in attrition of catalyst supports and catalysts, the relationship between strength and attrition rate for a given material, and test data can be used to great advantage in the design of attrition resistant catalysts. Several alternatives follow from the previous discussion for increasing attrition resistance: (1) increasing aggregate/agglomerate strength by means of advanced preparation methods, e.g., sol–gel granulation, spray drying, and carefully controlled precipitation methods (see Table 15 and Figure 29 for examples), (2) adding binders to improve strength and toughness, e.g., the addition of a polyvinylpyrrolidone binder to agglomerates of quartz sand increases agglomerate strength from 0.1 to 3 MPa [181], (3) coating aggregates with a porous but very strong material such as ZrO_2, e.g., embedding a fluidized-bed catalyst for partial oxidation of *n*-butane to maleic anhydride in a strong, amorphous matrix of zirconium hydrogen phosphate significantly improves its attrition resistance [182], and (4) chemical or thermal tempering of agglomerates to introduce compressive stresses that increase strength and attrition resistance, e.g., heating and cooling particles rapidly by passing them through a low-residence-time, high-temperature furnace to harden the agglomerate exterior, while preventing significant sintering of or phase changes in the porous interior. The subject of preventing mechanical degradation and other forms of catalyst deactivation is addressed in greater detail under Prevention of Catalyst Decay.

Summary of Deactivation Mechanisms for Solid Catalysts

Causes of solid (heterogeneous) catalyst deactivation are basically threefold: (1) chemical, (2) mechanical, and (3) thermal. Mechanisms of heterogeneous catalyst deactivation can be classified into five general areas: (1) chemical degradation including volatilization and leaching, (2) fouling, (3) mechanical degradation, (4) poisoning, and (5) thermal degradation. Poisoning and thermal degradation are generally slow processes, while fouling and some forms of chemical and mechanical degradation can lead to rapid, catastrophic catalyst failure. Some forms of poisoning and many forms of fouling are reversible; hence, reversibly poisoned or fouled catalysts are relatively easily regenerated. On the other hand, chemical, mechanical, and thermal forms of catalyst degradation are rarely reversible.

PREVENTION OF CATALYST DECAY

It is often easier to prevent rather than cure catalyst deactivation. Many poisons and foulants can be removed from feeds using guard beds, scrubbers, and/or filters. Fouling, thermal degradation, and chemical degradation can be minimized through careful control of process conditions, e.g., lowering temperature to lower sintering rate or adding steam, oxygen, or hydrogen to the feed to gasify carbon or coke-forming precursors. Mechanical degradation can be minimized by careful choice of carrier materials, coatings, and/or catalyst particle forming methods.

While treating or preventing catalyst deactivation is facilitated by an understanding of the mechanisms, additional perspectives are provided by examining the route by which each of the mechanisms causes loss of catalytic activity, *i.e.*, how it influences reaction rate [109]. Thus, catalytic activity can be defined in terms of the observed site-based rate constant k_{obs}, which is equal to the product of the active site density σ (number of sites per area of surface), the site-based intrinsic rate constant k_{intr}, and the effectiveness factor η, *i.e.*,

$$k_{obs} = \sigma k_{intr} \eta \tag{4}$$

Loss of catalytic activity may be due to a decrease in any of the three factors in Equation 4, whose product leads to k_{obs}. Thus, catalyst deactivation can be caused by (1) a decrease in the site density σ, (2) a decrease in intrinsic activity (*i.e.*, decrease in k_{intr}), and/or (3) lowered access of reactants to active sites (decrease in η). Poisoning, for example, leads to a loss of active sites, *i.e.*, $\sigma = \sigma_0(1 - \alpha)$, where α is the fraction of sites poisoned; sintering causes loss of active sites through crystallite growth and reduction of active surface area. Fouling can cause both loss of active sites due to blocking of surface sites as well as plugging of pores, causing a decrease in the effectiveness η. Moreover, poisoning, as discussed earlier, can also lead to a decrease in intrinsic activity by influencing the electronic structure of neighboring atoms. Thus, each of the deactivation mechanisms affects one or more of the factors comprising observed activity (see Table 16); all of the mechanisms, however, can effect a decrease in the number of catalytic sites.

Table 16: How Deactivation Mechanisms Affect the Rate of a Catalyzed Reaction and the Rapidity and Reversibility of Deactivation Process.

Deactivation mechanism	Effects on reaction rate: Decrease in number of active sites	Decrease in intrinsic activity (k_{intr})	Decrease in effectiveness factor (η)	Deactivation process: Fast or slow [a]	Reversible
Chemical degradation	×	×	× [b,c]	Varies	No
Fouling	×	×	-	Fast	Yes
Mechanical degradation	×	-	-	Varies	No
Poisoning	×	×	-	Slow	Usually
Thermal degradation/Sintering	×	× [b,d]	× [b,e]	Slow	Sometimes
Vaporization/leaching	×	× [b,f]	-	Fast	Sometimes

[a] Generally; [b] In some cases; [c] Chemical degradation can cause breakdown of support, pore plugging, and loss of porosity; [d] If the reaction is structure-sensitive, sintering could either increase or decrease intrinsic activity; [e] Sintering of the support may cause support collapse and loss of porosity; it may also increase average pore diameter. [f] Leaching of aluminum or other cations from zeolites can cause buildup of aluminum or other oxides in zeolite pores.

General Principles of Prevention

The age-old adage that says "an ounce of prevention is worth a pound of cure" applies well to the deactivation of catalysts in many industrial processes. The catalyst inventory for a large plant may entail a capital investment of tens of millions of dollars. In such large-scale processes, the economic return on this investment may depend on the catalyst remaining effective over a period of up to 3–5 years. This is particularly true of those processes involving irreversible or only partially reversible deactivation (e.g., sulfur poisoning or sintering). Some typical industrial catalysts, approximate catalyst lifetimes, and factors that determine their life are listed as examples in Table 17. It is evident that in many processes more than one mechanism limits catalyst life. Moreover, there is a wide variation in catalyst lifetimes among different processes, *i.e.*, from 10^{-6} to 15 years. While there is clearly greater interest in extending catalyst lifetimes in processes where life is short, it should be emphasized that great care must be exercised in protecting the catalyst in any process from process upsets (e.g., temperature runaway, short-term exposure to impure feeds, or changes in reactant composition) that might reduce typical catalyst life by orders of magnitude, e.g., from years to hours.

Table 17: Typical Lifetimes and Factors Determining the Life of Some Important Industrial Catalysts [a].

Reaction	Operating conditions	Catalyst	Typical life (years)	Process affecting life of catalyst charge	Catalyst property affected
Ammonia synthesis $N_2 + 3\,H_2 \rightarrow 2\,NH_3$	450–470 °C 200–300 atm	Fe with promoters (K_2O) and stabilizer (Al_2O_3)	10–15	Slow sintering	Activity
Methanation (ammonia and hydrogen plants) $CO/CO_2 + H_2 \rightarrow CH_4 + H_2O$	250–350 °C 30 atm	Supported nickel	5–10	Slow poisoning by S, As, K_2CO_3 from plant upsets	Activity and pore blockage
Acetylene hydrogenation ("front end") $C_2H_2 + H_2 \rightarrow C_2H_4$	30–150 °C 20–30 atm	Supported palladium	5–10	Slow sintering	Activity/selectivity and temperature
Sulfuric acid manufacturing $2\,SO_2 + O_2 \rightarrow 2\,SO_3$	420–600 °C 1 atm	Vanadium and potassium sulfates on silica	5–10	Inactive compound formation; pellet fracture; plugging by dust	Activity, pressure drop, and mass transfer
Methanol synthesis $CO + 2\,H_2 \rightarrow CH_3OH$	200–300 °C 50–100 atm	Copper on zinc and aluminum oxides	2–5	Slow sintering; poisoning by S, Cl, and carbonyls	Activity
Low temperature water gas shift $CO + H_2O \rightarrow CO_2 + H_2$	200–250 °C 10–30 atm	Copper on zinc and aluminum oxides	2–4	Slow poisoning and accelerated sintering by poisons	Activity
Hydrocarbon hydrodesulfurization $R_2S + 2\,H_2 \rightarrow H_2S + R_2$	300–400 °C 30 atm	Cobalt and molybdenum sulfides on aluminum oxide	1–10	Slow coking, poisoning by metal deposits in residuum	Activity, mass transfer, and pressure drop
High temperature water gas shift $CO + H_2O \rightarrow H_2 + CO_2$	350–500 °C 20–30 atm	Fe_3O_4 and chromia	1–4	Slow sintering, pellet breakage due to steam	Activity and pressure drop

Reaction	Operating conditions	Catalyst	Typical life (years)	Process affecting life of catalyst charge	Catalyst property affected
Steam reforming, natural gas $CH_4 + H_2O \rightarrow CO + 3\ H_2$	500–850 °C 30 atm	Nickel on calcium aluminate or α-alumina	1–3	Sintering, sulfur-poisoning, carbon formation, and pellet breakage due to plant upsets	Activity and pressure drop
Ethylene partial oxidation $2\ C_2H_4 + O_2 \rightarrow 2\ C_2H_4O$	200–270 °C 10–20 atm	Silver on α-alumina with alkali metal promoters	1–3	Slow sintering, poisoning by Cl, S	Activity and selectivity
Butane oxidation to maleic anhydride $C_4H_{10} + 3.5\ O_2 \rightarrow C_4H_2O_3 + 4\ H_2O$	400–520 °C 1–3 atm	Vanadium phosphorus oxide with transition metal additives	1–2	Loss of P; attrition or pellet breakage; S, Cl poisoning	Activity and selectivity
Reduction of aldehydes to alcohols $RCHO + H_2 \rightarrow RCH_2OH$	220–270 °C 100–300 atm	Copper on zinc oxide	0.5–1	Slow sintering, pellet breakage (depends on feedstock)	Activity or pressure drop
Ammonia oxidation $2\ NH_3 + 5/2\ O_2 \rightarrow 2\ NO + 3\ H_2O$	800–900 °C 1–10 atm	Pt–Rh alloy gauze	0.1–0.5	Surface roughness, loss of platinum	Selectivity, fouling by Fe
Oxychlorination of ethylene to ethylene dichloride $2\ C_2H_4 + 4\ HCl + O_2 \rightarrow 2\ C_2H_4Cl_2 + 2\ H_2O$	230–270 °C 1–10 atm	Copper chlorides on alumina (fluidized bed)	0.2–0.5	Loss by attrition and other causes resulting from plant upsets	Fluidized state and activity
Catalytic hydrocarbon reforming	460–525 °C 8–50 atm	Platinum alloys on treated alumina	0.01–0.5	Coking, frequent regeneration	Activity and mass transfer
Catalytic cracking of oils	500–560 °C 2–3 atm	Synthetic zeolites (fluidized bed)	0.000002	Very rapid coking, continuous regeneration	Activity and mass transfer

Adapted from Ref. [9].

While complete elimination of catalyst deactivation is not possible, the rate of damage can be minimized in many cases through understanding of the mechanisms, thereby enabling control of the deactivation process, *i.e.*, prevention is possible through control of catalyst properties, process conditions (*i.e.*, temperatures, pressures), feedstock impurities, methods of contacting, and process design. Figure 31 illustrates general approaches to eliminating or moderating deactivation through modifications in catalyst and/or process. Examples of how deactivation can be prevented are discussed below in connection with the most important causes of deactivation: chemical degradation, fouling by coke and carbon, poisoning, sintering, and mechanical degradation. Principles for preventing deactivation by these mechanisms are summarized in Table 18, while representative results from studies focusing on prevention or minimization of catalyst deactivation are summarized in Table 19.

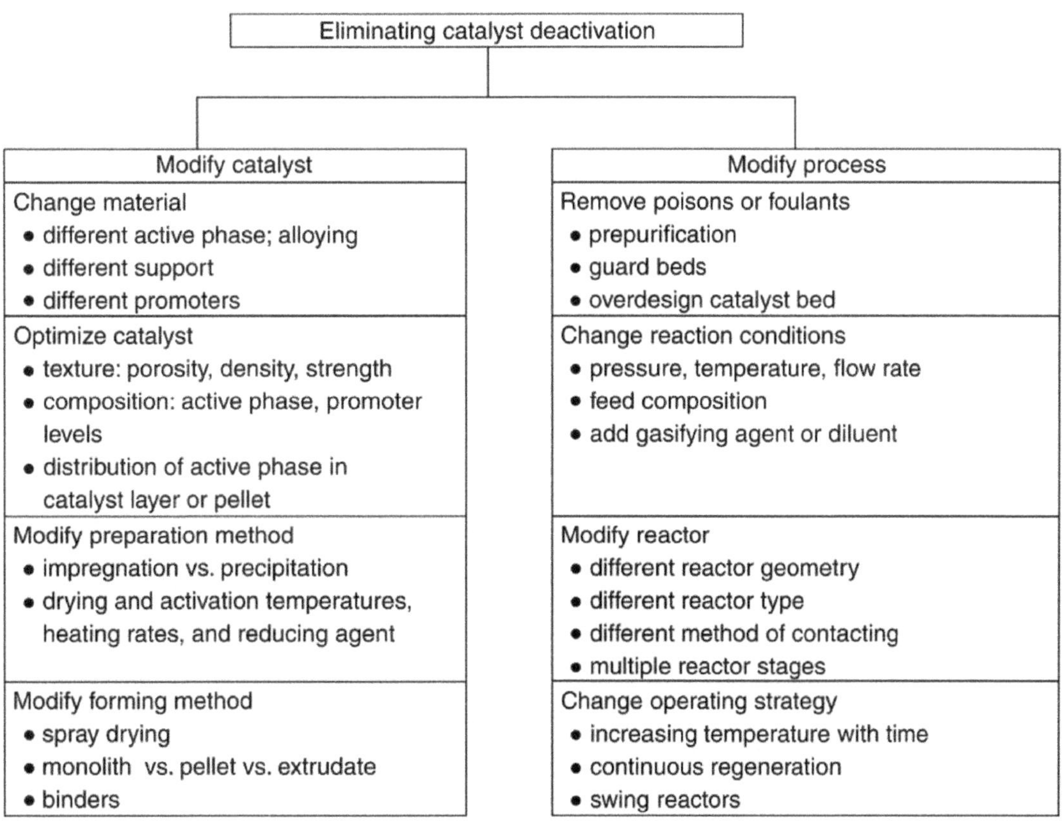

Figure 31: Approaches to eliminating catalyst deactivation.

Table 18: Methods for Preventing Catalyst Decay [8,41].

Basic mechanism	Problem	Cause	Methods of minimization
Chemical degradation	Oxidation of metal catalysts to inactive oxides	Oxidation of metal by contaminant O_2 or reactant/product water	(1) Purify feed of oxidants (2) Minimize reactant/product water by recycle/separation, staged reactors, and otherwise limiting conversion (3) Incorporate additives that facilitate resistance to oxidation
	Transformation of active phase to stable, inactive phase	Solid-state reaction of active phase with support or promoters	(1) Avoid conditions (e.g., oxidizing condition, high steam pressures, and high temperatures) that favor solid-state reactions (2) Select combinations of active phase and promoters/supports that are noninteracting
		Overreduction of active oxide phases	(1) Stabilize oxidation state using promoters that induce resistance to reduction or that serve as oxygen donors (2) Add steam to the reactants to prevent overreduction
Fouling by coke or carbon	Loss of catalytic surface sites due to formation of carbon or coke films	Free radical reactions in gas phase	(1) Avoid formation of free radicals, lower temperature (2) Minimize free space (3) Free radical traps, diluents (4) Add gasifying agents (e.g., H_2, H_2O)
		Free radical reactions at reactor walls	(1) Coat reactor with inert material
		Formation and growth on metal surfaces	(1) Avoid accumulation of coke precursors (e.g., atomic carbon, olefins) through careful choice of reactant conditions or membranes (2) Add gasifying agents (e.g., H_2, H_2O), diluents (3) Incorporate catalyst additives to increase rate of gasification or to change ensemble size (4) Passivate metal surfaces with sulfur (5) Decrease dispersion (6) Recycle inerts to flush surface of heavy oligomers and to moderate temperature
Fouling by coke or carbon (cont.)	Loss of catalytic surface sites due to formation of carbon or coke films	Formation and growth on metal oxides, sulfides	(1) Utilize measures 1, 2, 3, and 6 for metal surfaces (2) Design catalyst for optimum pore structure and acidity (3) Use shape-selective, coke-resistant molecular sieves
	Loss of catalyst effectiveness; plugging of pores; destruction of catalyst	Formation of gas phase coke, vermicular carbons, and liquid or solid cokes in massive quantities	(1) Minimize formation of free radicals or coke precursors as above (2) Use gasifying agents (3) Incorporate catalyst additives that lower solubility of carbon in metal or that change ensemble size (4) Use supports with large pores; large pellets
		Hot spots in pellet or bed	(1) Use wash coat or small pellets (2) Use slurry- or fluid-bed reactor, gas diluents
Mechanical failure	Crushing of granules, pellets, or monoliths in a fixed bed	Brittle fracture due to a mechanical load	(1) Minimize porosity of pellets or monoliths (2) Improve bonding of primary particles in agglomerates that make up pellets or monoliths using advanced forming methods, e.g., spray drying and controlled thermal treatments (3) Add binders such as carbon to the support material, which facilitate plastic deformation and thus protect against brittle fracture (4) Chemically or thermally temper agglomerates
	Attrition and/or erosion in fixed or moving beds	Abrasion of catalyst coatings or particles due to mechanical, thermal, or chemical stresses	(1) Avoid highly turbulent shear flows and/or cavitation, leading to high erosion rates (2) Avoid thermal stresses in the preparation and use of catalysts that lead to fracture or separation of coatings (3) Avoid formation of chemical phases of substantially different densities or growth of carbon filaments that cause fracture of primary particles and agglomerates. Choose supports, support additives, and coating materials, such as titanates, zirconia, and zirconates, having high fracture toughness
Poisoning	Loss of catalytic surface sites	Blockage of sites by strong adsorption of impurity	(1) Purify feed and/or use guard bed to adsorb poison (2) Employ additives that selectively adsorb poison (3) Choose reaction conditions that lower adsorption strength (4) Optimize pore structure and choose mass transfer regimes that minimize adsorption of poison on active sites (5) Apply coating that serves as diffusion barrier to poison
Thermal degradation, sintering	Loss of metal area	Metal particle or subparticle migration at high temperatures	(1) Lower or limit reaction temperature while facilitating heat transfer (2) Add thermal stabilizers to catalyst; and (3) avoid water
	Loss of support area	Crystallization and/or structural modification or collapse	Same as for avoiding loss of metal area

Table 19: Representative Results from Studies Focusing on Prevention/Minimization of Catalyst Deactivation.

Deactivation mechanism Process/Reaction Catalyst	Problem/cause	Method(s) of minimization	Ref.
		Chemical degradation	
Auto emissions control Pt– or Pd–Rh/Al_2O_3	In three-way catalyst, Rh is very active for NO reduction, but it forms a solid solution with Al_2O_3 that has no activity and alloys with Pt or Pd that reduce its activity	Place Rh in a separate catalyst layer from Pt or Pd to prevent alloying; support Rh on ZrO_2, which is a noninteracting support for Rh. In general, multilayer strategies (up to 6 layers) are used to prevent undesirable interactions between different components of the catalyst	[183–185]

Deactivation mechanism Process/Reaction Catalyst	Problem/cause	Method(s) of minimization	Ref.
		Chemical degradation	
Fischer–Tropsch synthesis Co supported on Al_2O_3, SiO_2, TiO_2, and Fe/Cu/K/SiO_2	Oxidation of active Co metal crystallites to inactive Co oxides, aluminates, and silicates and of active iron carbides to inactive Fe_3O_4 or Fe_3C in the presence of high pressure steam at high conversion	(1) Employ two- or three-stage process that enables lower conversion and lower concentrations of steam product in the first stage. Treat gaseous stream leaving the first or second stage to remove water and liquid hydrocarbons (2) Add noble metal promoters that facilitate and maintain high reducibility of the metal or metal carbide phases (3) Stabilize silica and alumina supports with coatings of hydrothermally stable materials such as ZrO_2 and $MgAl_2O_4$	[8,126,186,187
Partial oxidation of isobutene to methacrolein $Fe_2(MoO_4)_3$, $Mo_{12}Bi_xCe_yO_z$	Overreduction of the catalyst during reaction leads to activity decrease	(1) Stabilize reduction state of iron molybdate catalyst using an oxygen donor such as α-Sb_2O_4; the oxygen donor dissociates molecular oxygen to atomic oxygen that readily spills over to the catalyst (2) $Mo_{12}Bi_xCe_yO_z$ catalyst promoted with Co, Mg, Rb, and/or Cs oxides is highly resistant to reduction, highly selective to methacrolein, and long-lived	[154,156,188]
Steam reforming and steam-oxygen conversion of propane Pd/Al_2O_3	In the absence of steam, PdO is reduced to less active, less thermally stable Pd metal	Adding steam to the reactants inhibits oxidation of propane at lower reaction temperatures while preventing reduction of PdO at higher temperatures (up to 700–900 °C)	[189]

Prevention of Chemical Degradation (by Vapor–Solid and Solid–Solid Reactions)

The most serious problems of oxidation of metal catalysts, overreduction of oxide catalysts, and reaction of the active catalytic phase with carrier or promoter, can be minimized or prevented by careful catalyst and process design (as enumerated in Table 18 and illustrated in Table 19). For example, the loss of Rh due to solid-state reaction with alumina in the automotive three-way catalyst can be prevented by supporting Rh on ZrO_2 in a separate layer from Pt and/or Pd on alumina [215, 216,217,218,219,220,221,222] In Fischer–Tropsch synthesis, the oxidation of the active cobalt phase in supported cobalt catalysts to inactive oxides, aluminates, and silicates can be minimized by employing a two- or three-stage process in which product steam is moderated in the first stage by limiting conversion and in subsequent states by interstage removal of water [223] It can also be moderated by addition of noble metal promoters that facilitate and maintain high reducibility of the cobalt and by coating the alumina or silica support with materials such as ZrO_2 that are less likely to react with cobalt to form inactive phases.

Prevention of Fouling by Coke and Carbon

Rostrup-Nielsen and Trimm [57], Trimm [59], and Bartholomew [60] have discussed principles and methods for avoiding coke and carbon formation. General methods of preventing coke or carbon formation are summarized in Table 18. Most of these are based on one important fundamental principle: *carbon or coke results from a balance between the reactions that produce atomic carbon or coke precursors and the reactions of these species with H_2, H_2O, or O_2 that remove them from the surface.* If the conditions favor formation over gasification, these species accumulate on the surface and react further to form less active forms of carbon or coke, which either coat the surface with an inactive film or plug the pores, causing loss of catalyst effectiveness, pore plugging, or even destruction of the carrier matrix.

Methods to lower rates of formation of carbon or coke precursors relative to their rates of gasification vary with the mechanism of formation (*i.e.*, gas, surface, or bulk phase) and the nature of the active catalytic phase (e.g., metal or oxide). For example, gas phase formation can be minimized by choosing reaction conditions that minimize the formation of free radicals, by using free-radical traps, by introducing gasifying agents (e.g., H_2, H_2O) or gas diluents, and by minimizing the void space available for homogeneous reaction. Similarly, the formation and growth of carbon or coke species on metal surfaces is minimized by choosing reaction conditions that minimize the formation of atomic carbon or coke precursors and by introducing gasifying agents. Selective membranes or supercritical conditions can also be used to lower the gas-phase and surface concentrations of coke precursors. Since carbon or coke formation on metals apparently requires a critical ensemble of surface metal atoms and/or dissolution of carbon into the bulk metal, introduction of modifiers that change ensemble sizes (e.g., Cu or S in Ni or Ru) or that lower the solubility of carbon (e.g., Pt in Ni) can be effective in minimizing these forms of deactivation.

For example, in a detailed STM study of submonolayers of Au on Ni(111), Besenbacher and co-workers [71] found that the electron density of Ni atoms in the vicinity of Au atoms was increased; from density functional theory (DFT) calculations they concluded that the strength of carbon adsorption (and hence the tendency to form graphite) was decreased on next-nearest neighbor Ni atoms; from studies of the effects of S adsorption on methane activation and graphite formation on pure Ni, they were able to infer that the ensemble size needed for methane dissociation is smaller than that for graphite formation. These fundamental insights were used in the design of a 0.3% Au-promoted 16% Ni/$MgAl_2O_4$ catalyst that loses no activity over 4000 h during steam reforming of *n*-butane, while the corresponding unpromoted Ni catalyst loses about 5% of its initial activity (see Figure 32). In contrast to the moderating effects of noble metal additives, addition of 0.5% Sn to cobalt substantially increases the rate of carbon filament formation from ethylene [72], an effect desirable in the commercial production of carbon filament fibers.

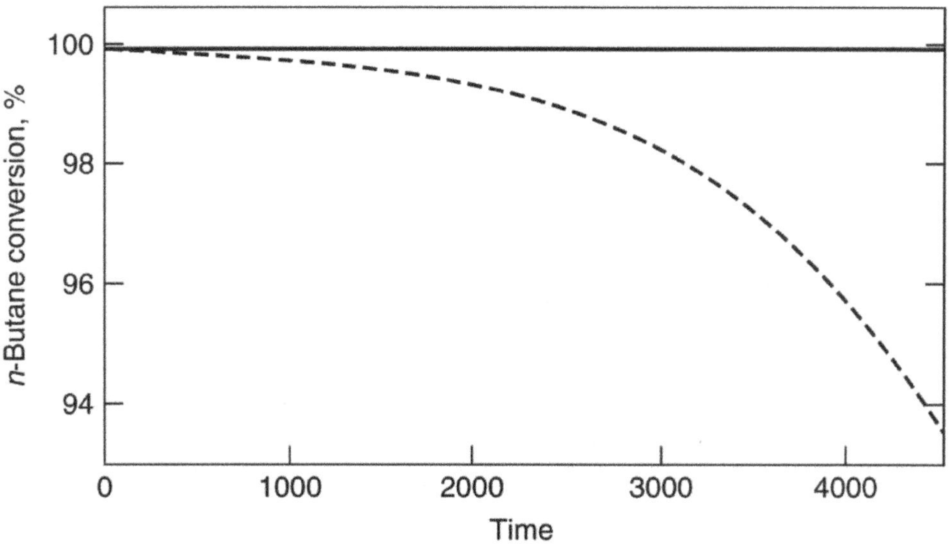

Figure 32: Conversion of *n*-butane as a function of time during steam reforming in a 3% *n*–butane–7% hydrogen–3% water in helium mixture at a space velocity of 1.2 h^{-1}. The dashed curve shows the *n*-butane conversion for the Ni catalyst (16.8% Ni) and the solid curve for the Au/Ni catalyst (16.4% Ni/0.3% Au). Reproduced from [71]. Copyright 1998, American Association for the Advancement of Science.

Coke deposition on oxide or sulfide catalysts occurs mainly on strongly acidic sites; accordingly the rate of coking can be lowered by decreasing the acidity of the support. For example, silanation of HY and HZSM-5 zeolites decreases their activities but improves catalyst life [236]. In steam reforming, certain catalyst additives, e.g., MgO, K_2O, or U_3O_8, facilitate H_2O or CO_2 adsorption and dissociation to oxygen atoms, which in turn gasify coke precursors [8,60,70].

Similarly, for steam reforming catalysts used for light alcohol and oxygenate conversion, the addition of partially reducible oxides, like ceria, in nickel perovskite ($La_{1-x}Ce_xNiO_3$) catalysts [237] or as a support for a cobalt catalyst [238], reduce the rate of carbon deposition. Alternatively, the reaction atmosphere may be modified to increase the gasification rate by adding oxidizing reactants (e.g., O_2 and/or CO_2) to reduce the rate of coke deposition [63]. This process is often described as autothermal reforming because it tends to balance the endothermic steam reforming reactions with exothermic reactions that make the process thermally neutral.

As in the case of poisoning (see below), there are certain reactor bed or catalyst geometries that minimize the effects of coking on the reaction. For example, specific film-mass transport or pore diffusion regimes favor coke or carbon deposition on either the outside or inside of the catalyst pellet [239,240]. Choosing supports with relatively large pores minimizes pore plugging; choice of large-diameter, mechanically-strong pellets avoids or delays reactor plugging. However, in view of the rapidity at which coke and carbon can deposit on, plug, and even destroy catalyst particles, the importance of preventing the onset of such formation cannot be overemphasized.

Reforming of naphtha provides an interesting case study of catalyst and process designs to avoid deactivation by coking [8,196,197,198,241]. The classical Pt/Al_2O_3 catalyst is bifunctional; that is, the metal catalyzes dehydrogenation, while the acid sites of the Al_2O_3 catalyze isomerization and hydrocracking. Together, the two functions catalyze dehydrocylization and aromatization. Addition of Re, Sn, or Ge, to Pt and sulfiding of the Pt–Re catalyst substantially reduce coke formation by diluting large Pt ensembles that would otherwise produce large amounts of coke, while addition of Sn and Ir improves selectivity for dehydrogenation relative to hydrogenolysis, the latter of which leads to coke formation. Naphtha reforming processes are designed for (1) high enough H_2 pressure to favor gasification of coke precursors while minimizing hydrocracking, (2) maintenance of Cl and S contents throughout the bed to ensure optimum acidity and coke levels, and (3) low enough overall pressure to thermodynamically and kinetically favor dehydrogenation and dehydrocylization. Accordingly, optimal process conditions are a compromise between case 1 and case 3. The above-mentioned improvements in catalyst technologies, especially resistance to coking, have enabled important process improvements, such as optimal operation at lower pressure; thus, processes have evolved over the past two to three decades from conventional fixed-bed reactors at high pressure (35 bar) using nonregenerative Pt catalysts to low pressure (3.5 bar), slowly moving-bed, continuously regenerated units with highly selective Pt/Sn catalysts, resulting in substantial economic benefits [198,241].

Prevention of Poisoning

Since poisoning is generally due to strong adsorption of feed impurities and since poisoned catalysts are generally difficult or impossible to regenerate, it is best prevented by removal of impurities from the feed to levels that will enable the catalyst to operate at its optimal lifetime. For example, it is necessary to lower the feed concentration of sulfur compounds in conventional methanation and Fischer–Tropsch processes involving base metal catalysts to less than 0.1 ppm in order to ensure a catalyst lifetime of 1–2 years. This is typically accomplished using a guard bed of porous ZnO at about 200 °C. In cracking or hydrocracking reactions on oxide catalysts, it is important to remove strongly basic compounds, such as ammonia, amines, and pyridines, from the feed; ammonia in some feedstocks, for example, can be removed by aqueous scrubbing. The poisoning of catalysts by metal impurities can be moderated by selective poisoning of the unwanted metal. For example, in catalytic cracking of nickel-containing petroleum feedstocks, nickel sites, which would otherwise produce copious amounts of coke, are selectively poisoned by antimony [242]. The poisoning of hydrotreating catalysts by nickel and vanadium metals can be minimized by (1) using a guard bed of inexpensive Mo catalyst or graded catalyst bed with inexpensive, low-activity Mo at the top (bed entrance) and expensive, high-activity catalyst at the bottom (see Figure 33) and (2) by depositing coke prior to the metals, since these metal deposits can be physically removed from the catalyst during regeneration [243].

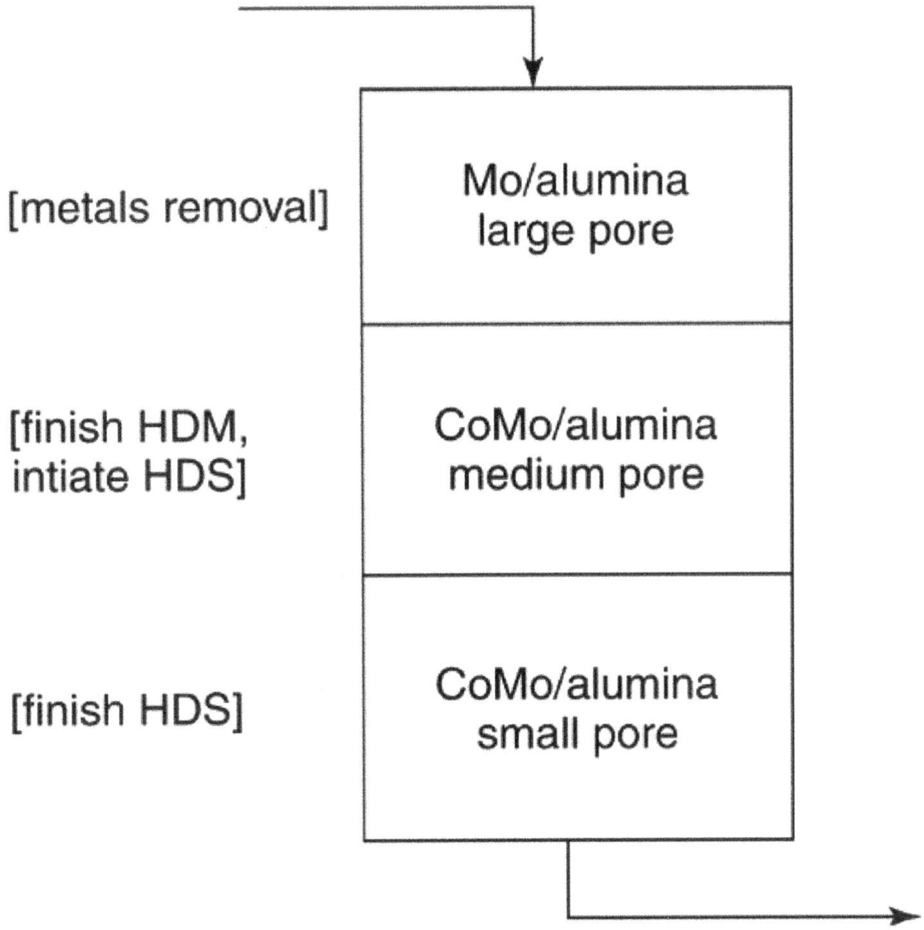

Figure 33: Staged reactor system with decreasing pore size strategy for hydrodemetalization (HDM)/hydrodesulfurization (HDS) of residuum. Reproduced from [214]. Copyright 1993, Marcel Dekker.

It may be possible to lower the rate of poisoning through careful choice of reaction conditions that lower the strength of poison adsorption [60] or by choosing mass-transfer-limiting regimes that limit deposits to the outer shell of the catalyst pellet, while the main reaction occurs uninterrupted on the interior of the pellet [239]. The manner in which the active catalytic material is deposited on a pellet (e.g., uniformly or in an eggshell or egg yolk pattern) can significantly influence the life of the catalyst [17,244].

An example of reducing catalyst poisoning (and oxidation) through process design has been reported in a process patent for staged hydrocarbon synthesis via the Fischer–Tropsch reaction [245]. While cobalt catalysts are favored because of their high activities and while it is desirable to achieve high conversions of CO in the process, the one-pass conversion for cobalt is limited by (1) its tendency to be oxidized at high partial pressures of product water observed at high CO conversions and (2) its tendency under these conditions to form the oxygenated products (e.g., alcohols and aldehydes) that poison or suppress its synthesis activity. One alternative is to separate products and recycle the unused CO and H_2, but this requires costly recompression and separation of the oxygenates. Costly separation and/or poisoning can be prevented by operating a first-stage reactor containing a cobalt catalyst to a moderately high conversion followed by reacting the remaining CO and H_2 in a second stage to above 95% conversion on an iron catalyst, which is not sensitive to the oxygenates and which shifts some of the product water to H_2 and CO_2, thus minimizing its hydrothermal degradation.

An example of reducing catalyst poisoning through catalyst design occurs in abatement of emissions for automotive and motorcycle engines [18,212]. Application of an alumina or zeolite coating, or alternatively preparing the active phase in a sublayer, provides a diffusion barrier that prevents or slows the access of poisons from the fuel or oil (e.g., phosphorus and/or zinc from lubricating oil or corrosion products) to the catalyst surface. The principle is to optimize the pore size distribution of the diffusion barrier to provide access to the catalytic phase of relatively small hydrocarbon, CO, NO, and O_2molecules, while preventing access of larger molecules, such as from lubricating oil and/or particulates.

Finally, another strategy that has been employed to reduce the impact of poisoning, particularly for sulfur, is the inclusion of traps or "getters" as part of the catalyst. These species, including rare earth oxides of thulium (Tm) [50] or Ce [51] and simple zinc oxide, essentially act as sacrificial stoichiometric reactants to protect the active metal by preferentially adsorbing the poison. These traps can extend the catalyst life, but because they are not catalytic as they perform, they are necessarily temporary agents if the poison remains in the feed to the process.

Prevention of Sintering

Since most sintering processes are irreversible or are reversed only with great difficulty, it is important to choose reaction conditions and catalyst properties that avoid such problems. Metal growth is a highly activated process; thus, by choosing reaction temperatures lower than 0.3–0.5 times the melting point of the metal, rates of metal sintering can be greatly minimized. The same principle holds true in avoiding recrystallization of metal oxides, sulfides, and supports. Of course, one approach to lowering reaction temperature is to maximize activity and surface area of the active catalytic phase.

Although temperature is the most important variable in the sintering process, differences in reaction atmosphere can also influence the rate of sintering. Water vapor, in particular, accelerates the crystallization and structural modification of oxide supports. Accordingly, it is vital to minimize the concentration of water vapor in high temperature reactions on catalysts containing high surface area supports.

Besides lowering temperature and minimizing water vapor, it is possible to lower sintering rates through addition of thermal stabilizers to the catalyst. For example, the addition of higher melting noble metals (such as rhodium or ruthenium) to a base metal (such as nickel) increases the thermal stability of the base metal [106]. Addition of Ba, Zn, La, Si, and Mn oxide promoters improves the thermal stability of alumina [246]. These additives can affect product selectivity, but generally positively toward desired products, and always through extending the productive life of the catalysts [8].

Designing thermally stable catalysts is a particular challenge in high temperature reactions, such as automotive emissions control, ammonia oxidation, steam reforming, and catalytic combustion. The development of thermally stable automotive catalysts has received considerable attention, thus providing a wealth of scientific and technological information on catalyst design (e.g., Refs. [8,215,216,217,218,219,220, 221,222]). The basic design principles are relatively simple: (1) utilize thermally and hydrothermally stable supports, e.g., high-temperature δ- or θ-aluminas or alkaline-earth or rare-earth oxides that form ultrastable spinels with γ-alumina; (2) use PdO rather than Pt or Pt–Rh for high temperature converters, since PdO is considerably more thermally stable in an oxidizing atmosphere because of its strong interaction with oxide supports; and (3) use multilayer strategies and/or diffusion barriers to prevent thermally induced solid-state reactions (e.g., formation of Rh aluminate) and to moderate the rate of highly exothermic CO and hydrocarbon oxidations. For example, a typical three-way automotive catalyst may contain alkaline-earth metal oxides (e.g., BaO) and rare-earth oxides (e.g., La_2O_3 and CeO_2), for stabilizing Pt and/or PdO on alumina, and ZrO_2 as a thermal stabilizer for the CeO_2 (an oxygen storage material) and as a noninteracting support for Rh in a separate layer or in a separate phase in a composite layer (see Figure 34).

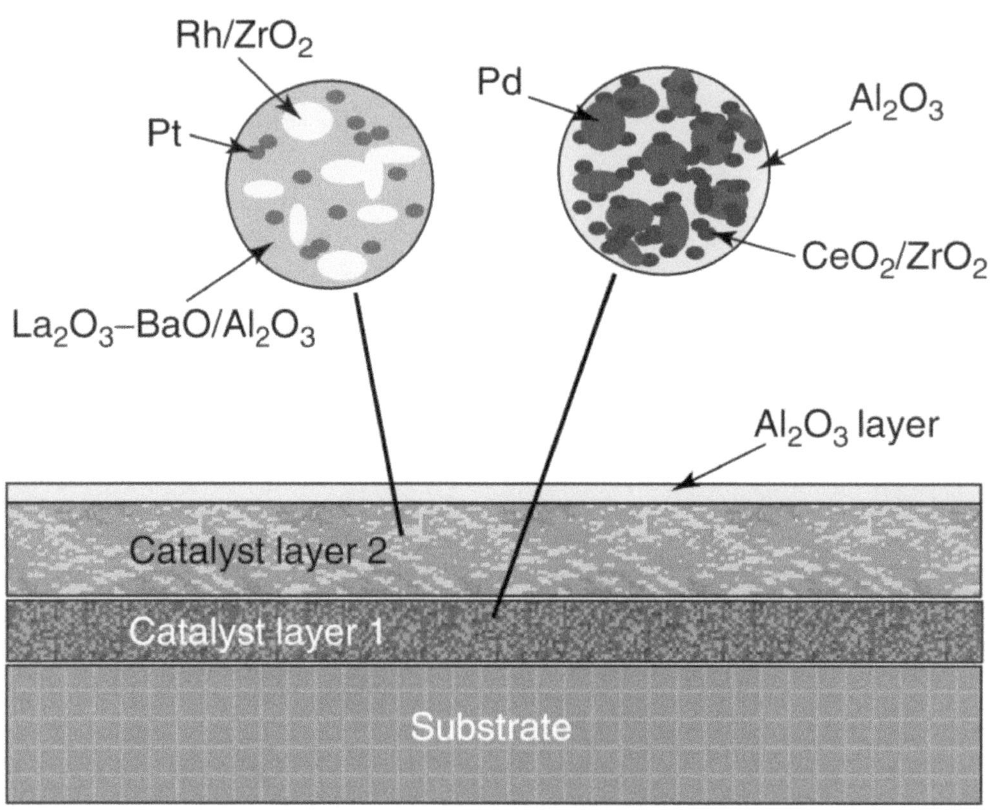

Figure 34: Conceptual design (by C. H. Bartholomew) of an advanced three-way catalyst for auto emissions control. Catalyst layer 1 is wash-coated first onto the monolithic substrate and consists of (**a**) well-dispersed Pd, which serves to oxidize CO/hydrocarbons and to reduce NO

and (**b**) CeO_2/ZrO_2 crystallites (in intimate contact with Pd), which store/release oxygen respectively, thereby improving the performance of the Pd. Catalyst layer 2 (added as a second wash coat) is a particle composite of Rh/ZrO_2 (for NO reduction) and Pt/La_2O_3–BaO/Al_2O_3 (with high to moderately-high activity for oxidation of CO and hydrocarbons). A thin (50–80 μm) coat of Al_2O_3, deposited over catalyst layer 2, acts as a diffusion barrier to foulants and/or poisons. Both the Al_2O_3 layer and catalyst layer 2 protect the sulfur-sensitive components of catalyst layer 1 from poisoning by SO_2.

Often, ideal metal dispersions require metal nanoparticles to be distributed closely together, but these particles are thermodynamically unstable on the surface and undergo rapid sintering, as described in Section 2.3 above. Recently, in an attempt to reduce sintering rates, researchers have attempted to stabilize the metal nanoparticles by first dispersing them on a support, encapsulating them in the same or another metal oxide, and then opening porosity to the particles (e.g., [247,248]). These approaches have met with varying degrees of success, but point to promising new areas of synthesis techniques that have the potential to reduce or to eliminate deactivation by sintering.

Prevention of Mechanical Degradation

While relatively few studies have focused on this topic, there are nevertheless principles that guide the design of processes and catalysts in preventing or minimizing mechanical degradation (see Table 19). In terms of catalyst design, it is important to (1) choose supports, support additives, and coatings that have high fracture toughness, (2) use preparation methods that favor strong bonding of primary particles and agglomerates in pellets and monolith coatings, (3) minimize (or rather optimize) porosity (thus maximizing density), and (4) use binders, such as carbon, to facilitate plastic deformation and thus protect against brittle fracture. Processes (and to some extent preparation procedures) should be designed to minimize (1) highly turbulent shear flows or cavitation that lead to fracture of particles or separation of coatings, (2) large thermal gradients or thermal cycling leading to thermal stresses, and (3) formation of chemical phases of substantially different densities or formation of carbon filaments leading to fracture of primary particles and agglomerates. Nevertheless, thermal or chemical tempering can be used in a controlled fashion to strengthen catalyst particles or agglomerates.

Examples of catalyst design to minimize attrition can be found in the recent scientific [230,231] and patent [232,233,234,235] literature focusing on the Fischer–Tropsch synthesis in slurry reactors. These studies indicate that (1) spray drying of particles improves their density and attrition resistance; (2) addition of silica and/or alumina into titania improves its attrition resistance, while addition of only 2000–3000 ppm of titania to γ-alumina improves alumina's attrition resistance; and (3) preformed alumina spheres promoted with La_2O_3 provide greater attrition resistance relative to silica. Increasing attrition resistance is apparently correlated with increasing density [230,231,235]. According to Singleton and co-workers [235], attrition resistance of Co/Al_2O_3 is improved when the γ-alumina support is (1) formed from synthetic boehmite having a crystallite diameter of 4–5 nm and (2) is pretreated in acidic solution having a pH of 1–3 (see Figure 35); moreover, attrition resistance decreases in the order $Co/Al_2O_3 > Co/SiO_2 > Co/TiO_2$ and is greater for catalysts prepared by aqueous*versus* nonaqueous impregnation.

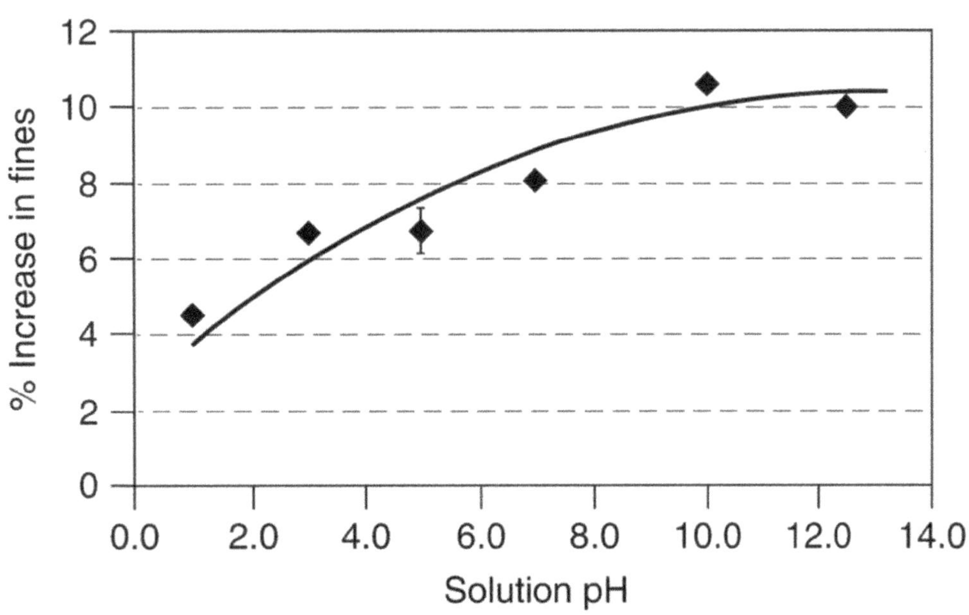

Figure 35: Effect of solution pH on the attrition resistance of 70-μm γ-Al_2O_3 particles measured in jet-cup tests [235]. The % increase in fines is defined at the % increase of particles of less than 11 μm.

REGENERATION OF DEACTIVATED CATALYSTS

Despite our best efforts to prevent it, the loss of catalytic activity in most processes is inevitable. When the activity has declined to a critical level, a choice must be made among four alternatives: (1) restore the activity of the catalyst, (2) use it for another application, (3) reclaim and recycle the important and/or expensive catalytic components, or (4) discard the catalyst. The first alternative (regeneration and reuse) is almost always preferred; catalyst disposal is usually the last resort, especially in view of environmental considerations.

The ability to reactivate a catalyst depends upon the reversibility of the deactivation process. For example, carbon and coke formation is relatively easily reversed through gasification with hydrogen, water, or oxygen. Sintering on the other hand is generally irreversible, although metal redispersion is possible under certain conditions in selected noble metal systems. Some poisons or foulants can be selectively removed by chemical washing, mechanical treatments, heat treatments, or oxidation [249,250]; others cannot be removed without further deactivating or destroying the catalyst.

The decision to regenerate/recycle or discard the entire catalyst depends largely on the rate of deactivation. If deactivation is very rapid, as in the coking of cracking catalysts, repeated or continuous regeneration becomes an economic necessity. Precious metals are almost always reclaimed where regeneration is not possible. Disposal of catalysts containing nonnoble heavy metals (e.g., Cr, Pb, or Sn) is environmentally problematic and should be a last resort; if disposal is necessary, it must be done with great care, probably at great cost. Accordingly, a choice to discard depends upon a combination of economic and legal factors [250]. Indeed, because of the scarcity of landfill space and an explosion of environmental legislation, both of which combine to make waste-disposal prohibitively expensive, there is a growing trend to regenerate or recycle spent catalysts [251,252]. A sizeable catalyst regeneration industry benefits petroleum refiners by helping to control catalyst costs and to limit liabilities [253,254]; it provides for *ex situ* regeneration of catalyst and recovery/recycling of metals, e.g., of cobalt, molybdenum, nickel, and vanadium from hydroprocessing catalysts [251].

Consistent with its importance, the scientific literature treating catalyst regeneration is significant and growing (includes nearly 1000 journal articles since 1990). Regeneration of sulfur-poisoned catalysts has been reviewed by Bartholomew and co-workers [28]. Removal of coke and carbon from catalysts has received attention in reviews by Trimm [59,250], Bartholomew [60], and Figueiredo [1]. Redispersion of sintered catalysts has been discussed by Ruckenstein and Dadyburjor [101], Wanke [102], and Baker and co-workers [103]. Useful case studies of regeneration of hydrotreating [255] and hydrocarbon-reforming catalysts [256] have also been reported. The proceedings of the 9th International Symposium on Catalyst Deactivation (2001) contains 12 papers treating catalyst regeneration [257]. Regeneration, recycling, and disposal of deactivated heterogeneous catalysts have been reviewed briefly by Trimm [250].

The patent literature treating catalyst regeneration/reactivation is enormous (more than 17,000 patents); the largest fraction of this literature describes processes for regeneration of catalysts in three important petroleum refining processes,*i.e.*, FCC, catalytic hydrocarbon reforming, and alkylation. However, a significant number of patents also claim methods for regenerating absorbents and catalysts used in aromatization, oligomerization, catalytic combustion, SCR of NO, hydrocracking, hydrotreating, halogenation, hydrogenation, isomerization, partial oxidation of hydrocarbons, carbonylations, hydroformylation, dehydrogenation, dewaxing, Fisher–Tropsch synthesis, steam reforming, and polymerization.

Conventional methods for regenerating (largely *in situ*) coked, fouled, poisoned, and/or sintered catalysts in some of these processes and representative examples thereof [258,259,260,261,262,263,264,265,266,267,268,269,270,271,272,273,274,275,276,277,278,279,280,281,282,283,284,285,286,287,288,289,290,291,292,293,294,295,296,297] are summarized in Table 20, while the basic principles and limitations involved in regeneration of coked, poisoned, and sintered catalysts are briefly treated in the subsections that follow.

Table 20: Conventional Methods for and Representative Examples of Catalyst Regeneration from Scientific and Patent Literatures.

Deactivation mechanism Process/Reaction Catalyst	Problem/cause	Method(s) of regeneration/phenomena studied/conclusions	Ref.
		Deactivation by coke, carbon	
Alkene aromatization oligomerization Zeolites, esp. ZSM-5, -22, -23, beta-zeolite, ferrierite	Catalyst fouling by condensation of heavy oligomers to coke	(1) ZSM-5 catalyst for light olefin oligomerization containing 2–3% coke is treated in 8–10% steam/air mixture (1300 kPa, 93 °C inlet) in a fluidized bed (2) A coked crystalline alumogallosilicate is contacted with oxygen at a concentration of 0.05–10 vol%, 420–580 °C, and 300–4000 h^{-1}	[258,259]
Alkylation of isoparaffins on solid catalysts Sulfated zirconia, USY[a], Nafion, silicalite, ZSM-5	Rapid catalyst deactivation due to coke formation; unacceptable product quality, and thermal degradation of catalyst during regeneration	(1) Coked zeolite is regenerated in liquid phase ($P > 3500$ kPa) fluid bed with H_2 in two steps: (a) at reaction temperature (20–50 °C) and (b) at 25 °C above reaction temperature (2) Coked Pd- and Pt/Y-zeolite catalysts containing 10–13% coke are regenerated in either air or H_2; H_2 treatment enables removal of most of the coke at low to moderate temperatures; higher temperatures are required for air (3) USY and other zeolites are regenerated in supercritical isobutane	[260–263]

Deactivation mechanism Process/Reaction Catalyst	Problem/cause	Method(s) of regeneration/phenomena studied/conclusions	Ref.
		Deactivation by coke, carbon	
Catalytic reforming of naphtha Pt/Al_2O_3 promoted with Re, Sn, Ge, or Ir	Poisoning and fouling by coke produced by condensation of aromatics and olefins	(1) Coke on Pt bimetallic reforming catalyst is removed off-stream in a fixed or moving bed at 300–600 °C, followed by oxychlorination (350–550 °C) (2) Coke on Pt/zeolite is removed in halogen-free oxygen-containing gas at $T < 415$ °C (3) Sintering during oxidation of coke on $Pt–Ir/Al_2O_3$ catalyst can be minimized by low regeneration temperatures (4) Study of influence of heating rate, temperature, and time on structural properties of regenerated $Pt–Sn/Al_2O_3$ (5) Study of effects of Cl, Sn content, and regeneration sequence on dispersion and selectivity of $Pt–Sn/Al_2O_3$ (6) Regenerated $Pt–Re/Al_2O_3$ is more stable than the fresh catalyst in *n*-heptane conversion and more selective for toluene	[264–269]
Dehydrogenation of propane and butane Cr_2O_3/Al_2O_3, Cr_2O_3/ZrO_2, FeO/K/MgO, Pt/Al_2O_3, $Pt–Sn/Al_2O_3$, Pt–Sn/KL-zeolite	Catalyst activity is low due to equilibrium limitations and build-up of product H_2; rapid loss of activity occurs due to coke formation	(1) Temperatures gradients were measured during burn off of coke formed on a chromia–alumina catalyst during butene dehydrogenation; data were used in developing a mathematical model for predicting temperatures and coke profiles (2) Coked supported palladium catalyst used in the dehydrogenation of dimethyltertrahydronaphthalenes to dimethylnaphthalenes is reactivated with an organic polar solvent at a temperature below 200 °C	[270,271]

Regeneration of Catalyst Deactivated by Coke or Carbon

Carbonaceous deposits can be removed by gasification with O_2, H_2O, CO_2, and H_2. The temperature required to gasify these deposits at a reasonable rate varies with the type of gas, the structure and reactivity of the carbon or coke, and the activity of the catalyst. Walker and co-workers [302] reported the following order (and relative magnitudes) for rates of uncatalyzed gasification at 10 kN/m^3 and 800 °C: O_2 (105) > H_2O (3) > CO_2 (1) > H_2 (3×10^{-3}). However, this activity pattern does not apply in general for other conditions and for catalyzed reactions [1]. Nevertheless, the order of decreasing reaction rate of $O_2 > H_2O > H_2$ can be generalized.

Rates of gasification of coke or carbon are greatly accelerated by the same metal or metal oxide catalysts upon which carbon or coke deposits. For example, metal-catalyzed coke removal with H_2 or H_2O can occur at a temperature as low as 400 °C [1]; β-carbon deposited in methanation can be removed with H_2 over a period of a few hours at 400–450 °C and with oxygen over a period of 15–30 min at 300 °C [60]. However, gasification of more graphitic or less reactive carbons or coke species in H_2 or H_2O may require temperatures as high as 700–900 °C [1], conditions, of course, that result in catalyst sintering.

Because catalyzed removal of carbon with oxygen is generally very rapid at moderate temperatures (e.g., 400–600 °C), industrial processes typically regenerate catalysts deactivated by carbon or coke in air. Indeed, air regeneration is used to remove coke from catalysts in catalytic cracking [81], hydrotreating processes [255], and catalytic reforming [256].

One of the key problems in air regeneration is avoiding hot spots or overtemperatures which could further deactivate the catalyst. The combustion process is typically controlled by initially feeding low concentrations of air and by increasing oxygen concentration with increasing carbon conversion [255,303]; nitrogen gas can be used as a diluent in laboratory-scale tests, while steam is used as a diluent in full-scale plant operations [303]. For example, in the regeneration of hydrotreating catalysts, McCulloch [255] recommends keeping the temperature at less than 450 °C to avoid the γ- to α-alumina conversion, MoO_3 sublimation, and cobalt or nickel aluminate formation, which occur at 815, 700, and 500–600 °C respectively.

Because coke burn-off is a rapid, exothermic process, the reaction rate is controlled to a large extent by film heat and mass transfer. Accordingly, burn-off occurs initially at the exterior surface and then progresses inward, with the reaction occurring mainly in a shrinking shell consistent with a "shell-progressive" or "shrinking-core" model, as illustrated inFigure 36 [304]. As part of this same work, Richardson [304] showed how experimental burn-off rate data can be fitted to various coking transport models, e.g., parallel or series fouling. Burn-off rates for coke deposited on SiO_2/Al_2O_3 catalysts were reported by Weisz and Goodwin [305]; the burning rate was found to be independent of initial coke level, coke type, and source of catalyst.

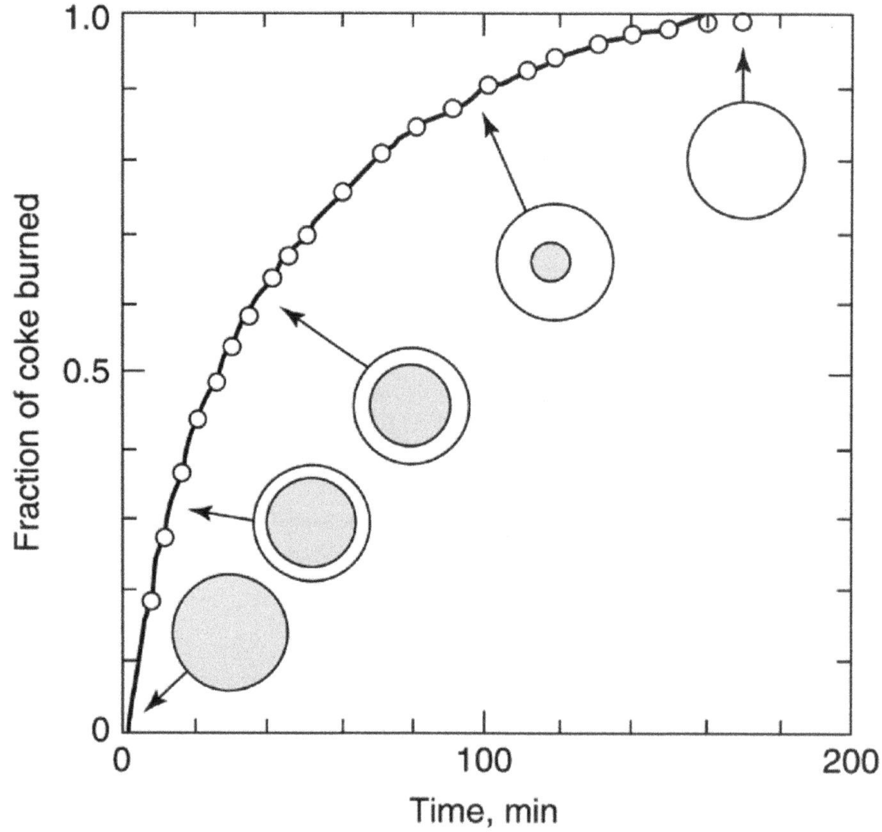

Figure 36: Shell-progressive regeneration of fouled pellet Reproduced from [304]. Copyright 1972, American Chemical Society.

Regeneration of Poisoned Catalysts

Much of the previous literature has focused on regeneration of sulfur-poisoned catalysts used in hydrogenations and steam reforming. Studies of regeneration of sulfur-poisoned Ni, Cu, Pt, and Mo with oxygen/air, steam, hydrogen, and inorganic oxidizing agents have been reported [28]. Rostrup-Nielsen [306] indicates that up to 80% removal of surface sulfur from Mg- and Ca-promoted Ni, steam reforming catalysts occurs at 700 °C in steam. The presence of both SO_2 and H_2S in the gaseous effluent suggests that the following reactions occur:

$$Ni\text{-}S + H_2O \rightarrow NiO + H_2S \quad (5)$$

$$H_2S + 2H_2O \rightarrow SO_2 + 3H_2 \quad (6)$$

Although this treatment is partially successful in the case of low-surface-area steam reforming catalysts, the high temperatures required for these reactions would cause sintering of most high-surface-area nickel catalysts.

Regeneration of sulfur-poisoned catalysts, particularly base metal catalysts, in air or oxygen has been largely unsuccessful. For example, the treatment of nickel steam-reforming catalysts in steam and air results in the formation of sulfates, which are subsequently reduced back to nickel sulfide upon contact with hydrogen. Nevertheless, sulfur can be removed as SO_2 at very low oxygen partial pressures, suggesting that regeneration is possible under carefully controlled oxygen atmospheres, including those provided by species such as CO_2 or NO that dissociate to oxygen. Apparently, at low oxygen pressures, the oxidation of sulfur to SO_2 occurs more rapidly than the formation of nickel oxide, while at atmospheric pressure the converse is true, *i.e.*, the sulfur or sulfate layer is rapidly buried in a nickel oxide layer. In the latter circumstance, the sulfur atoms diffuse to the nickel surface during reduction, thereby restoring the poisoned surface. Regeneration of sulfur-poisoned noble metals in air is more easily accomplished than with steam, although it is frequently attended by sintering. Regeneration of sulfur-poisoned nickel catalysts using hydrogen is impractical because (1) adsorption of sulfur is reversible only at high temperatures at which sintering rates are also high and (2) rates of removal of sulfur in H_2as H_2S are slow even at high temperature.

Inorganic oxidizing agents such as $KMnO_4$ can be used to oxidize liquid phase or adsorbed sulfur to sulfites or sulfates [16]. These electronically shielded structures are less toxic than the unshielded sulfides. This approach has somewhat limited application, *i.e.*, in partial regeneration of metal catalysts used in low temperature liquid-phase hydrogenation reactions or in liquid-phase destruction of chlorinated organic compounds. For example, Lowrey and Reinhard [286] reported successful regeneration in dilute hypochlorite solution of a Pd/Al_2O_3 catalyst deactivated during the aqueous-phase dechlorination of trichloroethylene (TCE) in the presence of sulfite or HS^- ions. These poisons are formed by sulfate-reducing bacteria present in natural groundwater and are apparently adsorbed on the alumina or Pd surfaces more strongly than sulfate ions. Figure 37 illustrates how readily the poisoned catalyst is regenerated by dilute hypochlorite solutions; indeed, it is evident in Figure 37b that regeneration every 5–10 days successfully maintains the catalytic conversion of TCE around 25% (a value only slightly less than that observed for reaction in distilled water).

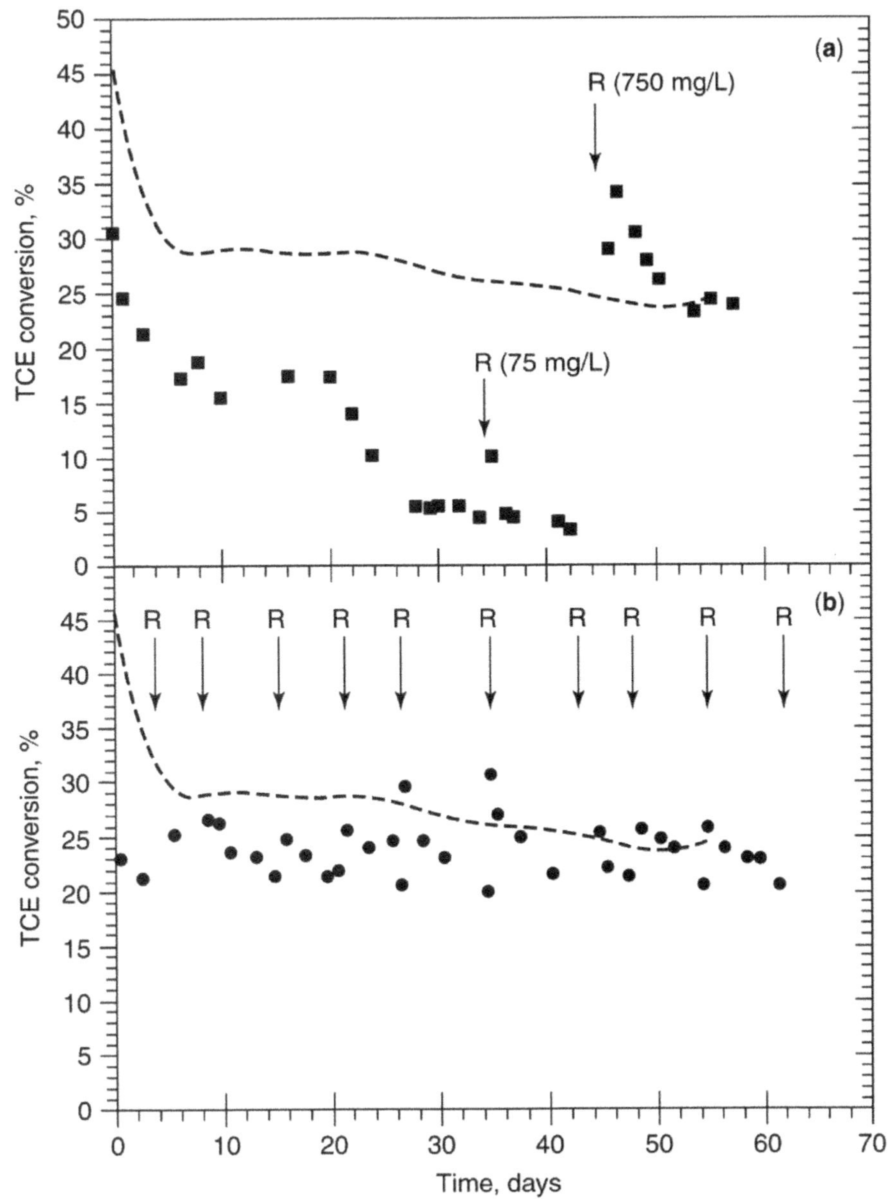

Figure 37: Effect of regeneration (R) with hypochlorite of Pd/Al_2O_3 catalysts used for aqueous phase dechlorination of trichloroethylene in the presence of HS^-/SO_3^{2}. Reproduced from [286]. Copyright 1992, American Chemical Society.

Detailed Case Study on Regeneration of Selective Catalytic Reduction (SCR) Catalysts

Introduction to SCR: Key to Abatement of NO_x from Coal Utility Boilers

NO_x, generally defined as NO and NO_2, emissions from coal utility boilers (approximately 30% of total NO_x emissions in the U.S.) contribute substantially to the formation of acid rain and photochemical smog, which in turn damage human health, property, agriculture, lakes, and forests. Selective catalytic reduction (SCR) technology has been used in utility boilers since the 1980s in Japan and Europe in response to stringent NO_x removal regulations. By 2000, SCR systems had been installed in coal-fired boilers totaling roughly 25 and 55 GW in Japan and Europe respectively [307,308]. Equivalent stringent NO_x abatement regulations were enacted later in the U.S. by the EPA, including

(1) the 1990 ARP and OTC mandates, requiring states to reduce NO_x emissions by 80%;

(2) the 1995 OTC-Phase 1 requiring Reasonably Available Control Technology (RACT);

(3) the 1998 NO_x SIP Call setting up a regional cap-and-trade program for 20 eastern states based on an equivalent NO_x emission rate of 0.15 lb/10^6-Btu; and

(4) the 2005 Clean Air Interstate Rule (CAIR) requiring all states to meet Best Available Retrofit Technology (BART) for existing plants, equivalent to emission rates of less than 0.05–0.10 lb/10^6-Btu [309,310].

By 2006, about 100 GW of coal-fired steam boilers in the U.S. used SCR. Presently, the U.S. has about 140 GW [309] of coal-boiler SCR capacity; world-wide, an estimated 300 GW of coal-boiler SCR is in operation.

Prior to the more recent stringent U.S. emissions regulations, boiler and engine manufacturers successfully reduced NO_x emissions by 30–60% using modifications to combustion processes, including reducing excess air, adding two-stage combustion features, altering burner design, *etc.* However, meeting the new reduction targets of 80–90% is, in general, only possible through catalytic after-treatment (SCR). Given ever more restrictive NO_x emission standards and the fact that worldwide power production from coal could double or triple in the next decade to an estimated 1500 GW [311], total installed SCR unit capacity is expected to grow commensurately, providing continued investment and design challenges in this area.

Selective Catalytic Reduction of NO_x

Reaction Chemistry and Preferred Catalysts

Selective catalytic reduction (SCR) is a process in which a reducing agent, typically NH_3, reacts selectively with the NO_x to produce N_2 without consumption of the excess O_2 present in the flue gas. Desirable stoichiometric reactions for SCR of NO and NO_2 (Equations 7 and 8) occur with high activity and selectivity to N_2 within a narrow temperature window of 300–400 °C on preferred commercial catalysts.

$$4NH_3 + 4NO + O_2 \rightarrow 4N_2 + 6H_2O \quad (7)$$

$$4NH_3 + 2NO_2 + O_2 \rightarrow 3N_2 + 6H_2O \quad (8)$$

Undesirable side reactions include oxidations of SO_2 (present in the flue gas) and the reducing agent NH_3. While only a small fraction of the SO_2 present in the flue gas is catalytically oxidized to SO_3, this acid precursor either corrodes downstream heat-exchange surfaces or reacts with NH_3 to form ammonium sulfates, which in turn can foul catalyst and/or heat exchange surfaces. Oxidation of NH_3 to either NO or N_2 may also occur at temperatures above 400 °C.

A typical commercial vanadia catalyst consists of 1 wt% V_2O_5 and 10 wt% WO_3 (alternatively 6 wt% MoO_3) supported on high-surface-area TiO_2 (mostly anatase, 60–80 m^2/g). TiO_2 has the decided advantage over Al_2O_3 as a support, since the former stabilizes the active vanadia species and does not form a bulk sulfate in the presence of SO_2-containing flue gases; thus TiO_2 promotes activity and extends catalyst life. WO_3 and MoO_3 prevent the transformation of anatase to rutile; they reside on basic sites of TiO_2, blocking adsorption of SO_3, thereby preventing sulfation of the support. Additionally, WO_3 and MoO_3 increase Brønsted acidity, promoting NO_x reduction while lowering SO_2 oxidation rate. Commercial vanadia-titania catalysts are typically supplied in the form of extruded monoliths or plates (see Figure 38), forms which minimize pressure drop [8].

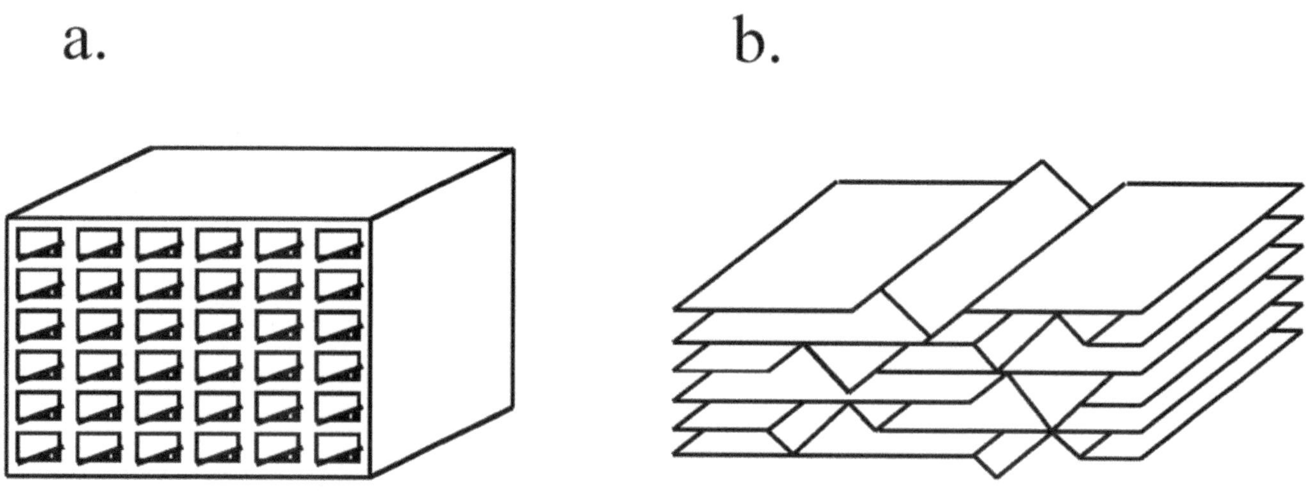

Figure 38: SCR catalyst support geometries: (**a**) extruded ceramic monolith; and (**b**) plate. Reproduced from [8]. Copyright 2006, Wiley-Interscience.

SCR Process Options

Two process options in terms of SCR reactant placement have found broad use for SCR units installed in coal-fired plants:

(1) the *high dust unit (HDU)* involving placement of the SCR unit after the economizer and prior to the air heater, particulate collector, and SO_2 scrubber; and

(2) the *tail end unit (TEU)* involving placement of the SCR unit following the SO_2 scrubber.

The HDU is used more widely in the U.S. and the TEU more frequently in Europe and Japan.

The HDU has the advantage of providing flue gas to the SCR unit at its ideal temperature range of 300–400 °C and disadvantages of

(1) deactivation of the catalyst due to erosion, fouling, and poisoning by fly ash thereby limiting its useful life to about 3–4 years;

(2) large monolith channel design to limit plugging by fly ash, but which also limits the amount of active catalyst per reactor volume; and

(3) requirement for a low activity catalyst to limit oxidation of SO_2 to SO_3 and the attendant formation of ammonium sulfates which foul and corrode downstream heat exchangers.

The TEU enables use of a smaller volume of high activity catalyst with small diameter channels, since particulates and SO_2 have already been removed upstream; moreover, since deactivation rate is much lower due to the absence of fly ash and other poisons, catalyst life is substantially extended (*i.e.*, to 15–20 years). A significant disadvantage is that the outlet scrubber gas, which is only about 120 °C, must be reheated to at least 200–250 °C for the SCR to occur at reasonable rates. The energy cost of reheating only 100 °C can be as much as 4–6% of the boiler capacity, unless a regenerative heat exchanger is used. In addition, the SCR catalyst must be designed to operate at significantly lower temperatures (200–290 °C relative to a typical 300–400 °C for an HDU).

Given the long life of the TEU catalyst, no regeneration is necessary. However, regeneration of the HDU catalyst is highly desirable, since the regeneration cost is significantly lower than the cost of a new catalyst. With this background, further discussion focuses on the deactivation and regeneration of the HDU catalyst.

Catalyst Deactivation, Rejuvenation, and Regeneration

Catalyst Deactivation

SCR catalysts have typical process lifetimes around 2–7 years, depending upon their application and placement in a power plant or other such facility. The principal causes of SCR catalyst deactivation [8,312] are fourfold:

`(1) fouling/masking of (deposition of solids on) catalyst surfaces, pores, and channels by fly ash components (e.g., sulfates and phosphates of Ca, K, and Na) or ammonium bisulfate;

(2) chemical poisoning of active sites by elements present in upstream lubricants or originating in the fuel such as As, Se, and P and alkali and alkaline earth metals;

(3) `hydrothermal sintering of the titania, especially as a result of high-temperature excursions; and

(4) abrasion or erosion by fly ash.

Erosion, fouling, and masking from fly ash and poisoning by As and alkali metals are specific to SCR catalysts installed near the hot, high-particulate side of a coal-fired boiler, accounting for the significantly lower catalyst life of 2–4 years for this configuration.

Formation of ammonium bisulfate depends on flue gas temperature, SO_3 concentration and NH_3 concentration [313]. Deposition of ammonium bisulfate is more likely to occur in catalyst pores at lower reactor temperatures in low-dust or tail-end (TEU) SCR units and on cooler surfaces of heat exchangers. Figure 39a shows typical activity loss *versus* time performance for a set of commercial V/Ti catalysts tested in a DOE pilot SCR unit installed in a slip-stream near the exit of a coal-fired boiler (HDU location) using high sulfur, Eastern U.S. coals; 20% of the initial catalyst activity is lost in about 14,000 h (1.6 years); however, the plant will not shut down until 50–60% of the initial activity has been lost (around 3–4 years). Activity and NH_3 slip are plotted against NH_3/NO ratio for the same catalysts in Figure 39b. To maintain NH_3 slip (exit NH_3 concentration) below a target maximum of 2–5 ppm (2 is highly preferred), the NH_3/NO ratio must be maintained near 0.8; under these conditions NO conversion is about 88%.

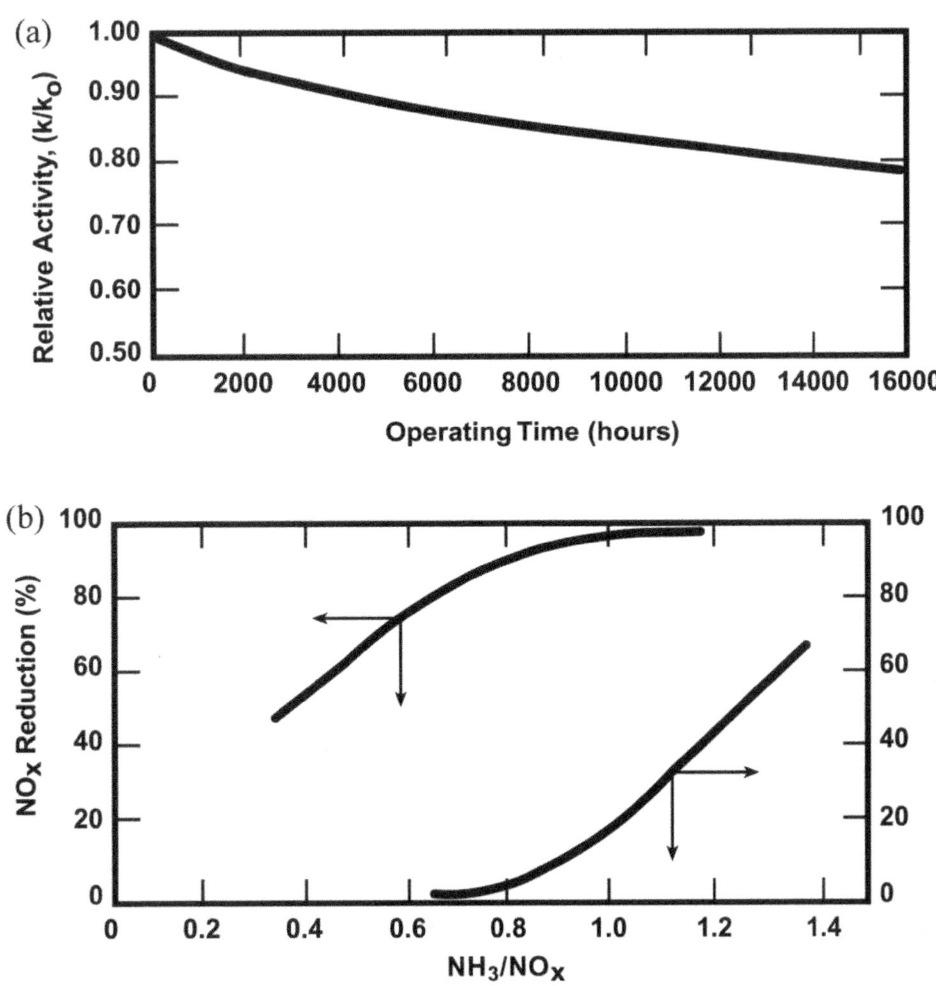

Figure 39. (**a**) Catalyst activity (k/k_o) *vs.* time; (**b**) Typical SCR performance. Reproduced from [313].

Prevention of deactivation requires optimal choices of catalyst design and process conditions. Abrasion, fouling, and/or poisoning by fly ash can be prevented by installation of a hot-side electrostatic precipitator or installing an active, low-temperature catalyst at the tail end of the process. Sintering is minimized by using catalyst promoters that enhance thermal stability and by maintaining reaction temperatures below critical values. The MoO_3 promoter extends catalyst life (in coal boilers) by preferentially adsorbing vapor-phase As which would otherwise adsorb on active V^{4+} sites. Free CaO in the fly ash (up to 3%) also scavenges As to low levels, forming calcium arsenide particles which are collected with the fly ash. Many U.S. coals contain adequate CaO; however, if the CaO content of the coal is too low, it can be added to the boiler or fuel. However, CaO levels above 3% of the fly ash are undesirable, since CaO reacts with SO_2 to form $CaSO_4$ which masks the exterior surface of the catalyst. Fouling by ammonium bisulfate is minimized by keeping exit SO_3 and NH_3 concentrations low and maintaining reaction temperatures above about 230 °C; SO_3 formation is minimized by keeping reaction temperatures below 350 °C or by using lower activity V_2O_5/TiO_2 or zeolite catalysts that have low selectivities for SO_3. Ultimately, however, extra catalyst volume is typically added to SCR reactors to extend periods between catalyst replacements.

For plants fueled by coal, substantial carry-over of inorganic ash occurs to HDU SCR units, a small, but significant fraction of which deposits on monolith walls, masks or blocks catalyst macropores, and plugs flow channels [314]. Extensive fouling necessitates the use of air lancing to purge the ash out of the catalyst channels. Figure 40 reveals the extent of serious channel plugging and erosion of an SCR catalyst in a pilot plant following several thousand hours of operation in flue gas containing coal fly ash. Plugging and excessive pressure drop are avoided by keeping monolith cell width at or above 7 mm.

Figure 40. Catalyst channel plugging (**left**) and damage due to erosion (**right**) during operation in an SCR facility. Reproduced from [313].

The type and extent of chemical deactivation depends on operating conditions, fuel type, catalyst geometry, shut-downs for boiler maintenance, *etc.* Mini-pilot tests and subsequent full-scale SCR operating experience have provided little evidence of poisoning by basic minerals from Western United States coals; rather they indicate that deactivation occurs principally by masking of catalyst layers and plugging of catalyst pores by $CaSO_4$ and other fly ash minerals. Moreover, laboratory analysis of catalysts exposed to power plant slip streams indicates that mineral poisons do not penetrate deep into catalyst pores [315,316] nor do they adsorb on Brønsted acid sites unless plant conditions cause moisture to condense on the catalyst.

Plant Operating Strategy to Maximize Catalyst Life

A typical SCR unit consists of a series of two to four catalyst layers (three is most common for coal boiler cleanup) through which the flue gas usually flows downward (see Figure 41). A layer of fresh catalyst can be added as catalyst performance declines over time [317]. Two general schemes are followed for replacing the spent catalyst, both of which take into consideration the *relative activity* or *design activity level*, a parameter that is usually defined as the ratio of NO_x conversion at any time divided by that produced by the fresh catalyst. Once the NO_x reduction performance declines to the minimum design activity level (typically 65–75% of fresh activity), the catalyst can either be replaced entirely (*simultaneous*replacement scheme) or one layer can be replaced at a time (*sequential* replacement scheme), usually beginning at the top and working down [313,318]. The sequential method results in increased overall catalyst life (on a per-volume-replaced basis), while annual replacement cost would be 60% lower for the simultaneous scheme (see Figure 42 [319]). Thus, optimal, cost-effective design of an SCR unit requires considering both the initial capital and annual costs.

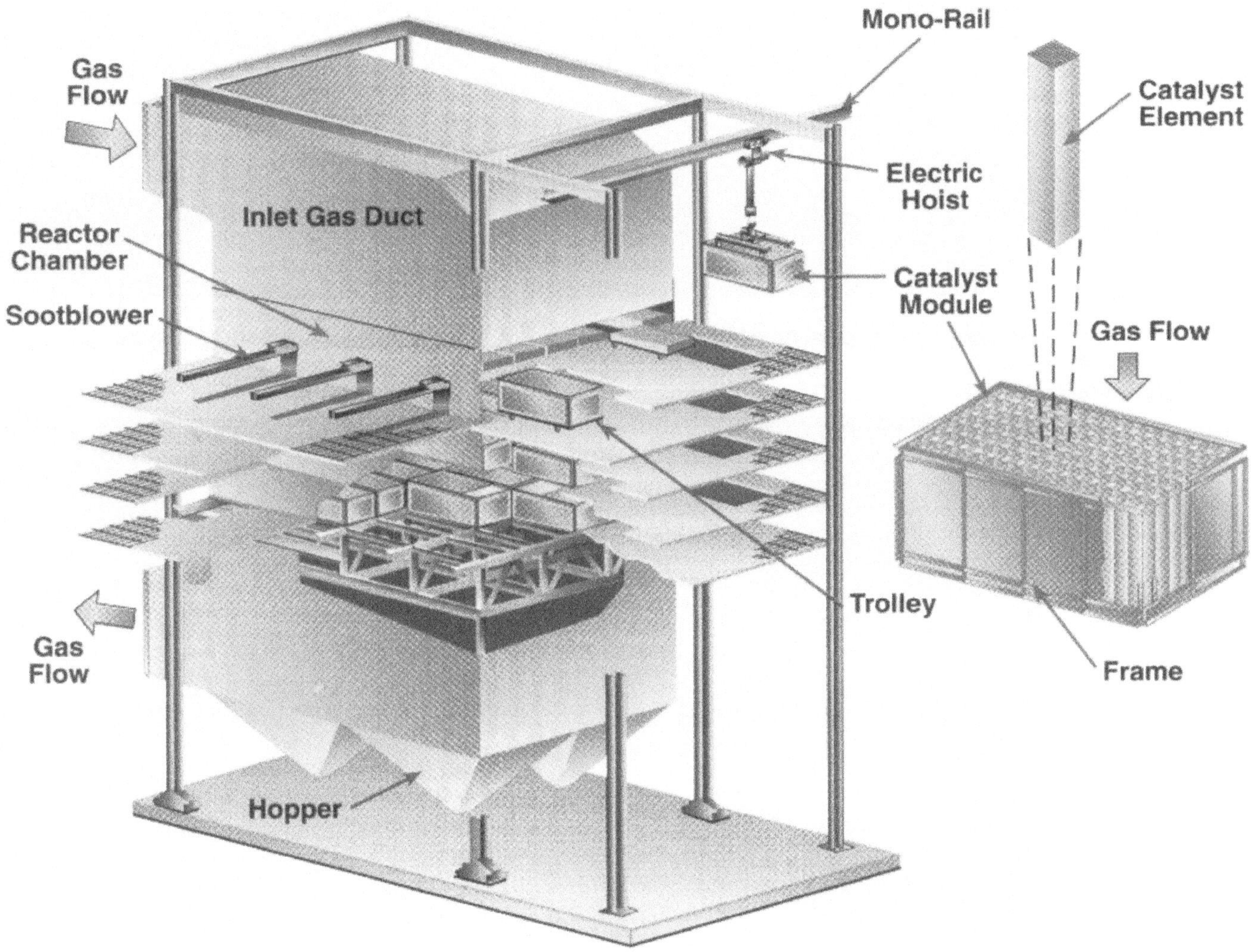

Figure 41: Vertical-flow fixed-bed SCR reactor. DOE SCR demonstration facility at Gulf Power Company's Plant Crist. Reproduced from [313].

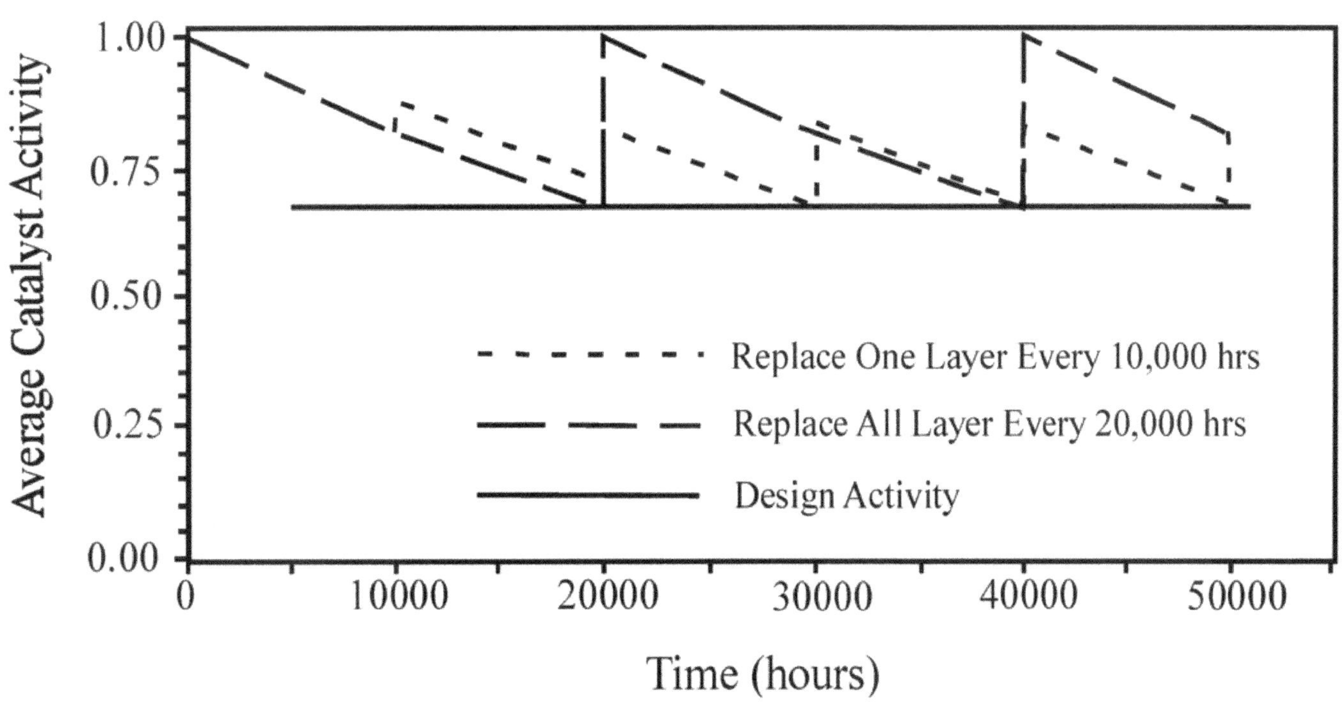

Figure 42: SCR replacement strategies: comparison of total replacement on a 20,000 h cycle relative to sequential replacement on a 10,000 h cycle while maintaining constant catalyst volume. Reproduced from [319].

Operating experience for commercial SCR installations has been better than anticipated. Catalyst lifetimes of 3–4 years at overall efficiencies of 75–90% for HDU's have been observed for electric boiler installations [312]. The principal contributors to operating cost include catalyst replacement cost, shutdown cost for catalyst replacement, and plant derating cost associated with catalyst pressure drop. Catalyst replacement or regeneration was typically required within 2–3 years and catalyst replacement times varied from 2–7 days. Pressure drop ranged from 0.8–15 cm of water for the various catalyst configurations and volumes. Pressure drops for plate type catalysts were significantly lower than for monolithic catalysts.

Catalyst Rejuvenation and Regeneration

While high-dust-catalyst life of 2–3 years is acceptable, advances in SCR catalyst regeneration technologies make it possible to extend life by several additional years. Recent experience indicates that even after long-term exposure to fly ash, foulants, and poisons, SCR catalysts may be successfully regenerated to the original performance or better [307,308,320,321,322].

Methods of Renewing Catalysts

Deactivated catalysts may be *cleaned, rejuvenated, and/or regenerated. Cleaning* commonly refers to removal of physical restrictions such as monolith channels plugged with fly ash or channel surfaces covered with a loose dust layer; these restrictions are easily removed *in situ* using compressed air, although cleaning will also be done as a first step in the other methods. *Rejuvenation* refers to relatively mild treatments that remove catalyst poisons or foulants inside the catalyst pores and restore part of the catalytic activity; these treatments are often done *in situ* or *on-site*. *Rejuvenation* involves removal of blinding layers and partial removal of some poisons; thus, activity is partly recovered, but none is added. *Regeneration*involves the *off-site*, complete restoration of catalytic activity through a series of relatively sophisticated treatments, some of which remove not only poisons and foulants, but also a part or much of the active catalytic materials from the support; hence, regeneration also involves restoration of the catalytically active materials bringing the catalyst to its original state or one of even higher activity. SCR catalysts are routinely and regularly cleaned or "blown out" during operation, while rejuvenation or regeneration is typically done after approximate 50–60% of the initial activity of the catalyst has been lost. *In situ rejuvenation* (ISR) *treatments* were practiced early (e.g., 1990s and early 2000s), while *off-site regeneration* (OSR) is now the*predominant practice* because of its greater effectiveness.

Rejuvenation or Regeneration?

According to McMahon [322], rejuvenating SCR catalyst may be more cost-effective than regenerating, if the catalyst is fairly new or the SCR system does not operate year around (as in the case of plants operating only during high pollutant levels, known as the "ozone season"). Otherwise, the choice between rejuvenation and regeneration depends largely on economics, *i.e.*,

(1) the plant's dispatch economics, including transportation costs;

(2) length of catalyst service;

(3) costs of removing and replacing the catalyst;

(4) the impact of the fuels combusted, *i.e.*, coal, oil, or gas; and

(5) the location of the catalyst in the plant, *i.e.*, HDU or TGU.

Examples of rejuvenation treatments are found in the scientific and patent literature. For example, work by Zheng and Johnsson [323] and others (e.g., [324,325]) indicates that activity of poisoned catalysts might be partially regenerated by washing with water, sulfuric acid, NH_4Cl, and/or catalyst precursor solutions (e.g., ammonium paratungstate and vanadyl sulfate), as well as a combination of washing and treatment with gaseous SO_2. The extent to which these rejuvenation methods are effective in restoring a significant fraction of the original catalyst activity varies significantly.

Rejuvenation: *On-site rejuvenation methods* generally include the following procedural types: (i) removal of dust in the monolith channels with compressed air followed by (ii) washing catalyst in a tank containing agitated, deionized water to remove the $CaSO_4$coating and alkali metal salts deposited by fly ash (the solution is generally mildly acidic due to impurities on the catalyst) or acidic aqueous solution (pH = 1–2 in either case) in a tank; (iii) rinsing vigorously with deionized water (usually in the same tank) to remove the dissolved and suspended deposits; and (iv) drying slowly in clean air at room temperature followed by drying gently in hot air. Examples of on-site regeneration methods include those developed and practiced in the time frame of 1995–2002 by SCR-Tech, SBW, Saar Energie, Steag, EnBW, HEW, BHK, and Integral [326,327,328]. The method described by Schneider and Bastuck [327] provided for adding catalytic materials, *i.e.*, vanadium and tungsten oxides (via impregnation of the V and W salts) to the cleaned catalyst.

The patent of Budin *et al.* [328] provides for more sophisticated treatments, including use of (i) nonionic surfactants and complex-forming or ion-exchange additives, (ii) washing with an acid or base, (iii) using acoustic radiation to remove fly-ash components, and (iv) addition of catalytic materials (oxides of V, W, Mo free of alkali and alkaline-earth metals, halogen, and sulfur) to restore activity, although few details or conditions of use are provided. In fact, no examples are provided in any of the patents cited directly above; accordingly, it is unclear to what extent and under what conditions the more sophisticated methods were used for on-site regeneration. The methods claimed by Budin *et al.* [328] are clearly more readily applied in off-site regeneration, as will be clear from the discussion below.

Regeneration: Bullock & Hartenstein [320], Cooper *et al.* [329], and McMahon [322] build a strong case for off-site regeneration and a comprehensive catalyst management program.

A Comprehensive Approach to Catalyst Management

The approach [320,322] includes

(1) strategies for extending catalyst life and reusability and planning for catalyst removal/rotation to coincide with power plant outages;

(2) catalyst inspection and testing before and following regeneration with replacement of badly damaged catalyst which is unregenerable;

(3) off-site regeneration using a series of robust washing and chemical treatments to remove channel blockages, deactivated catalyst metals, and poisons, followed by chemical treatments to restore active catalytic materials; and

(4) gentle drying/calcination in air to high temperatures to produce catalytically active oxides.

Common Regeneration Practices

Normal regeneration procedures [307,308,320,322,330,331,332,333] are designed to enhance removal of blockages, deactivated catalyst, and poisons and restore active catalytic material. These typically include the following steps:

(1) pressurized wet and dry treatments to remove channel blockages and outer dust layers;

(2) washing of catalyst units in tanks containing agitated water augmented with surfactants, dispersants, ion-exchange materials, emulsifiers, acid, base, and/or acoustic radiation to remove the outer $CaSO_4$ coating, alkali metal salts deposited in the catalyst pores, and deactivated (e.g., As-poisoned) catalyst;

(3) rinsing repeatedly in deionized water and repeating ultrasonic treatments between or in concert with chemical treatments, with a final rinse to finish removal of any catalyst or fouling residue;

(4) reimpregnation of the clean support with salts of the active catalytic materials (V, Mo, and W); and

(5) drying (calcining) at low heating rates to decompose the salts of the active catalytic materials to active metal oxides of V, Mo, and W.

Regeneration Process Profile: SCR-Tech Regeneration Process

SCR-Tech is the most prominent and experienced off-site regeneration company with 13 years of experience in the regeneration business and a documented record of research and development, going back to their German parent company ENVICA, who in 1997 began developing an offsite regeneration process. SCR-Tech was the first and until 2008 the only company in the U.S. to perform off-site regeneration. In September 2007, Evonik Energy Services (formerly Steag) opened an SCR catalyst regeneration facility in the U.S.

The SCR-Tech regeneration process involves a number of different process steps illustrated in Figure 43. Upon receipt of a shipment of catalyst, catalyst elements from several modules are inspected and analyzed; results of the analysis provide a basis for determining the precise protocol for treatment, *i.e.*, the number and order of processing steps [334,335]. A large catalyst module is then led through a protocol of soaking, washing, ultrasonic treatment, arsenic and/or phosphorus removal (as needed), replenishment of V and Mo, neutralization, and rinsing in various soaking pits, as shown in Figure 43; all of these wet chemical steps are performed at controlled pH and temperature. Finally, the catalyst is dried, inspected, and packaged for shipment. Performance guarantees are provided for complete removal of blinding layers, catalyst activity (typically higher after regeneration), SO_2 to SO_3 conversion rate (typically lower), mechanical stability (the same), and deactivation rate (the same) such that all properties of the regenerated catalyst are as good or better than the new catalyst.

A comparison of the physical appearances of SCR monoliths and plates before and after regeneration in Figure 44 reveals the rigor of the SCR-Tech cleaning process. The nearly complete removal of poisons originally in high concentrations by the regeneration process is demonstrated in Figure 45. Surface concentrations of CaO, P_2O_5, SiO_2, and SO_4 were also substantially reduced.

Table 21 compares the costs of regenerating *versus* buying a new catalyst [322]. This case is for a typical 500 MW unit with 650 m^3 of catalyst contained in 450 modules (150 modules in each of 3 layers). The purchase cost of new catalyst in 2006 was $3500 to 4500 per m^3. The cost to regenerate the catalyst is approximately 60% of this price. Thus, the purchase cost of one layer is $758,000 to $975,000 as compared to a regeneration cost of $455,000 to $585,000 resulting in savings per layer of $303,000 to $390,000 or $910,000 to $1.2 million for three layers. Assuming the SCR unit runs year around (as most do now) and catalyst life is three years, the annual savings due to regeneration is in the range of $300,000 to $600,000. The disposal cost for an SCR catalyst can range from $50 to $2,000/ton, the upper figure based on the cost of treating the vanadium as hazardous waste. Hence the disposal cost could be as high as $500,000 for a layer of catalyst. According to McMahon, SCR catalysts can be regenerated from 3 to 7 times.

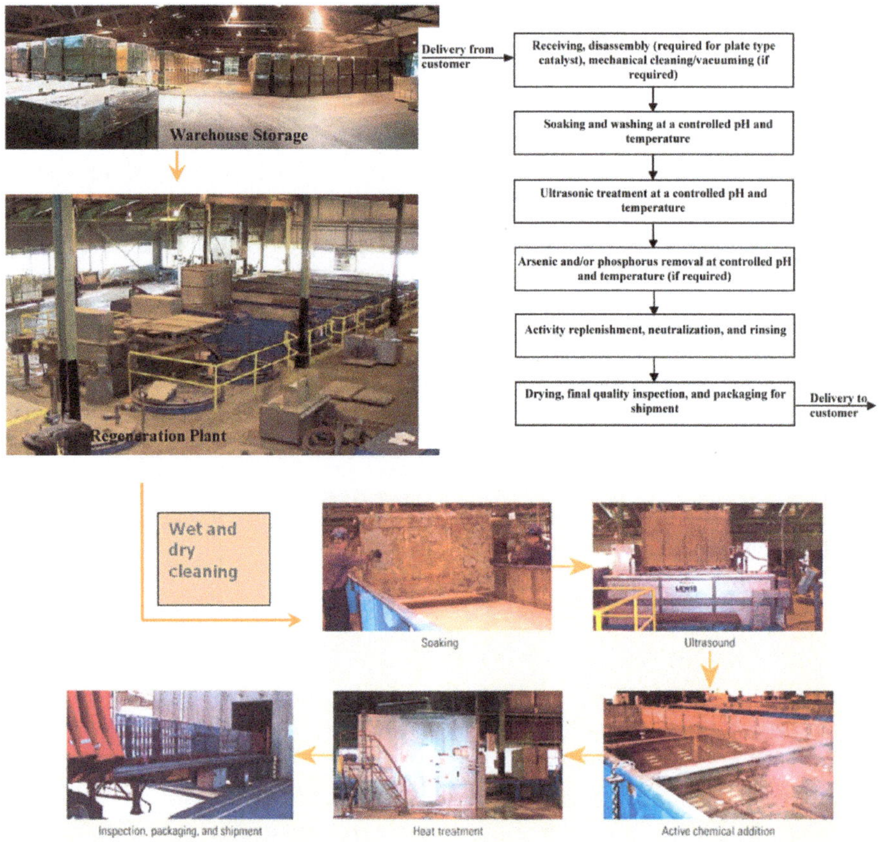

Figure 43: SCR-Tech catalyst regeneration process. Reproduced from [322,335,336,337]. Reproduced with permission of Electric Power and CoaLogix, Inc.

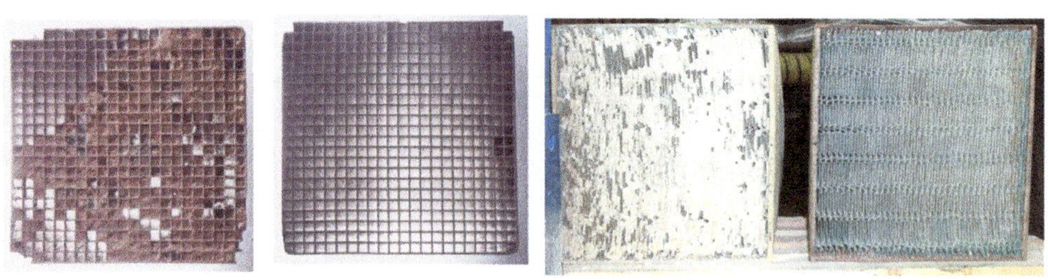

(a) (b)

Figure 44. (**a**) Monolith and (**b**) plate SCR catalysts before and after SCR-Tech regenerative treatment. Reproduced from [334]. Courtesy CoaLogix, Inc.

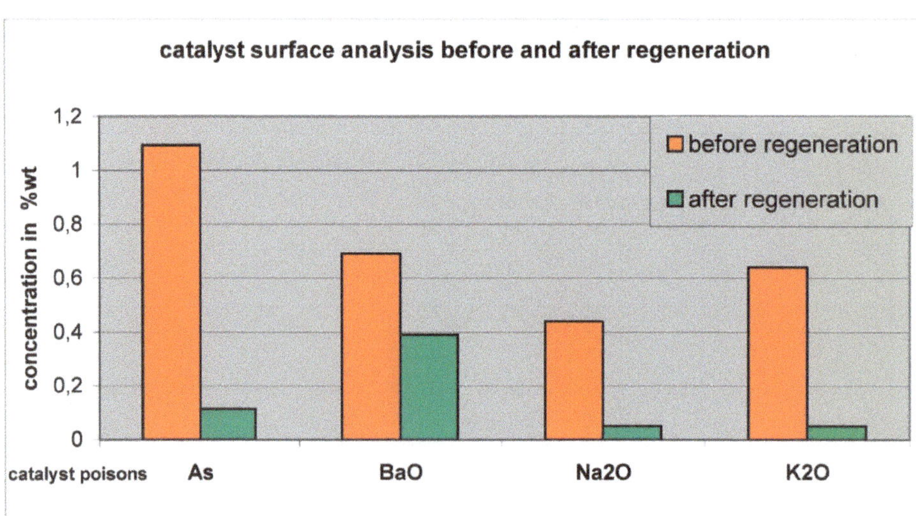

Figure 45: Concentration of principle poisons before and after regeneration. Reproduced from [320]. Courtesy CoaLogix, Inc.

Table 21: Cost per layer (217 m^3 or 150 modules) of new *versus* regenerated SCR catalyst. Adapted from [322]. Copyright 2006, Electric Power.

Catalyst Handling Step	New	Regenerated
Removal from SCR system	Comparable	Comparable
Transport out	Comparable	Comparable
Purchase price	\$758,000–\$975,000	\$455,000–\$585,000
Shipping	Comparable	Comparable
Installation	Comparable	Comparable
Net savings from regeneration	\$303,000–\$390,000 pls disposal cost	
Disposal cost	\$20,000–\$500,000	0

SCR Catalyst Case Study Summary Observations and Conclusions

Off-site regeneration processes are more sophisticated and demanding than on-site rejuvenation processes; the off-site regeneration processes provide significantly more efficient cleaning and reconstitution of the catalyst with full recovery of activity—sometimes greater than the fresh catalyst activity. Rejuvenation provides only partial (up to 85%) recovery of the original activity.

The development of offsite processes for regeneration of SCR catalysts is relatively new, having occurred largely over the past 10–15 years. SCR-Tech was the first and until 2008 the only company to operate an off-site regeneration facility in the U.S.

Because surface deposits are a primary deactivation mechanism, especially in HDU catalysts, extensive multi-step treatments are required, but rejuventation or regeneration appear to be a cost-effective method of catalyst management for SCR catalysts.

Redispersion of Sintered Catalysts

During catalytic reforming of hydrocarbons on platinum-containing catalysts, growth of 1-nm platinum metal clusters to 5–20-nm crystallites occurs. An important part of the catalyst regeneration procedure is the redispersion of the platinum phase by a high temperature treatment in oxygen and chlorine, generally referred to as "oxychlorination." A typical oxychlorination treatment involves exposure of the catalyst to HCl or CCl_4 at 450–550 °C in 2–10% oxygen for a period of 1–4 h (see details in Table 22). During coke burning, some redispersion occurs, e.g., dispersion (*D*) increases from 0.25 to 0.51, while during oxychlorination the dispersion is further increased, e.g., from 0.51 to 0.81 [256]. A mechanism for platinum redispersion by oxygen and chlorine is shown in Figure 46 [256]. It involves the adsorption of oxygen and chlorine on the surface of a platinum crystallite and formation of $AlCl_3$, followed by the formation of $PtCl_2(AlCl_3)_2$ complexes that dissociatively adsorb on alumina to oxychloro-platinum complexes. These latter complexes form monodisperse platinum clusters upon subsequent reduction.

Table 22: Typical Regeneration Procedure for Reforming Catalysts [a].

(*1*) *Preliminary operations:* cool the catalyst to about 200 °C and strip hydrocarbons and H_2 with N_2
(*2*) *Elimination of coke by combustion:* inject dilute air (0.5% O_2) at 380 °C and gradually increase oxygen content to about 2% by volume while maintaining temperature below 450–500 °C to prevent further sintering of the catalyst. To prevent excessive leaching of Cl_2, HCl or CCl_4 may be injected during the combustion step
(*3*) *Restoration of catalyst acidity:* restoration of acidity occurs at 500 °C by injection of a chlorinated compound in the presence of 100–200 ppm water in air
(*4*) *Redispersion of the metallic phase:* expose the catalyst to a few Torr of HCl or CCl_4 in 2–10% O_2 in N_2 at 510–530 °C for a period of about 4 h. After redispersion, O_2 is purged from the unit and the catalyst is reduced in H_2

[a] Ref. [255,256].

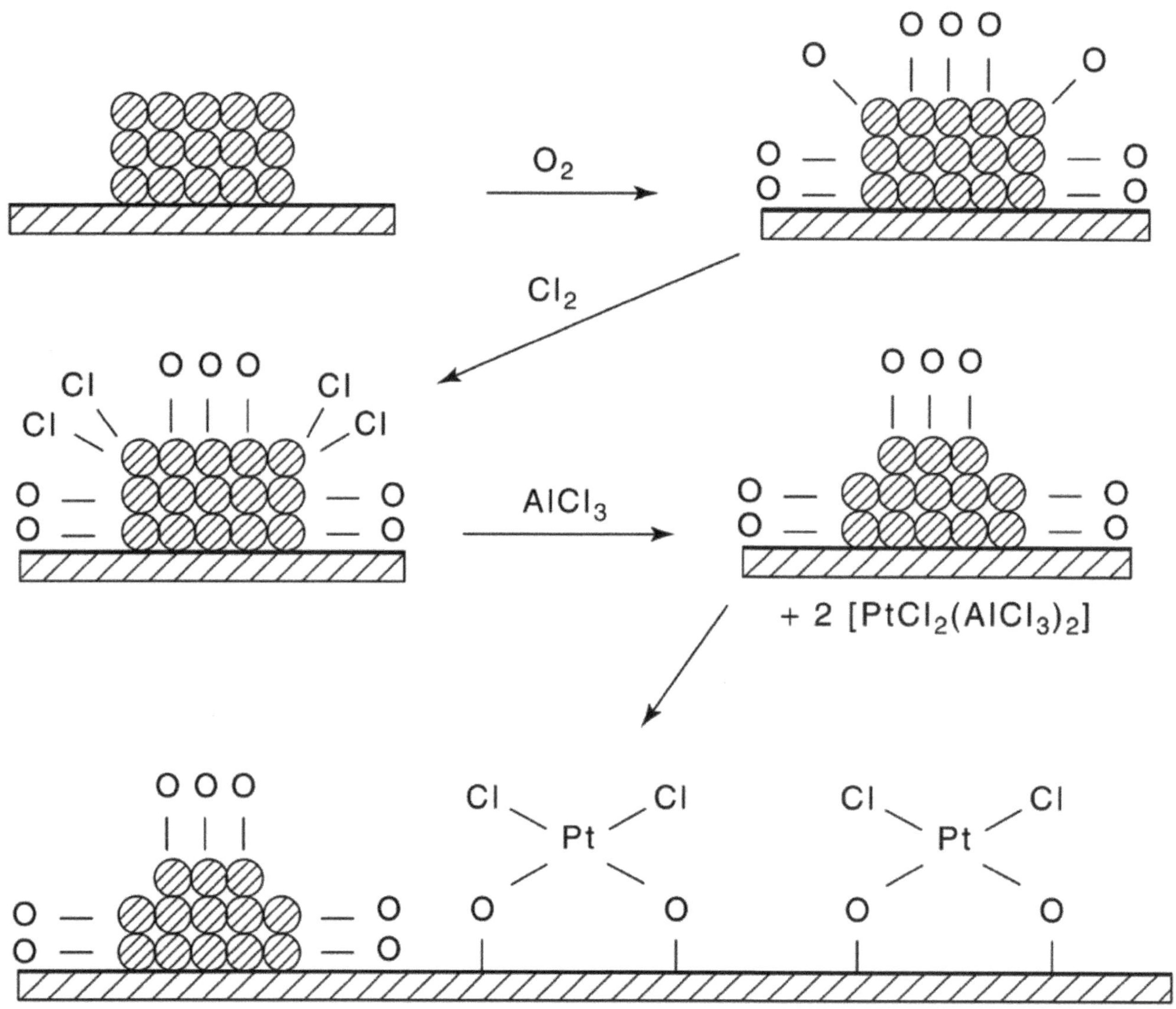

Figure 46: Proposed mechanism for redispersion by oxychlorination of alumina-supported platinum. Reproduced from [256]. Copyright 1982, Brill Nijhoff Publishers.

Some guidelines and principles regarding the redispersion process are worth enumerating:

(1) In cases involving a high degree of Pt sintering or poisoning, special regeneration procedures may be required. If large crystallites have been formed, several successive oxychlorinations are performed [256].

(2) Introducing oxygen into reactors in parallel rather than in series results in a significant decrease in regeneration time [101].

(3) Introduction of hydrocarbons present in the reactor recycle after regeneration is said to stabilize the catalyst; solvents such as ammonium acetate, dilute nitric acid containing lead nitrate, and EDTA and its diammonium salt are reported to dissolve out metal aggregates without leaching out the dispersed metal [101].

(4) The procedures for redispersion of Pt/alumina are not necessarily applicable to Pt on other supports or to other metals. For example, Pt/silica is redispersed at lower temperature and higher Cl_2 concentration (150–200 °C and 25% Cl_2). Pd/alumina can be redispersed in pure O_2 at 500 °C. While Pt–Re/alumina is readily redispersed by oxychlorination at 500 °C, Pt–Ir/alumina is not redispersed in the presence of O_2, unless the catalyst is pretreated with HCl [266].

An extensive scientific and patent literature of redisperson describes the use of chlorine, oxygen, nitric oxide, and hydrogen as agents for redispersion of sintered catalysts (summarized in Table 23). Most of the early literature shows positive effects for chlorine compounds in the presence of oxygen in redispersing alumina-supported platinum and other noble metals. Recent literature demonstrates the need for understanding the detailed surface chemistry in order to successfully develop and improve redispersion processes, especially in more complex catalyst systems such as alumina-supported bimetallics. For example, on the basis of a fundamental study of the redispersion surface chemistry, Fung [266] developed a redispersion procedure for Pt–Ir bimetallic catalysts using a wet HCl/air treatment, since the conventional oxychlorination is not effective for this catalyst.

Table 23: Representative Patents Prior to 1990 Treating Catalyst Redispersion.

Dispersing agent class	Dispersing agent	Metals/support	Patent No.	Ref.
Chlorine-Containing				
	Cl_2, Cl + halogen	Pt/zeolite	U.S. 4,645,751	[338]
	Cl, H_2O, O_2	Pt/zeolite	U.S. 4,657,874	[339]
	HCl, Cl–O	Ir	U.S. 4,491,636	[340]
	Cl, O_2	Pt–Ir, Ir	U.S. 4,467,045	[341]
	HCl, Cl	Pt–Ir–Re, Pt–Ir/zeolites	U.S. 4,359,400	[342]
	Cl, halogen	Ir, Pt–Ir/Al_2O_3	U.S. 4,480,046	[343]
	Cl–H_2O	Pt–Ir–Se/Al_2O_3	U.S. 4,492,767	[344]
	HCl–O–He	Pt–Ir–Se/Al_2O_3	U.S. 4,491,635	[345]
	Cl, O_2	Pt/zeolite	U.S. 4,855,269	[346]
	HCl, Cl, H_2O, O	Pt/zeolite	U.S. 4,925,819	[347]
	HCl, O	Ir, Pt–Ir/Al_2O_3	U.S. 4,444,896	[348]
	Cl, halogen	Ir, Pt–Ir/Al_2O_3	U.S. 4,444,895	[349]
	HCl	Ir, Pt–Ir/Al_2O_3	U.S. 4,517,076	[350]
Oxygen				
	O_2	Pt, Re/Al_2O_3	U.S. 4,482,637	[351]
Oxygen/N_2				
	O_2, N_2	Cu/Cr, Mn, Ru, Pd, Zn, Si, Mg, Ca, Sr, Ba	U.S. 4,855,267	[352]
Other				
	NO, NO + halogen	Pt, Pd/zeolite	Eu 0,306,170	[353]
	Halogen	Ru, Os, Rh, Pd/Al_2O_3	U.S. 4,891,346	[354]
	Halide	Ir, Pt–Ir/Al_2O_3	U.S. 4,447,551	[355]
	Halide, halogen/H_2O	Ir, Pt–Ir/Al_2O_3	U.S. 4,472,514	[356]
	Halogen	Ir, Pt–Ir/Al_2O_3	U.S. 4,473,656	[357]
	NO, NO + halogen, Cl	Group VIII metals/Al_2O_3, SiO_2, zeolites	U.S. 4,952,543	[358]
	H_2-halides, O_2	Ir, Pt–Ir/Al_2O_3	U.S. 4,444,897	[359]
	Halogen, H_2O	Ir, Pt–Ir/Al_2O_3	U.S. 4,472,515	[360]

Redispersion of alumina-supported platinum and iridium crystallites is also possible in a chlorine-free oxygen atmosphere, if chlorine is present on the catalyst. The extent of redispersion depends on the properties of the Pt/Al_2O_3catalyst and temperature; for example, the data in Figure 47 [102] for two different catalysts [catalyst 1 is a commercial Pt/Al_2O_3 (Engelhard); catalyst 2 is Pt/Al_2O_3 (Kaiser KA-201) impregnated with chloroplatinic acid] show that the maximum increases in dispersion occur at about 550 °C. The data also show that redispersion does not occur in a hydrogen environment. The question whether redispersion of platinum occurs only in oxygen without chlorine present on the catalyst remains controversial.

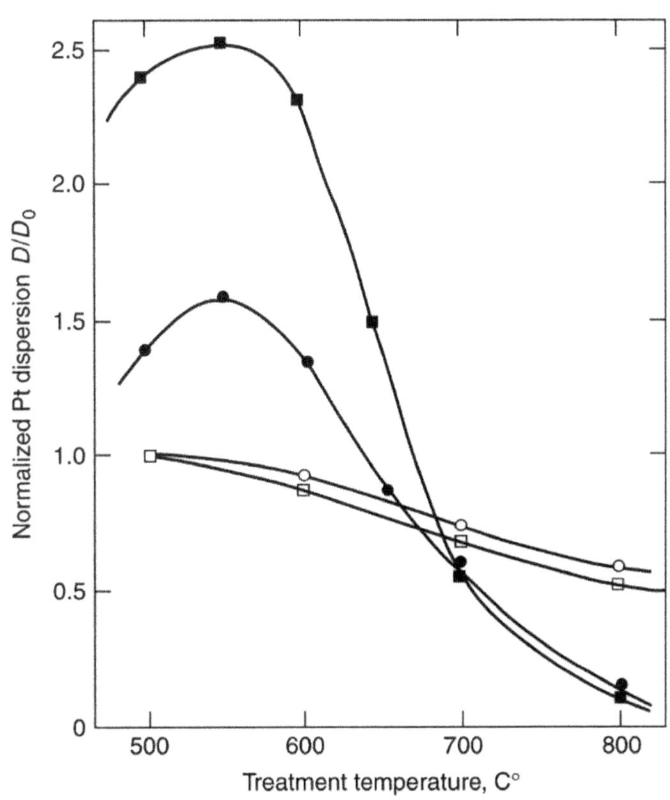

Figure 47: Effects of 1-h treatments in O_2 (closed symbols) and H_2 (open symbols) on the dispersion of Pt/Al_2O_3 catalysts: ○,● Pt/Al_2O_3 (Engelhard), □,■ Pt/KA-201 alumina (Kaiser). Reproduced from [102]. Copyright 1982, Brill Nijhoff Publishers.

Two models, "the thermodynamic redispersion model" and "the crystallite splitting model," have been advanced to explain the redispersion in oxygen [101,102,361]. The "thermodynamic" redispersion model hypothesizes the formation of metal oxide molecules that detach from the crystallite, migrate to active sites on the support, and form surface complexes with the support. Upon subsequent reduction, the metal oxide complexes form monodisperse metal clusters. In the "crystallite splitting" model, exposure of a platinum crystallite to oxygen at 500 °C leads to formation of a platinum oxide scale on the outer surface of the crystallite, which stresses and ultimately leads to splitting of the particle [361]. Dadyburjor hypothesizes that the crystallite splitting model is most applicable to the behavior of large crystallites and to all particles at relatively small regeneration times, while the thermodynamic migration model is useful for small particles and most particles after longer regeneration times.

Case Study: Cobalt based Fischer-Tropsch (FT) Catalyst Regeneration

Fischer-Tropsch (FT) synthesis is a catalytic process used to produce long chain hydrocarbons from synthesis gas consisting of carbon monoxide and hydrogen. Cobalt catalysts were initially developed by Franz Fischer and Hans Tropsch in the 1920s and similar cobalt-based catalysts are still in use today [8]. Although more expensive than iron based catalysts that are also used for FTS, supported cobalt FT catalysts are more active and selective for the desired liquid and wax products.

A recent review by the Davis group at the Center for Applied Energy Research at the University of Kentucky with Bukur at the University of Texas A&M in Qatar [362] focused on the results of studies using synchrotron radiation to characterize Co FT catalysts. The review includes a detailed consideration and analysis of the mechanisms and processes of sintering, oxidation, aluminate formation, and coking and carbide formation and under what operating conditions each is important. They summarize their and others' previous findings that oxidation primarily occurs on small (<2 nm) cobalt crystallites and at high partial pressures of water [362,363,364,365,366]. Further, they highlight the potentially complicated transformations between CoO and aluminates [362,364,367]. These complications highlight a complex mechanism that may be related to chemical-assisted sintering of Co FTS catalysts through a combination of the effect of CoO reduction during the initial activation of the catalysts and water exposure during operation. First, CoO, present either due to incomplete reduction of the catalysts [368] or oxidation of the small (<2 nm) crystallites as suggested by Davis' group [369,370] can apparently increase the sintering rate due to mobility that allows them to aggregate into larger CoO clusters that are subsequently reduced to metallic Co, as inferred from evidence presented in a number of studies [79,362,368,369,370,371]. Primarily, X-ray absorption near edge (XANES) analysis shows simultaneous increasing extent of reduction and increasing Co-Co coordination, due both to removal of oxygen and increases in particle size. Second, water is believed to cause chemical-assisted sintering [80,367,372,373,374], especially at high partial pressures that occur at CO conversions above about 65% [223], although the exact mechanisms are debated. Minor surface oxidation [373,374] and surface wetting [375] have been proposed, although Saib *et al.* have shown that cobalt oxidation is not an important deactivation route [79] in catalysts with Co particles >~8 nm, which are typical in commercial FTS catalysts.

A number of articles by researchers at Sasol, Eindhoven University of Technology, and the University of South Africa detailed the causes of deactivation and demonstrated the regenerability of alumina-supported cobalt FT catalysts [79,368,371,376,377,378,379,380,381,382]. Through a combination of studies on single crystal [377] and actual catalysts from pilot plants operated under industrial FT conditions [368,371], they concluded that contrary to prior hypotheses, neither formation of cobalt aluminates nor oxidation of the cobalt were significant deactivation mechanisms. In fact, extent of Co oxidation actually decreased with time on stream [371]. However, Co sintering and carbon deposition were identified as the primary means of deactivation. In unpublished presentations by these authors, the relative contributions of carbon deposition and sintering to the deactivation were reported as roughly equal. More interestingly, both of these deactivation mechanisms could be largely reversed through high pressure oxidation treatment [376,378], which removes both inactive carbon and redisperses the cobalt. Through high resolution transmission electron micrographs (HRTEM), the mechanism of redispersion of the cobalt was identified as the Kirkendall effect, which results in the formation of spherical shells of cobalt oxide that during subsequent reduction disperse into smaller crystallites of cobalt (see Figure 48). Bezemer *et al.* have previously shown that unpromoted Co FT catalysts require Co crystallites of at least 6 nm in diameter to achieve maximum turnover frequency, but this is the optimum size because larger crystallites display the same surface activity as the 6 nm particles [383]. The oxidative regeneration and reduction process described by Hauman *et al.* [376] and Weststrate *et al.* [377] recovers ~95% of the fresh catalyst activity by removing the carbon deposits and returning the sintered cobalt particles to near the optimum 6 nm size. While the rate per mass of catalyst is nearly constant following regeneration, some smaller particles are produced on model catalysts because the rate on a turnover frequency basis decreases by roughly 1/3 compared to the fresh catalysts [376].

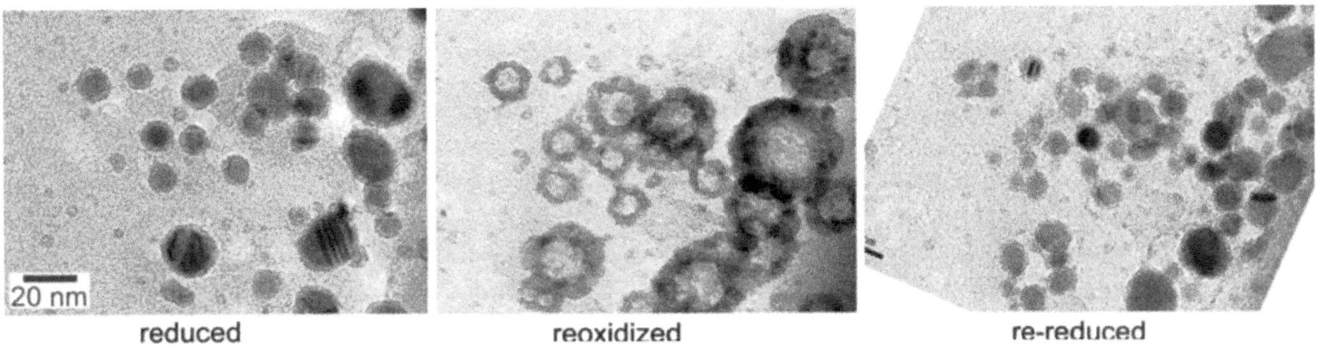

Figure 48: Bright field TEM images showing redispersion of cobalt particles supported on a flat model silica by oxidative treatment. The center image shows hollow spheres created by the Kirkendall effect, which form dispersed smaller particles upon re-reduction in the right hand image. Reproduced from [378]. Copyright 2011, Springer.

These results are significant because they show the power of careful evaluation of the root causes of deactivation in an important catalytic system and then show how proper choice of regeneration conditions can extend the life of the catalysts by redispersion of the active metal. However, promoters may not be redispersed as completely as the cobalt during repeated regeneration. Although traditional promoters, like Pt and Ru, appear to remain with the Co and maintain their effect, some promoters like Au tend to segregate and lose their promotion effect, as indicated by TPR peaks shifting to higher temperatures [384].

SUMMARY

This article focuses on the causes, mechanisms, prevention, modeling, and treatment (experimental and theoretical) of deactivation. Several general, fundamental principles are evident:

(1) The causes of deactivation are basically of three kinds: chemical, mechanical, and thermal. The five intrinsic mechanisms of catalyst decay, (a) poisoning, (b) fouling, (c) thermal degradation, (d) chemical degradation, and (e) mechanical failure, vary in their reversibility and rates of occurrence. Poisoning and thermal degradation are generally slow, irreversible processes, while fouling with coke and carbon is generally rapid and reversible by regeneration with O_2 or H_2.

(2) Catalyst deactivation is more easily prevented than cured. Poisoning by impurities can be prevented through careful purification of reactants or mitigated to some extent by adding traps or "getters" as components of the catalyst. Carbon deposition and coking can be prevented by minimizing the formation of carbon or coke precursors through gasification, careful design of catalysts and process conditions, and by controlling reaction rate regimes, e.g., mass transfer regimes, to minimize effects of carbon and coke formation on activity. Sintering is best avoided by minimizing and controlling the temperature of reaction, although recent developments have focused on encapsulating metal crystallites to eliminate mobility, while still allowing access for reactants and products.

(3) Catalyst regeneration is feasible in some circumstances, especially to recover activity loss due to rapid coking or longer term deactivation associated with loss of active metal dispersion. Typically, regeneration or rejuvenation strategies are dictated by process or economic necessity to obtain desired process run lengths. Life cycle operating strategies are important considerations when evaluating catalyst regeneration/rejuvenation *versus* replacement decisions. Rejuvenation treatments can extend the useful life of catalysts. Selective catalytic reduction catalysts provide an example of rejuvenation practiced in a commercial process.

ACKNOWLEDGMENTS

The authors wish to acknowledge Brigham Young University for its support and the employees of MDPI, for their untiring assistance in publishing this review.

CONFLICTS OF INTEREST

The authors declare no conflict of interest.

REFERENCES

1. Figueiredo, J.L. Carbon formation and gasification on nickel. In *Progress in Catalyst Deactivation; (NATO Advanced Study Institute Series E, No. 54)*; Martinus Nijhoff Publishers: Boston, MA, USA, 1982; pp. 45–63.
2. Hughes, R. *Deactivation of Catalysts*; Academic Press: London, UK, 1984.
3. Oudar, J.; Wise, H. *Deactivation and Poisoning of Catalysts*; Marcel Dekker: New York, NY, USA, 1985.
4. Butt, J.B.; Petersen, E.E. *Activation, Deactivation, and Poisoning of Catalysts*; Academic Press: San Diego, CA, USA, 1988.
5. Denny, P.J.; Twigg, M.V. Factors determining the life of industrial heterogeneous catalysts. In *Catalyst Deactivation 1980 (Studies in Surface Science and Catalysis)*; Delmon, B., Froment, G.F., Eds.; Elsevier: Amsterdam, The Nerthelands, 1980; Volume 6, pp. 577–599.
6. Bartholomew, C.H. Catalyst Deactivation. *Chem. Eng.* 1984, *91*, 96–112.
7. Butt, J.B. Catalyst deactivation and regeneration. In *Catalysis—Science and Technology*; Anderson, J.R., Boudart, M., Eds.; Springer-Verlag: New York, NY, USA, 1984; Volume 6, pp. 1–63.

8. Bartholomew, C.H.; Farrauto, R.J. *Fundamentals of Industrial Catalytic Processes*, 2nd ed.; Wiley-Interscience: Hoboken, NJ, USA, 2006.
9. Delmon, B.; Froment, G.F. *Catalyst Deactivation 1980 (Studies in Surface Science and Catalysis)*; Elsevier: Amsterdam, The Nerthelands, 1980; Volume 6.
10. Delmon, B.; Froment, G.F. *Catalyst Deactivation 1987 (Studies in Surface Science and Catalysis)*; Elsevier: Amsterdam, The Nerthelands, 1987; Volume 34.
11. Bartholomew, C.H.; Butt, J.B. *Catalyst Deactivation 1991 (Studies in Surface Science and Catalysis)*; Elsevier: Amsterdam, The Nerthelands, 1991; Volume 68.
12. Delmon, B.; Froment, G.F. *Catalyst Deactivation 1994 (Studies in Surface Science and Catalysis)*; Elsevier: Amsterdam, The Nerthelands, 1994; Volume 88.
13. Bartholomew, C.H.; Fuentes, G.A. *Catalyst Deactivation 1997 (Studies in Surface Science and Catalysis)*; Elsevier: Amsterdam, The Nethelands, 1997; Volume 111.
14. Delmon, B.; Froment, G.F. *Catalyst Deactivation 1999 (Studies in Surface Science and Catalysis)*; Elsevier: Amsterdam, The Nethelands, 1999; Volume 126.
15. Moulijn, J.A. Catalyst Deactivation. *Appl. Catal. A* 2001, *212*, 1–255.
16. Maxted, E.B. The poisoning of metallic catalysts. *Adv. Catal.* 1951, *3*, 129–177.
17. Hegedus, L.L.; McCabe, R.W. Catalyst poisoning. In *Catalyst Deactivation 1980 (Studies in Surface Science and Catalysis)*; Delmon, B., Froment, G.F., Eds.; Elsevier: Amsterdam, The Nethelands, 1980; Volume 6, pp. 471–505.
18. Hegedus, L.L.; McCabe, R.W. *Catalyst Poisoning*; Marcel Dekker: New York, NY, USA, 1984.
19. Butt, J.B. Catalyst poisoning and chemical process dynamics. In *Progress in Catalyst Deactivation (NATO Advanced Study Institute Series E, No. 54)*; Figueiredo, J.L., Ed.; Martinus Nijhoff Publishers: Boston, MA, USA, 1982; pp. 153–208.
20. Barbier, J. Effect of poisons on the activity and selectivity of metallic catalysts. In *Deactivation and Poisoning of Catalysts*; Oudar, J., Wise, H., Eds.; Marcel Dekker: New York, NY, USA, 1985; pp. 109–150.
21. Bartholomew, C.H. Mechanisms of nickel catalyst poisoning. In *Catalyst Deactivation 1987 (Studies in Surface Science and Catalysis)*; Delmon, B., Froment, G.F., Eds.; Elsevier: Amsterdam, The Nethelands, 1987; Volume 34, pp. 81–104.
22. Rostrup-Nielsen, J.R. Promotion by poisoning. In *Catalyst Deactivation 1991 (Studies in Surface Science and Catalysis)*; Bartholomew, C.H., Butt, J.B., Eds.; Elsevier: Amsterdam, The Nethelands, 1991; Volume 68, pp. 85–101.
23. Inga, J.; Kennedy, P.; Leviness, S. A1 Fischer-tropsch process in the presence of nitrogen contaminants. WIPO Patent WO 2005/071044, 4 August 2005.
24. Völiter, V.J.; Hermann, M. Katalytische Wirksamkeit van reinem und von CO-vergiftetem Platin bei der p-H_2-Umwandlung. *Z. Anorg. Allg. Chem.* 1974, *405*, 315.
25. Baron, K. Carbon monoxide oxidation on platinum-lead films. *Thin Solid Films* 1978, *55*, 449–462.
26. Clay, R.D.; Petersen, E.E. Catalytic activity of an evaporated platinum film progressively poisoned with arsine. *J. Catal.*1970, *16*, 32–43.
27. Madon, R.J.; Seaw, H. Effect of Sulfur on the Fischer-Tropsch Synthesis. *Catal. Rev.—Sci. Eng.* 1977, *15*, 69–106.
28. Bartholomew, C.H.; Agrawal, P.K.; Katzer, J.R. Sulfur poisoning of metals. *Adv. Catal.* 1982, *31*, 135–242.
29. Rostrup-Nielsen, J.R. Sulfur Poisoning. In *Progress in Catalyst Deactivation (NATO Advanced Study Institute Series E, No. 54)*; Figueiredo, J.L., Ed.; Martinus Nijhoff Publishers: Boston, MA, USA, 1982; pp. 209–227.
30. Wise, H.; McCarty, J.; Oudar, J. Sulfur and carbon interactions with metal surfaces. In *Deactivation and Poisoning of Catalysts*; Oudar, J., Wise, H., Eds.; Marcel Dekker: New York, NY, USA, 1985; pp. 1–50.
31. Rostrup-Nielsen, J.R.; Nielsen, P.E.H. Catalyst deactivation in synthesis gas production, and important syntheses. In*Deactivation and Poisoning of Catalysts*; Oudar, J., Wise, H., Eds.; Marcel Dekker: New York, NY, USA, 1985; pp. 259–323.
32. Grossmann, A.; Erley, W.; Ibach, H. Adsorbate-induced surface stress and surface reconstruction: oxygen, sulfur and carbon on Ni(111). *Surf. Sci.* 1995, *337*, 183–189.
33. Ruan, L.; Stensgaard, I.; Besenbacher, F.; Lægsgaard, E. Observation of a missing-row structure on an fcc (111) surface: The (5 √3 ×2)S phase on Ni(111) studied by scanning tunneling microscopy. *Phys. Rev. Lett.* 1993, *71*, 2963–2966.
34. Kitajima, Y.; Yokoyama, T.; Ohta, T.; Funabashi, M.; Kosugi, N.; Kuroda, H. Surface EXAFS and XANES studies of (5v3× 2)S/Ni(111). *Surf. Sci.* 1989, *214*, L261–L269.
35. Perdereau, M.; Oudar, J. Structure, mécanisme de formation et stabilité de la couche d'adsorption du soufre sur le nickel. *Surf. Sci.* 1970, *20*, 80–98.
36. Oudar, J. Sulfur adsorption and poisoning of metallic catalysts. *Catal. Rev.—Sci. Eng.* 1980, *22*, 171–195.
37. McCarroll, J.J.; Edmonds, T.; Pitkethly, R.C. Interpretation of a complex low energy electron diffraction pattern: Carbonaceous and sulphur-containing structures on Ni(111). *Nature* 1969, *223*, 1260–1262.
38. Edmonds, T.; McCarroll, J.J.; Pitkethly, R.C. Surface structures formed during the interaction of sulphur compounds with the (111) face of nickel. *J. Vac. Sci. Technol.* 1971, *8*, 68–74.
39. Bartholomew, C.H. Mechanisms of catalyst deactivation. *Appl. Catal. A* 2001, *212*, 17–60.
40. Ruan, L.; Besenbacher, F.; Stensgaard, I.; Lægsgaard, E. Atom-resolved studies of the reaction between H_2S and O on Ni(110). *Phys. Rev. Lett.* 1992, *69*, 3523–3526.
41. Hepola, J.; McCarty, J.; Krishnan, G.; Wong, V. Elucidation of behavior of sulfur on nickel-based hot gas cleaning catalysts. *Appl. Catal. B* 1999, *20*, 191–203.

42. Erley, W.; Wagner, H. Sulfur poisoning of carbon monoxide adsorption on Ni(111). *J. Catal.* 1978, *53*, 287–294.
43. Rendulic, K.D.; Winkler, A. The initial sticking coefficient of hydrogen on sulfur- and oxygen-covered polycrystalline nickel surfaces. *Surf. Sci.* 1978, *74*, 318–320.
44. Goodman, D.W.; Kiskinova, M. Chemisorption and reactivity studies of H_2 and CO on sulfided Ni(100). *Surf. Sci.* 1981, *105*, L265–L270.
45. Kiskinova, M.; Goodman, D.W. Modification of chemisorption properties by electronegative adatoms: H_2 and CO on chlorided, sulfided, and phosphided Ni(100). *Surf. Sci.* 1981, *108*, 64–76.
46. Johnson, S.; Madix, R.S. Desorption of hydrogen and carbon monoxide from Ni(100), Ni(100)p(2 × 2)S, and Ni(100)c(2 × 2)S surfaces. *Surf. Sci.* 1981, *108*, 77–98.
47. Madix, R.J.; Thornberg, M.; Lee, S.B. CO-sulfur interaction on Ni(110); evidence for local interactions, not long range electronic effects. *Surf. Sci.* 1983, *133*, L447–L451.
48. Hardegree, E.L.; Ho, P.; White, J.M. Sulfur adsorption on Ni(100) and its effect on CO chemisorption: I. TDS, AES and work function results. *Surf. Sci.* 1986, *165*, 488–506.
49. Erekson, E.J.; Bartholomew, C.H. Sulfur poisoning of nickel methanation catalysts: II. Effects of H_2S concentration, CO and H_2O partial pressures and temperature on reactivation rates. *Appl. Catal.* 1983, *5*, 323–336.
50. Jacobs, G.; Ghadiali, F.; Pisanu, A.; Padro, C.L.; Borgna, A.; Alvarez, W.E.; Resasco, D.E. Increased Sulfur Tolerance of Pt/KL Catalysts Prepared by Vapor-Phase Impregnation and Containing a Tm Promoter. *J. Catal.* 2000, *191*, 116–127.
51. Jongpatiwut, S.; Sackamduang, P.; Rirksomboon, T.; Osuwan, S.; Alvarez, W.E.; Resasco, D.E. Sulfur- and water-tolerance of Pt/KL aromatization catalysts promoted with Ce and Yb. *Appl. Catal. A* 2002, *230*, 177–193.
52. Jacobs, G.; Ghadiali, F.; Pisanu, A.; Borgna, A.; Alvarez, W.E.; Resasco, D.E. Characterization of the morphology of Pt clusters incorporated in a KL zeolite by vapor phase and incipient wetness impregnation. Influence of Pt particle morphology on aromatization activity and deactivation. *Appl. Catal. A* 1999, *188*, 79–98.
53. McVicker, G.B.; Kao, J.L.; Ziemiak, T.J.J.; Gates, W.E.; Robbins, J.L.; Treacy, M.M.J.; Rice, S.B.; Vanderspurt, T.H.; Cross, V.R.; Ghosh, A.K. Effect of Sulfur on the Performance and on the Particle Size and Location of Platinum in Pt/KL Hexane Aromatization Catalysts. *J. Catal.* 1993, *139*, 48–61.
54. Farbenindustrie, I.G. Improvements in the Manufacture and Production of Unsaturated Hydrocarbons of Low Boiling Point. British Patent 322,284, 5 December 1929.
55. Kritzinger, J.A. The role of sulfur in commercial iron-based Fischer–Tropsch catalysis with focus on C_2-product selectivity and yield. *Catal. Today* 2002, *71*, 307–318.
56. Sparks, D.E.; Jacobs, G.; Gnanamani, M.K.; Pendyala, V.R.R.; Ma, W.; Kang, J.; Shafer, W.D.; Keogh, R.A.; Graham, U.M.; Gao, P.; *et al.* Poisoning of cobalt catalyst used for Fischer–Tropsch synthesis. *Catal. Today* 2013, *215*, 67–72.
57. Rostrup-Nielsen, J.R.; Trimm, D.L. Mechanisms of carbon formation on nickel-containing catalysts. *J. Catal.* 1977, *48*, 155–165.
58. Trimm, D.L. The Formation and Removal of Coke from Nickel Catalyst. *Catal. Rev.—Sci. Eng.* 1977, *16*, 155–189.
59. Trimm, D.L. Catalyst design for reduced coking (review). *Appl. Catal.* 1983, *5*, 263–290.
60. Bartholomew, C.H. Carbon deposition in steam reforming and methanation. *Catal. Rev.—Sci. Eng.* 1982, *24*, 67–112.
61. Albright, L.F.; Baker, R.T.K. *Coke Formation on Metal Surfaces (ACS Symposium Series 202)*; American Chemical Society: Washington, DC, USA, 1982.
62. Menon, P.G. Coke on catalysts—harmful, harmless, invisible and beneficial types. *J. Mol. Catal.* 1990, *59*, 207–220.
63. Rostrup-Nielsen, J.R. Conversion of hydrocarbons and alcohols for fuel cells. *Phys. Chem. Chem. Phys.* 2001, *3*, 283–288.
64. Trane-Restrup, R.; Resasco, D.E.; Jensen, A.D. Steam reforming of light oxygenates. *Catal. Sci. Technol.* 2013, *3*, 3292–3302.
65. De Lima, S.M.; da Silva, A.M.; da Costa, L.O.O.; Assaf, J.M.; Jacobs, G.; Davis, B.H.; Mattos, L.V.; Noronha, F.B. Evaluation of the performance of Ni/La_2O_3 catalyst prepared from $LaNiO_3$ perovskite-type oxides for the production of hydrogen through steam reforming and oxidative steam reforming of ethanol. *Appl. Catal. A* 2010, *377*, 181–190.
66. Deken, J.D.; Menon, P.G.; Froment, G.F.; Haemers, G. On the nature of carbon in Niα-Al_2O_3 catalyst deactivated by the methane-steam reforming reaction. *J. Catal.* 1981, *70*, 225–229.
67. Durer, W.G.; Craig, J.H., Jr.; Lozano, J. Surface carbon and its effects on hydrogen adsorption on Rh(100). *Appl. Surf. Sci.* 1990, *45*, 275–277.
68. Moeller, A.D.; Bartholomew, C.H. Deactivation by carbon of nickel and nickel-molybdenum methanation catalysts.*Prepr.—Am. Chem. Soc., Div. Fuel Chem.* 1980, *25*, 54–70.
69. Marschall, K.-J.; Mleczko, L. Short-contact-time reactor for catalytic partial oxidation of methane. *Ind. Eng. Chem. Res.*1999, *38*, 1813–1821.
70. Rostrup-Nielsen, J.R. Catalytic steam reforming. In *Catalysis—Science and Technology*; Anderson, J.R., Boudart, M., Eds.; Springer-Verlag: New York, NY, USA, 1984; Volume 5, pp. 1–117.
71. Besenbacher, F.; Chorkendorff, I.; Clausen, B.S.; Hammer, B.; Molenbroek, A.M.; Nørskov, J.K.;

Stensgaard, I. Design of a surface alloy catalyst for steam reforming. *Science* 1998, *279*, 1913–1915.

72. Nemes, T.; Chambers, A.; Baker, R.T.K. Characteristics of carbon filament formation from the interaction of Cobalt–Tin particles with ethylene. *J. Phys. Chem. B* 1998, *102*, 6323–6330.

73. Bartholomew, C.H.; Strasburg, M.V.; Hsieh, H. Effects of support on carbon formation and gasification on nickel during carbon monoxide hydrogenation. *Appl. Catal.* 1988, *36*, 147–162.

74. Vance, C.K.; Bartholomew, C.H. Hydrogenation of carbon dioxide on group viii metals: III, Effects of support on activity/selectivity and adsorption properties of nickel. *Appl. Catal.* 1983, *7*, 169–177.

75. Baker, R.T.K.; Chludzinski, J.J., Jr. Filamentous carbon growth on nickel-iron surfaces: The effect of various oxide additives. *J. Catal.* 1980, *64*, 464–468.

76. Brown, D.E.; Clark, J.T.K.; Foster, A.I.; McCarroll, J.J.; Sims, M.L. Inhibition of coke formation in ethylene steam cracking. In *Coke Formation on Metal Surfaces (ACS Symposium Series 202)*; Albright, L.F., Baker, R.T.K., Eds.; American Chemical Society: Washington, DC, USA, 1982; pp. 23–43.

77. Bitter, J.H.; Seshan, K.; Lercher, J.A. Deactivation and coke accumulation during CO_2/CH_4 reforming over Pt catalysts.*J. Catal.* 1999, *183*, 336–343.

78. Rostrup-Nielsen, J.R. Coking on nickel catalysts for steam reforming of hydrocarbons. *J. Catal.* 1974, *33*, 184–201.

79. Saib, A.M.; Moodley, D.J.; Ciobica, I.M.; Hauman, M.M.; Sigwebela, B.H.; Weststrate, C.J.; Niemanstsverdriet, J.W.; van de Loosdrecht, J. Fundamental understanding of deactivation and regeneration of cobalt Fischer–Tropsch synthesis catalysts. *Catal. Today* 2010, *154*, 271–282.

80. Tsakoumis, N.E.; Rønning, M.; Borg, O.; Rytter, E.; Holmen, A. Deactivation of cobalt based Fischer–Tropsch catalysts: A review. *Catal. Today* 2010, *154*, 162–182.

81. Gates, B.C.; Katzer, J.R.; Schuit, G.C.A. *Chemistry of Catalytic Processes*; McGraw-Hill: New York, NY, USA, 1979.

82. Naccache, C. Deactivation of acid catalysts. In *Deactivation and Poisoning of Catalysts*; Oudar, J., Wise, H., Eds.; Marcel Dekker: New York, NY, USA, 1985; pp. 185–203.

83. Appleby, W.G.; Gibson, J.W.; Good, G.M. Coke Formation in Catalytic Cracking. *Ind. Eng. Chem. Process Des. Dev.* 1962,*1*, 102–110.

84. Beuther, H.; Larson, O.H.; Perrotta, A.J. The mechanism of coke formation on catalysts. In *Catalyst Deactivation 1980 (Studies in Surface Science and Catalysis)*; Delmon, B., Froment, G.F., Eds.; Elsevier: Amsterdam, The Nerthelands, 1980; Volume 6, pp. 271–282.

85. Gayubo, A.G.; Arandes, J.M.; Aguayo, A.T.; Olazar, M.; Bilbao, J. Deactivation and acidity deterioration of a silica/alumina catalyst in the isomerization of cis-butene. *Ind. Eng. Chem. Res.* 1993, *32*, 588–593.

86. Augustine, S.M.; Alameddin, G.N.; Sachtler, W.M.H. The effect of Re, S, and Cl on the deactivation of Pt γ-Al_2O_3reforming catalysts. *J. Catal.* 1989, *115*, 217–232.

87. Guisnet, M.; Magnoux, P. Coking and deactivation of zeolites: Influence of the Pore Structure. *Appl. Catal.* 1989, *54*, 1–27.

88. Bauer, F.; Kanazirev, V.; Vlaev, C.; Hanisch, R.; Weiss, W. Koksbildung in ZSM-5-Katalysatoren. *Chem. Tech.* 1989, *41*, 297–301.

89. Grotten, W.A.; Wojciechowski, B.W.; Hunter, B.K. On the relationship between coke formation chemistry and catalyst deactivation. *J. Catal.* 1992, *138*, 343–350.

90. Bellare, A.; Dadyburjor, D.B. Evaluation of modes of catalyst deactivation by coking for cumene cracking over zeolites. *J. Catal.* 1993, *140*, 510–525.

91. Uguina, M.A.; Serrano, D.P.; Grieken, R.V.; Vènes, S. Adsorption, acid and catalytic changes induced in ZSM-5 by coking with different hydrocarbons. *Appl. Catal. A* 1993, *99*, 97–113.

92. Li, C.; Chen, Y.-W.; Yang, S.-J.; Yen, R.-B. *In-situ* FTIR investigation of coke formation on USY zeolite. *Appl. Surf. Sci.*1994, *81*, 465–468.

93. Buglass, J.G.; de Jong, K.P.; Mooiweer, H.H. Analytical studies of the coking of the zeolite ferrierite. *J. Chem. Soc. Abstr.*1995, *210*. 105-PETR.

94. Chen, D.; Rebo, H.P.; Moljord, K.; Holmen, A. Effect of coke deposition on transport and adsorption in zeolites studied by a new microbalance reactor. *Chem. Eng. Sci.* 1996, *51*, 2687–2692.

95. Guisnet, M.; Magnoux, P.; Martin, D. Roles of acidity and pore structure in the deactivation of zeolites by carbonaceous deposits. In *Catalyst Deactivation 1997 (Studies in Surface Science and Catalysis)*; Bartholomew, C.H., Fuentes, G.A., Eds.; Elsevier: Amsterdam, The Nethelands, 1997; Volume 111, pp. 1–19.

96. Masuda, T.; Tomita, P.; Fujikata, Y.; Hashimoto, K. Deactivation of HY-type zeolite catalyst due to coke deposition during gas-oil cracking. In *Catalyst Deactivation 1999 (Studies in Surface Science and Catalysis)*; Delmon, B., Froment, G.F., Eds.; Elsevier: Amsterdam, The Nethelands, 1999; Volume 126, pp. 89–96.

97. Cerqueira, H.S.; Magnoux, P.; Martin, D.; Guisnet, M. Effect of contact time on the nature and location of coke during methylcyclohexane transformation over a USHY zeolite. In *Catalyst Deactivation 1999 (Studies in Surface Science and Catalysis)*; Delmon, B., Froment, G.F., Eds.; Elsevier: Amsterdam, The Nethelands, 1999; Volume 126, pp. 105–112.

98. Wanke, S.E.; Flynn, P.C. The sintering of supported metal catalysts. *Catal. Rev.—Sci. Eng.* 1975, *12*, 93–135.

99. Wynblatt, P.; Gjostein, N.A. Supported metal crystallites. *Prog. Solid State Chem.* 1975, *9*, 21–58.

100. Ruckenstein, E.; Pulvermacher, B. Kinetics of crystallite sintering during heat treatment of supported metal catalysts.*AIChE J.* 1973, *19*, 356–364.

101. Ruckenstein, E.; Dadyburjor, D.B. Sintering and redispersion in supported metal catalysts. *Rev. Chem. Eng.* 1983, *1*, 251–356.

102. Wanke, S.E. Sintering of commercial supported platinum group metal catalysts. In *Progress in Catalyst Deactivation (NATO Advanced Study Institute Series E, No. 54)*; Figueiredo, J.L., Ed.; Martin Nijhoff Publishers: Boston, MA, USA, 1982; pp. 315–328.

103. Baker, R.T.; Bartholomew, C.H.; Dadyburjor, D.B. Sintering and Redispersion: Mechanisms and Kinetics. In *Stability of Supported Catalysts: Sintering and Redispersion*; Horsley, J.A., Ed.; Catalytica: Mountain View, CA, USA, 1991; pp. 169–225.

104. Bartholomew, C.H. Model catalyst studies of supported metal sintering and redispersion kinetics. In *Catalysis (Specialist Periodical Report)*; Spivey, J.J., Agarwal, S.K., Eds.; Royal Society of Chemistry: Cambridge, UK, 1993; Volume 10, pp. 41–82.

105. Bartholomew, C.H. Sintering kinetics of supported metals: new perspectives from a unifying GPLE treatment. *Appl. Catal. A* 1993, *107*, 1–57.

106. Bartholomew, C.H. Sintering kinetics of supported metals: perspectives from a generalized power law approach. In*Catalyst Deactivation 1994 (Studies in Surface Science and Catalysis)*; Delmon, B., Froment, G.F., Eds.; Elsevier: Amsterdam, The Nethelands, 1994; Volume 88, pp. 1–18.

107. Bartholomew, C.H. Sintering and redispersion of supported metals: Perspectives from the literature of the past decade. In *Catalyst Deactivation 1997 (Studies in Surface Science and Catalysis)*; Bartholomew, C.H., Fuentes, G.A., Eds.; Elsevier: Amsterdam, The Nethelands, 1997; Volume 111, pp. 585–592.

108. Bartholomew, C.H.; Sorenson, W. Sintering kinetics of silica- and alumina-supported nickel in hydrogen atmosphere.*J. Catal.* 1983, *81*, 131–141.

109. Moulijn, J.A.; van Diepen, A.E.; Kapteijn, F. Catalyst deactivation: is it predictable?: What to do? *Appl. Catal. A* 2001,*212*, 3–16.

110. Bridger, G.W.; Spencer, M.S. Methanol synthesis. In *Catalyst Handbook*, 2nd ed.; Twigg, M.V., Ed.; Manson Publishing: London, UK, 1996; pp. 441–468.

111. Fuentes, G.A. Catalyst deactivation and steady-state activity: A generalized power-law equation model. *Appl. Catal.*1985, *15*, 33–40.

112. Fuentes, G.A.; Ruiz-Trevino, F.A. Towards a better understanding of sintering phenomena in catalysis. In *Catalyst Deactivation 1991 (Studies in Surface Science and Catalysis)*; Bartholomew, C.H., Butt, J.B., Eds.; Elsevier: Amsterdam, The Nethelands, 1991; Volume 68, pp. 637–644.

113. Bournonville, J.P.; Martino, G. Sintering of Alumina Supported Platinum. In *Catalyst Deactivation 1980 (Studies in Surface Science and Catalysis)*; Delmon, B., Froment, G.F., Eds.; Elsevier: Amsterdam, The Nethelands, 1980; Volume 6, pp. 159–166.

114. Somorjai, G.A. Small-Angle X-Ray Scattering and Low Energy Electron Diffraction Studies on Catalyst Surfaces. In *X-ray and Electron Methods of Analysis (Progress in Analytical Chemistry)*; Van Olphen, H., Parrish, W., Eds.; Plenum Press: New York, NY, USA, 1968; Volume 1, pp. 101–126.

115. Seyedmonir, S.R.; Strohmayer, D.E.; Guskey, G.J.; Geoffroy, G.L.; Vannice, M.A. Characterization of supported silver catalysts: III. Effects of support, pretreatment, and gaseous environment on the dispersion of Ag. *J. Catal.* 1985, *93*, 288–302.

116. Trimm, D.L. Thermal stability of catalyst supports. In *Catalyst Deactivation 1991 (Studies in Surface Science and Catalysis)*; Bartholomew, C.H., Butt, J.B., Eds.; Elsevier: Amsterdam, The Nethelands, 1991; Volume 68, pp. 29–51.

117. Shastri, A.G.; Datye, A.K.; Schwank, J. Influence of chlorine on the surface area and morphology of TiO_2. *Appl. Catal.*1985, *14*, 119–131.

118. Oberlander, R.K. Aluminas for catalysts: Their preparations and properties. In *Applied Industrial Catalysis*; Leach, B.E., Ed.; Academic Press: Orlando, FL, USA, 1984; Volume 3, pp. 64–112.

119. Wefers, K.; Misra, C. *Oxides and Hydroxides of Aluminum*; Alcoa Technical Paper No. 19; Alcoa Laboratories: Pittsburg, PA, USA, 1987.

120. Hegedus, L.L.; Baron, K. Phosphorus accumulation in automotive catalysts. *J. Catal.* 1978, *54*, 115–119.

121. Summers, J.; Hegedus, L.L. Modes of catalyst deactivation in stoichiometric automobile exhaust. *Ind. Eng. Chem. Prod. Res. Dev.* 1979, *18*, 318–324.

122. Peter-Hoblyn, J.D.; Valentine, J.M.; Sprague, B.N.; Epperly, W.R. Methods for reducing harmful emissions from a diesel engine. U.S. Patent 6,003,303, 21 December 1999.

123. Manson, I. Self-regenerating diesel exhaust particulate filter and material. U.S. Patent 6,013,599, 11 January 2000.

124. Deeba, M.; Lui, Y.K.; Dettling, J.C. Four-way diesel exhaust catalyst and method of use. U.S. Patent 6,093,378, 25 July 2000.

125. Dry, M.E. The fischer-tropsch synthesis. In *Catalysis—Science and Technology*; Anderson, J., Boudart, M., Eds.; Springer-Verlag: New York, NY, USA, 1981; pp. 159–218.

126. Huber, G.W.; Guymon, C.G.; Stephenson, B.C.; Bartholomew, C.H. Hydrothermal stability of Co/SiO_2 Fischer-Tropsch synthesis catalysts. In *Catalyst Deactivation 2001 (Studies in Surface Science and Catalysis)*; Spivey, J.J., Roberts, G.W., Davis, B.H., Eds.; Elsevier: Amsterdam, The Netheland, 2001; Volume 139, pp. 423–430.

127. Busca, G.; Lietti, L.; Ramis, G.; Berti, F. Chemical and mechanistic aspects of the selective catalytic reduction of NO_x by ammonia over oxide catalysts: A review. *Appl. Catal. B* 1998, *18*, 1–36.

128. Kobylinski, T.P.; Taylor, B.W.; Yong, J.E. Stabilized ruthenium catalysts For NO_x reduction. Proc. SAE, 1974. Paper 740250.

129. Shelef, M.; Gandhi, H.S. The reduction of nitric oxide in automobile emissions: Stabilisation of catalysts containing ruthenium. *Platinum Met. Rev.* 1974, *18*, 2–14.

130. Gandhi, H.S.; Stepien, H.K.; Shelef, M. Optimization of ruthenium-containing, stabilized, nitric oxide reduction catalysts. *Mat. Res. Bull.* 1975, *10*, 837–845.

131. Bartholomew, C.H. Reduction of nitric oxide by monolithic-supported palladium-nickel and palladium-ruthenium alloys. *Ind. Eng. Chem. Prod. Res. Dev.* 1975, *14*, 29–33.

132. Clark, R.W.; Tien, J.K.; Wynblatt, P. Loss of palladium from model platinum-palladium supported catalysts during annealing. *J. Catal.* 1980, *61*, 15–18.

133. Shen, W.M.; Dumesic, J.A.; Hill, C.G. Criteria for stable Ni particle size under methanation reaction conditions: Nickel transport and particle size growth via nickel carbonyl. *J. Catal.* 1981, *68*, 152–165.

134. Pannell, R.B.; Chung, K.S.; Bartholomew, C.H. The stoichiometry and poisoning by sulfur of hydrogen, oxygen and carbon monoxide chemisorption on unsupported nickel. *J. Catal.* 1977, *46*, 340–347.

135. Lohrengel, G.; Baerns, M. Determination of the metallic surface area of nickel and its dispersion on a silica support by means of a microbalance. *Appl. Catal.* 1981, *1*, 3–7.

136. Qamar, I.; Goodwin, J.G. Fischer-Tropsch Synthesis over Composite Ru Catalysts. In Proceedings of 8th North American Catalysis Society Meeting, Philadelphia, PA, USA; 1983; Paper C-22.

137. Goodwin, J.G.; Goa, D.O.; Erdal, S.; Rogan, F.H. Reactive metal volatilization from Ru/Al_2O_3 as a result of ruthenium carbonyl formation. *Appl. Catal.* 1986, *24*, 199–209.

138. Watzenberger, O.; Haeberle, T.; Lynch, D.T.; Emig, G. Deactivation of Heteropolyacid Catalysts. In *Catalyst Deactivation 1991 (Studies in Surface Science and Catalysis)*; Bartholomew, C.H., Butt, J.B., Eds.; Elsevier: Amsterdam, The Nethelands, 1991; Volume 68, pp. 441–448.

139. Agnelli, M.; Kolb, M.; Mirodatos, C. Co hydrogenation on a nickel catalyst: 1. Kinetics and modeling of a low-temperature sintering process. *J. Catal.* 1994, *148*, 9–21.

140. Lee, H.C.; Farrauto, R.J. Catalyst deactivation due to transient behavior in nitric acid production. *Ind. Eng. Chem. Res.*1989, *28*, 1–5.

141. Farrauto, R.J.; Lee, H.C. Ammonia oxidation catalysts with enhanced activity. *Ind. Eng. Chem. Res.* 1990, *29*, 1125–1129.

142. Sperner, F.; Hohmann, W. Rhodium-platinum gauzes for ammonia oxidation. *Platinum Met. Rev.* 1976, *20*, 12–20.

143. Hess, J.M.; Phillips, J. Catalytic etching of Pt/Rh gauzes. *J. Catal.* 1992, *136*, 149–160.

144. Kuo, C.L.; Hwang, K.C. Does morphology of a metal nanoparticle play a role in ostwald ripening processes? *Chem. Mater.* 2013, *25*, 365–371.

145. Bartholomew, C.H. Hydrogen adsorption on supported cobalt, iron, and nickel. *Catal. Lett.* 1990, *7*, 27–51.

146. Wu, N.L.; Phillips, J. Catalytic etching of platinum during ethylene oxidation. *J. Phys. Chem.* 1985, *89*, 591–600.

147. Wu, N.L.; Phillips, J. Reaction-enhanced sintering of platinum thin films during ethylene oxidation. *J. Appl. Phys.* 1986,*59*, 769–779.

148. Wu, N.L.; Phillips, J. Sintering of silica-supported platinum catalysts during ethylene oxidation. *J. Catal.* 1988, *113*, 129–143.

149. Bielanski, A.; Najbar, M.; Chrzaoszcz, J.; Wal, W. Deactivation of the V_2O_5-MoO_3, catalysts in the selective oxidation of eenzene to maleic anhydride and the changes in its morphology and chemical composition. In *Catalyst Deactivation 1980 (Studies in Surface Science and Catalysis)*; Delmon, B., Froment, G.F., Eds.; Elsevier: Amsterdam, The Nethelands, 1980; Volume 6, pp. 127–140.

150. Burriesci, N.; Garbassi, F.; Petrera, M.; Petrini, G.; Pernicone, N. Solid state reactions in Fe-Mo oxide catalysts for methanol oxidation during aging in industrial plants. In *Catalyst Deactivation 1980 (Studies in Surface Science and Catalysis)*; Delmon, B., Froment, G.F., Eds.; Elsevier: Amsterdam, The Nethelands, 1980; Volume 6, pp. 115–126.

151. Xiong, Y.L.; Castillo, R.; Papadopoulou, C.; Dada, L.; Ladriere, J.; Ruiz, P.; Delmon, B. The protecting role of antimony oxide against deactivation of iron molybdate in oxidation catalysts. In *Catalyst Deactivation 1991 (Studies in Surface Science and Catalysis)*; Bartholomew, C.H., Butt, J.B., Eds.; Elsevier: Amsterdam, The Nethelands, 1991; Volume 68, pp. 425–432.

152. Farrauto, R.J.; Hobson, M.; Kennelly, T.; Waterman, E. Catalytic chemistry of supported palladium for combustion of methane. *Appl. Catal. A* 1992, *81*, 227–237.

153. Gai-Boyes, P.L. Defects in Oxide Catalysts: Fundamental Studies of Catalysis in Action. *Catal. Rev.—Sci. Eng.* 1992, *34*, 1–54.

154. Delmon, B. Solid-state reactions in catalysts during ageing: Beneficial role of spillover. In *Catalyst Deactivation 1994 (Studies in Surface Science and Catalysis, Volume 88)*; Delmon, B., Froment, G.F., Eds.; Elsevier: Amsterdam, The Nethelands, 1994; pp. 113–128.

155. Erickson, K.M.; Karydis, D.A.; Boghosian, S.; Fehrmann, R. Deactivation and compound formation in sulfuric-acid catalysts and model systems. *J. Catal.* 1995, *155*, 32–42.

156. Delmon, B. Solid state reactions in catalysts: An approach to real active systems and their deactivation. In *Catalyst Deactivation 1997 (Studies in Surface Science and Catalysis)*; Bartholomew, C.H., Fuentes, G.A., Eds.; Elsevier: Amsterdam, The Nethelands, 1997; Volume 111, pp. 39–51.

157. Jackson, N.B.; Datye, A.K.; Mansker, L.; O'Brien, R.J.; Davis, B.H. Deactivation and attrition of iron catalysts in synthesis gas. In *Catalyst Deactivation 1997 (Studies in Surface Science and Catalysis)*; Bartholomew,

C.H., Fuentes, G.A., Eds.; Elsevier: Amsterdam, The Nethelands, 1997; Volume 111, pp. 501–516.

158. Eliason, S.A.; Bartholomew, C.H. Temperature-programmed reaction study of carbon transformations on iron fischer-tropsch catalysts during steady-state synthesis. In *Catalyst Deactivation 1997 (Studies in Surface Science and Catalysis)*; Bartholomew, C.H., Fuentes, G.A., Eds.; Elsevier: Amsterdam, The Nethelands, 1997; Volume 111, pp. 517–526.

159. Baranski, A.; Dziembaj, R.; Kotarba, A.; Golebiowski, A.; Janecki, Z.; Pettersson, J.B.C. Deactivation of iron catalyst by water-potassium thermal desorption studies. In *Catalyst Deactivation 1999 (Studies in Surface Science and Catalysis)*; Delmon, B., Froment, G.F., Eds.; Elsevier: Amsterdam, The Nethelands, 1999; Volume 126, pp. 229–236.

160. Querini, C.A.; Ravelli, F.; Ulla, M.; Cornaglia, L.; Miró, E. Deactivation of Co, K catalysts during catalytic combustion of diesel soot: Influence of the support. In *Catalyst Deactivation 1999 (Studies in Surface Science and Catalysis)*; Delmon, B., Froment, G.F., Eds.; Elsevier: Amsterdam, The Nethelands, 1999; Volume 126, pp. 257–264.

161. Wilson, J.; de Groot, C. Atomic-scale restructuring in high-pressure catalysis. *J. Phys. Chem.* 1995, *99*, 7860–7866.

162. Parkinson, G.S.; Novotny, Z.; Argentero, G.; Schmid, M.; Pavelec, J.; Kosak, R.; Blaha, P.; Diebold, U. Carbon monoxide-induced adatom sintering in a Pd–Fe_3O_4 model catalyst. *Nat. Mater.* 2013, *12*, 724–728.

163. Pham, H.N.; Reardon, J.; Datye, A.K. Measuring the strength of slurry phase heterogeneous catalysts. *Powder Technol.*1999, *103*, 95–102.

164. Kalakkad, D.S.; Shroff, M.D.; Kohler, S.; Jackson, N.; Datye, A.K. Attrition of precipitated iron Fischer-Tropsch catalysts. *Appl. Catal. A* 1995, *133*, 335–350.

165. Callister, W.D. *Materials Science and Engineering: An Introduction*; John Wiley, & Sons, Inc.: New York, NY, USA, 2000.

166. Coble, R.L.; Kingery, W.D. Effect of porosity on physical properties of sintered alumina. *J. Am. Ceram. Soc.* 1956, *39*, 377–385.

167. Deng, S.G.; Lin, Y.S. Granulation of sol-gel-derived nanostructured alumina. *AIChE J.* 1997, *43*, 505–514.

168. Thoma, S.G.; Ciftcioglu, M.; Smith, D.M. Determination of agglomerate strength distributions: Part 1. Calibration via ultrasonic forces. *Powder Technol.* 1991, *68*, 53–61.

169. Bankmann, M.; Brand, R.; Engler, B.H.; Ohmer, J. Forming of high surface area TiO_2 to catalyst supports. *Catal. Today*1992, *14*, 225–242.

170. Kenkre, V.M.; Endicott, M.R. A theoretical model for compaction of granular materials. *J. Am. Ceram. Soc.* 1996, *79*, 3045–3054.

171. Song, H.; Evans, J.R.G. A die pressing test for the estimation of agglomerate strength. *J. Am. Ceram. Soc.* 1994, *77*, 806–814.

172. Werther, J.; Xi, W. Jet attrition of catalyst particles in gas fluidized beds. *Powder Technol.* 1993, *76*, 39–46.

173. Bhatt, B.L.; Schaub, E.S.; Hedorn, E.C.; Herron, D.M.; Studer, D.W.; Brown, D.M. Liquid phase Fischer-Tropsch synthesis in a bubble column. *Proceedings of Liquefaction Contractors Review Conference*; Stiegel, G.J., Srivastava, R.D., Eds.; U.S. Department of Energy: Pittsburgh, PA, USA, 1992; pp. 402–423.

174. Pham, H.N.; Datye, A.K. The synthesis of attrition resistant slurry phase iron Fischer-Tropsch catalysts. *Catal. Today*2000, *58*, 233–240.

175. Bemrose, C.R.; Bridgewater, J. A review of attrition and attrition test methods. *Powder Technol.* 1987, *49*, 97–126.

176. Ghadiri, M.; Cleaver, J.A.S.; Tuponogov, V.G.; Werther, J. Attrition of FCC powder in the jetting region of a fluidized bed. *Powder Technol.* 1994, *80*, 175–178.

177. Weeks, S.A.; Dumbill, P. Method speeds FCC catalyst attrition resistance determinations. *Oil Gas J.* 1990, *88*, 38–40.

178. Zhao, R.; Goodwin, J.G.; Jothimurugesan, K.; Spivey, J.J.; Gangwal, S.K. Comparison of attrition test methods: ASTM standard fluidized bed vs jet cup. *Ind. Eng. Chem. Res.* 2000, *39*, 1155–1158.

179. Doolin, P.K.; Gainer, D.M.; Hoffman, J.F. Laboratory testing procedure for evaluation of moving bed catalyst attrition.*J. Test. Eval.* 1993, *21*, 481.

180. Oukaci, R.; Singleton, A.H.; Wei, D.; Goodwin, J.G. Attrition resistance of Al_2O_3-supported cobalt F-T catalysts. In Proceedings of the 217th National Meeting, ACS Division of Petroleum Chemistry, Anaheim, CA, USA, 21–25 March 1999; p. 91.

181. Adams, M.J.; Mullier, M.A.; Seville, J.P.K. Agglomerate strength measurement using a uniaxial confined compression test. *Powder Technol.* 1994, *78*, 5–13.

182. Emig, G.; Martin, F.G. Development of a fluidized bed catalyst for the oxidation of *n*-butane to maleic anhydride. *Ind. Eng. Chem. Res.* 1991, *30*, 1110–1116.

183. Silver, R.G.; Summers, J.C.; Williamson, W.B. *Catalysis and Automotive Pollution Control II*; Elsevier: Amsterdam, The Nethelands, 1991; p. 167.

184. Fisher, G.B.; Zammit, M.G.; LaBarge, J. *Investigation of Catalytic Alternatives to Rhodium in Emissions Control*; SAE Report 920846; SAE International: Warrendale, PA, USA, 1992.

185. Farrauto, R.J.; Heck, R.M. Catalytic converters: state of the art and perspectives. *Catal. Today* 1999, *51*, 351–360.

186. Beer, G.L. Extended catalyst life two stage hydrocarbon synthesis process. U.S. Patent 6,169,120, 2 January 2001.

187. Bartholomew, C.H.; Stoker, M.W.; Mansker, L.; Datye, A. Effects of pretreatment, reaction, and promoter on microphase structure and Fischer-Tropsch activity of precipitated iron catalysts. In *Catalyst Deactivation 1999 (Studies in Surface Science and Catalysis)*; Delmon, B., Froment, G.F., Eds.; Elsevier: Amsterdam, The Nethelands, 1999; Volume 126, pp. 265–272.

188. Nagano, O.; Watanabe, T. Method for producing methacrolein. U.S. Patent 5,728,894, 17 March 1998.

189. Maillet, T.; Barbier, J.; Duprez, D. Reactivity of steam in exhaust gas catalysis III. Steam and oxygen/steam conversions of propane on a Pd/Al_2O_3 catalyst. *Appl. Catal. B* 1996, *9*, 251–266.

190. Mathys, G.M.K.; Martens, L.R.M.; Baes, M.A.; Verduijn, J.P.; Huybrechts, D.R.C. Alkene oligomerization. WIPO Patent WO 1993/016020 (A3), 16 September 1993.

191. Mathys, G.M.K.; Martens, L.R.M.; Baes, M.A.; Verduijn, J.P.; Huybrechts, D.R.C. Alkene oligomerization. U.S. Patent 5,672,800, 30 September 1997.

192. Stine, L.O.; Muldoon, B.S.; Gimre, S.C.; Frame, R.R. Process for oligomer production and saturation. U.S. Patent 6,080,903, 26 June 2000.

193. Subramaniam, B.; Arunajatesan, V.; Lyon, C.J. Coking of solid acid catalysts and strategies for enhancing their activity. In *Catalyst Deactivation 1999 (Studies in Surface Science and Catalysis)*; Delmon, B., Froment, G.F., Eds.; Elsevier: Amsterdam, The Nethelands, 1999; Volume 126, pp. 63–77.

194. Ginosar, D.M.; Fox, R.V.; Kong, P.C. Solid catalyzed isoparaffin alkylation at supercritical fluid and near-supercritical fluid conditions. WIPO Patent WO 1999/033769 (A1), 8 July 1999.

195. Ribeiro, F.H.; Bonivardi, A.L.; Kim, C. Transformation of platinum into a stable, high-temperature, dehydrogenation-hydrogenation catalyst by ensemble size reduction with rhenium and sulfur. *J. Catal.* 1994, *150*, 186–198.

196. Ginosar, D.; Subramaniam, B. Coking and activity of a reforming catalyst in near-critical and dense supercritical reaction mixtures. In *Catalyst Deactivation 1994 (Studies in Surface Science and Catalysis)*; Delmon, B., Froment, G.F., Eds.; Elsevier: Amsterdam, The Nethelands, 1994; Volume 88, pp. 327–334.

197. Petersen, E.E. Catalyst deactivation: Opportunity amidst woe. In *Catalyst Deactivation 1997 (Studies in Surface Science and Catalysis)*; Bartholomew, C.H., Fuentes, G.A., Eds.; Elsevier: Amsterdam, The Nethelands, 1997; Volume 111, pp. 87–98.

198. Gosselink, J.W.; Veen, J.A.R.V. Coping with catalyst deactivation in hydrocarbon processing. In *Catalyst Deactivation 1999 (Studies in Surface Science and Catalysis)*; Delmon, B., Froment, G.F., Eds.; Elsevier: Amsterdam, The Nethelands, 1999; Volume 126, pp. 3–16.

199. Lin, L.; Zao, T.; Zang, J.; Xu, Z. Dynamic process of carbon deposition on Pt and Pt–Sn catalysts for alkane dehydrogenation. *Appl. Catal.* 1990, *67*, 11–23.

200. Resasco, D.E.; Haller, G.L. Catalytic dehydrogenation of lower alkanes. In *Catalysis (Specialist Periodiodical Report*; Spivey, J.J., Agarwal, S.K., Eds.; Royal Society of Chemistry: Cambridge, UK, 1994; Volume 11, pp. 379–411.

201. Cortright, R.D.; Dumesic, J.A. Microcalorimetric, Spectroscopic, and Kinetic Studies of Silica Supported Pt and Pt/Sn Catalysts for Isobutane Dehydrogenation. *J. Catal.* 1994, *148*, 771–778.

202. Weyten, H.; Keizer, K.; Kinoo, A.; Luyten, J.; Leysen, R. Dehydrogenation of propane using a packed-bed catalytic membrane reactor. *AIChE J.* 1997, *43*, 1819–1827.

203. Praserthdam, P.; Mongkhonsi, T.; Kunatippapong, S.; Jaikaew, B.; Lim, N. Determination of coke deposition on metal active sites of propane dehydrogenation catalysts. In *Catalyst Deactivation 1997 (Studies in Surface Science and Catalysis)*; Bartholomew, C.H., Fuentes, G.A., Eds.; Elsevier: Amsterdam, The Nethelands, 1997; Volume 111, pp. 153–158.

204. Rose, B.H.; Kiliany, T.R. Dual catalyst system. WIPO Pat. App. WO 2000/069993 (A1), 12 May 2000.

205. Guerrero-Ruiz, A.; Sepulveda-Escribano, A.; Rodriguez-Ramos, I. Cooperative action of cobalt and MgO for the catalysed reforming of CH_4 with CO_2. *Catal. Today* 1994, *21*, 545–550.

206. Qin, D.; Lapszewicz, J. Study of mixed steam and CO_2 reforming of CH_4 to syngas on MgO-supported metals. *Catal. Today* 1994, *21*, 551–560.

207. Stagg, S.; Resasco, D. Effects of promoters and supports on coke formation on Pt catalysts during CH_4 reforming with CO_2. In *Catalyst Deactivation 1997 (Studies in Surface Science and Catalysis)*; Bartholomew, C.H., Fuentes, G.A., Eds.; Elsevier: Amsterdam, The Nethelands, 1997; Volume 111, pp. 543–550.

208. Fujimoto, K.; Tomishige, K.; Yamazaki, O.; Chen, Y.; Li, X. Development of catalysts for natural gas reforming: Nickel-magnesia solid solution catalyst. *Res. Chem. Intermed.* 1998, *24*, 259–271.

209. Marshall, C.L.; Miller, J.T. Process for converting methanol to olefins or gasoline. U.S. Patent 5,191,142, 2 March 1993.

210. Gayubo, A.G.; Aguayo, A.T.; Campo, A.E.S.D.; Benito, P.L.; Bilbao, J. The role of water on the attenuation of coke deactivation of a SAPO-34 catalyst in the transformation of methanol into olefins. In *Catalyst Deactivation 1999 (Studies in Surface Science and Catalysis)*; Delmon, B., Froment, G.F., Eds.; Elsevier: Amsterdam, The Nethelands, 1999; Volume 126, pp. 129–136.

211. Barger, P.T. SAPO catalysts and use thereof in methanol conversion processes. U.S. Patent 5,248,647, 28 Sepetember 1993.

212. Cox, J.P.; Evans, J.M. Exhaust gas catalyst for two-stroke engines. WIPO Patent WO 1999/042202 (A1), 20 February 1998.

213. Leviness, S.C.; Mart, C.J.; Behrmann, W.C.; Hsia, S.J.; Neskora, D.R. Slurry hydrocarbon synthesis process with increased catalyst life. WIPO Patent WO 1998/050487 (A1), 2 May 1997.

214. Bartholomew, C.H. Catalyst deactivation in hydrotreating of residua: A review. In *Catalytic Hydroprocessing of Petroleum and Distillates*; Oballa, M., Shih, S., Eds.; Marcel Dekker: New York, NY, USA, 1993; pp. 1–32.

215. Summers, J.; Williamson, W.B. Palladium-only catalysts for closed-loop control. In *Environmental Catalysis 1993*; Armor, J.N., Ed.; American Chemical Society: Washington, DC, USA, 1993; Volume 552, pp. 94–113.

216. Dettling, J.; Hu, Z.; Lui, Y.K.; Smaling, R.; Wan, C.Z.; Punke, A. Smart Pd TWC technology to meet stringent

standards. In *Studies in Surface Science and Catalysis*; Elsevier: Amsterdam, The Nethelands, 1995; Volume 96, pp. 461–472.

217. Berndt, M.; Ksinsik, D. Catalyst and process for its preparation. U.S. Patent 4,910,180, 20 March 1990.
218. Prigent, M.; Blanchard, G.; Phillippe, P. Catalyst supports/catalysts for the treatment of vehicular exhaust gases. U.S. Patent 4,985,387, 15 January 1991.
219. Williamson, W.B.; Linden, D.G.; Summers, J.C., II. High-temperature three-way catalyst for treating automotive exhaust gases. U.S. Patent 5,041,407, 20 August 1991.
220. Williamson, W.B.; Linden, D.G.; Summers, J.C., II. High durability and exhuast catalyst with low hydrogen sulfide emissions. U.S. Patent 5,116,800, 26 May 1992.
221. Narula, C.K.; Watkins, W.L.H.; Chattha, M.S. Binary La-Pd oxide catalyst and method of making the catalyst. U.S. Patent 5,234,881, 10 August 1993.
222. Wan, C.-Z.; Tauster, S.J.; Rabinowitz, H.N. Catalyst composition containing platinum and rhodium components. U.S. Patent 5,254,519, 19 October 1993.
223. Argyle, M.D.; Frost, T.S.; Bartholomew, C.H. Modeling cobalt fischer tropsch catalyst deactivation with generalized power law expressions. *Top. Catal.* 2014, *57*, 415–429.
224. Furuya, T.; Yamanaka, S.; Hayata, T.; Koezuka, J.; Yoshine, T.; Ohkoshi, A. Hybrid catalytic combustion for stationary gas turbine—Concept and small scale test results. In Proceedings of Gas Turbine Conference and Exhibition, Anaheim, CA, USA, 31 May–4 June 1987. ASME Paper 87-GT-99.
225. Kawakami, T.; Furuya, T.; Sasaki, Y.; Yoshine, T.; Furuse, Y.; Hoshino, M. Feasibility study on honeycomb ceramics for catalytic combustor. In Proceedings of Gas Turbine and Aeroengine Congress and Exposition, Toronto, ON, Canada, 4–8 June 1989. ASME Paper 89-GT-41.
226. Betta, R.D.; Ribeiro, F.; Shoji, T.; Tsurumi, K.; Ezawa, N.; Nickolas, S. Catalyst structure having integral heat exchange. U.S. Patent 5,250,489, 5 October 1993.
227. Fujii, T.; Ozawa, Y.; Kikumoto, S. High pressure test results of a catalytic combustor for gas turbine. *J. Eng. Gas Turbines Power* 1998, *120*, 509–513.
228. Borio, D.O.; Schbib, N.S. Simulation and optimization of a set of catalytic reactors used for dehydrogenation of butene into butadiene. *Comput. Chem. Eng.* 1995, *19*, S345–S350.
229. Fiato, R.A. Two stage process for hydrocarbon synthesis. U.S. Patent 5,028,634, 2 July 1991.
230. Zhao, R.; Goodwin, J.G., Jr.; Jothimurugesan, K.; Gangwal, S.K.; Spivey, J.J. Spray-dried Iron Fischer-Tropsch catalysts. 1. Effect of structure on the attrition resistance of the catalysts in the calcined state. *Ind. Eng. Chem. Res.* 2001, *40*, 1065–1075.
231. Zhao, R.; Goodwin, J.G., Jr.; Jothimurugesan, K.; Gangwal, S.K.; Spivey, J.J. Spray-dried iron Fischer-Tropsch catalysts. 2. Effect of carburization on catalyst attrition resistance. *Ind. Eng. Chem. Res.* 2001, *40*, 1320–1328.
232. Singleton, A.H.; Oukaci, R.; Goodwin, J.G. Processes and catalysts for conducting fischer-tropsch synthesis in a slurry bubble column reactor. U.S. Patent 5,939,350, 17 August 1999.
233. Plecha, S.; Mauldin, C.H.; Pedrick, L.E. Titania catalysts, their preparation and use in Fischer-Tropsch synthesis. U.S. Patent 6,087,405, 11 July 2000.
234. Plecha, S.; Mauldin, C.H.; Pedrick, L.E. Titania catalysts, their preparation and use in fischer-tropsch synthesis. U.S. Patent 6,124,367, 26 September 2000.
235. Singleton, A.H.; Oukaci, R.; Goodwin, J.G. Reducing fischer-tropsch catalyst attrition losses in high agitation reaction systems. WIPO Patent WO 2000/071253 (A3), 30 November 2000.
236. Seitz, M.; Klemm, E.; Emig, G. Poisoning and regeneration of NOx adsorbing catalysts for automotive applications. In*Catalyst Deactivation 1999 (Studies in Surface Science and Catalysis)*; Delmon, B., Froment, G.F., Eds.; Elsevier: Amsterdam, The Nethelands, 1999; Volume 126, pp. 211–218.
237. de Lima, S.M.; da Silva, A.M.; da Costa, L.O.O.; Assaf, J.M.; Mattos, L.V.; Sarkari, R.; Venugopal, A.; Noronha, F.B. Hydrogen production through oxidative steam reforming of ethanol over Ni-based catalysts derived from $La_{1-x}Ce_xNiO_3$ perovskite-type oxides. *Appl. Catal. B* 2012, *121–122*, 1–9.
238. Song, H.; Ozkan, U.S. Ethanol steam reforming over Co-based catalysts: Role of oxygen mobility. *J. Catal.* 2009, *261*, 66–74.
239. Masamune, S.; Smith, J.M. Performance of fouled catalyst pellets. *AIChE J.* 1966, *12*, 384–394.
240. Murakami, Y.; Kobayashi, T.; Hattori, T.; Masuda, M. Effect of intraparticle diffusion on catalyst fouling. *Ind. Eng. Chem. Fundam.* 1968, *7*, 599–605.
241. Stork, W.H.J. Molecules, catalysts and reactors in hydroprocessing of oil fractions. In *Hydrotreatment and Hydrocracking of Oil Fractions (Studies in Surface Science and Catalysis)*; Froment, G.F., Delmon, B., Grange, P., Eds.; Elsevier: New York, NY, USA, 1997; Volume 106, pp. 41–67.
242. Parks, G.D.; Schaffer, A.M.; Dreiling, M.J.; Shiblom, C.M. Surface Studies on the Interaction of Nickel and Antimony on Cracking Catalysts. *Prepr.—Am. Chem. Soc., Div. Petr. Chem.* 1980, *25*, 335.
243. Trimm, D.L. Introduction to Catalyst Deactivation. In *Progress in Catalyst Deactivation (NATO Advanced Study Institute Series E, No. 54)*; Figueiredo, J.L., Ed.; Martinus Nijhoff Publishers: Boston, MA, USA, 1982; pp. 3–22.
244. Becker, E.R.; Wei, J.J. Nonuniform distribution of catalysts on supports: I. Bimolecular Langmuir reactions. *J. Catal.*1977, *46*, 365–381.
245. Long, D.C. Staged hydrocarbon synthesis process. U.S. Patent 5,498,638, 12 March 1996.
246. Powell, B.R., Jr. Stabilitzation of high surface area aluminas. In Proceedings of the Materials Research

Society Annual Meeting, Boston, MA, USA, 16–21 November 1980. Paper H9.

247. Dai, Y.; Lim, B.; Yang, Y.; Cobley, C.M.; Li, W.; Cho, E.C.; Grayson, B.; Fanson, P.T.; Campbell, C.T.; Sun, Y.; *et al.* A sinter-resistant catalytic system based on platinum nanoparticles supported on TiO_2 nanofibers and covered by porous silica. *Angew. Chem. Int. Ed.* 2010, *49*, 8165–8168.

248. Shang, L.; Bian, T.; Zhang, B.; Zhang, D.; Wu, L.-Z.; Tung, C.-H.; Yin, Y.; Zhang, T. Graphene-supported ultrafine metal nanoparticles encapsulated by mesoporous silica: Robust catalysts for oxidation and reduction reactions. *Angew. Chem. Int. Ed.* 2014, *53*, 250–254.

249. Heck, R.; Farrauto, R. *Catalytic Air Pollution Control: Commercial Technology*; Van Nostrand Reinhold: New York, NY, USA, 1995.

250. Trimm, D.L. The regeneration or disposal of deactivated heterogeneous catalysts. *Appl. Catal. A* 2001, *212*, 153–160.

251. Berrebi, G.; Dufresne, P.; Jacquier, Y. Recycling of spent hydroprocessing catalysts: EURECAT technology. *Environ. Prog.* 1993, *12*, 97–100.

252. D'Aquino, R.L. For catalyst regeneration, incentives anew. *Chem. Eng.* 2000, *107*, 32–33, 35.

253. Chang, T. Regeneration industry helps refiners control costs, limit liabilities. *Oil Gas J.* 1998, *96*, 49.

254. Blashka, S.R.; Duhon, W. Catalyst regeneration. *Int. J. Hydrocarbon Eng.* 1998, *4*, 60–64.

255. McCulloch, D.C. Catalytic hydrotreating in petroleum refining. In *Applied Industrial Catalysis*; Leach, B.E., Ed.; Academic Press: New York, NY, USA, 1983; pp. 69–121.

256. Franck, J.P.; Martino, G. Deactivation and Regeneration of Catalytic-Reforming Catalysts. In *Progress in Catalyst Deactivation (NATO Advanced Study Institute Series E., No. 54)*; Figueiredo, J.L., Ed.; Martinus Nijhoff Publishers: Boston, MA, USA, 1982; pp. 355–397.

257. Spivey, J.J.; Roberts, G.W.; Davis, B.H. *Catalyst Deactivation 2001 (Studies in Surface Science and Catalysis*; Elsevier: Amsterdam, The Nethelands, 2001; Volume 139.

258. Haddad, J.H.; Harandi, M.N.; Owen, H. Upgrading light olefin fuel gas in a fluidized bed catalyst reactor and regeneration of the catalyst. U.S. Patent 5,043,517, 27 August 1991.

259. Ueda, M.; Murakami, T.; Shibata, S.; Hirabayashi, K.; Kondoh, T.; Adachi, K.; Hoshino, N.; Inoue, S. Process for the regeneration of coke-deposited, crystalline silicate catalyst. U.S. Patent 5,306,682, 26 April 1994.

260. Zhang, S.Y.-F.; Gosling, C.D.; Sechrist, P.A.; Funk, G.A. Dual regeneration zone solid catalyst alkylation process. U.S. Patent 5,675,048, 7 October 1997.

261. Panattoni, G.; Querini, C.A. Isobutane alkylation with C_4 olefins: regeneration of metal-containing catalysts. In*Catalyst Deactivation 2001 (Studies in Surface Science and Catalysis)*; Spivey, J.J., Roberts, G.W., Davis, B.H., Eds.; Elsevier: Amsterdam, The Nethelands, 2001; Volume 139, p. 181.

262. Thompson, D.N.; Ginosar, D.M.; Burch, K.C. Regeneration of a deactivated USY alkylation catalyst using supercritical isobutene. *Appl. Catal. A* 2005, *279*, 109–116.

263. Petkovic, L.M.; Ginosar, D.M.; Burch, K.C. Supercritical fluid removal of hydrocarbons adsorbed on wide-pore zeolite catalysts. *J. Catal.* 2005, *234*, 328–339.

264. Dufresne, P.; Brahma, N. Offsite regeneration process for a catalyst containing at least one precious metal. U.S. Patent 5,854,162, 29 December 1998.

265. Innes, R.A.; Holtermann, D.L.; Mulaskey, B.F. Low temperature regeneration of coke deactivated reforming catalysts. U.S. Patent 5,883,031, 16 March 1999.

266. Fung, S.C. Regenerating a reforming catalyst—this bimetallic noble-metal catalyst is used in a refining process to convert saturated-hydrocarbons to aromatics - simply burning off the carbon was not enough—a procedure had to be found for redispersing the metals on the alumina support as well. *Chemtech* 1994, *24*, 40–44.

267. Alfonso, J.C.; Aranda, D.A.G.; Schmal, M.; Frety, R. Regeneration of a Pt-Sn/Al_2O_3 catalyst: Influence of heating rate, temperature and time of regeneration. *Fuel Proc. Technol.* 1997, *50*, 35–48.

268. Arteaga, G.J.; Anderson, J.A.; Rochester, C.H. Effects of Catalyst Regeneration with and without Chlorine on Heptane Reforming Reactions over Pt/Al_2O_3 and Pt–Sn/Al_2O_3. *J. Catal.* 1999, *187*, 219–229.

269. Pieck, C.L.; Vera, C.R.; Parera, J.M. Study of industrial and laboratory regeneration of Pt-Re/Al_2O_3 catalysts. In*Catalyst Deactivation 2001 (Studies in Surface Science and Catalysis)*; Spivey, J.J., Roberts, G.W., Davis, B.H., Eds.; Elsevier: Amsterdam, The Nethelands, 2001; Volume 139, pp. 279–286.

270. Acharya, D.R.; Hughes, R.; Kennard, M.A.; Liu, Y.P. Regeneration of fixed beds of coked chromia—alumina catalyst.*Chem. Eng. Sci.* 1992, *47*, 1687–1693.

271. Sikkenga, D.L.; Zaenger, I.C.; Williams, G.S. Palladium catalyst reactivation. U.S. Patent 4,999,326, 12 March 1991.

272. Ekstrom, A.; Lapszewicz, J. On the role of metal carbides in the mechanism of the Fischer-Tropsch reaction. *J. Phys. Chem.* 1984, *88*, 4577–4580.

273. Ekstrom, A.; Lapszewicz, J. The reactions of cobalt surface carbides with water and their implications for the mechanism of the Fischer-Tropsch reaction. *J. Phys. Chem.* 1987, *91*, 4514–4619.

274. Pedrick, L.E.; Mauldin, C.H.; Behrmann, W.C. Draft tube for catalyst rejuvenation and distribution. U.S. Patent 5,268,344, 7 December 1993.

275. Owen, H.; Schipper, P.H. Process and apparatus for regeneration of FCC catalyst with reduced NOx and or dust emissions. U.S. Patent 5,338,439, 16 August 1994.

276. Raterman, M.F. Two-stage fluid bed regeneration of catalyst with shared dilute phase. U.S. Patent 5,198,397, 30 March 1993.

277. Apelian, M.R.; Fung, A.S.; Hatzikos, G.H.; Kennedy, C.R.; Lee, C.-H.; Kiliany, T.R.; Ng, P.K.; Pappal, D.A. Regeneration of noble metal containing zeolite catalysts

via partial removal of carbonaceous deposits. U.S. Patent 5,393,717, 28 February 1995.

278. Clark, D.E. Hydrocracking process using a reactivated catalyst. U.S. Patent 5,340,957, 23 August 1994.

279. Yoshimura, Y.; Sato, T.; Shimada, H.; Matsubayashi, N.; Imamura, M.; Nishijima, A.; Yoshitomi, S.; Kameoka, T.; Yanase, H. Oxidative regeneration of spent molybdate and tungstate hydrotreating catalysts. *Energy Fuels* 1994, *8*, 435–445.

280. Oh, E.-S.; Park, Y.-C.; Lee, I.-C.; Rhee, H.-K. Physicochemical changes in hydrodesulfurization catalysts during oxidative regeneration. *J. Catal.* 1997, *172*, 314–321.

281. Aquavo, A.T.; Gayubo, A.G.; Atutxa, A.; Olazar, M.; Bilbao, J. Regeneration of a catalyst based on a SAPO-34 used in the transformation of methanol into olefins. *J. Chem. Tech. Biotech.* 1999, *74*, 1082–1088.

282. Forbus, N.P.; Wu, M.M.-S. Regeneration of methanol/methyl ether conversion catalysts. U.S. Patent 4,777,156, 11 October 1988.

283. Krishna, A.; Hsieh, C.; English, A.E.; Pecoraro, T.; Kuehler, C. Additives improve FCC process; increase catalyst life, lower regeneration temperatures and improve yield and quality of products cost effectively. *Hydrocarbon Processing*1991, 59–66.

284. Altomare, C.; Koermer, G.; Schubert, P.; Suib, S.; Willis, W. A designed fluid cracking catalyst with vanadium tolerance. *Chem. Mater.* 1989, *1*, 459–463.

285. Aguinaga, A.; Montes, M. Regeneration of a nickel/silica catalyst poisoned by thiophene. *Appl. Catal. A* 1992, *90*, 131–144.

286. Lowry, G.V.; Reinhard, M. Pd-Catalyzed TCE Dechlorination in Groundwater: Solute Effects, Biological Control, and Oxidative Catalyst Regeneration. *Environ. Sci. Technol.* 2000, *34*, 3217–3223.

287. Trinh, D.C.; Desvard, A.; Martino, G. Process for regenerating used catalysts by means of hydrogen peroxide aqueous solution stabilized with an organic compound. U.S. Patent 4,830,997, 16 May 1989.

288. Sherwood, D.E. Process for the reactivation of spent alumina-supported hydrotreating catalysts. U.S. Patent 5,230,791, 27 July 1993.

289. Fung, S.C. Deactivation and regeneration/redispersion chemistry of Pt/KL-zeolite. In *Catalyst Deactivation 2001 (Studies in Surface Science and Catalysis)*; Spivey, J.J., Roberts, G.W., Davis, B.H., Eds.; Elsevier: Amsterdam, The Nethelands, 2001; Volume 139, pp. 399–406.

290. Didillon, B. Catalyst regeneration process and use of the catalyst in hydrocarbon conversion processes. U.S. Patent 5,672,801, 30 September 1997.

291. Tsao, Y.-Y.P.; von Ballmoos, R. Reactivating catalysts containing noble metals on molecular sieves containing oxides of aluminum and phosphorus. U.S. Patent 4,929,576, 29 May 1990.

292. Dufresne, P.; Brahma, N.; Girardier, F. Off-site regeneration of hydroprocessing catalysts. *Rev. Inst. Fr. Pet.* 1995, *50*, 283–293.

293. Clark, F.T.; Hensley, A.L., Jr. Process for regenerating a spent resid hydroprocessing catalyst using a group IIA metal. U.S. Patent 5,275,990, 4 January 1994.

294. Brito, A.; Arvelo, R.; Gonzalez, A.R. Variation in structural characteristics of a hydrotreatment catalyst with deactivation/regeneration cycles. *Ind. Eng. Chem. Res.* 1998, *37*, 374–379.

295. Teixeira-da-Silva, V.L.S.; Lima, F.P.; Dieguez, L.C. Regeneration of a deactivated hydrotreating catalyst. *Ind. Eng. Chem. Res.* 1998, *37*, 882–886.

296. Snape, C.E.; Diaz, M.C.; Tyagi, Y.R.; Martin, S.C.; Hughes, R. Characterisation of coke on deactivated hydrodesulfurisation catalysts and a novel approach to catalyst regeneration. In *Catalyst Deactivation 2001 (Studies in Surface Science and Catalysis)*; Spivey, J.J., Roberts, G.W., Davis, B.H., Eds.; Elsevier: Amsterdam, The Nethelands, 2001; Volume 139, pp. 359–365.

297. Sherwood, D.E., Jr.; Hardee, J.R., Jr. Method for the reactivation of spent alumina-supported hydrotreating catalysts. U.S. Patent 5,445,728, 29 August 1995.

298. Maholland, M.K.; Fu, C.-M.; Lowery, R.E.; Kubicek, D.H.; Bertus, B.J. Demetallization and passivation of spent cracking catalysts. U.S. Patent 5,021,377, 4 June 1991.

299. Kubicek, D.H.; Fu, C.-M.; Lowery, R.E.; Maholland, K.M. Reactivation of spent cracking catalysts. U.S. Patent 5,141,904, 25 August 1992.

300. Fu, C.-M.; Maholland, M.; Lowery, R.E. Reactivation of spent, metal-containing cracking catalysts. U.S. Patent 5,151,391, 29 September 1992.

301. Hu, Y.; Luo, B.; Sun, K.; Yang, Q.; Gong, M.; Hu, J.; Fang, G.; Li, Y. Dry regeneration-demetalization technique for catalyst for residuum and/or heavy oil catalytic cracking. U.S. Patent 6,063,721, 16 May 2000.

302. Walker, P.L., Jr.; Rusinko, F., Jr.; Austin, L.G. Gas Reactions of Carbon. *Adv. Catal.* 1959, *11*, 133–221.

303. Fulton, J.W. Regenerating Spent Catalyst. *Chem. Eng.* 1988, *96*, 111–114.

304. Richardson, J.T. Experimental determination of catalyst fouling parameters. Carbon profiles. *Ind. Eng. Chem. Process Des. Dev.* 1972, *11*, 8–11.

305. Weisz, P.B.; Goodwin, R.B. Combustion of carbonaceous deposits within porous catalyst particles: II. Intrinsic burning rate. *J. Catal.* 1966, *6*, 227–236.

306. Rostrup-Nielsen, J.R. Some principles relating to the regeneration of sulfur-poisoned nickel catalyst. *J. Catal.* 1971, *21*, 171–178.

307. Hartenstein, H. *Feasibility of SCR Technology for NO_x Control Technology for the Milton R. Young Station, Center, North Dakota*; Expert Report on behalf of the United States Department of Justice: Washington, DC, USA; July; 2008.

308. Hartenstein, H.; Ehrnschwender, M.; Sibley, A.F. SCR Regeneration-10 Years of R&D and Commercial Application. In Proceedings of Power Plant Air Pollution Control Mega Symposium, Baltimore, MD, USA, 24–28 August 2008. Paper No. 104.

309. EPA; Office of Air and Radiation. Update on Cap and Trade Programs for SO_2 and NO_x. In Presented at the Enviornmental Markets Association, 12 Annual Spring Conference, Miami, FL, USA, 29 April–2 May 2008.

310. EPA. NO_x Budget Trading Program—Basic Information. Available online: http://www.epa.gov/airmarkt/progsregs/nox/docs/NBPbasicinfo.pdf (accessed on 31 October 2014).

311. McIlvaine Company. News Release: Coal to Play Bigger Role in Future Power Generation With World Capacity of 1450 GW by 2012. Available online: http://www.mcilvainecompany.com/brochures/Alerts_for_Internet/s9%20moni/1104coal%20to%20play%20bigger%20role%20in%20future%20power%20generation.htm (accessed on 31 October 2014).

312. Hjalmarsson, A.-K. *NO_x Control Technologies for Coal Combustion*; IEACR/24; IEA Coal Research: Lexington, KY, USA, 1990.

313. U.S. Department of Energy; Southern Company Services. Control of Nitrogen Oxide Emissions: Selective Catalytic Reduction. *Clean Coal Technology*; Topical Report 9. 1997.

314. Janssen, F.; Meijer, R. Quality control of $DeNO_x$ catalysts: Performance testing, surface analysis and characterization of $DeNO_x$ catalysts. *Catal. Today* 1993, *16*, 157–185.

315. Ashton, J.; Nackos, A.; Bartholomew, C.H.; Hecker, W.C.; Baxter, L. Poisoning/Deactivation of Vanadia/Titanium Dioxide SCR Catalysts in Coal Systems. In Presented at the 19th Annual Western States Catalysis Club Symposium, Albuquerque, NM, USA, 25 February 2005.

316. Guo, X.; Nackos, A.; Ashton, J.; Bartholomew, C.H.; Hecker, W.C.; Baxter, L. Poisoning Study of V_2O_5-WO_3/TiO_2Catalysts by Na, K, and Ca. In Presented at 19th Annual Western States Catalysis Club Symposium; Albuquerque, NM, USA: 25 February 2005.

317. Forzatti, P.; Lietti, L.; Tronconi, E. Nitrogen Oxides Removal–Industrial. In *Encyclopedia of Catalysis*; Horváth, T.I., Ed.; Wiley: Hoboken, NJ, USA, 2003.

318. Rubin, E.S.; Kalagnanam, J.R.; Frey, C.H.; Berkenpas, M.B. Integrated Environmental Control Modeling of Coal-Fired Power Systems. *Journal of the Air and Waste Management Association* 1997, *47*, 1180–1188.

319. Cochran, J.R.; Ferguson, A.W. Selective Catalytic Reduction for NO_x Emission Control. In Proceedings of the First International Conference on Air Pollution, Monterrey, Mexico, 16-18 Feb. 1993; Zannetti, P., Brebbia, C.A., Garcia Gardea, J.E., Ayala Milian, G., Eds.; Volume 1, pp. 703–718.

320. Bullock, P.E.; Hartenstein, H.U. O&M Cost Reduction of a Coal-Fired US Merchant Plant Through an Optimized SCR Management Strategy Involving Catalyst Regeneration. In Proceedings of the Conference on Selective Catalytic Reduction and Selective Non-Catalytic Reduction for NO_x Control, Pittsburgh, PA, USA, 15–17 May 2002.

321. Babcock Hitachi's Rejuvenation of Mehrum SCR Catalyst. In *FGD and $DeNO_x$ Newsletter*; No. 315; The McIlvaine Company: Northfield, IL, USA, 2004.

322. McMahon, B. Catalyst regeneration: The business case. *Power* 2006, *150*, 36–39.

323. Zheng, Y.; Jensen, A.; Johnsson, J. Laboratory investigation of selective catalytic reduction catalysts: Deactivation by potassium compounds and catalyst regeneration. *Ind. Eng. Chem. Res.* 2004, *43*, 941–947.

324. Khodayari, R.; Odenbrand, C. Regeneration of commercial SCR catalysts by washing and sulphation: effect of sulphate groups on the activity. *Appl. Catal. B* 2001, *33*, 277–291.

325. Khodayari, R.; Odenbrand, C. Regeneration of commercial TiO_2-V_2O_5-WO_3 SCR catalysts used in bio fuel plants.*Appl. Catal. B* 2001, *30*, 87–89.

326. Dörr, H.-K.; Koch, G.; Bastuck, W. Method for renewed activation of honeycomb-shaped catalyst elements for denitrating flue gases. U.S. Patent 6,387,836 (B1), 14 May 2002.

327. Schneider, P.; Bastuck, W. Method to re-enable honeycomb catalyst elements constructed for the denitrification of flue gases. DE Patent 10222915 (B4), 15 January 2004.

328. Budin, R.; Krotla, K.; Rabitsch, H. Process for regenerating used $deNO_x$ or dedioxin catalytic converters. U.S. Patent 6,484,733 (B2), 26 November 2002.

329. Cooper, M.; Harrison, K.; Lin, C. An Experimental Program to Optimize SCR Catalyst Regeneration for Lower Oxidation of SO_2 to SO_3. In Proceedings of the 2006 Environmental Controls Conference, Pittsburgh, PA, USA, 16–18 May 2006.

330. Patel, N.; Hartenstein, H.-U.; Wenz, F. Process for decoating a washcoat catalyst substrate. U.S. Patent 7,559,993 (B1), 14 July 2009.

331. Patel, N.; Hartenstein, H.-U.; Wenz, F. Process for decoating a washcoat catalyst substrate. U.S. Patent 6,929,701, 16 August 2005.

332. Tate, A.; Skipper, J.; Wenz, F. Environmentally sound handling of deactivated SCR catalyst. Coal Power, 31 July 2008.

333. Hartenstein, H.U.; Hoffman, T. Method of regeneration of SCR catalyst. U.S. Patent 7,723,251, 25 May 2010.

334. Servatius, P.; Schluttig, A. Lifetime extension of SCR-$DeNO_x$ catalysts using SCR-Tech's high efficiency ultrasonic regeneration process. In Proceedings of 2001 Conference of Selective Catalytic Reduction and Selective Non-Catalytic Reduction for NO_x Control, Pittsburgh, PA, USA, 16–18 May 2001.

335. Hartenstein, H.; Gutberlet, H. Catalyst Regeneration—An Integral Part of Proper Catalyst Management. In Proceedings of the 2001 Workshop on Selective Catalytic Reduction, Baltimore, MD, USA, 13–15 November 2001.

336. Wenz, F.; Deneault, R.; Franklin, H.N. The Goals, Challenges and Successes of Regenerating Selective Catalytic Reduction Catalyst. In Proceedings of the Electric Power 2006 Conference–NO_x Control III–SCR Technologies, Atlanta, GA, USA, 2–4 May 2006.

337. CoaLogix. Available online: http://www.coalogix.com (accessed on 31 October 2014).

338. McCullen, S.B.; Wong, S.S.; Huang, T.J. Regeneration of noble metal-highly siliceous zeolite with sequential hydrogen halide and halogen or organic-halogen compound treatment. U.S. Patent 4,645,751, 24 February 1987.

339. Borghard, W.S.; Huang, T.J.; McCullen, S.B.; Schoennagel, H.J.; Tsao, Y.-.P.; Wong, S.S. Redispersion of agglomerated noble metals on zeolite catalysts. U.S. Patent 4,657,874, 14 April 1987.

340. Fung, S.C.; Rice, R.W. Process using halogen/oxygen for reactivating iridium and selenium containing catalysts. U.S. Patent 4,491,636, 1 January 1985.

341. Fung, S.C. Redispersion of Ir catalysts by low temperature reduction step. U.S. Patent 4,467,045, 21 August 1984.

342. Landolt, G.R.; McHale, W.D.; Schoennagel, H.J. Catalyst regeneration procedure. U.S. Patent 4,359,400, 16 November 1982.

343. Fung, S.C.; Weissman, W.; Carter, J.L. Reactivation process for iridium-containing catalysts using low halogen flow rates. U.S. Patent 4,480,046, 30 October 1984.

344. Fung, S.C. Low temperature decoking process for reactivating iridium and selenium containing catalysts. U.S. Patent 4,492,767, 8 January 1985.

345. Fung, S.C.; Weissman, W.; Carter, J.L.; Kmak, W.S. Reactivating iridium and selenium containing catalysts with hydrogen halide and oxygen. U.S. Patent 4,491,635, 1 January 1985.

346. Mohr, D.H. Process for regenerating a monofunctional large-pore zeolite catalyst having high selectivity for paraffin dehydrocyclization. U.S. Patent 4,855,269, 8 August 1989.

347. Fung, S.C.; Tauster, S.J.; Koo, J.Y. Method of regenerating a deactivated catalyst. U.S. Patent 4,925,819, 15 May 1990.

348. Fung, S.C.; Kmak, W.S. Reactivation of iridium-containing catalysts by halide pretreat and oxygen redispersion. U.S. Patent 4,444,896, 24 April 1984.

349. Fung, S.C.; Weissman, W.; Carter, J.L. Reactivation process for iridium-containing catalysts using low halogen flow rates. U.S. Patent 4,444,895, 24 April 1984.

350. Boyle, J.P.; Gilbert, J.B. Reactivation of iridium-containing catalysts. U.S. Patent 4,517,076, 14 May 1985.

351. Buss, W.D.; Hughes, T.R. *In situ* hydrocarbon conversion catalyst regeneration and sulfur decontamination of vessels communicating with catalyst reactor. U.S. Patent 4,482,637, 13 November 1984.

352. Cheng, W.-H. Regeneration of methanol dissociation catalysts. U.S. Patent 4,855,267, 8 August 1989.

353. Huang, Y.-Y.; LaPierre, R.B.; McHale, W.D. Process for dispersing or redispersing a group VIII noble metal species on a porous inorganic support. European Patent 0,306,170 (B1), 8 March 1989.

354. Hucul, D.A. Redispersal of Ru, Os, Rh and Pd catalysts and processes therewith. U.S. Patent 4,891,346, 2 January 1990.

355. Fung, S.C.; Rice, R.W. Process for reactivating iridium-containing catalysts. U.S. Patent 4,447,551, 8 May 1984.

356. Fung, S.C. Process for reactivating iridium-containing catalysts in series. U.S. Patent 4,472,514, 18 September 1984.

357. Fung, S.C.; Rice, R.W. Process for reactivating iridium-containing catalysts. U.S. Patent 4,473,656, 25 September 1984.

358. Huang, Y.-Y.; LaPierre, R.B.; McHale, W.D. Process for dispersing or redispersing a Group VIII noble metal species on a porous inorganic support. U.S. Patent 4,952,543, 28 August 1990.

359. Fung, S.C.; Weissman, W.; Carter, J.L.; Kmak, W.S. Reactivating iridium-containing catalysts with hydrogen halide and oxygen. U.S. Patent 4,444,897, 24 April 1984.

360. Fung, S.C. Low temperature decoking process for reactivating iridium containing catalysts. U.S. Patent 4,472,515, 18 September 1984.

361. Dadyburjor, D.B. Regions of validity for models of regeneration of sintered supported metal catalysts. In *Catalyst Deactivation (Studies in Surface Science and Catalysis)*; Delmon, B., Froment, G.F., Eds.; Elsevier: Amsterdam, The Nethelands, 1980; Volume 6, pp. 341–351.

362. Jacobs, G.; Ma, W.; Gao, P.; Todic, B.; Bhatelia, T.; Bukur, D.B.; Davis, B.H. The application of synchrotron methods in characterizing iron and cobalt Fischer–Tropsch synthesis catalysts. *Catal. Today* 2013, *214*, 100–139.

363. Van Steen, E.; Claeys, M.; Dry, M.E.; van de Loosdrecht, J.; Viljoen, E.L.; Visagie, J.L. Stability of nanocrystals: Thermodynamic analysis of oxidation and re-reduction of cobalt in water/hydrogen mixtures. *J. Phys. Chem. B* 2005,*109*, 3575–3577.

364. Ma, W.; Jacobs, G.; Ji, Y.; Bhatelia, T.; Bukur, D.B.; Khalid, S.; Davis, B.H. Fischer–Tropsch synthesis: Influence of CO conversion on selectivities, H_2/CO usage ratios, and catalyst stability for a Ru promoted Co/Al_2O_3 catalyst using a slurry phase reactor. *Top. Catal.* 2011, *54*, 757–767.

365. Jacobs, G.; Sarkar, A.; Ji, Y.; Patterson, P.M.; Das, T.K.; Luo, M.; Davis, B.H. Fischer-Tropsch synthesis: Characterization of interactions between reduction promoters and Co for Co/Al_2O_3–based GTL catalysts. In Proceedings of AIChE Annual Meeting, San Francisco, CA, USA, 12–17 November 2006.

366. Das, T.K.; Jacobs, G.; Patterson, P.M.; Conner, W.A.; Davis, B.H. Fischer–Tropsch synthesis: characterization and catalytic properties of rhenium promoted cobalt alumina catalysts. *Fuel* 2003, *82*, 805–815.

367. Jacobs, G.; Das, T.K.; Patterson, P.M.; Luo, M.; Conner, W.A.; Davis, B.H. Fischer–Tropsch synthesis: Effect of water on Co/Al_2O_3 catalysts and XAFS characterization of reoxidation phenomena. *Appl. Catal. A* 2004, *270*, 65–76.

368. Moodley, D.J.; Saib, A.M.; van de Loosdrecht, J.; Welker-Nieuwoudt, C.A.; Sigwebela, B.H.; Niemanstsverdriet, J.W. The impact of cobalt aluminate formation on the deactivation of cobalt-based Fischer–Tropsch synthesis catalysts.*Catal. Today* 2011, *171*, 192–200.

369. Jermwongratanachai, T.; Jacobs, G.; Shafer, W.D.; Ma, W.; Pendyala, V.R.R.; Davis, B.H.; Kitiyanan, B.; Khalid, S.; Cronauer, D.C.; Kropf, A.J.; Marshall, C.L. Fischer–Tropsch synthesis: Oxidation of a fraction of cobalt crystallites in research catalysts at the onset of FT at partial pressures mimicking 50% CO conversion. *Top. Catal.* 2014, *57*, 479–490.

370. Azzam, K.; Jacobs, G.; Ma, W.; Davis, B.H. Effect of cobalt particle size on the catalyst intrinsic activity for Fischer–Tropsch synthesis. *Catal. Lett.* 2014, *144*, 389–394.

371. Saib, A.M.; Borgna, A.; van de Loosdrecht, J.; van Berge, P.J.; Niemanstsverdriet, J.W. XANES study of the susceptibility of nano-sized cobalt crystallites to oxidation during realistic Fischer–Tropsch synthesis. *Appl. Catal. A*2006, *312*, 12–19.

372. Dalai, A.K.; Davis, B.H. Fischer–Tropsch synthesis: A review of water effects on the performances of unsupported and supported Co catalysts. *Appl. Catal. A* 2008, *348*, 1–15.

373. Sadeqzadeh, M.; Chambrey, S.; Piché, S.; Fongarland, P.; Luck, F.; Curulla-Ferré, D.; Schweich, D.; Bousquet, J.; Khodakov, A.Y. Deactivation of a Co/Al_2O_3 Fischer–Tropsch catalyst by water-induced sintering in slurry reactor: Modeling and experimental investigations. *Catal. Today* 2013, *215*, 52–59.

374. Sadeqzadeh, M.; Hong, J.; Fongarland, P.; Curulla-Ferre, D.; Luck, F.; Bousquet, J.; Schweich, D.; Khodakov, A.Y. Mechanistic modeling of cobalt based catalyst sintering in a fixed bed reactor under different conditions of Fischer–Tropsch synthesis. *Ind. Eng. Chem. Res.* 2012, *51*, 11955–11964.

375. Soled, S.L.; Kiss, G.; Kliewer, C.; Baumgartner, J.; El-Malki, E.-M. Learnings from Co Fischer-Tropsch catalyst studies. In Present at 245th ACS National Meeting & Exposition, New Orleans, LA, USA, 7–11 April 2013.

376. Hauman, M.M.; Saib, A.M.; Moodley, D.J.; Plessis, E.D.; Claeys, M.; van Steen, E. Re-dispersion of cobalt on a model fischer–tropsch catalyst during reduction–oxidation–reduction cycles. *Chem. Catal. Chem.* 2012, *4*, 1411–1419.

377. Weststrate, C.J.; Kizilkaya, A.C.; Rossen, E.T.R.; Verhoeven, M.W.G.M.; Ciobica, I.M.; Saib, A.M.; Niemanstsverdriet, J.W. Atomic and Polymeric Carbon on Co(0001): Surface Reconstruction, Graphene Formation, and Catalyst Poisoning. *J. Phys. Chem. C* 2012, *116*, 11575–11583.

378. Weststrate, C.J.; Hauman, M.M.; Moodley, D.J.; Saib, A.M.; van Steen, E.; Niemanstsverdriet, J.W. Cobalt Fischer–Tropsch catalyst regeneration: The crucial role of the kirkendall effect for cobalt redispersion. *Top. Catal.* 2011, *54*, 811–816.

379. Moodley, D.J.; van de Loosdrecht, J.; Saib, A.M.; Overett, M.J.; Datye, A.K.; Niemanstsverdriet, J.W. Carbon deposition as a deactivation mechanism of cobalt-based Fischer–Tropsch synthesis catalysts under realistic conditions. *Appl. Catal. A* 2009, *354*, 102–110.

380. van de Loosdrecht, J.; Balzhinimaev, B.; Dalmon, J.-A.; Niemanstsverdriet, J.W.; Tsybulya, S.V.; Saib, A.M.; van Berge, P.J.; Visagie, J.L. Cobalt Fischer-Tropsch synthesis: Deactivation by oxidation? *Catal. Today* 2007, *123*, 293–302.

381. Saib, A.M.; Borgna, A.; van de Loosdrecht, J.; van Berge, P.J.; Niemanstsverdriet, J.W. In Situ surface oxidation study of a planar $Co/SiO_2/Si(100)$ model catalyst with nanosized cobalt crystallites under model Fischer–Tropsch synthesis conditions. *J. Phys. Chem. B* 2006, *110*, 8657–8664.

382. Moodley, D.J.; van de Loosdrecht, J.; Saib, A.M.; Niemanstsverdriet, J.W. The formation and influence of carbon on cobalt-based Fischer-Tropsch synthesis catalysts: An integrated review. In *Advances in Fischer-Tropsch Synthesis, Catalysts, and Catalysis*; Davis, B.H., Occelli, M.L., Eds.; CRC Press Taylor & Francis Group: Boca Raton, FL, USA, 2010; pp. 49–81.

383. Bezemer, G.L.; Herman, J.H.; Kuipers, P.C.E.; Oosterbeek, H.; Holewign, J.E.; Xu, X.; Kapteijn, F.; van Dillen, A.J.; de Jong, K.P. Cobalt particle size effects in the Fischer–Tropsch reaction studied with carbon nanofiber supported catalysts. *J. Am. Chem. Soc.* 2006, *128*, 3956–3964.

384. Jacobs, G.; Ribeiro, M.C.; Ma, W.; Ji, Y.; Khalid, S.; Sumodjo, P.T.A.; Davis, B.H. Group 11 (Cu, Ag, Au) promotion of $15\%Co/Al_2O_3$ Fischer–Tropsch synthesis catalysts. *Appl. Catal. A* 2009, *361*, 137–151.

CHAPTER 2

DEACTIVATION AND REGENERATION OF COMMERCIAL TYPE FISCHER-TROPSCH CO-CATALYSTS

Erling Rytter and Anders Holmen

Keywords: Fischer-Tropsch; deactivation; regeneration; Co-catalysts

INTRODUCTION

In a gas-to-liquid (GTL) plant the high H_2/CO ratio obtained from reforming of natural gas to synthesis gas (syngas) obviates the need for shifting CO with steam to yield more hydrogen (and CO_2) for the FT unit. This is one main reason for using a cobalt catalyst instead of the much cheaper iron alternative as catalytic metal for the FT reaction. In addition, the cobalt catalyst is more active and has a simpler product slate of mainly paraffins and some α–olefins. However, both cobalt metal in itself, precious metal promoters as well as advanced overall formulations, make the catalyst inherently costly. Further, Co catalysts typically lose about half their activity within a few months. Assuming an economically acceptable catalyst lifetime of 2–3 years, this means that the catalyst cost will add several USD to the price per bbl of produced synthetic crude. Therefore, improving catalyst stability is a major focus among technology providers and plant operators. It follows logically that a basic understanding of the mechanisms involved in the deactivation process is vital to improving catalyst stability. Fortunately, it appears that, at least for most commercial Co catalysts, rejuvenation of catalyst activity is possible.

A comprehensive review of deactivation of Co FT catalysts appeared a few years ago [1]. This review discusses a wide variety of deactivation mechanisms comprising sintering; re-oxidation of cobalt, including surface oxidation; formation of stable compounds between cobalt and the support, e.g., cobalt aluminate; surface reconstruction; formation of carbon species on the cobalt surface; carbiding; and poisoning. However, less focus is given in the review to long-term deactivation under commercial conditions.

Historically, details of deactivation mechanisms and rates have been scarce particularly as only a few plants are operated commercially. Nevertheless, some data can be found in the patent literature, mainly based on operation of pilot or demonstration plants, and in conference presentations. A complicating factor is that an industrial process is typically operated at constant global production, *i.e.*, deactivation is counteracted by a steady increase in operating temperature. Only a few reports on deactivation under these most relevant conditions could be found. Fortunately, Sasol and its collaborator, Eindhoven University of Technology, have published extensive data on mechanisms and rates of deactivation at industrial conditions for their Co/Pt/alumina catalyst. Several of their papers have focused on long term deactivation due to polycarbon deposition [2].

We would also like to draw attention to an extensive report to US-DOE where long term experiments are reported focusing mainly on the effect of water [3]. Catalysts with the formulations Co(15 wt.%)/Re(0.2; 0.5; 1.0 wt.%)/γ-Al_2O_3, Co(10 wt.%)/Ru(0.2 wt.%)/TiO_2, Co(15 wt.%)/Pt(0;0.5 wt.%)/γ-Al_2O_3, Co(15 wt.%)/Ru(0.2;0.5;1.0 wt.%)/ γ-Al_2O_3 and Co(12.4 wt.%)/SiO_2 were tested for up to 3500 h in a 1L autoclave (CSTR-reactor). The authors claim that carbon deposition may be minimized by careful temperature control, and that deactivation caused by sintering and oxidation are the major concerns. These conclusions are controversial and have been disputed by several investigations; see later in this review and in our previous review [1].

Argyle *et al.* have fitted previously published activity *versus* time data to first or second order general power law rate

Citation: Erling Rytter and Anders Holmen, Deactivation and Regeneration of Commercial Type Fischer-Tropsch Co-Catalysts, doi: 10.3390/catal5020478

expressions incorporating a limiting activity and have shown how parallel routes, e.g., sintering and carbon deposition deactivation, can be modeled separately. For example, their model predicts that during a 60 day run under typical FTS conditions a commercial Co catalyst loses about 30% activity within 10–15 days due to rapid sintering, while an additional 30% activity is lost gradually over the 60 day period due to carbon [4].

Causes of deactivation may depend on catalyst material and properties, e.g., support, promoters, dispersion, extent of reduction, *etc.*; reactor type; and especially operating conditions. It appears that after an initial break-in period during which cobalt is equilibrated with its reactor environment in terms of crystallite size, possibly crystal structure, and degree of reduction, a slow long term deactivation is observed. The origin of this latter deactivation period is discussed in terms of carbon formation and/or re-oxidation of the metal.

CATALYST ACTIVITY

To understand catalyst deactivation, it is first necessary to understand the factors that determine initial catalyst activity. Activity is largely dependent on the degree of reduction of the cobalt metal precursor and the average size of the cobalt crystallites, which together determine the surface density of catalytically active sites, *i.e.*, the dispersion of the metal. It has been verified that the turn-over-frequency (TOF) is rather constant for Co crystallites larger than 6–8 nm [5]. As activity falls off rapidly below this threshold, methods for making very high dispersion catalysts have limited relevance for FT-synthesis. Cobalt crystallite size and degree of reduction depend on several factors, including cobalt precursor, support material and its pretreatment; pore diameter, pore volume and available total surface area; method of impregnation or deposition; drying and calcination conditions; reduction conditions; *etc.* It is especially important to calcine and reduce the catalyst at optimum conditions, *i.e.*, optimal gas flow rates, temperature ramp and final temperature [6]. For example, overly high calcination and reduction temperatures result in large cobalt oxide and cobalt metal crystallites and, therefore, undesirably low dispersion due to over-sintering. For a given degree of reduction and crystallite size, the activity per kg catalyst is proportional to the cobalt loading. The loading of a commercial type catalyst may vary from 12 to 30 wt.% and is a compromise between several catalyst properties. For instance, a lower surface area and pore volume support will be able to accommodate less cobalt, but might be considerably more attrition resistant.

To optimize cobalt crystallite size is not particularly challenging as long as one is able to control preparation conditions. It is industrial practice to add a metal promoter that enhances the degree of reduction and maintains a targeted dispersion [7,8]. The literature provides no solid evidence that such metal promoters are able to enhance reaction rates or surface concentration of intermediates. Promoters used at a commercial or demo scale include platinum, rhenium and ruthenium. The promoter will add significantly to the cost of the catalyst; catalyst grade Re is today priced at *ca.* 3000 USD/kg and Pt at 45,000 USD/kg [9]. Typical loadings are up to 0.5 and 0.05 wt.% for Re and Pt, respectively. It can also be mentioned that a possible effect of cobalt being in the *fcc* or *hcp* crystallographic phase, with the latter being more active, has been reported [10]. However, studies on the actual configuration of an active cobalt crystal are needed to be able to correlate activity with atomic arrangement on the surface of a working catalyst. It is well documented that the catalyst support has a strong influence on the selectivity to C_{5+} of the process, but as long as known impurities like alkali, alkaline earth metals and sulfur are eliminated it is less clear how support chemistry and pore structure influence activity of a catalyst for a given cobalt crystallite size. However, different supports have varying interactions with cobalt oxide and therefore influence reducibility.

It is well known that reaction rate greatly depends on process conditions. Generally, rate increases with increasing temperature and overall pressure. Furthermore, indications are that the rate increases with H_2/CO ratio, possibly due to higher methane make, and decreases with increasing conversion as the partial pressure of syngas is reduced and a high level of product water may block active sites [11].

Of the CH_x monomers generated on the surface of a catalyst, CH_2 is probably the most abundant intermediate and is probably readily incorporated in the chain during polymerization. A smaller portion of the monomer will be hydrogenated all the way to methane. The growing chain can terminate by β–hydrogen abstraction and leave the surface as an olefin or be hydrogenated to an alkane. Olefins can also be hydrogenated in a secondary reaction. There is evidence from experiments at low conversion and small catalyst particle size that the primary product is dominated by olefins, but for practical purposes the olefin to paraffin ratio is well above two for C_3 and then diminishes rapidly with chain length [12].

FISCHER-TROPSCH REACTORS

Apart from the type of FT-catalyst, selection of the FT-reactor, as well as how it is operated and incorporation in the XTL flowsheet, is the principal factor influencing catalyst deactivation. Comparison of properties of the main reactor types for low-temperature Fischer-Tropsch synthesis is given in Table 1. Conversion per path will vary, mainly with the propensity for heat removal and temperature control, whereas the temperature typically is between 200 and 250 °C and the pressure will be in the range 15–30 bar. H_2/CO ratio in the make-up gas from the syngas generator will be slightly below 2, whereas the outlet ratio of the FT-reactor will be considerable lower; down to 1.2, in some cases even below 1.0. Evidently, high temperature and low H_2/CO ratio are expected to promote deactivation, but reports on these effects are not available.

Table 1: Properties for different reactor types for Low-Temperature Fischer-Tropsch synthesis *.

Reactor	Conversionper path(%)	Capacity per reactor (bbl/day)	Characteristics
Tubular fixed-bed	30–35	≤6000	≤ 30,000 tubes with catalysts pellets or extrudates.
Slurry bubble column	55–65	≤25,000	Internal heat exchanger and optional product filter
Micro-channel	65–75	≤1000	Metal block with < 2mm diameter channels

* Based on open literature and patents for commercial Low Temperature Fischer-Tropsch (LTFT) synthesis with cobalt catalysts [13,14]. See also references in Table 2.

In a comparative study of reactor types for LTFT synthesis Guettel and Turek conclude that the productivity per reactor volume of a slurry bubble-column reactor or monolith reactor is up to one order of magnitude higher than for a fixed-bed or micro-channel reactor [15]. However, this conclusion is at variance with other work, reported for micro-channel reactors for which superior productivities have been reported [16].

Fixed-bed reactor. Due to the necessity of controlling the heat evolved during reaction, the design of a fixed-bed FT reactor is based on a multi-tube heat-exchange type of reactor where catalyst pellets are loaded into the tube bundles and the shell contains evaporating water. In this type of reactor axial temperature typically increases through a maximum of 5–20°°C, and it is imperative to avoid hot-spots which cause sintering. In order to minimize deactivation due to temperature effects several measures are taken. Once-through CO conversion is limited to 30–35%. Tube diameter is typically 2.5–5 cm and the size of the catalyst pellets or extrudates is relatively small, in the range of 1–3 mm. Small extrudates size is also important to secure good radial mixing and minimize diffusion limitations, thus maintaining high selectivity to liquids. It has been shown that above *ca.* 200 µm particle size the higher effective H_2/CO ratio in the inner part of the pellets significantly reduces C_{5+} yield [12]. Therefore, an egg-shell catalyst design is preferred where only the outer parts are impregnated with active metal. Another factor limiting the applicable superficial gas velocity and the global rate is pressure drop. In general, a challenge related to deactivation in a fixed-bed reactor is variations in the partial pressures of reactants and products along the tube length and within the catalyst particles.

H_2/CO ratio depends on several factors, but the make-up gas typically has a ratio slightly below 2 for maximizing C_{5+}selectivity. With recycle of the product gas to the FT-section the feed ratio to the reactor may be significantly lower, possibly 1.6–1.7, and there may be a gradual reduction from the inlet to around 1.4–1.5 at the outlet. Compared to a slurry reactor, the average gas composition may be richer in hydrogen and due to less efficient temperature control, the overall operating temperature usually lower. A consequence is lower reaction rates, but this is at least in part compensated for by a lower average partial pressure of product water and thereby a higher syngas pressure. The overall effect on deactivation, in particular carbon deposition, is complex and challenging to predict.

A distinct advantage of the tubular fixed-bed reactor is a well proven commercial design. Several tens of thousands of tubes can be incorporated within the reactor shell. Scale-up is comparatively easy, and optimization can be done in a single tube laboratory reactor. Operational experience with catalyst fouling or attrition and resulting difficulties with loading and unloading tubes are trade secrets, but it can be expected that an experienced operator is able to control these factors. Minimizing catalyst deactivation or being able to perform *in situ* regeneration is critical in order to reduce catalyst consumption and avoid an extensive unloading-reloading sequence. The liquid product is inherently separated from the catalyst and any need for removing residual particles and metal components will be low.

Slurry bubble-column reactor (SBCR). Catalyst particles are suspended in the liquid hydrocarbon product of the FT process and synthesis gas is bubbled through the slurry. Depending on the density of the catalysts particles, their diameter and the superficial gas velocity, there is a profile of solid concentration diminishing from bottom to top of the reactor. Gaseous components leave from the top of the reactor. Higher boiling products have to be removed from the reactor as liquid, and separated from the catalyst. Several methods for this purpose have been patented, both *in situ* and *ex situ* techniques. Broadly they can be classified as employment of filters, settling devices, magnetic separation and hydrocyclones. Sasol uses internal filters combined with secondary polishing filters of the product [14].

Settling of the catalyst should be avoided as overheating and consequently catalyst deactivation will occur. Particularly critical are the gas distribution system and, depending on design measures to prevent particle settling in stagnant regions below the nozzles. One way to improve overall liquid circulation, and thereby avoid settling, is to install so-called internal down-comers. A serious threat to the catalyst in a slurry operation is any upset in production, like a sudden reduction in syngas flow. Without adequate back-up systems such events will lead to settling and serious overheating in the catalyst mass due to continuous FT-synthesis with residual syngas. Similar conditions may occur in slurry separation devices like filters, but no public information is available on any effects on catalyst deactivation in these devices.

An SBCR operates preferentially in the churn turbulent flow regime for best distribution of catalyst particles as well as minimizing mass and heat transport restrictions. In the churn turbulent flow regime there is a mixture of smaller and larger bubbles that undergo frequent beak-up and coalescence. This mechanism prevents serious film transport restrictions on the catalyst slurry interphase, and with catalyst particles below 200 µm the H_2/CO ratio as well as water vapor pressure can be assumed constant over the entire reactor volume. Thus, deactivation should be more easily controlled compared to other reactor configurations. Further details on operation of SBCRs can be found in the book on Fischer-Tropsch technology by Steinberg and Dry [17].

Reasons for selection of a slurry bubble column reactor for Fischer-Tropsch synthesis include: (1) a comparatively simple construction; (2) high space-time-yield and catalyst efficiencies; (3) high heat transfer coefficients; and (4) isothermal conditions. Continuous catalyst regeneration of a slip stream is a viable option. Challenges include minimizing catalyst particle attrition and efficiently separating catalyst from the products. Efficient liquid and gas back-mixing and a high exit water concentration lead to high selectivity; the high exit water concentration is beneficial in reducing coke deposition. On the other hand reactant concentrations are lower than the average of a fixed-bed reactor resulting in comparatively lower global rates. Single pass conversion is typically in the range of 55–65%, significantly higher than for fixed-bed. Conversion is limited by the feasible height of the reactor, but there is also an upper conversion limit above about 75–80% for which the water-gas-shift activity will lead to possible catalyst oxidation and a steep increase in CO_2 yield [18]. Naturally, extensive recycle of syngas in the FT-section of the plant is necessary to obtain a very high overall CO conversion.

Micro-channel reactor. Micro-channel reactor technology holds great promise for process intensification due to outstanding heat and mass transfer rates [19,20]. Combined with highly active and stable catalysts, micro-channel reactors can achieve high volume based productivities. In some cases very high conversions (~ 90%) can be obtained while maintaining high C_5+ selectivity. Detailed studies of flow and temperature behavior have shown that a micro-channel reactor can operate isothermally and with very low pressure drop [19]. Except for the normal initial deactivation, the catalyst in the micro-channel reactor is remarkably stable even at very high conversions [21]. Observed rates of deactivation appear to be lower in the micro-channel reactor compared with the fixed bed laboratory reactor at similar conditions. Velocys and CompactGTL are operating microchannel demonstration plants [20].

COMMERCIAL CATALYST FORMULATIONS

Scale-up to commercial catalyst production is demanding, and little public information on the industrial manufacturing processes is available. Great care must be taken to obtain a homogeneous catalyst material, but the targeted distribution of cobalt on the support depends on the actual process. For slurry catalysts with diameters typically in the range 40–120 μm pore diffusion resistance of the syngas is negligible ensuring full utilization of the available surface area [12]. For micro-channel reactors the catalyst either will resemble a slurry catalyst or be impregnated onto special trays that are inserted into the channels. On the other hand, a fixed-bed catalyst is typically designed as an egg-shell or rim catalyst in which only the outer few hundred micrometers contain the active phase. As to the degree of reduction it has been shown by Sasol that the initial value is not critical, as the syngas will reduce the catalyst further under the first months of operation, thereby increasing CO conversion [2].

Cobalt FT catalysts can be classified according to supports and promoters used. Table 2 lists several commercial type catalysts. Precious metal promoters like Pt and Ru may facilitate hydrogenation activity and hence reduce propensity for carbon deposition as indicated, for example, by Exxon Mobil for ruthenium.

Some relative activities have been estimated and are included in the table. The values are based on fixed-bed data reported in the patent literature. As the process conditions vary considerably, an attempt to normalize the activities is made by using a simplified kinetic expression with activation energy of 110 kJ/mol, and partial pressures are average pressures in the reactor [22]. It is recognized that this comparison is only approximate, but still a guide. Activities are generally lower for a given catalyst operated in slurry compared to fixed-bed in spite of limited apparent diffusion limitations. The origin of this effect is not understood. Particularly low activities have been reported by Nippon and ENI/IFP, perhaps an indication of large cobalt crystals in the working catalyst.

It appears that Shell favors a titania based support over their previous zirconia modified silica system. Promoters are either Mn or V, the latter claimed to lower CO_2 make [23]. Titania has relatively large pores and moderately low surface areas, but is known to facilitate high selectivities to C_{5+} products. Fixed-bed catalysts contain modifiers like citric acid added prior to the forming step. In addition to Shell, it should be mentioned that BP is promoting its fixed-bed technology and claims that a CO_2 resistant support is vital [24].

Velocys/Oxford Catalyst and Compact GTL are offering micro-channel fixed-bed technologies with catalyst diameter or thickness of 0.1–0.5 mm. Velocys' carbon combustion preparation technique may very well take the edge off initial sintering during FT synthesis by optimizing cobalt crystallite size already at the catalyst manufacturing stage. Loading and off-loading the catalyst can be particular challenge for these systems, but efficient methods for this purpose are claimed by these companies.

Table 2: Formulation of commercial type cobalt catalysts and their application *.

Technology provider	Support/ modifier	Promoter	Reactor type	Reactor scale (bbl/d)	Relative activity	Reference
Sasol	γ-Alumina/ Si **	Pt	Slurry	16,000		[25]
Shell	Titania	Mn; V	Fixed	6000	0.3	[26]
GTL.F1	Ni-aluminate/ α-Alumina	Re	Slurry	1000	0.9	[27]
ENI/IFP/ Axens	γ-alumina/SiO_2/spinel	?	Slurry	20	0.19	[28]
Nippon Oil	Silica/ Zirconia	Ru	Slurry	500	0.16	[29]
Syntroleum	γ-Alumina/ Si **; La	Ru	Slurry	80		[30]
BP	ZnO	?	Fixed			[24]
Exxon Mobil	Titania/ γ-Alumina	Re	Slurry	200		[31]
Conoco-Phillips	γ-Alumina/ Boron	Ru/Pt/Re	Slurry	400	0.68	[32]
Compact GTL	Alumina?	Re?	Micro	500		[33]
Oxford cat. /Velocys	Titania-silica	Re	Micro	1000		[34]

* Deduced from open literature and patents. Actual commercial formulation may vary. ** Si from TEOS; tetraethoxy silane.

The platinum promoted Sasol catalyst is prepared on γ-alumina (Puralox SCCa-2/150 or -5/150: pore volume 0.5 mL/g; surface area 150 m^2/g) and stabilized by impregnation with tetra-ethoxy-silane (TEOS) followed by calcination to give a surface concentration of *ca.* 2.5 Si atoms/nm^2. This procedure apparently modifies the surface so that the support becomes less prone to dissolution in the acidic product water. The GTL.F1/Statoil catalyst is based on a larger pore diameter γ-alumina modified with nickel and fired at high temperature to produce a nickel-aluminate (spinel)/α-alumina mixture. The pore properties resemble titania-based supports, but with very high attrition resistance. Also the ENI/IFP catalyst is supported on Si-modified γ-alumina, but probably strengthened by silanation and calcination giving a final SiO_2 content of 6–7 wt.%. Other support modification methods have been described by ENI/IFP in earlier patents, including formation of spinel compounds. It is unclear whether the catalyst formulation contains a reduction or other type of promoter. In their slurry catalyst development, Nippon has apparently adopted a silica based support formulation similar to that of Shell's previous fixed-bed catalyst, but using ruthenium as a promoter. No information on attrition resistance has been revealed, as is the case for most other slurry catalysts as well. Syntroleum used a catalyst similar to Sasol's, but also with a ruthenium promoter. Exxon Mobil, that pioneered titania as a support with an alumina binder, BP, Conoco-Phillips and Syntroleum have terminated their developments in FT-technology. However, BP is still licensing their FT technology and Exxon Mobil has announced that the technology is ready for commercialization should the right project be prioritized.

CAUSES OF DEACTIVATION

From our previous review on deactivation during LTFT synthesis the main causes of deactivation are sintering, re-oxidation of cobalt, formation of stable compounds between cobalt and the support, surface reconstruction, formation of carbon species on the cobalt surface, carbiding and poisoning [1]. In addition there can be a loss of catalyst material from the reaction zone due to attrition. The chemical environment is challenging with a number of reactive chemical species generated including significant amounts of water. In addition, the exothermicity of the reaction may lead to hot spots in the catalyst.

There appear to be two main "schools" for describing long-term deactivation mechanisms based on demo slurry operations, one favoring re-oxidation [35,36], and one poly-carbon formation on the surface [2]. It should be realized that both catalyst system and process conditions can affect the results. In addition, an initial sintering stage may be observed if the fresh catalyst contains crystallites in the range of 6–12 nm. The consequences of severe deactivation can be a significant decline in the activity over a typical design period for catalyst life of two years to an estimated 25–30% of the initial value. In addition, slurry catalyst loss due to attrition can be significant. All previous experience considered catalyst replacement due to deactivation will contribute significantly to the operational costs of an FT plant.

In a study on the effect of impurities it was found [37], by impregnating 400 ppm of the impurity element from nitrate precursors, a poisoning effect which decreases in the following order:

Na > Ca > K > Mg > P,

Mn, Fe and Cl showing minimal effects. The latter is surprising as chlorine causes a 25% reduction in hydrogen chemisorption. Even alkali concentrations of less than 100 ppm have a large effect on the rate (site time yield) [38]. However, no effect of alkali on the hydrogen chemisorption was observed. The impact of alkali and alkaline earth elements is far stronger than any stoichiometric blocking of surface sites, and might be related to the strong electronegativity of the elements leading to blocking of steps on cobalt thus preventing CO dissociation [39]. Particular care must be taken to avoid alkali and alkaline earth elements in impregnation fluids and washing water as well as contaminants in the catalyst support. Sulfur as H_2S or $(CH_3)_2S$ added to the syngas gives a deactivation consistent with stoichiometric blocking of cobalt surface sites as *in situ* measurement of cobalt dispersion by H_2S is consistent with hydrogen derived dispersion of a fresh catalyst. The effect of ammonia appears to be strongly catalyst dependent, and reports vary from negligible influence to strong negative consequences.

Contributions to attrition of catalysts for three-phase slurry bubble column operation include mechanical abrasion and breakage of catalyst particles; chemical dissolution; and synergisms between these mechanisms. It appears that Sasol focuses more on avoiding chemical attack on their γ-alumina based support [40], whereas Statoil/GTL.F1 [41], IFP/ENI/Axens [28], and probably Exxon [42], have developed more mechanically robust slurry catalysts.

Sasol has reported deactivation profiles in several publications for their Co/Pt/modified γ-alumina catalyst [2]. We refer to the section below on carbon deposition and to the previous review for further details on the Sasol work and their extensive documentation of carbon deposits [1].

Statoil/GTL.F1 have disclosed deactivation profiles at several conferences for their attrition resistant catalyst of Co/Re/aluminate spinel catalyst. In a CSTR slurry reactor test over 3000 h the temperature was adjusted regularly, typically in 2–3 weeks intervals, to keep conversion reasonable constant [43]. Somewhat surprisingly, temperature was *decreased* from 222 °C to 215 °C during the operation meaning that the catalyst activity increased regularly. This is in line with the reported enhanced reduction during first months of slurry FT-synthesis. More surprisingly, the C_{5+} selectivity increased simultaneously by *ca.* 5%, significantly more than expected given the reduction in temperature. In another presentation on CSTR results, the rate of hydrocarbon formation increased up to 800 h time-on-stream (TOS) followed by decline towards end of test at 1600 h [44]. Characterization of a commercial catalyst after *ca.* one month operation in a commercial scale slurry-bubble column confirms an increase in degree of reduction [45]. This is concurrent with sintering probably facilitated by a high steam partial pressure [46].

DEACTIVATION BY CARBON DEPOSITION

Higher hydrocarbons (waxes) are desired products from LTFT synthesis on Co catalysts. The hydrocarbons can accumulate on the surface and can slowly be converted to carbon or coke that blocks the active sites. By using TPR Lee *et al.*[47] could distinguish between several forms of carbon on the catalyst surface from CO disproportionation; they suggested that carbon was present in two forms: atomic and polymeric carbon. Support for stabilization of graphitic carbon on an *fcc*-Co(111) surface has been obtained by quantum-mechanical calculations [48]. DFT data indicate chemical bonding between graphene and cobalt, as also supported by other studies [49]. Direct STM evidence for formation of graphene on Co(0001) has been demonstrated through decomposition of ethylene [50]. It was found that carbon on the surface induces cobalt reconstruction and weakens CO and H_2 adsorptions.

Thus, there are ample investigations showing that carbon in different forms can interact with and block cobalt surfaces. Support for this deactivation mechanism comes from a few long term studies using commercial catalysts in pilot reactors. Sasol studied catalyst deactivation by periodically removing samples from a pilot slurry bubble column reactor operated for 6 months [2]. Wax was removed by inert solvent extraction before the catalyst samples were characterized by temperature programmed hydrogenation and oxidation, chemisorption, TEM and LEIS. Polymeric carbon was found both on the alumina support and on cobalt. This carbon is resistant to hydrogen treatment at temperatures above the FT synthesis temperature. The amount of polymeric carbon correlated well with observed long term deactivation. From XANES data they ruled out oxidation of cobalt during the run, but there was significant sintering taking place during the first 10–15 days on stream. Moodley *et al.* [2] concluded that accumulation of polymeric carbon was responsible for at least a part of long term catalyst deactivation.

Build-up of graphitic or polymeric carbon as deactivation mechanism was recognized by Syntroleum [51]. By TGA-MS they estimated that about 1% carbon was deposited on the catalyst. Carbon deposition on cobalt/ZnO has also been proposed by BP based on results from a demonstration plant and laboratory studies [52].

DEACTIVATION BY RE-OXIDATION

Schanke *et al.* investigated the influence of water on deactivation of unpromoted or Re promoted alumina supported cobalt catalysts. Adding 20–28% steam to 50% syngas in the feed of a lab-scale fixed-bed reactor resulted in significant deactivation due to oxidation of highly dispersed cobalt crystals and surface cobalt atoms [53]. Although these experiments clearly show oxidation of cobalt, the conditions represent very high conversion levels (> 80%). Oxidation takes place within the stability range for bulk cobalt metal, and is presumed to be a consequence of surface reactivity of small crystallites. It is also evident that the effect of water depends critically on the support material used, e.g., samples of γ–alumina from different sources behave very differently. Even a positive effect of water on activity has been reported; see [1,3] for further discussion. From activity tests over a range of conversions in a slurry CSTR reactor Co/Re/γ-alumina catalyst activity is observed to slightly increase with conversion up to 85%; above 85% conversion it drops rapidly [18]. At high conversions high partial pressures of H_2O may oxidize small cobalt crystallites and promote aluminate formation thus enhancing WGS activity as shown by a significant CO_2 make. Similar results were found for a Pt promoted catalyst [54].

Very recently, the group of A.Y. Kodakov in cooperation with Total published evidence for surface oxidation of cobalt nano-crystallites in alumina supported, Pt promoted catalysts during FT synthesis [55]. However, surface oxidation was only clearly evident by STEM-EELS after an excursion to 340 °C and 100% CO conversion. B.H. Davis and coworkers exposed a freshly reduced catalyst directly to a water vapor pressure equivalent to 50% conversion [56]. They observed oxidation of a fraction of the smaller cobalt crystallites when supported on alumina or activated carbon, and recommended that careful crystallite size management is required for a commercial catalyst. Kliewer *et al.* studied redox transformations of cobalt catalysts by TEM in terms of agglomeration of the metal, mixed-oxide formation with the support and reversible oxidation followed by reduction under mild hydrogen treatment [57]. They claim that reactor and TEM studies show that nanoscale Co crystallites can oxidize to CoO during commercially relevant FT synthesis conditions in spite of bulk thermodynamic data that suggest otherwise [58]. The propensity for oxidation is enhanced by small Co crystallites and high CO conversion with attendant high water vapor pressure and a high H_2O to CO ratio in the reactor. The oxide crystallites thus formed can be fully reduced by hydrogen at standard FT temperature and pressure. Reported TEM images from Exxon show that the oxidized cobalt metal wets the support surface and thereby facilitates contact between nearby crystallites. This has also been illustrated in a presentation from Statoil in Figure 1 [59]. Both images show a thickness of a cobalt oxide crystallite of *ca.* 5 nm wetting the surface. Further, the TEM image indicates mixed orientation of several cobalt oxide crystallites and an amorphous layer at the support interphase. Upon re-reduction metal crystallites may agglomerate depending on the initial spatial distribution on the catalyst support.

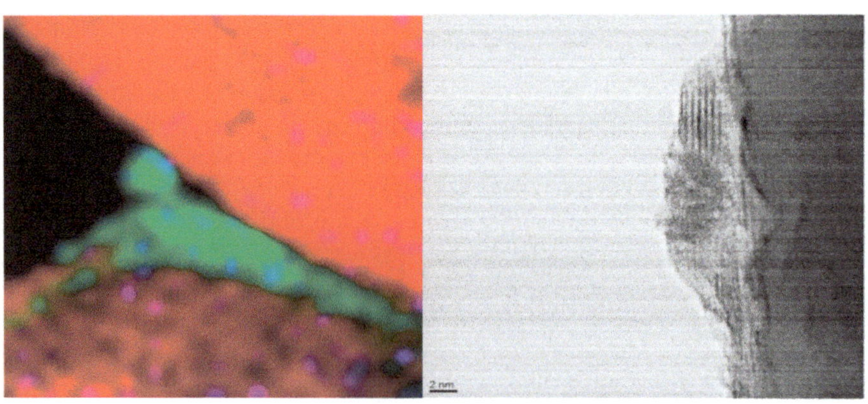

Figure 1: Wetting of cobalt oxide on a support. Left: EELS spectra; alumina: red; cobalt: green. Right: TEM.

That cobalt distribution can vary significantly with catalyst preparation procedure is illustrated in Figure 2 for a Co/Re/γ-alumina catalyst [43]. Cobalt dispersions are comparable, but we see that the distribution of clusters of the oxide varies significantly. Although it can be imagined that small well dispersed clusters are less prone to deactivation, this needs to be verified.

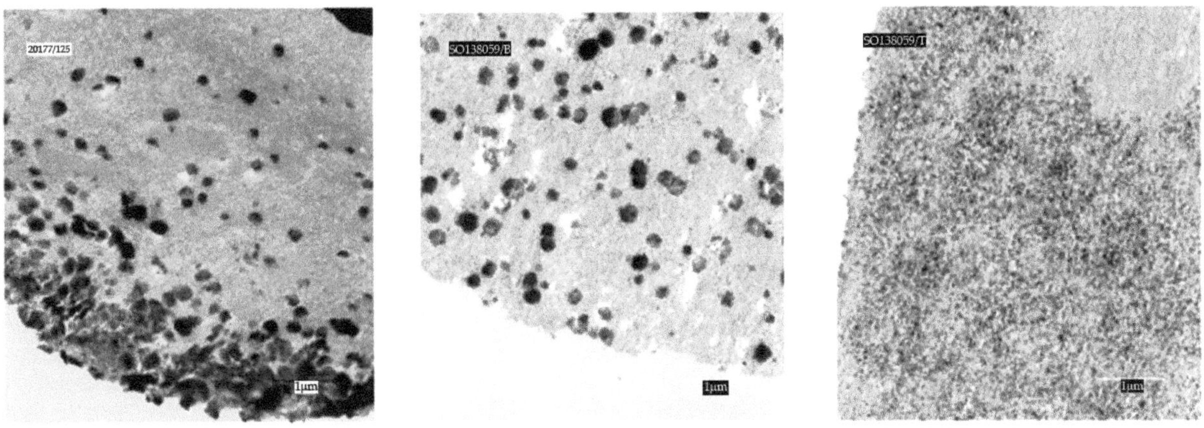

Figure 2: Cobalt oxide clusters on a γ-alumina. Bar shows 1 μm for all images.

It is well known that mixed-oxides can be formed between cobalt and silica, alumina and titania. In part, a surface layer of mixed-oxide is formed during catalyst preparation from water solution followed by drying and calcination. It has also been claimed that enhanced mixed oxide formation takes place during FT-reaction at high conversion levels (> 70%) concurrent with oxidation of cobalt to Co^{2+}. In the case of silica supported catalysts, well defined crystalline needles of cobalt silicate are rapidly formed at higher conversions [60].

Redox reactions of cobalt have been used deliberately to enhance the catalytic properties of FT catalysts by employing a reduction-oxidation-reduction (ROR) procedure [61,62]. An ROR treatment can increase Co dispersion presumably by forming hollow oxide domes during controlled oxidation that break up into smaller Co crystallites during re-reduction [63]. Whether the resulting Co crystals that will be in close proximity with each other experience agglomeration during the first months of commercial FT synthesis has apparently not been reported.

CATALYST REJUVENATION AND REGENERATION

Regeneration of cobalt LTFT synthesis catalysts is largely described in the patent literature. The options involve treatment of the catalyst with air (oxygen), hydrogen and/or CO and variations thereof in addition to procedures for removing produced wax. Therefore, regeneration addresses reversing the main deactivation processes of carbon deposition, metal oxidation and sintering by combustion, reduction and re-dispersion, respectively. A review of early reports on regeneration covering 1930–1952 has been presented [64]. It was concluded that there is no universal process for regenerating Co FT-catalysts; the art over this relatively short time period comprising conflicting results with patents covering a wide range of processes involving oxidation, reduction, combined oxidation-reduction, steam-reduction, operating at elevated temperatures and solvent extraction.

Commercial regeneration processes are either *in situ* in the FT-reactor itself or *ex situ* after removal of part of or the entire catalyst inventory. Indications are that Shell successfully regenerates their catalyst regularly, but it is undisclosed whether this takes place inside the tubes of the fixed-bed reactor or in a separate unit, whereas Sasol apparently removes part of the slurry from the reactor continuously for regeneration and re-deployment of the catalyst into the reactor. The latter approach allows continuous operation of the LTFT synthesis. Micro-channel reactors pose special challenges depending on the catalyst configuration. *In situ* regeneration is an option, or the catalyst can be removed for external treatment either by unloading the catalyst particles or removing multi-channel trays with catalyst attached.

A summary of regeneration procedures from some of the main industrial companies that are or recently have been involved in Fischer-Tropsch technology development is given in Table 3. The table is representative of available information but should not be assumed to be complete or up-to-date; nevertheless it illustrates what is probably the preferred approach of each individual company. In the first column the type of commercial regeneration process and primary FT-reactor type are listed and whether the regeneration is intermittent or continuous, while data in the other columns refer to the actual test protocol and results described in the patent literature. Note that reactor type can be different in columns one and three.

It is clearly possible to regenerate a deactivated catalyst to a level close to the original activity by different combinations of wax removal, hydrogenation and combustion of carbonaceous deposits; followed by re-reduction if needed. There is unfortunately little information available on the long-term performance of regenerated catalysts.

Sasol has published a brief summary of their procedure for removing most of the wax, followed by combustion of the remaining carbonaceous species and complete reactivation [65]. Specifically, Sasol describes dewaxing by hydrogenolysis with pure hydrogen for 2 h at 220 °C and reduction for 2 h at 350 °C [66]. After passivation with CO_2, the catalyst is subjected to oxidation with air in a fluidized-bed calcination unit at 250 °C for 6 h under a pressure preferably of *ca.* 10

bar. Re-reduction is performed at 425 °C and 98% of the original activity is regained. In cooperation with Eindhoven University, Sasol has investigated mechanisms of deactivation and regeneration for both model and commercial catalysts [67]. After a heptane wash and reoxidation, several FT cycles were demonstrated with no apparent permanent loss in activity. The oxidation step is described in terms of the Kirkendall effect involving spreading of a Co oxide film during oxidation followed by breaking up of the film to form small re-dispersed Co crystallites.

Table 3: Summary of regeneration concepts and procedures based on patents and presentations from LTFT technology companies.

Technology owner Regeneration configuration*	Catalyst TOS	FT test reactor	Wax removal step	Primary hydrogenation	Calcination/ Oxidation/ Re-dispersion	Regeneration effect (activity)
Sasol [66,68] Slurry; continuous *Ex situ*	Co/Pt/ alumina -	Slurry**	H_2 strip at 220 °C or xylene wash	H_2 at 350 °C	Air in fluid-bed	98%
Shell [69] Fixed-bed; intermittent *in situ*	Co/Mn/ Titania- Years	Fixed-bed Full scale	Gas oil wash	Diluted H_2	Diluted O_2 at 270 °C; NH_3/CO_2	> 100% < 100% (800 h TOS)
GTL.F1 [70] Slurry; continuous *Ex situ*	Co/Re/ aluminate -	Slurry**	Draining or cyclohexane wash	No	Diluted air in fluid-bed	98–115%
ExxonMobil [71] Slurry; continuous *Ex situ*	Co/Re/titania - A few days	Fixed-bed (lab)	Filtration + H_2 strip	No	Diluted O_2 in fixed-bed	~ 100%
Nippon Oil [72] Slurry; continuous *Ex situ*	Co/ zirconia-silica -	Slurry (lab)	"de-oiling"	No	Steaming at 200 °C; 25 barin fixed-bed	95%
ConocoPhillips [73] Slurry; continuous *Ex situ*	Co/Ru or Re/ (F-)alumina - 1014 h	Fixed-bed (lab) or slurry.**	No	7% H_2/steam at 300 °C/ 3 bar	No	80–95%
Syntroleum [74] Slurry; continuous *Ex situ*	Co/alumina - 4000 h	Slurry**	N_2 at 316–343 °C	No	Diluted O_2 in fluid-bed	"Good performance"
Oxford Catalyst/ Velocys [75] Micro-channel; *in situ* intermittent	Co/Re/ titania-silica?	Micro-channel	N_2 flow?	H_2 *in situ*	O_2 *in situ*	~ 100%

* Estimated. ** Size of slurry reactor is undisclosed, but the technology providers have had both lab and pilot/demo units in operation.

It appears that Shell is aiming at *in situ* regeneration in the tubes of their fixed-bed FT reactor, although an external post FT-reaction step is part of their described procedure [69]. Regeneration is based on a procedure that most likely comes from their Bintulu plant in Malaysia. After wax removal a mild hydrogenation and oxidation is conducted. The catalyst is then taken out of the FT-reactor, treated with a concentrated aqueous ammonia solution and subsequently with CO_2, giving Co ammonium carbonate. Most likely the latter procedure gives cobalt ammine carbonate complexes suitable for re-dispersing cobalt [76].

Regeneration may re-disperse cobalt, possibly even to a level higher than for a freshly reduced catalyst [70]. A promoted cobalt catalyst on a spinel support was subjected to an extended run in a slurry-bubble column reactor as described by Schanke *et al.* [14]. Actual time on stream was not reported, but samples from four different TOS's were analyzed and regenerated. Data from the patent have been plotted in Figure 3. All activity data are from standard runs in a laboratory fixed-bed reactor at 210 °C, 20 bar pressure and H_2/CO ratio of 2:1, and after a conventional reduction protocol with hydrogen at 350 °C. Fresh and used catalysts, all embedded in wax, were drained to remove excessive wax at 85 °C before activity testing. A successive reduction in activity to *ca.* 50% was experienced. If the spent catalyst was calcined at 300 °C to burn off excessive wax and carbon deposits before testing, the activity is only slightly reduced compared to the fresh catalyst. Interestingly, even an enhanced activity was experienced if the wax was removed by cyclohexane/n-heptane solvent extraction before calcination. Pore volumes and surface areas of freshly calcined catalysts and used catalysts subject to wax draining and oxidation are unchanged. From these data it appears that hydrogenation by itself has moderate or no effect on regeneration, but that burning off coke deposits completely rejuvenates the catalyst.

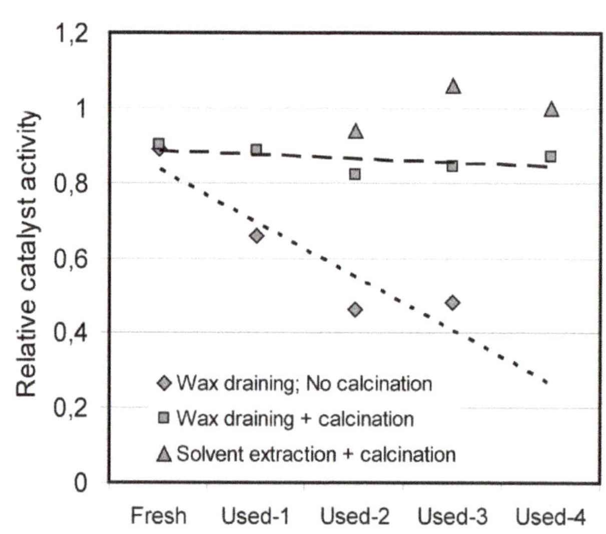

Figure 3: Catalyst activity before and after oxidation of carbon deposits of a catalyst used in a slurry bubble column reactor (data from ref. [70]).

In an early patent, Exxon researchers describe successful rejuvenation by atmospheric hydrogen treatment at typical FT synthesis temperatures of 200–230 °C of spent Co/Ru/titania catalysts [77]. The promoter is needed to facilitate the rejuvenation and is claimed also to inhibit carbon deposits [78]. However, examples are only for FT runs of 10–30 days at 50–60% CO conversion in lab-scale, fixed-bed reactors, and therefore may only addresses hydrogenolysis of very heavy waxes that partially block pores and possibly hydrogenation of oligomeric carbons on the surface of the catalyst. In a later patent, as shown in the table above, Exxon Mobil has demonstrated that reduction with hydrogen after oxidative regeneration can be performed in the slurry FT-reactor itself at mild process conditions of 200 °C and 20 bar, thus resembling the FT synthesis conditions. The company has a number of patents describing regeneration procedures, including adding more active metal after combustion of carbon deposits [79].

As pointed out above, - deposits of heavy hydrocarbon waxes reside in the pores of a used catalyst that should be largely removed before regeneration. Hydrogen treatment may reduce part of the wax through hydrogenation, but may also leave residual components at the surface. ConocoPhillips claims to have designed a suitable reactivation procedure that both removes heavy hydrocarbons and reduces the active metal [73]. For a 19 wt.% Co/0.1 wt.% Ru on alumina catalyst run for 40 days at standard FT-conditions, best regeneration results were obtained in a fixed-bed using a mixture of 93% steam/7% H_2at 300 °C and 3 bar where 95% of initial activity was regained. Similar experiments for a catalyst composition of 19 wt.% Co/0.1 wt.% Re on fluorinated alumina in a slurry reactor were less successful as 71% of initial activity was obtained, up from 33% after the FT period.

In their continuous regeneration of a spent slurry catalyst, Syntroleum focuses on removing as much wax as possible by nitrogen stripping at 300–350 °C and 3 bar, followed by calcination. Improved cobalt dispersion and reducibility are claimed, but no actual performance data are reported for the regenerated catalyst [74]. Nippon Oil has followed an alternative approach where they hydrothermally treat the spent catalyst with steam at elevated pressure [72]. Much of the activity is regained, although it appears that the TOS between regenerations is relatively short in view of the high activity level of *ca.* 75% before regeneration, thus moderating build-up of polymeric carbonaceous deposits.

Oxford Catalyst/Velocys have presented interesting long-term operation and regeneration performances for their micro-channel reactor [75]. Velocys' catalyst and reactor operate at much higher space velocities and productivities than conventional catalyst/large-scale FT reactors. To compensate for deactivation, temperature is increased from 205 °C to 232 °C after *ca.* 650 days TOS, accompanied by a slight increase in CO conversion from 71.2 to 73.6%. Naturally, the temperature increase is accompanied by a reduction in C_{5+} selectivity, in this case by as much as from 87.9% to 75.8%. Both activity and selectivity are fully restored after *in situ* hydrogen treatment followed by calcination with oxygen and re-reduction. The first step of hydrogenation probably re-reduces smaller cobalt crystallites and partly removes some wax and deposits. The hydrogenation step can be carried out under conditions resembling FT-synthesis and can thus be carried out at the plant location and with the catalyst loaded in the reactor. However, burning off the carbonaceous deposits is far more demanding and probably requires removing the catalyst from the reactor. These results are qualitatively in agreement with what can be expected from large scale fixed-bed and slurry bubble column operations although the rate of deactivation for the Velocys' catalyst is quantitatively smaller than those reported in available literature for large scale reactors. The alternative micro-channel LTFT provider, CompactGTL, appears to have only a regeneration patent directed at removing accumulated ammonia deposits [80].

CONCLUSIONS

An attempt to visualize the main catalyst deactivation mechanisms that impact FT catalyst activity during commercial operation is shown in Figure 4. The regions shown are approximate with respect to TOS and conversion, and there certainly will be some overlap. The following takes place:

In the initial phase, (a few weeks) there will be driving forces towards both an increase in activity by reduction as well as a decrease due to sintering. The net effect can be positive or negative and will to a large degree depend on the catalyst formulation, pretreatment and FT reactor environment. These mechanisms are well documented for slurry operations, but are less evident for fixed-bed and micro-reactors.

Sintering is favored by small Co crystallites and high conversion levels due to enhanced water activity, whereas reduction is facilitated by low conversion and water vapor pressure.

To a large extent oxidation of cobalt depends on operating conditions and may proceed at high conversion conditions, particularly for smaller cobalt crystals and may involve interaction with the support material.

Polycarbon deposition is the principal long-term cause of deactivation for all LTFT catalysts and reactor types since commercial catalysts are stabilized against oxidation by operation at realistic conversions, while reduction and sintering are short term phenomena which determine steady-state activity after just 1–2 months of TOS.

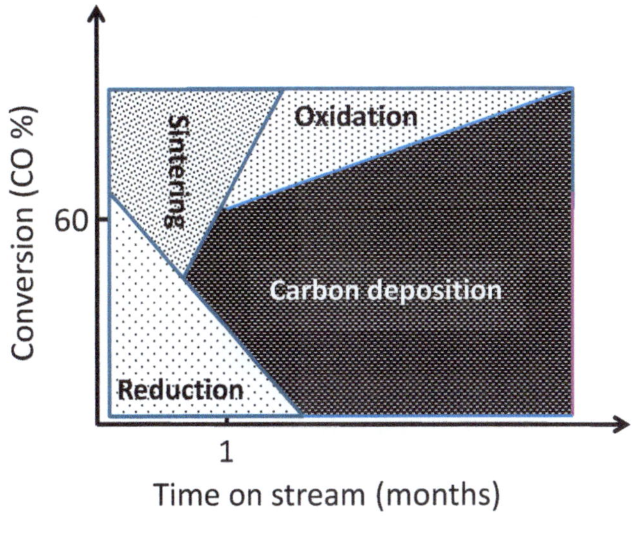

Figure 4: Main reactions taking place during equilibration and deactivation of Co Fischer-Tropsch catalysts

Although deactivation during the first 6–18 months of operation can cause 30–60% loss in activity in low-temperature Fischer-Tropsch synthesis with cobalt based catalysts, fortunately it appears possible to regenerate the catalyst to approach the activity of a freshly equilibrated catalyst. In a slurry reactor system a slip stream can be taken from the reactor for continuous rejuvenation. For a fixed-bed reactor, including micro-channel reactors, an intermittent *in situ* rejuvenation and*in situ* regeneration procedures can presumably be implemented as long as removal of wax and combustion of carbonaceous deposits can be controlled to prevent large temperature excursions. Nevertheless, based on patent literature, *ex situ*regeneration procedures appear to be the norm for most FT processes.

ACKNOWLEDGEMENT

The financial support from inGAP centre of research-based innovation, supported by the Norwegian research council under contract no. 174893, is acknowledged with gratitude.

REFERENCES AND NOTES

1. Tsakoumis, N.E.; Rønning, M.; Borg, Ø.; Rytter, E.; Holmen, A. Deactivation of cobalt based Fischer-Tropsch catalysts. A review. *Catal. Today* 2010, *154*, 162–182e.
2. Moodley, D.J.; van de Loosdrecht, J.; Saib, A.M.; Overett, M.J.; Datye, A.K.; Niemantsverdriet, J.W. Carbon deposition as a deactivation mechanism of cobalt-based Fischer–Tropsch synthesis catalysts under realistic conditions. *Appl. Catal.* 2009, *354*, 102–110.
3. Davis, B.H.; Iglesia, E. Technology Development for Iron and Cobalt Fischer-Tropsch Catalysts, Final Technical Report. Available online: http://www.fischer-tropsch.org/DOE/DOE_reports/40308/FC26-98FT40308-f/FC26-98FT40308-f_toc.htm (accessed on 13 March 2015).
4. Argyle, M.D.; Frost, T.S.; Bartholomew, C.H. Cobalt Fischer–Tropsch Catalyst Deactivation Modeled Using Generalized Power Law Expressions. *Top. Catal.* 2014, *57*, 415–429.
5. Bezemer, G.L.; Bitter, J.H.; Kuipers, H.P.C.E.; Oosterbeek, H.; Holewijn, J.E.; Xu, X.; Kapteijn, F.; van Dillen, A.J.; de Jong, K.P. Cobalt particle size effects in the Fischer-Tropsch reaction studied with carbon nanofiber supported catalysts. *J. Am. Chem. Soc.* 2006, *128*, 3956–3964.
6. Borg, Ø. Role of Alumina Support in Cobalt Fischer-Tropsch Synthesis. Ph.D. Theses, NTNU, Norway, April 2007.
7. Diehl, F.; Khodakov, A.Y. Promotion of Cobalt Fischer-Tropsch Catalysts with Noble Metals: A Review. *Oil Gas Sci. Technol. Rev. IFP.* 2009, *64*, 11–24.
8. Ma, W.; Jacobs, G.; Keogh, R.A.; Bukur, D.B.; Davis, B.H. Fischer-Tropsch synthesis: Effect of Pd, Pt, Re, and Ru noble metal promoters on the activity and selectivity of a 25%Co/Al_2O_3 catalyst. *Appl. Catal. A* 2012, *437–438*, 1–9.
9. Information could be found at http://www.metal-pages.com/metalprices/rhenium/ and http://www.metal-pages.com/metals/platinum/metal-prices-news-information/.
10. Ducreux, O.; Rebours, B.; Lynch, J.; Roy-Auberger, M.; Bazin, D. Microstructure of Supported Cobalt Fischer-Tropsch Catalysts. *Oil Gas Sci. Technol. Rev. IFP.* 2009, *64*, 49–62.
11. Blekkan, E.A.; Borg, Ø.; Frøseth, V.; Holmen, A. Fischer-Tropsch synthesis on cobalt catalysts: The effect of water. In*Catalysis*; Royal Society of Chemistry: Cambridge, UK, 2007; Volume 20, pp. 1–21e.
12. Rytter, E.; Eri, S.; Skagseth, T.H.; Schanke, D.; Bergene, E.; Myrstad, R.; Lindvåg, A. Catalyst Particle Size of Cobalt/Rhenium on Porous Alumina and the Effect on Fischer–Tropsch Catalytic Performance. *Ind. Eng. Chem. Res.*2007, *46*, 9032–9036.
13. Zennaro, R. The ENI-IFP/Axens GTL technology for the conversion of natural gas. In Proceedings of 6th World GTL Summit, London, UK, 17–19 May 2006.
14. Schanke, D.; Wagner, M.; Taylor, P. *Proceedings of 1st annual gas processing symposium*; Doha, Qatar, 10–12 January 2009, Alfadala, H., Reklaitis, G.V.R., El-Halwagi, M.M., Eds.; Elsevier: Amsterdam, The Netherlands, 2009.
15. Guettel, R.; Turek, T. Comparison of different reactor types for low temperature Fischer–Tropsch synthesis: A simulation. *Chem. Eng. Sci.* 2009, *64*, 955–964.
16. LeViness, S. presentation, Energy Frontiers International, 22nd October 2012. Available online: http://www.velocys.com/arcv/press/ppt/EFI%202012%20Presentation%20121019_1_rev2.pdf (accessed on 3 December 2014).
17. Steynberg, A.P.; Dry, M.E.; Davis, B.H.; Breman, B.B. Fischer-Tropsch Reactors. *Stud. Surf. Sci. Catal.* 2004, *152*, 64–195.
18. Schanke, D.; Lian, P.; Eri, S.; Rytter, E.; Sannæs, B.H.; Kinnari, K.J. Optimisation of Fischer-Tropsch Reactor Design and Operation in GTL Plants. *Stud. Surf. Sci. Catal.* 2001, *136*, 239–244.
19. Myrstad, R.; Eri, S.; Pfeifer, P.; Rytter, E.; Holmen, A. Fischer-Tropsch synthesis in a microstructured reactor. *Catal. Today* 2009, *147S*, S301–S304.
20. Holmen, A.; Myrstad, R.; Venvik, H.J.; Zhu, J.; Chen, D. Monolithic, microchannel and carbon nanofibers/carbon felt reactors for syngas conversion by Fischer-Tropsch synthesis. *Catal. Today* 2013, *216*, 150–157.
21. Yang, J.; Eiras, S.B.; Myrstad, R.; Pfeifer, P.; Venvik, H.J.; Holmen, A. Fischer-Tropsch synthesis over high-loading Co-based catalysts in a microreactor. *Prep. ACS Div. Energy Fuels* 2014, *59*, 828–830.
22. Rytter, E.; Ochoa-Fernández, E.; Fahmi, A. *Catalytic Process Development for Renewable Materials*; Imhof, P., van der Waal, J.-K., Eds.; Wiley: Weinheim, Germany, 2013; pp. 265–308.
23. Creyghton, E.J.; Mesters, C.M.A.M.; Reynhout, M.J.; Verbist, G.L.M.M. Process for preparing a catalyst. Patent WO2008071640 A2, June 2008.

24. Freide, J.J.H.M.F.; Collins, J.P.; Nay, B.; Sharp, C. A history of the BP Fischer-Tropsch catalyst from laboratory to full scale demonstration in Alaska and beyond. *Stud. Surf. Sci. Catal.* 2007, *163*, 37–44.

25. Van Berge, P.J.; Barradas, S. Catalysts. U.S. Patent 7365040 B2, March 2008.

26. Dogterom, R.J.; Mesters, C.M.A.M.; Reynhout, M.J. Process for preparing a hydrocarbon synthesis catalyst. Patent WO2007068731 A1, June 2007.

27. Rytter, E.; Skagseth, T.H.; Wigum, H.; Sincadu, N. Enhanced strength of alumina based Co Fischer-Tropsch catalyst. Patent WO 2005072866, August 2005.

28. Bellussi, G.; Carluccio, L.C.; Zennaro, R.; del Piero, G. Process for the preparation of Fischer-Tropsch catalysts with a high mechanical, thermal and chemical stability. Patent WO2007009680 A1, January 2007.

29. Ikeda, M.; Waku, T.; Aoki, N. Catalyst for Fisher-Tropsch synthesis and method for producing hydrocarbon. Patent WO2005099897 A1, October 2005.

30. Inga, J.; Kennedy, P.; LeViness, S. Fischer-Tropsch process in the presence of nitrogen contaminants. Patent WO2005071044, August 2005.

31. Soled, S.L.; Fiato, R.A.; Iglesia, E. Cobalt-ruthenium catalyst for Fischer-Tropsch synthesis. Eur. Pat. Appl. 87310896, June 1989.

32. Srinivasan, N.; Espinoza, R.L.; Coy, K.L.; Jothimurugesan, K. Fischer-Tropsch catalysts using multiple precursors. U.S. Patent 6822008 B2, November 2004.

33. Compact GTL. The modular gas solution. Available online: http://www.compactgtl.com (accessed on 24 February 2015).

34. Velocys. Think Smaller. Available online: http://www.velocys.com/resources/Velocys_Booklet.pdf (accessed on 24 February 2015).

35. Kliewer, C.E.; Soled, S.L.; Kiss, G. Characterizing Intrinsic Deactivation in Cobalt-Catalyzed Fischer-Tropsch Synthesis. Available online: http:///www.nycsweb.org/files/37637807.pdf (accessed on 24 February 2015).

36. Kliewer, C.E.; Kiss, G.; Soled, S.L. Characterizing Intrinsic Deactivation in Cobalt-Catalyzed Fischer-Tropsch Synthesis. *Microsc. Microanal.* 2010, *16*, 1258–1259.

37. Borg, Ø.; Hammer, N.; Enger, B.C.; Myrstad, R.; Lindvåg, O.A.; Eri, S.; Skagseth, T.H.; Rytter, E. Effect of biomass-derived synthesis gas impurity elements on cobalt Fischer–Tropsch catalyst performance including *in situ* sulphur and nitrogen addition. *J. Catal.* 2011, *279*, 163–173.

38. Balonek, C.M.; Rane, A.H.L.S.; Schmidt, E.R.L.; Holmen, A. Effect of alkali metal impurities on Co-Re catalysts for Fischer-Tropsch synthesis from biomass-derived syngas. *Catal. Lett.* 2010, *138*, 8–13.

39. Lillebø, A.H. Conversion of biomass derived synthesis gas into liquid fuels via the Fischer-Tropsch synthesis process: Effect of alkali and alkaline earth metal impurities and CO conversion levels on cobalt based catalysts. Ph.D. Thesis, NTNU, Norway, September 2014.

40. Van Berge, P.J.; van de Loosdrecht, J.; Barradas, S. Method of treating an untreated catalyst support and forming a catalyst precursor and catalyst from the teated support. U.S. Patent 6875720 B2, October 2005.

41. Rytter, E.; Eri, S.; Schanke, D.; Wigum, H.; Skagseth, T.H.; Borg, Ø.; Bergene, E. Development of an Attrition Resistant Fischer–Tropsch Catalyst for Slurry Operation. *Topics Catal.* 2011, *54*, 801–810.

42. Behrmann, W.C.; Davis, S.M.; Mauldin, C.H. Catalyst preparation. Eur. Patent 535790 A1, August 1992.

43. Rytter, E.; Eri, S.; Schanke, D.; Wigum, H.; Skagseth, T.H.; Sincadu, N.; Bergene, E. Proceedings of 9th Novel Gas Conversion Symposium. Lyon, France, 30 May–3 June 2010.

44. Rytter, E.; Schanke, D.; Eri, S.; Wigum, H.; Skagseth, T.H.; Lian, P.; Sincadu, N. Optimization of Statoil's Fischer-Tropsch Co/Re/alumina catalyst. *Prep. ACS Div. Petro. Chem.* 2004, *49*, 182–183.

45. Tsakoumis, N.E.; Dehghan-Niri, R.; Rønning, M.; Walmsley, J.C.; Borg, Ø.; Rytter, E.; Holmen, A. X-ray absorption, X-ray diffraction and electron microscopy study of spent cobalt based catalyst in semi-commercial scale Fischer–Tropsch synthesis. *Appl. Catal. A* 2014, *479*, 59–89.

46. Bartholomew, C.H. Sintering kinetics of supported metals: New perspectives from a unifying GPLE treatment. *Appl. Catal. A* 1993, *107*, 1–57.

47. Lee, D.; Lee, J.-H.; Ihm, A.-K. Effect of Carbon Deposits on Carbon Monoxide Hydrogenation over Alumina Supported Cobalt Catalyst. *Appl. Catal.* 1988, *36*, 199–207.

48. Swart, J.C.W.; van Steen, E.; Ciobícã, I.M.; van Santen, R.A. Interaction of graphene with FCC-Co(111). *Phys. Chem. Chem. Phys.* 2009, *11*, 803–807.

49. Tan, K.F.; Xu, J.; Chang, J.; Borgna, A.; Saeys, M.A. Carbon deposition on Co catalysts during Fischer–Tropsch synthesis: A computational and experimental study. *J. Catal.* 2010, *274*, 121–129.

50. Weststrate, C.J.; Kızılkaya, A.C.; Rossen, E.T.R.; Verhoeven, M.W.G.M.; Ciobîcă, I.M.; Saib, A.M.; Niemantsverdriet, J.W. Atomic and Polymeric Carbon on Co(0001): Surface Reconstruction, Graphene Formation, and Catalyst Poisoning. *J. Phys. Chem. C* 2012, *116*, 11575–11583.

51. Gruver, V.; Young, R.; Engman, J.; Robota, H.J. The Role of Accumulated Carbon in Deactivating Cobalt Catalysts during FT Synthesis in a Slurry-Bubble-Column Reactor. *Prep. Am. Chem. Soc., Div. Petrol. Chem.* 2005, *50*, 164–169.

52. Freide, J.J.H.M.F.; Gamlin, T.D.; Hensman, J.R.; Nay, B.; Sharp, C. Deactivation of Cobalt-Based Catalyst in the F-T Process. *J. Nat. Gas. Chem.* 2004, *13*, 1–9.

53. Schanke, D.; Hilmen, A.M.; Bergene, E.; Kinnari, K.; Rytter, E.; Ådnanes, E.; Holmen, A. Study of the deactivation mechanism of Al_2O_3-supported cobalt Fischer-Tropsch catalyst. *Catal. Lett.* 1995, *34*, 269–284.

54. Li, J.; Zhan, X.; Zhang, Y.; Jacobs, G.; Das, T.; Davis, B.H. Fischer–Tropsch synthesis: effect of water on the deactivation of Pt promoted Co/Al_2O_3 catalysts. *Appl. Catal. A* 2002, *228*, 203–212.
55. Lancelot, C.; Ordomsky, V.V.; Stéphan, O.; Sadeqzadeh, M.; Karaca, H.; Lecroix, M.; Curulla-Ferré, D.; Luck, F.; Fongarland, P.; Griboval-Constant, A.; Kodakov, A.Y. Direct Evidence of Surface Oxidation of Cobalt Nanoparticles in Alumina-Supported Catalysts for Fischer-Tropsch Synthesis. *ACS Catal.* 2014, *4*, 4510–4515.
56. Jermwongratanachai, T.; Jacobs, G.; Shafer, W.D.; Ma, W.; Pendyala, V.R.R.; Davis, B.H.; Kitiyanan, B.; Khalid, S.; Cronauer, D.C.; Kropf, A.J.; Marshall, C.L. Fischer–Tropsch Synthesis: Oxidation of a Fraction of Cobalt Crystallites in Research Catalysts at the Onset of FT at Partial Pressures Mimicking 50% CO Conversion. *Top. Catal.* 2014, *57*, 479–490.
57. Kliewer, C.E.; Kiss, G.; DeMartin, G.J. Ex situ transmission electron microscopy: A fixed-bed approach. *Microsc. Microanal.* 2006, *12*, 135–144.
58. Kubaschewski, O.; Alcock, C.B. *Metallurgical Thermochemistry*, 5th ed.; Pergaman: New York, NY, USA, 1979.
59. Rytter, E.; Schanke, D.; Eri, S.; Wigum, H.; Skagseth, T.H.; Bergene, E. Some QA and optimization issues during development of the Statoil FT-catalyst. Available online: http://www.globalspec.com/reference/43550/203279/qa-and-optimization-issues-during-development-of-the-statoil-ft-catalyst (accessed on 24 February 2015).
60. Kiss, G.; Kliever, C.E.; DeMartin, G.J.; Culross, C.C.; Baumgartner, J.E. Hydrothermal deactivation of silica-supported cobalt catalysts in Fischer–Tropsch synthesis. *J. Catal.* 2003, *217*, 127–140.
61. Eri, S.; Goodwin, J.G., Jr.; Marcelin, G.; Riis, T. Catalyst for Production of Hydrocarbons. U.S. Patent 4801573, January 1987.
62. Kobylinski, T.P.; Kibby, C.L.; Pannell, R.B.; Leigh, E.L. Synthesis Gas Conversion using ROR Activated Catalyst. Eur. Patent 253924, July 1988.
63. Ernst, B.; Bensaddik, A.; Hilaire, L.; Chaumette, P.; Kiennemann, A. Study on a cobalt silica catalyst during reduction and Fischer-Tropsch reaction: In situ EXAFS compared to XPS and XRD. *Catal. Today* 1998, *39*, 329–341.
64. Arcuri, K.B.; LeViness, S.C. The Regeneration of Hydrocarbon Synthesis Catalyst—A Partial Review of the Related Art Published during 1930 to 1952. Available online: http://www.fischer-tropsch.org/primary_documents/presentations/AIChE%202003%20Spring%20National%20Meeting/Paper%2086a%20Arcuri%20Regeneration.pdf (accessed on 24 February 2015).
65. Weststrate, C.; Hauman, M.; Moodley, D.; Saib, A.; Steen, E.; Niemantsverdriet, J. Cobalt Fischer–Tropsch Catalyst Regeneration: The Crucial Role of the Kirkendall Effect for Cobalt Redispersion. *Topics Catal.* 2011, *54*, 811–815.
66. Van de Loosdrecht, J.; Saib, A.M. Catalysts. Patent WO 2008/139407, November 2008.
67. Saib, A.M.; Gauché, J.L.; Weststrate, C.J.; Gibson, P.; Boshoff, J.H.; Moodley, D.J. Fundamental Science of Cobalt Catalyst Oxidation and Reduction Applied to the Development of a Commercial Fischer-Tropsch Regeneration Process. *Ind. Eng. Chem. Res.* 2014, *53*, 1816–1824.
68. Van de Loosdrecht, J.; Booysen, W.A. Catalysts. U.S. Patent 2011/0245355, June 2011.
69. Bezemer, G.L.; Nkrumah, S.; Smits, J.T.M. Process for Regenerating a Catalyst. U.S. Patent 2012/0165417, December 2011.
70. Rytter, E.; Eri, S.; Skagseth, T.H.; Borg, Ø. Fischer-Tropsch Catalyst Regeneration. U.S. Patent 0210939, August 2013.
71. Soled, S.L.; Baumgartner, J.E.; Kiss, G. *In situ* catalyst regeneration/activation process. U.S. Patent 6900151, November 2003.
72. Ono, H.; Nagayasu, Y.; Hayasaka, K. Method for Manufacturing a Regenerated Fischer-Tropsch Synthesis Catalyst and Hydrocarbon Manufacturing Method. U.S. Patent 2012/0322899, December 2012.
73. Wright, H.A.; Raje, A.P.; Espinoza, R.L. Pressure Swing Catalyst Reneration Procedure for Fischer-Tropsch Catalyst. U.S. Patent 6962947, November 2005.
74. Huang, J.-R.; Arcuri, K.; Agee, K.; Schubert, P.F. Process for Regenerating a Slurry Fischer-Tropsch Catalyst. U.S. Patent 6989403, April 2004.
75. LeViness, S. 245th ACS National Meeting, April 2013, New Orleans. Available online: http://www.velocys.com/arcv/press/ppt/AIChE%202013%20Presentation%2004–30–13%20-%20LeViness%20v1_rev1.pdf (accessed on 10 March 2105).
76. Bailey, S.; Gray, G.; Lok, M.C. Methods for the production of cobalt catalysts supported on silicon dioxide and their use. Patent WO 2001/062381, August 2001.
77. Iglesia, E.; Soled, S.L.; Fiato, R.A. Cobalt-ruthenium catalysts for Fischer-Tropsch synthesis and process for their preparation. U.S. Patent 4738948, April 1988.
78. Iglesia, E.; Soled, S.L.; Fiato, R.A.; Via, G.H. Bimetallic Synergy in Cobalt Ruthenium Fischer-Tropsch Synthesis Catalysts. *J. Catal.* 1993, *143*, 345–368.
79. Daage, M.; Koveal, R.J.; Chang, M. Catalyst Regeneration. U.S. Patent 6800579, January 2002.
80. Minnie, O.R.; Ndifor, E.N.; Kavian, S.; Smith, B.; Knowles, R. Process for the regeneration of a Fischer-Tropsch Catalyst. Patent WO 2013/093423, June 2013.

CHAPTER 3

STRUCTURED PEROVSKITE-BASED CATALYSTS AND THEIR APPLICATION

Sylvain Keav, Santhosh Kumar Matam, Davide Ferri and Anke Weidenkaff

Keywords: Perovskite-type oxides; honeycombs; structured catalysts; three-way catalysts; CO oxidation; hydrocarbon combustion; NO_x reduction; self-regenerative properties; washcoat

INTRODUCTION

Unburnt hydrocarbons (UHC), carbon monoxide (CO), and nitrogen oxides (NO_x) are the major noxious gases present in the exhaust of internal combustion engines. In order to comply with the stringent emission levels imposed by the legislation, these compounds have to be eliminated before they are released into the surrounding environment. For this purpose, gasoline-powered vehicles are equipped with a Three-Way Catalytic (TWC) converter. This emission control device, constituted of a mixture of catalytically active components deposited onto a monolithic honeycomb substrate, allows the simultaneous conversion of UHC, CO, and NO_x into harmless CO_2, H_2O, and N_2 (Equations (1)–(4)).

$$UHC + O_2 \rightarrow CO_2 + H_2O \quad (1)$$

$$2CO + O_2 \rightarrow 2CO_2 \quad (2)$$

$$UHC + NO \rightarrow \frac{1}{2}N_2 + CO_2 + H_2O' \quad (3)$$

$$CO + NO \rightarrow \frac{1}{2}N_2 + CO_2 \quad (4)$$

Because oxidation and reduction reactions must occur simultaneously, TWC are most efficient when operated under conditions close to the stoichiometric point (*i.e.*, air-to-fuel ratio of 14.7 *w/w* for gasoline vehicles). The composition of the exhaust is continuously determined by an oxygen sensor and the air-to-fuel ratio is adjusted accordingly by the fuel-control feedback system. A time lag between the analysis and the adjustment produces slight redox fluctuations of the exhaust gas composition around the stoichiometry [1].

The current TWC technology used in gasoline vehicles is based on noble metals. Platinum and palladium are used as oxidation components. Both can also contribute to NO_x reduction but are much less efficient than rhodium. The performance of the noble metals is increased by dispersion on a high specific surface area (*ca.* 100 $m^2 \cdot g^{-1}$) γ-Al_2O_3 support. Ceria-zirconia, a high Oxygen Storage Capacity (OSC) material was introduced to minimize the air-to-fuel ratio fluctuations in the exhaust [2]. Other common additives include BaO (NO_x trap), zeolites (UHC or NO_x adsorbent), La_2O_3 (thermal stability enhancer for γ-Al_2O_3), TiO_2, NiO and SiO_2. Natural gas vehicles (NGV) generate exhaust emissions containing higher concentrations of methane, a potent greenhouse gas [3] and the most difficult hydrocarbon to oxidize, with respect to their gasoline counterparts. For this reason, TWC for NGV are mainly based on Pd, the most active catalyst for methane oxidation, and contain a threefold higher noble metal loading (*ca.* 3 $mg \cdot cm^{-3}$) than their gasoline equivalents (*ca.* 1 $mg \cdot cm^{-3}$) [4].

TWC are more exposed to high temperatures (up to 1000 °C) than diesel oxidation catalysts as a consequence of their vicinity to the engine or of the higher fuel load during operation. Consequently, thermal ageing can cause noble metal sintering processes or diffusion into the alumina support with mileage [5]. Because of the progressive loss of catalytic activity upon ageing and of the price and scarcity of noble metals, there is a need to further develop this technology for limiting their amount in the catalytic converter and for increasing their tolerance to the harsh operating conditions.

Already, in 1971, $LaCoO_3$ perovskite-type oxide was proposed as an economic substitute for noble metal-based exhaust after treatment catalysts [6]. Co-, Fe-, Mn-, and Ru-based perovskite-type oxides exhibited promising activity for

Citation: Sylvain Keav, Santhosh Kumar Matam, Davide Ferri and Anke Weidenkaff, Structured Perovskite-Based Catalysts and Their Application, doi:10.3390/catal4030226

the oxidation of CO and the reduction of NO_x [7,8,9,10,11]. $La_{0.8}Sr_{0.2}CoO_3$ was shown to compare well with 1 wt% Pt/Al_2O_3 for hydrocarbon oxidation [12]. The promising oxygen storage capacity of $LaMn_{0.976}Rh_{0.024}O_{3+\delta}$ was also pointed out [13]. However, the high levels of sulfur in the fuels greatly limited their use in real applications. Recently, the use of $LaFe_{0.95}Pd_{0.05}O_3$ [14] for the exhaust after treatment of gasoline engines and the demonstration that $La_{0.9}Sr_{0.1}CoO_3$ can rival with noble metal-based catalysts for the treatment of diesel vehicle exhaust [15] confirmed the suitability of and rejuvenated the interest for this class of catalysts. Over the past several years, a continuous effort was undertaken to understand and improve the catalytic properties of perovskite-type oxides in automotive applications. In this review, we briefly describe the redox properties of perovskite-type oxides and the relevant mechanisms involved in oxidation and reduction reactions peculiar of TWC operation. Because of our interest in catalytic converters for NGV applications, the main part of the review is dedicated to the practical aspects of the use of structured perovskite-type oxide catalysts with a special emphasis on TWC. The advantages and limits will be discussed with the aim of identifying the most suitable types of catalysts for automotive applications.

PEROVSKITE-TYPE OXIDES

Perovskite-type oxides with the general formula ABO_3 can crystallise in cubic structure with a space group Pmm or in distorted orthorhombic, rhombohedral, tetragonal, monoclinic or triclinic symmetry. The content of oxygen and defects can vary depending on the composition due to a large stability range of the structure. The larger A-site cation is often a rare earth (La, Sm, Pr), an alkaline earth (Sr, Ba, Ca), or an alkali (Na, K) metal cation coordinated to 12 oxygen anions. The B-site cation is typically a smaller transition metal cation occupying octahedral interstices in the oxygen framework. The stability of the perovskite phases is largely governed by geometric considerations that are best summarized by the Goldschmidt tolerance factor, t (Equation (5)):

$$t = \frac{(r_A + r_O)}{\sqrt{2}(r_B + r_O)} \quad (5)$$

where r_A, r_B, and r_O are the respective radii of A, B, and oxygen ions. The value of the Goldschmidt tolerance factor must be in the range from 0.75 to 1.00 in order to obtain the perovskite structure [16]. Several combinations of A- and B-site cations can form a stable perovskite-like structure. Additionally, A- and B-site cations, as well as oxygen anions can be partially substituted by other suitable elements [17,18]. Partial substitution allows the control of the valence state of A- and B-site cations as well as the non-stoichiometry (cationic or anionic vacancies) in the mixed oxide based on electroneutrality arguments. The presence of a multiplicity of oxidation states is the major responsible for the catalytic properties of perovskite-type catalysts [19,20]. Due to this wide diversity, perovskite-type oxides also offer a large array of properties [21] (e.g., ferroelectricity, piezzoelectricity, pyroelectricity, thermolectricity [22], magnetism, superconductivity), which can be finely tuned for intended applications. For these reasons, more than 2000 scientific papers are devoted to perovskite-type materials every year and several review articles specifically focus on their preparation, structure elucidation, and applications [23,24,25,26,27].

Perovskite-type oxides possess adsorption, acid-base, as well as redox properties, which lead to attractive catalytic properties [25]. The very first studies concerning their use as catalysts were conducted by Parravano *et al.* [28] and Dickens *et al.* [29] in 1952 and 1965, respectively. Since then, they were evaluated in various catalytic (e.g., total or partial oxidation of hydrocarbons, oxygenated compounds or halocompounds; hydrogenation of CO or CO_2; hydrogenolysis of hydrocarbons), photocatalytic and electrocatalytic processes [25].

Parameters Influencing the Catalytic Activity

The catalytic activity of perovskite-type oxides is largely controlled by the nature of the B-site cation. Amongst noble metal-free perovskites, those based on Co and Mn perform the best in the oxidation of hydrocarbons and CO (Figure 1) [30,31,32]. The role of the A-site cation was found to be less decisive in the studied systems. Nitadori *et al.* [31], as well as Panich *et al.* [32], did not observe significant activity differences between Y-, La-, Pr-, Nd-, Sm-, Eu-, and Gd-based cobaltates tested in the oxidation of propane, methanol, CO, ethylbenzene, and propylene (Figure 1). However, partial substitution at the A-site by a cation of different valence (e.g., substitution of La^{3+} by Sr^{2+} or K^+) can form oxygen vacancies and/or change the oxidation state of the B-site cation, which enhances substancially the catalytic properties of the material [31,33]. Detailed review concerning the effects of substitution on the CO oxidation activity of perovskite-type oxides can be found in reference [27].

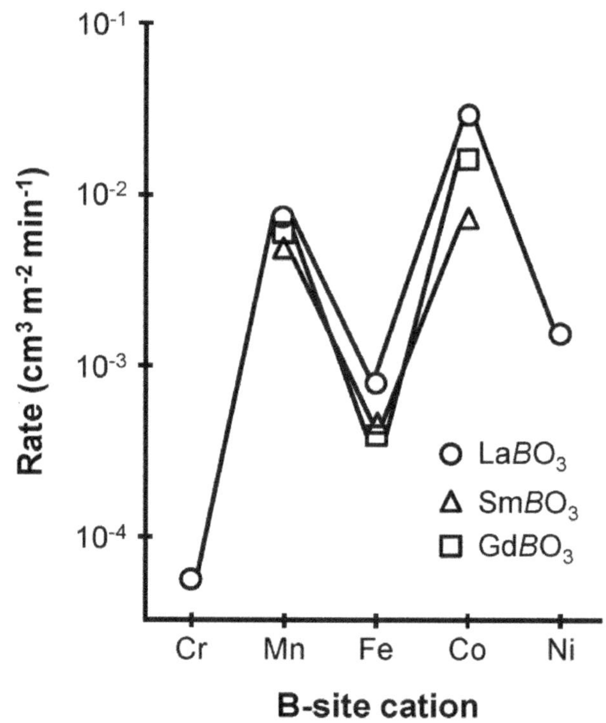

Figure 1: Influence of A- and B-site cations on the catalytic activity of perovskite-type oxides for propane oxidation at 227 °C (Remark: B in $LaBO_3$, $SmBO_3$ and $GdBO_3$ refers to the B-site metal cation and not to boron element). Adapted with permission from Nitadori *et al.* [31]. Copyright 1988, The Chemical Society of Japan.

The performance does not only depend on the phase composition of the mixed oxide but also on the preparation procedure and its parameters, as these affect the crystallinity, texture and morphology of the final material [34,35,36,37,38,39]. For example, the TWC activity of $La_{0.7}Sr_{0.3}Mn_{0.7}Cu_{0.3}O_{3-\delta}$ prepared by complexation of nitrate salts in the presence of citric acid and ammonia is strongly affected by the acid-to-ammonia ratio employed for the synthesis [40] since it affects the polymerization of the precursor molecules. The catalytic performance of perovskite-type oxides is well-known to increase with their Specific Surface Area (SSA) [11,23] which can often be very low. The main reason for this is that perovskite-type oxides are essentially high temperature phases and most of their preparation procedures require heating up at relatively high temperature. Solution chemistry methods (sol-gel, freeze-drying, spray-drying, co-precipitation, *etc.*) can generally produce highly crystalline materials at lower temperature (700–800 °C) than solid-state reaction (900–1000 °C). Consequently, materials prepared by such means are usually preferred for catalytic applications because they have a larger SSA (up to 30 $m^2 \cdot g^{-1}$ compared to less than 2 $m^2 \cdot g^1$). As an exception to this rule, a solid state reaction method designated as reactive grinding, which was developpped by the group of Kaliaguine in the late 1990s [41,42,43,44,45], proved to be efficient in producing perovskite-type oxides with high SSA. This low-temperature mechano-synthesis procedure consists in the high-energy ball-milling of oxide precursors for several hours at a temperature lower than 40 °C. The so-formed low SSA perovskite-type phase is subsequently ball-milled in the presence of a non-reactive additive (*i.e.*, ZnO) which is leached at the end of the synthesis [46] to obtain a high SSA (sometimes above 100 $m^2 \cdot g^{-1}$ [42]) material.

Reaction Mechanisms

The high TWC performance of perovskite-type oxides is related to their oxygen mobility and storage capacity. Certain compounds, such as $LaMn_{1-x}Ni_xO_3$, even demonstrate significantly higher OSC than commercial ceria-zirconia [47]. Oxygen uptake and release is associated with the existence of structural defects and the change of the oxidation state of the B-site cation. Some perovskite-type materials are employed as supports for high redox potential oxides which are chemically incompatible with conventional γ-Al_2O_3. For example, MnO_x/$LaAlO_3$ is characterized by a promisingly high OSC [48]. Because two types of catalytically active oxygen species co-exist in a perovskite-type oxide (Figure 2) [49], two redox mechanisms have been proposed. Suprafacial oxygen species (α) are adsorbed on the oxide surface and are available at low temperature, generally below 600 °C. At higher temperature, bulk oxygen from the lattice, referred to as intrafacial oxygen (β), is activated and takes part in the catalytic reaction according to a Mars-van-Krevelen mechanism. In both cases, dissociation of molecular oxygen (or NO) from the gas phase must replenish oxygen species from the material, which are consumed during the reaction, so that the latter can proceed [13].

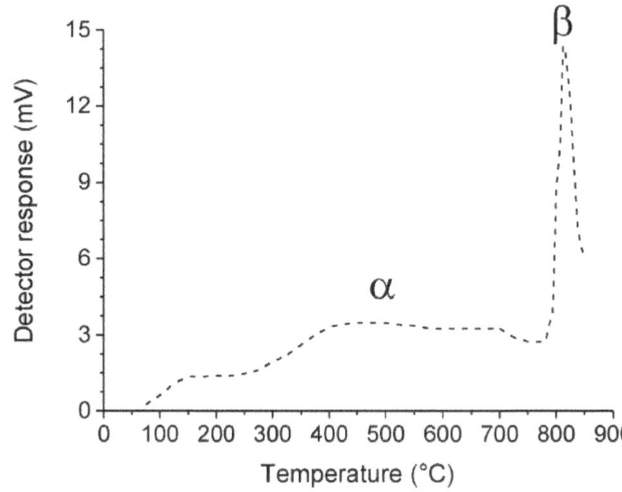

Figure 2: Oxygen temperature-programmed desorption on $La_{0.6}Sr_{0.4}CoO_3$. The broad low-temperature desorption peak corresponds to α-oxygen species and the sharp high-temperature one to β-oxygen species. Adapted with permission from Yamazoe *et al.*[49]. Copyright 1981, The Chemical Society of Japan.

Oxidation of hydrocarbons by molecular oxygen is governed either by a suprafacial or an intrafacial mechanism. The combustion of propylene and iso-butene on $LaBO_3$ (B = Cr, Mn, Fe, Co, and Ni) at 250 °C occurs through a suprafacial process [50]. Methane oxidation takes place according to a suprafacial mechanism below 800 °C and according to an intrafacial one at higher temperature [51]. The combustion of toluene on $LaFeO_3$ and $YFeO_3$ is probably a suprafacial reaction, although the scenario of an intrafacial mechanism cannot be ruled out in the case of the latter oxide [52].

CO oxidation by molecular oxygen is recognized to occur through a suprafacial process [26,53,54]. The reaction (Equations (8)–(10)) takes place between oxygen species dissociatively adsorbed on surface B metal cations (Equation (6)) and CO species non-dissociatively adsorbed on O^{2-} anions (Equation (7)) [55,56]. In agreement with its limited influence on activity, the A-site cation does not seem to be involved in adsorption and reaction.

$$O_{2(g)} \rightarrow O_{2(ads)} \rightarrow 2O_{(ads)} \quad (6)$$

$$CO_{(g)} \rightarrow CO_{(ads)} \quad (7)$$

$$CO_{(ads)} + 2O_{(ads)} \rightarrow CO_{3(ads)} \text{ (rate limiting step)} \quad (8)$$

$$CO_{3(ads)} \rightarrow CO_{2(ads)} + O_{(ads)} \quad (9)$$

$$CO_{2(ads)} \rightarrow CO_{2(g)} \quad (10)$$

NO reduction takes place following an intrafacial mechanism [9,25,33,53,54]. NO dissociatively adsorbs on a surface oxygen vacancy (□) (Equation (11)). The recombination of two adsorbed nitrogen species leads to the formation of N_2(Equation (12)) while parallel pathways produce undesired N_2O (Equation (13)) and isocyanate species (Equation (14)) [57]. The oxygen vacancy is restored by the reducing species in the gas phase (e.g., CO (Equation (15))). It has to be noted that, on perovskite-type oxides containing redox-active B-site cations (e.g., Co, Ni), the consumption of O_2 proceeds preferentially to that of NO when both oxidizing species are present in the gas phase [58].

$$NO_{(g)} + \text{B-}\square\text{-B} \rightarrow N_{(ads)} + \text{B-O-B} \quad (11)$$

$$2\,N_{(ads)} \rightarrow N_{2(g)} \quad (12)$$

$$N_{(ads)} + NO_{(g)} \rightarrow N_2O_{(g)} \quad (13)$$

$$N_{(ads)} + CO_{(g)} \rightarrow NCO_{(ads)} \quad (14)$$

$$CO_{(g)} + B\text{-}O\text{-}B \rightarrow CO_{2(ads)} + B\text{-}\square\text{-}B \quad (15)$$

Incorporation of a Noble Metal

Despite the promising TWC performance of perovskite-type oxides, their very low SSA and their reactivity with typical components of three-way catalytic converters (e.g., γ-Al_2O_3) have to be overcome by the development of appropriate strategies. These issues will be developed in more details further in the article. The low price of noble metal-free perovskite-type oxides is counterbalanced by some of their limits, which appear critical for their implementation in real TWC. Most of these materials have a very low resistance against sulfur-containing poisons and those that do not (e.g., the $LaCrO_3$ class) generally provide insufficient overall performance (Figure 3) [10,59,60]. The partial substitution of the B-site cation by a noble metal dramatically increases the sulfur tolerance of perovskite-type oxides [61]. Tzimpilis *et al.* reported the activation of a $La_{1.034}Mn_{0.966}Pd_{0.05}O_z$ catalyst in the presence of 8 ppm of SO_2 [62,63]. Regardless of their promising performance, noble metal-free perovskite type oxides still remain less active than noble metal supported catalysts [64]. Their operating window (range of exhaust compositions around the stoichiometry at which high conversions of UHC, CO, and NO_x are simultaneously reached) is also much narrower (Figure 4) [65]. The incorporation of small amounts of a noble metal in the perovskite significantly promotes the reduction of NO_x (which is otherwise limited) and improves the low temperature oxidation of UHC and CO (Figure 5) [13,66,67] (note the difference of reaction temperature and the inversion of the x axis between Figure 4 and Figure 5). Some noble metal-based perovskite-type oxides present additional advantages (e.g., better selectivity [64], activity improvement in the presence of water [68]) compared to conventional TWC. Furthermore, the incorporation of the noble metal in the perovskite matrix protects it against sintering and reduces its losses by volatilization under oxidizing atmosphere [9,69].

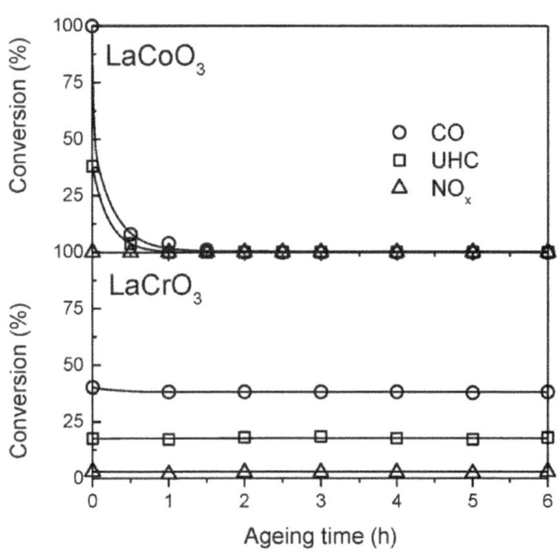

Figure 3: TWC performance of $LaCoO_3$ and $LaCrO_3$ at 550 °C and 32,500 h^{-1} during ageing under an oxygen-rich sulfur-containing atmosphere. Adapted with permission from Jovanovic *et al.* [60]. Copyright 1991, Elsevier.

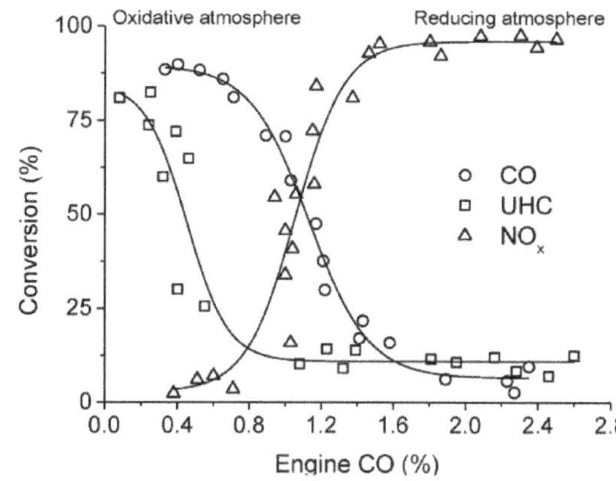

Figure 4: TWC activity of $LaCoO_3$ at 655–685 °C and 18,000–21,500 h^{-1} as a function of CO concentration in the exhaust gases (the percent of CO in the exhaust is indicative of the air-to fuel ratio). Adapted with permission from Sorenson *et al.* [65]. Copyright 1974, The American Ceramic Society.

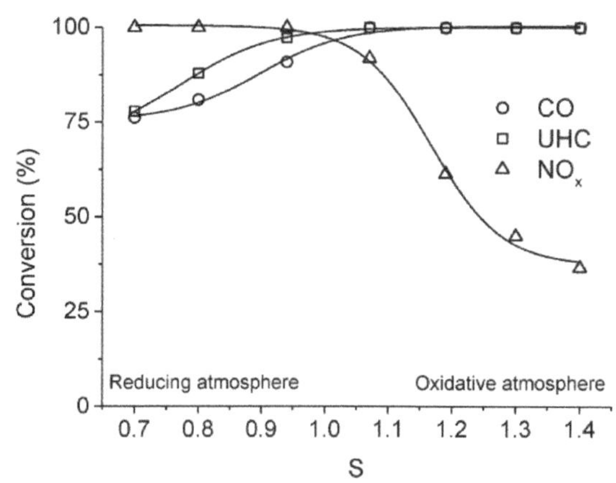

Figure 5: TWC activity of $LaMn_{0.976}Rh_{0.024}O_{3+\delta}$ at 400 °C and 13,000 h^{-1} as a function of the stoichiometric factor $\left(S = \frac{2 \cdot C_{O_2} + C_{NO}}{C_{CO} + 9 \cdot C_{C_3H_6}}\right)$. Adapted with permission from Guilhaume *et al.* [13]. Copyright 1997, Elsevier.

The interest in perovskite-type oxides for TWC applications was intensified in the early 2000s with the discovery of the so-called self-regenerative property [70,71]. The high temperature (up to 1000 °C) and the redox environment of exhaust gases are known to provoke the progressive deactivation of conventional supported noble metal catalysts. This loss of activity, which results from the agglomeration and growth of metal particles, is claimed to be suppressed in self-regenerative perovskite-based catalysts. The self-regenerative function was first reported on $LaFe_{0.57}Co_{0.38}Pd_{0.05}O_3$ [70] in which palladium (but also a small fraction of cobalt) oscillates between two oxidation and coordination states depending on the redox environment. Under slightly oxidative atmosphere, it is incorporated in the perovskite lattice as Pd^{3+} cations occupying B-sites. Under slightly reducing environment, it segregates out of the perovskite crystal in the form of metallic nanoparticles of 1–3 nm. Therefore, palladium reversibly moves in and out of the perovskite matrix in response to the cyclic redox fluctuations of the exhaust gas composition. These reversible structural changes, which are fast (only a few seconds at 600 °C [14]) and occur even at low temperature (100–400 °C) [72,73], suppress the growth of Pd particles compared to Pd/Al_2O_3. To illustrate this phenomenon, the behaviors of two catalysts of similar initial activity are compared after ageing in a real exhaust at 900 °C for 100 h. The reference $Pd/\gamma\text{-}Al_2O_3$ catalyst loses about 10% of its activity (Figure 6) owing to the sintering of noble metal particles (D > 100 nm). On the contrary, $LaFe_{0.57}Co_{0.38}Pd_{0.05}O_3$ maintains a high Pd dispersion (D = 1–3 nm) and its activity remains unchanged (Figure 6) [70,74]. Following the discovery of the self-regenerative property of $LaFe_{0.57}Co_{0.38}Pd_{0.05}O_3$, several other "intelligent" catalysts based on platinum or rhodium hosted in $LaFeO_3$, $LaAlO_3$, $CaTiO_3$, $CaZrO_3$, $SrTiO_3$, $SrZrO_3$, $BaTiO_3$, or $BaZrO_3$ structure were identified [75]. Each combination of noble metal and host structure is characterized by the proportion of precious metal likely to form a solid solution with the perovskite-type oxide and, therefore, to be involved in the self-regeneration process (Table 1) [75]. The stronger metal-support interaction of the noble metal with a perovskite-type oxide support, as compared to a conventional alumina support, significantly increases its resistance to thermal sintering, allowing to enhance the catalytic performance while minimizing the noble metal content [76,77].

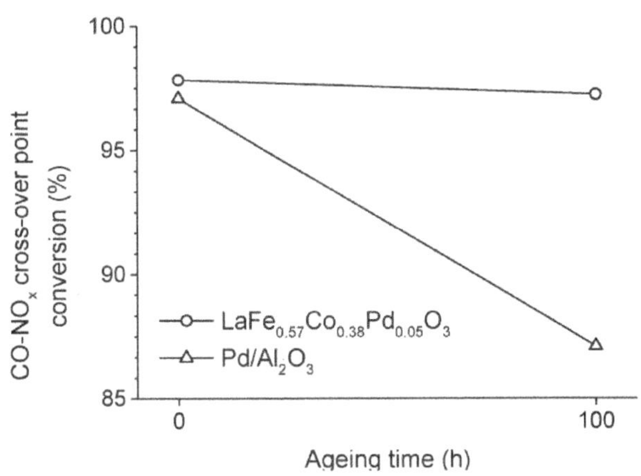

Figure 6: Change in the CO-NO_x cross-over point conversion (which is indicative of the catalytic activity) of $LaFe_{0.57}Co_{0.38}Pd_{0.05}O_3$ and Pd/Al_2O_3 catalysts upon ageing in a real exhaust gas at 900 °C for 100 h. Adapted with permission from Nishihata *et al.*[70]. Copyright 2002, Nature Publishing Group.

Table 1: Proportion of noble metal in $AB_{0.95}B'_{0.05}O_3$ (B' = Pd, Pt, Rh) actually forming a solid solution with the host structure under reducing (lower value) or oxidative (upper value) atmosphere ("-" indicates that no data is available; "x" indicates that the precious metal and the perovskite do not form a solid solution). Adapted from Tanaka*et al.* [75].

Host structure	Noble metal		
	Pd	Rh	Pt
$LaFeO_3$	0%–100%	63%–100%	25%–83%
$LaAlO_3$	-	69%–100%	-
$CaTiO_3$	x	28%–100%	20%–100%
$CaZrO_3$	-	53%–100%	27%–100%
$SrTiO_3$	-	37%–100%	50%–100%
$SrZrO_3$	-	x	x
$BaTiO_3$	-	0%–100%	25%–93%%
$BaZrO_3$	-	47%–100%	x

Recent HAADF-STEM observations on Pd-$LaFeO_3$, Pt-$CaTiO_3$, Rh-$CaTiO_3$, and Pt-$BaCeO_3$ systems question the self-regenerative function [78,79]. The microscopy study evidenced that a fraction of the fully reversible formation-dissolution of noble metal particles under redox cycling predominantly takes place within the bulk of the host structure. This implies that most of the metallic clusters formed under reducing atmosphere are not directly accessible to the catalytic reaction. However, it should be kept in mind that the easy $A^0 \leftrightarrow A^{3+}$ (A = Pd, Pt, Rh) transformation is accompanied by an oxygen uptake-release and that a defective perovskite structure is formed when the noble metal segregates out of the oxide. Both phenomena undoubtedly play a role in the activity of the material. The formation and dissolution of noble metal clusters at the surface of the perovskite particles was found to be much more limited than expected [78]. The surface noble metal clusters formed after reduction in 10 vol% H_2/N_2 at 800 °C for 1 h tended to coarsen rather than return into the perovskite matrix after oxidation in 20 vol% O_2/N_2 at 800 °C for 1 h [79]. At least, the partial embedding of these particles in the support surface confirms that they strongly interact with the host structure (Figure 7) [78]. In fact, contrary to what happens during the relatively long pretreatment procedure prior to HAADF-STEM observation (1 h under constant atmosphere [79] or alternance between reducing and oxidizing environments every 10 min [78]), one can expect that this interaction should be constantly revived under the high frequency (0.5–5 Hz) redox fluctuations of a real exhaust. This should limit the mobility of the precious metal particles on the surface of the host material and prevent their agglomeration. Therefore, it is likely that the durability of "self-regenerative" materials does not originate from a fully reversible phenomenon but from a strong interaction between surface precious metal particles and the perovskite surface maintaining them in a highly dispersed state under real operating conditions.

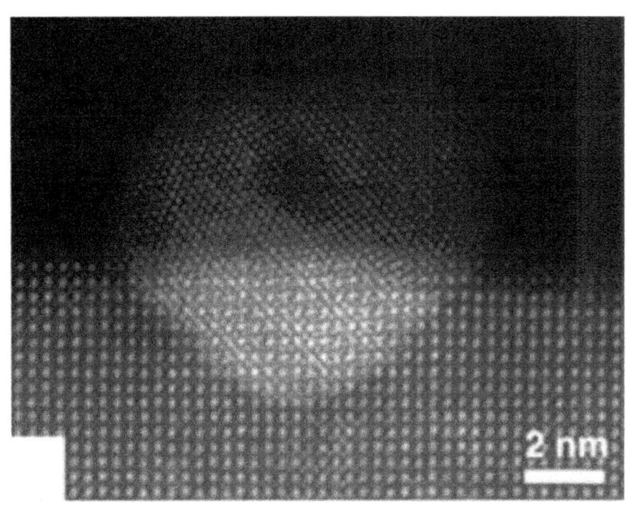

Figure 7: TEM image of a Pd particle partially embedded in the surface of a $LaFeO_3$support. Reproduced with permission from Katz *et al.* [78]. Copyright 2011, American Chemical Society.

A last point worth mentioning is that the incorporation of the noble metal in the perovskite matrix is not beneficial for all applications. Indeed, perovskite-supported noble metal catalysts remain more active than noble metal-substituted perovskites [80]. As a consequence, Pd-based perovskite-type oxides tested in the treament of NGV exhaust were found to activate upon thermal treatment at high temperature (850–950 °C) due to the migration of Pd out of the perovskite lattice [63] and the growth of noble metal particles [81].

PREPARATION OF STRUCTURED CATALYSTS

TWC for automotive applications are traditionally supported on a structured carrier. Structured catalysts offer several advantages compared to their powder and pellet counterparts. These include (i) a very low pressure drop; (ii) an increased efficiency due to an improved mass transfer; (iii) a high safety (self-draining reactor, no risk of clogging); (iv) an easy separation from the reaction medium and (v) cost reduction (low production costs, low catalyst loadings) [82,83]. The requirements, preparation procedures and performance of structured catalysts have been discussed in several review articles [82,83,84,85]. The most common structured carriers are monoliths, also referred to as honeycombs. These single block structures possess numerous small parallel straight channels which can be of square, triangular, hexagonal or round shape. Channel density can vary from 25 to 1600 channels per square inch (cpsi) and has to be carefully chosen depending on the specific requirements of the process. A higher cell density implies a higher geometric surface area but also more complex manufacturing and washcoating procedures as well as higher pressure drops [82]. Monolith substrates used in automotive applications generally have square channels with a cell density of 300–600 cpsi [86,87]. More than 90% are made of cordierite ($2MgO{\cdot}2Al_2O_3{\cdot}5SiO_2$) [83], which offers high mechanical strength, very low thermal expansion coefficient and resistance to high temperatures and thermal shocks [82,85,86]. Other ceramic materials (mullite, silicon carbide, alumina,*etc.*) or metals (Al, Ni, stainless steel, FeCrAl alloy, *etc.*) are also used in practice [82,83,85].

In the previous sections of this review, we have demonstrated that perovskite-type materials show promising activities in the combustion of hydrocarbons, oxidation of CO and reduction of NO_x. In order to confirm their potential for real application in TWC converters, it is necessary to evaluate their performance as active components of structured catalysts. Amongst the extensive literature related to perovskite-type catalysts, studies concerning specifically structured ones are still relatively scarce. Furthermore, the great majority is dedicated to the oxidation of CO or various hydrocarbons under lean atmosphere and at relatively low gas hourly space velocity (GHSV < 10,000 h^{-1}). It is therefore difficult, from these data, to assess the performance of perovskite-type catalysts in TWC converters, whose operating conditions (stoichiometric atmosphere, GHSV up to 100,000 h^{-1}) significantly deviate from those previously mentioned.

In the following, we will refer to the two main categories of structured catalysts described by Avila *et al.* [85] (Figure 8), namely incorporated and coated catalysts.

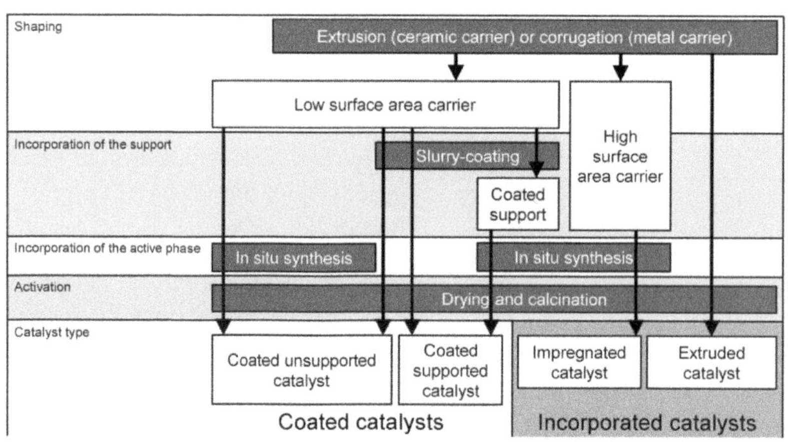

Figure 8: Paths for the preparation of structured perovskite catalysts. Adapted with permission from a more general scheme by Avila *et al.* [85]. Copyright 2005, Elsevier.

Incorporated Catalysts

In incorporated catalysts, the active phase is an integral part of the structured body. Consequently, a high mass of the active component is present and a first requirement is therefore that the catalytic material should not be too expensive. Because of their nature, they become relevant to conditions where catalyst erosion can occur and diffusion limitations are negligible. They are produced either by direct extrusion of a plastic paste containing the catalytic material or by impregnation of precursors on a high surface area substrate. To our knowledge, only perovskite-containing structured catalysts of the first category have been studied thus far. Most of available articles have been produced by the same group: Isupova, Sadykov *et al.* [88,89,90,91,92,93] prepared extruded monoliths from a wide range of perovskite catalyst compositions (*A* = La, Sr, Ce, Dy, Y; *B* = Fe, Mn, Ni, Co, Cu).

Perovskite-based monoliths can be produced in a reduced number of steps including the extrusion of a plastic paste containing the catalytic powder, a drying step and a calcination step. Even in the absence of additives, catalyst powders with sufficient surface reactivity (*i.e.*, high surface density of acid and hydroxyl groups) easily form a water-based paste from which high-strength extrudates can be produced [88]. In other cases, it is necessary to use additives in order to, on the one hand, adjust the rheological properties of the paste to make it suitable for the extrusion process and, on the other hand, optimize the mechanical strength and porosity of the final monolith. The use of an alumina binder enhances the contacts between perovskite particles, resulting in increased SSA, mechanical strength and resistance to thermal shocks [88,89,90]. In some cases, there can be counterproductive effects of employing a binder. A decrease of the catalytic activity can result from the partial blockage of active centers by alumina particles [89] or from the formation of less active hexa-aluminate surface micro-phases due to chemical interactions between the binder and the perovskite-type catalyst at high temperature [90]. In order to force the conversion of the alumina binder into a hexa-aluminate phase before it can alter the perovskite, sacrificial A- and/or B-site cations can be added to the extrusion paste as nitrate salts [91]. While small amounts of a surfactant (ethylene glycol, glycerol, carboxymethyl hydro cellulose) improve the rheological properties of the extrusion paste and limit the formation of cracks during drying, too high amounts are responsible for a significant increase of the mean pore size (Figure 9) and consequently a decrease of the mechanical strength [88,89]. The addition of ceramic fibers improves the resistance to thermal shocks [88] and modifies the pore structure of the monolith [91,92]. Organic or mineral acids can be used as peptizing agents. Despite that acetic acid is more environment-friendly than nitric acid (HNO_3 releases NO_x during the calcination step), monoliths with higher mechanical strength were produced with the latter [88,89].

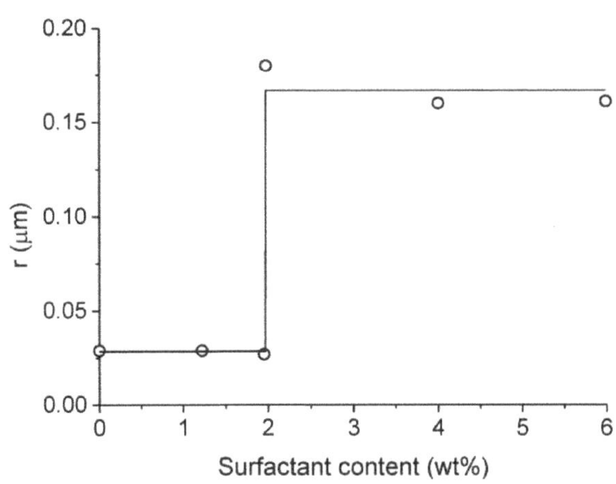

Figure 9: Influence of surfactant (ethylene glycol) content on the mean pore radius of monolithic $LaMnO_3$. Adapted with permission from Isupova *et al.* [88]. Copyright 1996, Elsevier.

The drying step has to be performed slowly under controlled humidity conditions in order to avoid the formation of cracks [89,90]. Extruded monoliths can contain up to 70–80 wt% of active phase. Appropriate formulation of the extrusion paste can yield high SSA (30–75 $m^2 \cdot g^{-1}$), porosity (up to 0.3 $cm^3 \cdot g^{-1}$), and mechanical strength (up to 10–20 $MPa \cdot cm^{-2}$), even after calcination at 900 °C. The extrusion procedure was not reported to alter the phase composition [91,92].

The catalytic performance of extruded perovskite catalysts was found to be mostly determined by their SSA [91,92]. Isupova *et al.* calculated that the specific rates of methane disappearance per gram of active component were comparable to those reported by other authors on powder and coated perovskite-type catalysts [92]. Therefore, an extruded catalyst, which, by its nature, contains a higher amount of active phase than a coated catalyst of comparable geometry (volume, density of channels, *etc.*) will show a higher overall activity in the kinetic regime [94]. Nevertheless, as active sites are distributed within the walls of the structured body, some of them are not easily accessible to the reactants and mass transfer limitations commonly occur at high temperature (e.g., above 560 °C for methane oxidation [91,93]). High temperature (up to 1300 °C) tests on pilot installations during several months showed the better stability in terms of activity, mechanical strength, monolith integrity and resistance to poisons (HF, HCl, SO_2) of extruded perovskite-type catalysts compared to the coated oxide catalysts conventionally used for the same applications [88,89,90].

Coated Catalysts

Coated catalysts are obtained when a thin layer (*ca.* 100 µm) of active components is deposited on the walls of a low surface area structured carrier. Although perovskite-type coatings can be obtained through plasma spraying [95], chemical vapor deposition [96], flame-assisted vapor deposition [97], physical vapor deposition [98] or even atomic layer deposition [99], these advanced techniques are not easily scalable, based on cost considerations, and generally produce dense layers which are not suitable for catalytic applications. We will, therefore, limit our discussion to more conventional and cost effective methods. Precursor techniques where the active material is synthesized directly on the substrate wall will be distinguished from procedures where a slurry of the powder catalyst is washcoated.

Precursor-Derived "On-Substrate-Wall" Synthesis of Catalysts

Immersion of the structured carrier in an aqueous solution of salt precursors in the absence of any additive [100,101] or in the presence of citric acid [102] is employed to prepare perovskite-type active phases directly on the substrate. Citric acid acts as a chelating agent and enables the preparation of highly concentrated precursor solutions. The increased viscosity and concentration of such solutions result in increased amounts of deposited material. A variation of this preparation protocol uses a polymerized solution of nitrate precursors, citric acid and ethylene glycol prepared by the Pechini method [101,103,104,105]. After withdrawal of the impregnated carrier, the excess solution is removed by blowing pressurized air. Drying can be carried out in a conventional furnace (80–200 °C) [102,103,104], under air-flow at room temperature [101], or in a microwave oven [106]. The latter option allows a homogeneous drying, preventing the redistribution of the active material and therefore ensuring its good distribution on the carrier [107]. Finally, calcination at 700–1100 °C [101,102,104] converts the precursors into the desired perovskite-type oxide. Phase pure materials are generally obtained but secondary phases are sometimes detected [101]. However, the method presents some drawbacks. First, it is long, because several dip-dry-calcine cycles are required to obtain the desired active phase loading. Then, low SSA (below 6 $m^2 \cdot g^{-1}$) of the synthesized perovskite-type materials is generally obtained [101,103,104]. In addition, the partial dissolution of the substrate in the acidic and chelating solution used for impregnation leads to the formation of less active secondary phases (e.g., spinels, partially substituted perovskites) [101,103,104]. This phenomenon can be limited by precoating a thin layer of ZrO_2 or Ln_2O_3primer (Ln = mixture of La, Ce, Pr, Nd, Sm) on the structured carrier prior to the deposition of the perovskite phase [103].

Polymerized solutions prepared by the Pechini method produce a distinct active phase layer covering the walls of the structured carrier [101]. The coating has a thickness of 2–160 μm [101,102,103] and shows a coherent interface with the substrate which was interpreted as a sign of good adherence [101,102]. However, large cracks (up to 30 μm; Figure 10a) were observed in the coating and no adherence assessment test was performed. Traditional wet impregnation with a solution of metal nitrate precursors does not produce a separate catalytic layer but the active component is uniformly distributed across the walls of the structured carrier [100,101]. Despite these differences in morphology, similar activities are measured on both types of samples [101]. However, in the presence of mass transfer limitations, samples prepared following the Pechini method exhibit higher performance due to the better accessibility to their active sites (active phase distributed at the surface of the monolith walls rather than within them).

In situ solution combustion synthesis [4,108,109,110,111,112,113,114,115,116] seems to be a good alternative to the previously described procedures. It allows the production of a well-adhered and porous perovskite layer in a reduced number of steps. After being dipped into an aqueous solution containing nitrate precursors (oxidant), urea (fuel) and ammonium nitrate (combustion booster), the structured carrier is placed in an oven at 600–650 °C. This rapidly brings the aqueous phase to boil and the precursor mixture to ignite: within few minutes, the heat released by urea combustion allows the transformation of the nitrate precursors into a phase pure perovskite-type oxide [4,108,109,110,111]. The generation of large amounts of gaseous products in a very short period of time during the combustion produces a 40–100 μm thick, highly porous and spongy coating (Figure 10b) with high SSA (4–30 $m^2 \cdot g^{-1}$) [4,108,109,110,112,113]. Such a structure minimizes the pressure drops and improves the mass transfer [113]. Although catalytic powders prepared by solution combustion synthesis were reported to easily crumble [109,114], coatings obtained by the same procedure show excellent adherence. The weight loss of samples submitted to a vibration test consisting in a 5 h treatment in an ultrasonic bath is lower than 1% [4,110,112,113,115]. Thermal ageing at temperatures as high as 850 °C in the presence of SO_2 [115] or water [109] did not induce significant deactivation and no chemical interactions between the active phase and the structured carrier were evidenced [108,109,110,112,114,115,116].

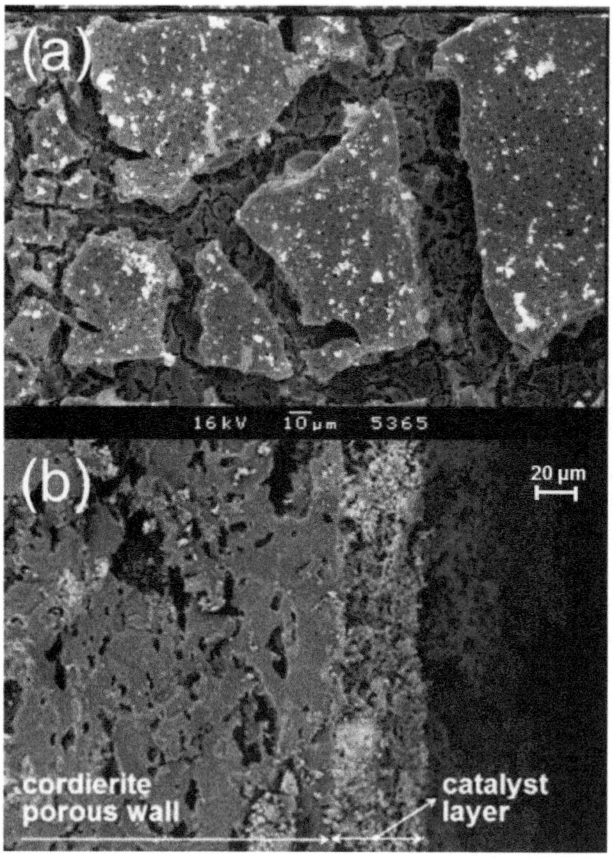

Figure 10: SEM images of (**a**) a $LaMnO_3$ film prepared by CM (top-view) and (**b**) a $La_{0.8}Li_{0.2}Cr_{0.8}O_3$ layer deposited via solution combustion synthesis (cross-section). Reproduced with permission from Isupova *et al.* [101] and Fino *et al.* [112]. Copyright 2002 and 2006, Elsevier.

Washcoated Catalysts

Washcoated catalysts are prepared from an already formed active phase powder which is used to produce a slurry and applied to the structured carrier surface by dip-coating. This technique consists in dipping the dry substrate into the catalyst suspension for a short period of time. After subsequent withdrawal, the excess slurry remaining in the channels/pores of the substrate is removed by blowing pressurized air. The deposited catalyst layer is then left to dry between room temperature and 200 °C and finally fixed to the carrier by calcination at 500–700 °C.

Washcoated Unsupported Catalysts

Perovskite-type materials used to prepare washcoated catalysts are generally obtained by conventional techniques such as the amorphous citrate method (CM) [106,117,118], co-precipitation [94,100,119] or flame-assisted synthesis (e.g., flame hydrolysis [118,120,121,122]). The solvent of the slurry is typically water [39,117,120,121,122,123,124] but organic solvents, such as *n*-butanol, are sometimes employed [62,63]. Similarly to extrusion pastes, various additives can be used to tune the rheological properties of the slurry and improve the quality of the coating. An alumina or zirconia binder increases the mechanical strength and adherence of the coated layer [106]. A temporary binder, such as Tylose (a cellulose derivative), can sometimes be employed [39,94,100,119]. This additive, which acts in the same way as a permanent binder, prevents damage that may occur during handling of the catalyst and is eliminated during the calcination step. A peptizer, such as nitric [117,120] or citric acid [125,126] improves the dispersion of suspended particles in the slurry. A critical parameter known to influence the stability of a slurry is the size of the suspended particles. If these particles are sufficiently small (*i.e.*, if the selected synthesis protocol produces very fine catalyst particles [121] or if an appropriate size fraction has formerly been collected through grinding and sieving [126]), they can be easily dispersed by low energy ball-milling [118,120,121,122] or even simple stirring at room temperature [117]. Otherwise, the slurry must be submitted to high energy ball-milling in order to attain catalyst particles of the desired size, typically below 5 µm [106].

Active phase layers deposited by dip-coating are generally homogeneously distributed over the surface of the carrier but cracks sometimes appear during drying, calcination [117] and operation. In the worst cases, the formation of deep cracks upon thermal ageing leads to the partial exfoliation of the active phase [122]. Therefore, besides catalytic activity, both the quality and adherence of the washcoat are important issues to be considered. The mechanical properties of the coated layer primarily depend on the characteristics of the initial catalyst powder. Perovskite-type materials with high SSA and good crystallinity can be obtained by CM after calcination at relatively low temperature (700–750 °C). Contrary to flame-made powders, such materials never experienced high temperatures and are characterized by a poor thermal stability. As a consequence, a coated CM-made perovskite powder is significantly deactivated after exposure to a temperature of 800 °C [120] and large cracks appear between the active phase and the carrier [118]. The composition of the mixed oxide is also of great importance because it influences the surface chemistry (e.g., nature and density of surface groups) of the oxide as well as the way the latter bonds to the carrier. For example, Kucharczyk *et al.* observed that the adherence of washcoated $La_{1-x}Ag_xMnO_3$ decreases with increasing the amount of Ag incorporated into the perovskite (Figure 11 [126]). Adherence obviously depends also on the nature of the substrate: the bonding of the catalyst is weaker on metal carriers, which have a much lower porosity and a smoother surface than ceramic ones [82,83,85,117].

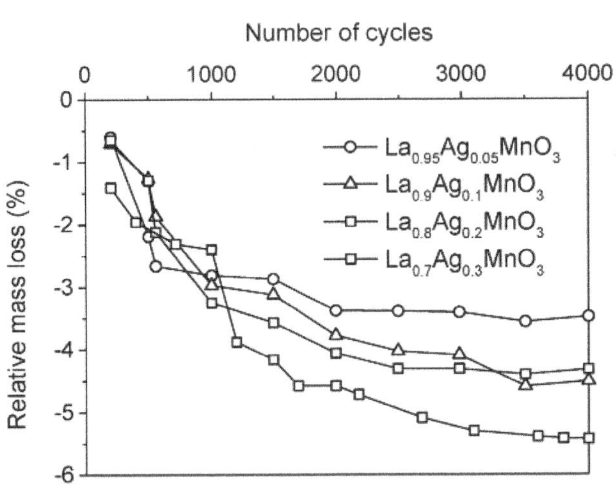

Figure 11: Influence of La substitution by Ag on the resistance of washcoated $La_{1-x}Ag_xMnO_3$ catalysts to 4000 cycles of heating to 1000 °C and rapid cooling to room temperature. Adapted with permission from Kucharczyk *et al.* [126]. Copyright 2008, Elsevier.

As previously mentioned, it is possible to improve the mechanical properties of the coated layer by means of additives. Qi*et al.* [106] observed that the mechanical strength of a coating consisting only of perovskite-based particles was so poor that it could simply be blown away by a strong nitrogen flow. Adherence was greatly improved through the addition of a binder. The grafting properties of the substrate, whether the latter is made of metal [125,126] or ceramic [120,121,122,126], can be greatly enhanced through the use of a primer. This very thin layer constituted of one or several oxide(s) (e.g., Al_2O_3, TiO_2, La_2O_3) is applied to the carrier prior to washcoat. It is generally obtained by dipping the substrate in a water-based sol [118,120,125,126] or an aqueous solution containing precursors [118]. The nanoparticles in the sol or the ions in the solution easily interact with the surface of the structured carrier, producing a well-adhered, dense oxide layer upon calcination. The increased roughness of this primer, as compared to the original substrate, improves the bonding with the active phase [120]. Fabbrini *et al.* concluded that despite its much lower SSA, La_2O_3 constitutes a better primer than Al_2O_3 [121]. In fact, besides being prone to chemical interactions with the perovskite catalyst, the latter oxide has a poor thermal stability and its thermal expansion coefficient significantly differs from that of cordierite, favouring crack formation (Figure 12a,b) [118]. Lanthanum nitrate was found to be a better precursor of La_2O_3 than lanthanum acetate, the difference resting on their distinct decomposition behavior [121]. Lanthanum acetate tends to form platelet-shaped deposits, a morphology which is much less favorable to a good grafting of coated particles than the needle-like structures obtained when lanthanum nitrate is employed. Upon exposure to high temperature, lanthanum acetate agglomerates into larger plates, which can easily detach from the substrate, while lanthanum nitrate produces a compact layer ensuring the good anchoring of the active phase [121]. Catalytic layers deposited by slurry-coating were reported to be less adherent than those obtained by direct impregnation [106]. Therefore, a well-adhered perovskite coating obtained by *in situ* on-substrate-wall synthesis can also be used as an efficient primer [100]. Since this primer can have exactly the same formulation as the slurry-coated particles, a better adherence can be achieved while no deactivation due to chemical interactions can occur. From the available literature, it is unfortunately difficult to clearly identify the optimum parameters (slurry composition, active phase content, nature and amount of binder/primer, *etc.*) necessary to produce a well-adhered washcoat of a perovskite-type oxide. Beside the scarce quantification of the adherence properties in the available literature (Figure 11 [126]), the lack of a reference adherence assessment test makes it arduous to compare results obtained by different research groups.

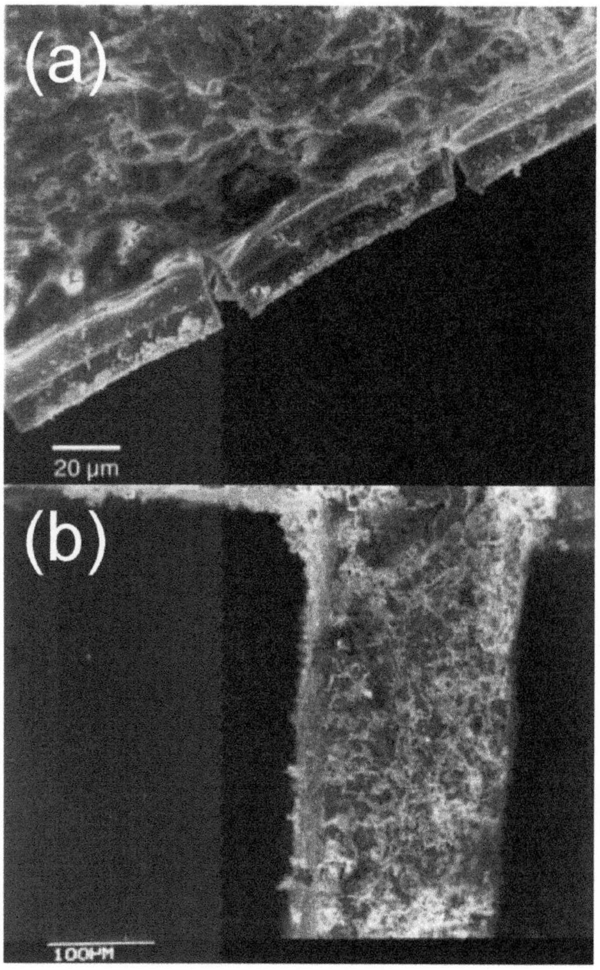

Figure 12: SEM images of washcoated $La_{0.9}Ce_{0.1}CoO_3$ after accelerated ageing at 800 °C for 1 h: (**a**) Al_2O_3 primer and (**b**) La_2O_3 primer. Reproduced with permission from Fabbrini *et al.* [118]. Copyright 2003, Elsevier.

In addition to mechanical aspects, the chemical stability of the catalyst layer is also of great importance. In most cases, no changes of texture, crystallinity or composition occur during the dip-coating procedure [117]. Although surface modifications of the active phase (decrease of crystallinity, formation of surface hydroxyl groups) may take place during the preparation of the slurry, they are usually not critical for activity and thermal resistance [122]. As previously observed on extruded and *in situ* synthesized structured catalysts, chemical interactions between the active component and the substrate/binder/primer lead to the partial destruction of the perovskite structure and subsequent deactivation. For example, solid state reaction between the perovskite washcoat and the cordierite substrate [123], between La_2NiO_4 and the La_2O_3primer [122], between $La_{0.8}Ce_{0.2}NiO_3$ and the ZrO_2 binder [106] or between $LaCoO_3$ and the Al_2O_3 primer [118] were reported.

In the absence of chemical incompatibilities, the main origin for the deactivation of washcoated unsupported perovskite-type catalysts is the reduction of their SSA [123,126]. Kucharczyk *et al.* reported that after 24 h of ageing at 900 °C, the methane conversion at 750 °C on monolith-carried $La_{0.8}Ag_{0.2}MnO_3$ decreased from 98.9% to 67.0% as SSA dropped from 29.4 to 1.7 $m^2 \cdot g^{-1}$ [126]. Sufficient stability can be obtained through the use of appropriate synthesis procedure and binder/primer additives [120,126]: more than 90% of the initial activity of a $La_{0.9}Ce_{0.1}CoO_3$ washcoated catalyst could be maintained after ageing at 800 °C for 1 h [118,121]. Arendt *et al.* [117] reported that, at low temperature, perovskite-type catalysts were more active when carried on a metallic than on a ceramic substrate. The opposite behavior was observed at high temperature. Authors concluded that this could originate from the different thermal conductivity of the two carrier materials. However, since the catalyst particles adhered to the metal substrate (0.02–0.06 μm aggregates) were much smaller than those deposited on the ceramic one (3 μm aggregates), particle size effects cannot be excluded. The same authors observed that washcoated catalysts had increased activity compared to the corresponding powders. They suggested that the carrier material was able to activate molecular oxygen into strongly oxidizing species capable of limiting the accumulation of deactivating coke. On the contrary, Fabbrini, Forni *et al.* [118,120] calculated that, in the absence of a primer, the specific activity (per gram of active component) of carried catalysts, was lower than that of the corresponding powders, whereas identical activities were obtained when a primer was used.

Washcoated Supported Catalysts

Washcoated supported perovskite-type catalysts can be prepared by impregnation or deposition-precipitation of the active phase on an oxide support powder (e.g., γ-Al_2O_3, ZrO_2, MgO) either before [39] or after [127,128,129,130,131,132] that the latter is washcoated on the structured carrier. This produces a homogeneously distributed active phase [127] on a porous support firmly anchored to the walls of the substrate [66,128].

Similarly to other types of coated catalysts, exposure to high temperature can cause fractures in the washcoat. Nevertheless, the latter were not reported to strongly reduce adherence [127,128]. ZrO_2 [128] and MgO [129] supports are chemically inert with respect to most of active phases. On the contrary, extensive interaction exists on γ-Al_2O_3 supported systems, which results, as earlier mentioned, in the formation of less active spinel structures, simple oxides and aluminate phases [66,130]. $LaAl_{11}O_{18}$ and $LaCr_{1-x}Al_xO_3$ form upon solid-state reaction between $LaCrO_3$ and γ-Al_2O_3 at high temperature [131] while Mn from $LaMnO_3$ is incorporated in the alumina lattice even at low temperature [129]. Such chemical interactions can be limited by precoating a La_2O_3 layer on alumina prior to the application of the perovskite phase [66].

Deposition on a porous oxide support significantly increases the dispersion of the perovskite material, which enhances the catalytic performance [128,129,131]. Cimino *et al.* compared fresh unsupported and ZrO_2-supported $LaMnO_3$ monolithic catalysts. A three times higher specific rate of methane combustion per gram of catalyst (corresponding to a 14-fold increase in the specific rate per gram of active phase) was obtained on the supported system [128]. Activity enhancement is however not always so spectacular [119] and has a limit: large perovskite crystals form upon increase of the perovskite-type oxide loading and exposure to high temperatures, which leads to pore blocking and loss of dispersion [39,128,129,130,131]. The decrease of activity was reported to be proportional to the loss of specific surface area: a reduction of both these parameters by a factor four was observed after treating a monolithic γ-Al_2O_3 supported $LaMnO_3$ catalyst for 9 h at 1100 °C [127]. Although no support-active phase interactions were evidenced in the case of ZrO_2 supported catalysts, the latter more largely deactivated than γ-Al_2O_3 supported ones due to strong sintering. The SSA of the washcoat decreased from 47 to 3 $m^2 \cdot g^{-1}$ after exposure to 1100 °C for 8 h, thus providing an increase of T_{50} (temperature of 50% methane conversion) by 175–190 °C [128].

IMPLEMENTATION IN TWC CONVERTERS

In the previous sections, we have demonstrated the promising TWC performance of powdered perovskite-type oxides and the possibility to prepare structured catalysts based on these materials. The activity of perovskite-type oxides logically increases with increasing amount of accessible active phase. Consequently, a higher efficiency is obtained on extruded and coated structured catalysts containing a higher loading of active component [104,120]. Accordingly, extruded catalysts with 70–80 wt% of active phase provide better conversion efficiencies in the kinetic regime than the coated unsupported catalysts that contain 5–30 wt% of active phase [101]. For real scale applications (e.g., automotive exhaust treatment), where high velocity streams have to be treated, washcoated catalysts are more advantageous than extruded ones due to minimized diffusion limitations and, hence, a larger fraction of accessible active phase.

Structured catalysts can be obtained by different synthesis methods. Non perovskite oxide layers prepared from precursors directly on the surfaces of a structured carrier are generally denser and have a lower SSA than those obtained by slurry-coating [84]. The difference is not so pronounced in the case of perovskite-type oxides due to the sudden release of large amounts of combustion related gases (NO_x, CO_2, H_2O) upon calcination. Thus, the precursor solution based synthesis produces a relatively porous catalytic perovskite-type oxide layer with a SSA, thickness and catalytic performance which are comparable to those of washcoated layers. However, the significant release of heat and combustion related corrosive gases during the calcination restricts large scale production. Perovskite-type materials synthesized directly on the substrate surface by soft chemistry have a relatively low stability at high temperatures which results in sintering and deactivation. For these reasons, washcoated structured catalysts prepared from high temperature stable powders (synthesized, for example by flame hydrolysis) are more appropriate for TWC applications [118]. Amongst all types of structured catalysts, the best performance is obtained on washcoated supported ones containing a low amount (up to 6 wt% referred to the whole structured catalyst) of a highly dispersed perovskite-type oxide phase. Labhsetwar *et al.* [66] demonstrated the difference between supported and unsupported perovskite-type oxides coated on monoliths for TWC applications (Figure 13).

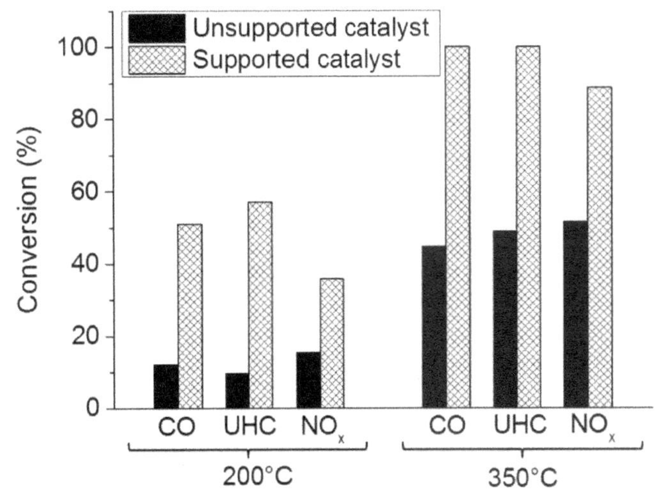

Figure 13: Comparison of CO, UHC, and NO_x conversions obtained on unsupported and supported $La_{0.7}Sr_{0.3}Mn_{0.95}Pt_{0.05}O_3$ washcoate d cordierite honeycomb monoliths. Adapted from Labhsetwar *et al.* [66].

The authors also studied the suitability of these catalysts for real scale application by testing them on an engine dynamometer. With supported perovskite type oxides coated structured catalysts, high conversions of CO (82%), UHC (86%) and NO_x (71%) were obtained. Despite these promising results, the oxide support (typically γ-Al_2O_3) used to disperse the active phase has a lower thermal resistance than the perovskite-type oxide. The sintering of the oxide support results in the significant growth of supported perovskite-type oxide particles. Consequently, unsupported systems perform better than the supported ones after exposure to very high temperatures (*i.e.*, 1100 °C [101]).

The major advantage of perovskite-type oxides is that a large array of elements can be accomodated into the structure. This makes it possible to tailor catalytic properties of the materials for specific applications. However, this strength of perovskite-type oxides can be a drawback when it reacts with the substrate, binder or primer materials that are present in a TWC converter. This leads to decreased TWC activity. In extruded catalysts, such chemical interactions are not a serious issue owing to the high active phase content. Sadykov *et al.* have even reported advantages of such interactions: the conversion of the alumina binder into a hexa-aluminate phase upon solid state reaction with the perovskite-type oxide improves the mechanical strength of the extruded catalyst [90]. The strongest interactions are observed for supported structured catalysts due to the high dispersion of the active phase component [119]. In such catalysts, any alteration to the relatively low amount of active phase implies considerable activity loss. To prevent this, several authors have suggested that the surface of the oxide support should be thermally treated to achieve an inert upper layer prior to the deposition of the active phase [131]. In agreement with this, α-Al_2O_3 was found to be less reactive than γ-Al_2O_3 to perovskite-type oxides [101]. Despite its reduced reactivity towards the active phase, α-Al_2O_3 has a significantly low SSA, which limits the dispersion of the supported perovskite-type oxide. Therefore, the deposition of a primer layer, e.g., La_2O_3, on a high SSA support appears to be the best option to protect the active perovskite-type oxide phase from chemical modifications while ensuring its sufficient dispersion.

Additional issues have to be taken into account for perovskite-type oxides to be implemented in TWC converters. For health and environmental reasons, certain elements cannot be included in the composition of a TWC due to their oxidation and volatilization under the high temperature exhaust environment comprising water vapor and corrosive gases. The volatile metal-based species are then potentially released in the environment through high space velocity exhaust gases. German clean air regulations identified that cobalt-based species formed under these conditions were potentially carcinogenic. On the same line, the "self-regenerative" perovskite-type oxide catalyst $LaFe_{0.95}Pd_{0.05}O_3$ was commercialized as a TWC component whereas $LaFe_{0.57}Co_{0.38}Pd_{0.05}O_3$ was not [14]. Though the thermal stability of perovskite-type oxides is high, their sulfur resistance is limited. The low sulfur tolerance of perovskite-type oxides can be overcome by appropriately designing the structured catalyst. A double layer configuration where the active perovskite-type oxide phase is located in the inner layer can avoid interactions between the active phase and sulfur-containing compounds and hence catalyst deactivation [14]. TWC converters containing a noble metal-based perovskite-type oxide with the double layer configuration shows better performance, over a wide range of space velocities, than a standard catalyst with a 1.5 times higher noble metal loading. As a result, self-regenerative Co-free perovskite-type catalysts were implemented in commercial TWC converters by Daihatsu Motor Co. Ltd. Around 3.2 million TWC converters based on this technology have been installed between 2002 and 2007 in Japan and in Europe [133].

Recently, solid foams have attracted particular attention as potential alternatives to honeycomb monoliths [134]. These sponge-like substrates contain interconnected and ramdomly-organized open pores in the millimeter-range. These foams can typically be made of the same constituting materials as monoliths and are characterized by their pore density, which can vary from 5 to 100 pores per inch (ppi). Compared to monoliths, the structure of foams creates higher pressure drops but allows radial mixing and enhanced turbulence [135]. Such features improve mass and heat transfers, which are crucial parameters in exhaust after-treatment applications. There are very few studies on perovskite-type oxides based foam catalysts [119,130] and none of them evaluates the performance of the foams in relation to the conventional honeycomb monoliths.

The multifunctional properties of perovskite-type oxides could also impart new possibilities to the TWC converters. For example, by exploiting the dielectric properties of perovskite-type oxides, TWC converters can be dielectrically-heated to minimize exhaust emissions during the cold start of a gasoline engine [87] A cordierite monolith coated with $LaCoO_3$exhibits rapid heat-up under microwave irradiation allowing to attain the 50% UHC conversion temperature within 15 s. Similarly, $La_{0.8}Ce_{0.2}MnO_3$ was reported as a dielectric material for the periodical regeneration of monolithic soot filters in diesel exhaust after-treatment systems [124]. Perovskite-type oxides are also studied for thermoelectric applications [22]: manganites, cuprates and cobaltites, which are known for their suitable catalytic performance, can be used to build thermoelectric oxide modules for the direct conversion of heat into electricity [136]. As a consequence, an appropriately designed TWC converter made of a suitable perovskite-type oxide could simultaneously remove noxious compounds from the exhaust gases and recycle the waste heat into electric energy for the vehicle.

CONCLUSIONS

Perovskite-type oxides are reviewed in relation to their application in exhaust after-treatment technologies as three-way catalysts (TWC). The main advantage of these materials is their robust crystal structure that can be used to catalyze redox reactions due to their flexible oxygen content. The possibility to accommodate simultaneously different metal cations at A- and B-sites allows to tune the catalytic properties for a specific application such as TWC, which requires redox properties with high thermal stability. As such, perovskite-type oxides exhibit good oxidation activity which can be improved by including a fraction of noble metals like Pd, Pt, or Rh at the B-site. However, significant improvement in the oxidation activity of the materials is only observed when the noble metal is dislodged from the B-site of the crystal and forms metal nanoparticles of few nanometers at the surface. Recent studies demonstrate that the dislodged noble metal is not completely reintegrated into the perovskite crystal, questioning the self-regenerative property of the materials. The potential use of these materials for real scale TWC applications is studied by converting powder catalysts into structured ones such as monoliths or foams.

ACKNOWLEDGMENTS

The authors are grateful to the COFUND program (EMPAPOSTDOCS project PCOFUND-GA-2010-267161) and the National Research Program SNF-NRP 62 "Smart Materials" (project no. 406240-126127) for their financial support. We also thank Gion Pirovino for his support.

CONFLICTS OF INTEREST

The authors declare no conflict of interest.

REFERENCES

1. Taylor, K.C. Automobile Catalytic Converters. *Stud. Surf. Sci. Catal.* 1987, *30*, 97–116.
2. Trovarelli, A. *Catalysis by Ceria and Related Materials*, 1st ed.; Imperial College Press: London, UK, 2002.
3. Forster, P.; Ramaswamy, V.; Artaxo, P.; Berntsen, T.; Betts, R.; Fahey, D.W.; Haywood, J.; Lean, J.; Lowe, D.C.; Myhre, G.; *et al.* Changes in Atmospheric Constituents and in Radiative Forcing. In *Climate Change 2007: The Physical Science Basis. Contribution of Working Group I to the Fourth Assessment Report of the Intergovernmental Panel on Climate Change*; Solomon, S., Qin, D., Manning, M., Chen, Z., Marquis, M., Averyt, K.B., Tygnor, M., Miller, H.L., Eds.; Cambridge University Press: Cambridge, UK; New York, NY, USA, 2007; pp. 129–234.

4. Fino, D.; Russo, N.; Saracco, G.; Specchia, V. Supported Pd-perovskite catalyst for CNG engines' exhaust gas treatment. *Prog. Solid State Chem.* 2007, *35*, 501–511.

5. Yao, H.C.; Japar, S.; Shelef, M. Surface Interactions in the System Rh/Al_2O_3. *J. Catal.* 1977, *50*, 407–418.

6. Libby, W.F. Promising Catalyst for Auto Exhaust. *Science* 1971, *171*, 499–500.

7. Voorhoeve, R.J.H.; Remeika, J.P.; Freeland, P.E.; Matthias, B.T. Rare-Earth Oxides of Manganese and Cobalt Rival Platinum for the Treatment of Carbon Monoxide in Auto Exhaust. *Science* 1972, *177*, 353–354.

8. Voorhoeve, R.J.H.; Remeika, J.P.; Johnson, D.W., Jr. Rare-Earth Manganites: Catalysts with Low Ammonia Yield in the Reduction of Nitrogen Oxides. *Science* 1973, *180*, 62–64.

9. Voorhoeve, R.J.H.; Remeika, J.P.; Trimble, L.E. Perovskites containing ruthenium as catalysts for nitric oxide reduction.*Mater. Res. Bull.* 1974, *9*, 1393–1404.

10. Voorhoeve, R.J.H.; Trimble, L.E. Exploration of perovskite-like catalysts: Ba_2CoWO_6 and Ba_2FeNbO_6 in NO reduction and CO oxidation. *Mater. Res. Bull.* 1974, *9*, 655–666.

11. Gallagher, P.K.; Johnson, D.W., Jr.; Schrey, F. Studies of some supported perovskite oxidation catalysts. *Mater. Res. Bull.*1974, *9*, 1345–1352.

12. Tabata, K.; Misono, M. Elimination of pollutant gases—Oxidation of CO, reduction and decomposition of NO. *Catal. Today* 1990, *8*, 249–261.

13. Guilhaume, N.; Primet, M. Three-Way Catalytic Activity and Oxygen Storage Capacity of Perovskite $LaMn_{0.976}Rh_{0.024}O_{3+\delta}$. *J. Catal.* 1997, *165*, 197–204.

14. Tanaka, H. An intelligent catalyst: The self-regenerative palladium-perovskite catalyst for automotive emissions control. *Catal. Surv. Asia* 2005, *9*, 63–74.

15. Kim, C.H.; Qi, G.; Dahlberg, K.; Li, W. Strontium-Doped Perovskites Rival Platinum Catalysts for Treating NO_x in Simulated Diesel Exhaust. *Science* 2010, *327*, 1624–1627.

16. Goldschmidt, V.M. Die Gesetze der Krystallochemie. *Die Naturwissenschaften* 1926, *8*, 477–485.

17. Ebbinghaus, S.G.; Abicht, H.P.; Dronskowski, R.; Müller, T.; Reller, A.; Weidenkaff, A. Perovskite-related oxynitrides—Recent developments in synthesis, characterisation and investigations of physical properties. *Progr. Solid State Chem.*2009, *37*, 173–205.

18. Yoon, S.; Maegli, A.E.; Karvonen, L.; Matam, S.K.; Shkabko, A.; Riegg, S.; Großmann, T.; Ebbinghaus, S.G.; Pokrant, S.; Weidenkaff, A. Bandgap tuning in $SrTi(N,O,F)_3$ by anionic-lattice variation. *J. Solid State Chem.* 2013, *206*, 226–232.

19. Weidenkaff, A.; Ebbinghaus, S.G.; Lippert, T.; Montenegro, M.J.; Soltmann, C.; Wessicken, R. Phase formation and phase transition of $Ln_{1-x}Ca_xCoO_{3-\delta}$ (Ln = La, Er) applied for bifunctional air electrodes. *Cryst. Eng.* 2002, *5*, 449–457.

20. Weidenkaff, A. Preparation and Application of Nanostructured Perovskite Phases. *Adv. Eng. Mater.* 2004, *6*, 709–714.

21. Raveau, B. The perovskite history: More than 60 years of research from the discovery of ferroelectricity to colossal magnetoresistance via high TC superconductivity. *Prog. Solid State Chem.* 2007, *35*, 171–173.

22. Weidenkaff, A.; Robert, R.; Aguirre, M.; Bochet, L.; Lippert, T.; Canulescu, S. Development of thermoelectric oxides for renewable energy conversion technologies. *Renew. Energy* 2008, *33*, 342–347.

23. Voorhoeve, R.J.H.; Johnson, D.W., Jr.; Remeika, J.P.; Gallagher, P.K. Perovskite Oxides: Materials Science in Catalysis.*Science* 1977, *195*, 827–833.

24. Bhalla, A.S.; Guo, R.; Roy, R. The perovskite structure—A review of its role in ceramic science and technology. *Mater. Res. Innov.* 2000, *4*, 3–26.

25. Peña, M.A.; Fierro, J.L.G. Chemical Structures and Performance of Perovskite Oxides. *Chem. Rev.* 2001, *101*, 1981–2017.

26. Tejuca, L.G.; Fierro, J.L.G.; Tascón, J.M.D. Structure and Reactivity of Perovskite-Type Oxides. *Adv. Catal.* 1989, *36*, 237–328.

27. Royer, S.; Duprez, D. Catalytic Oxidation of Carbon Monoxide over Transition Metal Oxides. *ChemCatChem* 2011, *3*, 24–65.

28. Parravano, G. Ferroelectric Transitions and Heterogeneous Catalysis. *J. Chem. Phys.* 1952, *20*, 342–343.

29. Dickens, P.G.; Whittingham, M.S. Recombination of oxygen atoms on oxide surfaces. Part 2.—Catalytic activities of the alkali metal tungsten bronzes. *Trans. Faraday Soc.* 1965, *61*, 1226–1231.

30. Tascón, J.M.D.; Tejuca, L.G. Catalytic activity of perovskite-type oxides $LaMeO_3$. *React. Kinet. Catal. Lett.* 1980, *15*, 185–191.

31. Nitadori, T.; Ichiki, T.; Misono, M. Catalytic Properties of Perovskite-Type Mixed Oxides (ABO_3) Consisting of Rare Earth and 3d Transition Metals. The Roles of the A- and B-Site Ions. *Bull. Chem. Soc. Jpn.* 1988, *61*, 621–626.

32. Panich, N.M.; Pirogova, G.N.; Korosteleva, R.I.; Voronin, Y.V. Oxidation of CO and hydrocarbons over perovskite-type complex oxides. *Russ. Chem. Bull.* 1999, *48*, 694–697.

33. Voorhoeve, R.J.H.; Remeika, J.P.; Trimble, L.E.; Cooper, A.S.; Disalvo, F.J.; Gallagher, P.K. Perovskite-like $La_{1-x}K_xMnO_3$ and related compounds: Solid state chemistry and the catalysis of the reduction of NO by CO and H_2.*J. Sol. State. Chem.* 1975, *14*, 395–406.

34. Hackenberger, M.; Stephan, K.; Kießling, D.; Schmitz, W.; Wendt, G. Influence of the preparation conditions on the properties of perovskite-type oxide catalysts. *Solid State Ionics* 1997, *101–103*, 1195–1200.

35. Giannakas, A.E.; Ladavos, A.K.; Pomonis, P.J. Preparation, characterization and investigation of catalytic activity for NO + CO reaction of $LaMnO_3$ and $LaFeO_3$ perovskites prepared via microemulsion method. *Appl. Catal. B* 2004, *49*, 147–158.

36. Lu, Y.; Eyssler, A.; Otal, E.H.; Matam, S.K.; Brunko, O.; Weidenkaff, A.; Ferri, D. Influence of the synthesis

method on the structure of Pd-substituted perovskite catalysts for methane oxidation. *Catal. Today* 2013, *208*, 42–47.

37. Campagnoli, E.; Tavares, A.; Fabbrini, L.; Rossetti, I.; Dubitsky, Y.A.; Zaopo, A.; Forni, A. Effect of preparation method on activity and stability of $LaMnO_3$ and $LaCoO_3$ catalysts for the flameless combustion of methane. *Appl. Catal. B*2005, *55*, 133–139.

38. Oliva, C.; Bonoldi, L.; Cappelli, S.; Fabbrini, L.; Rossetti, I.; Forni, L. Effect of preparation parameters on $SrTiO_{3\pm\delta}$catalyst for the flameless combustion of methane. *J. Mol. Catal. A* 2005, *226*, 33–40.

39. Stephan, K.; Hackenberger, M.; Kiessling, D.; Wendt, G. Supported perovskite-type oxide catalysts for the total oxidation of chlorinated hydrocarbons. *Catal. Today* 1999, *54*, 23–30.

40. Ran, R.; Wu, X.; Weng, D. Effect of complexing species in a sol-gel synthesis on the physicochemical properties of $La_{0.7}Sr_{0.3}Mn_{0.7}Cu_{0.3}O_{3+\lambda}$ catalyst. *J. Alloys Compd.* 2006, *414*, 169–174.

41. Kaliaguine, S.; van Neste, A. Process for Synthesizing Perovskites Using High Energy Milling. U.S. Patent 6,017,504, 25 January 2000.

42. Kaliaguine, S.; van Neste, A.; Szabo, V.; Gallot, J.E.; Bassir, M.; Muzychuk, R. Perovskite-type oxides synthesized by reactive grinding—Part I. Preparation and characterization. *Appl. Catal. A* 2001, *209*, 345–358.

43. Royer, S.; Bérubé, F.; Kaliaguine, S. Effect of the synthesis conditions on the redox and catalytic properties in oxidation reactions of $LaCo_{1-x}Fe_xO_3$. *Appl. Catal. A* 2005, *282*, 273–284.

44. Zhang, R.; Villanueva, A.; Alamdari, H.; Kaliaguine, S. Catalytic reduction of NO by propene over $LaCo_{1-x}Cu_xO_3$perovskites synthesized by reactive grinding. *Appl. Catal. B* 2006, *64*, 220–223.

45. Levasseur, B.; Kaliaguine, S. Methanol oxidation on $LaBO_3$ (B = Co, Mn, Fe) perovskite-type catalysts prepared by reactive grinding. *Appl. Catal. A* 2008, *343*, 29–38.

46. Shu, J.; Kaliaguine, S. Well-dispersed perovskite-type oxidation catalysts. *Appl. Catal. B* 1998, *16*, L303–L308.

47. Ran, R.; Wu, X.; Weng, D.; Fan, J. Oxygen storage capacity and structural properties of Ni-doped $LaMnO_3$ perovskites.*J. Alloys Compd.* 2013, *577*, 288–294.

48. Chang, Y.F.; McCarty, J.G. Novel oxygen storage components for advanced catalysts for emission control in natural gas fueled vehicles. *Catal. Today* 1996, *30*, 163–170.

49. Yamazoe, N.; Teraoka, Y.; Seiyama, T. TPD and XPS study on thermal behavior of adsorbed oxygen in $La_{1-x}Sr_xCoO_3$.*Chem. Lett.* 1981, *10*, 1767–1770.

50. Kremenić, G.; Nieto, J.M.L.; Tascón, J.M.D.; Tejuca, L.G. Chemisorption and catalysis on $LaMO_3$ oxides. *J. Chem. Soc. Faraday Trans. 1* 1985, *81*, 939–949.

51. Arai, H.; Yamada, T.; Eguchi, K.; Seiyama, T. Catalytic combustion of methane over various perovskite-type oxides.*Appl. Catal.* 1986, *26*, 265–276.

52. Markova-Velichkova, M.; Lazarova, T.; Tumbalev, V.; Ivanov, G.; Kovacheva, D.; Stefanov, P.; Naydenov, A. Complete oxidation of hydrocarbons on $YFeO_3$ and $LaFeO_3$ catalysts. *Chem. Eng. J.* 2013, *231*, 236–244.

53. Voorhoeve, R.J.H.; Remeika, J.P.; Trimble, L.E. Defect chemistry and catalysis in oxidation and reduction over perovskite-type oxides. *Ann. N. Y. Acad. Sci.* 1976, *272*, 3–21.

54. Fierro, J.L.G. Structure and composition of perovskite surface in relation to adsorption and catalytic properties. *Catal. Today* 1990, *8*, 153–174.

55. Tascón, J.M.D.; Fierro, J.L.G.; Tejuca, L.G. Kinetics and Mechanism of CO Oxidation on $LaCoO_3$. *Z. Phys. Chem.* 1981,*124*, 249–257.

56. Tascón, J.M.D.; Tejuca, L.G. Adsorption of CO_2 on the perovskite-type oxide $LaCoO_3$. *J. Chem. Soc. Faraday Trans. 1*1981, *77*, 591–602.

57. Forni, L.; Oliva, C.; Barzetti, T.; Selli, E.; Ezerets, A.M.; Vishniakov, A.V. FT-IR and EPR spectroscopic analysis of $La_{1-x}Ce_xCoO_3$ perovskite-like catalysts for NO reduction by CO. *Appl. Catal. B* 1997, *13*, 35–43.

58. Harada, T.; Teraoka, Y.; Kagawa, S. Perovskite-type oxides as catalysts for selective reduction of nitric oxide by ethylene. *Appl. Surf. Sci.* 1997, *121–122*, 505–508.

59. Yu Yao, Y.-F. The oxidation of hydrocarbons and CO over metal oxides: IV. Perovskite-type oxides. *J. Catal.* 1975, *36*, 266–275.

60. Jovanovic, D.; Dondur, V.; Terlecki-Baricevic, A.; Grbic, B. Three-way activity and sulfur tolerance of single phase perovskites. *Stud. Surf. Sci. Catal.* 1991, *71*, 371–379.

61. Gallagher, P.K.; Johnson, D.W., Jr.; Vogel, E.M.; Schrey, F. Effect of the Pt content of $La_{0.7}Pb_{0.3}MnO_3$ on its catalytic activity for the oxidation of CO in the presence of SO_2. *Mat. Res. Bull.* 1975, *10*, 623–628.

62. Tzimpilis, E.; Moschoudis, N.; Stoukides, M.; Bekiaroglou, P. Ageing and SO_2 resistance of Pd containing perovskite-type oxides. *Appl. Catal. B* 2009, *87*, 9–17.

63. Tzimpilis, E.; Moschoudis, N.; Stoukides, M.; Bekiaroglou, P. Preparation, active phase composition and Pd content of perovskite-type oxides. *Appl. Catal. B* 2008, *84*, 607–615.

64. Mondragón Rodríguez, G.C.; Kelm, K.; Heikens, S.; Grünert, W.; Saruhan, B. Pd-integrated perovskites for TWC applications: Synthesis, microstructure and N_2O-selectivity. *Catal. Today* 2012, *184*, 184–191.

65. Sorenson, S.C.; Wronkiewicz, J.A.; Sis, L.B.; Wirtz, G.P. Properties of $LaCoO_3$ as a Catalyst in Engine Exhaust Gases.*Am. Ceram. Soc. Bull.* 1974, *53*, 446–449.

66. Labhsetwar, N.K.; Watanabe, A.; Biniwale, R.B.; Kumar, R.; Mitsuhashi, T. Alumina supported, perovskite oxide based catalytic materials and their auto-exhaust application. *Appl. Catal. B* 2001, *33*, 165–173.

67. Tanaka, H.; Mizuno, N.; Misono, M. Catalytic activity and structural stability of $La_{0.9}Ce_{0.1}Co_{1-x}Fe_xO_3$ perovskite catalysts for automotive emission control. *Appl. Catal. A* 2003, *244*, 371–382.

68. Bradow, R.; Jovanović, D.; Petrović, S.; Jovanović, Ž.; Terlecki-Baričević, A. Ruthenium Perovskite Catalysts for Lean NO_x Automotive Emission Control. *Ind. Eng. Chem. Res.* 1995, *34*, 1929–1932.

69. Shelef, M.; Gandhi, H.S. The Reduction of Nitric Oxide in Automobile Emissions—Stabilisation of catalysts containing ruthenium. *Platin. Met. Rev.* 1974, *18*, 2–14.

70. Nishihata, Y.; Mizuki, J.; Akao, T.; Tanaka, H.; Uenishi, M.; Kimura, M.; Okamoto, T.; Hamada, N. Self-regeneration of a Pd-perovskite catalyst for automotive emissions control. *Nature* 2002, *418*, 164–167.

71. Tanaka, H.; Tan, I.; Uenishi, M.; Kimura, M.; Dohmae, K. Regeneration of palladium subsequent to solid solution and segregation in a perovskite catalyst: An intelligent catalyst. *Top. Catal.* 2001, *16/17*, 63–70.

72. Uenishi, M.; Taniguchi, M.; Tanaka, H.; Kimura, M.; Nishihata, Y.; Mizuki, J.; Kobayashi, T. Redox behavior of palladium at start-up in the Perovskite-type $LaFePdO_x$ automotive catalysts showing a self-regenerative function.*Appl. Catal. B* 2005, *57*, 267–273.

73. Uenishi, M.; Tanaka, H.; Taniguchi, M.; Tan, I.; Sakamoto, Y.; Matsunaga, S.; Yokota, K.; Kobayashi, T. The reducing capability of palladium segregated from perovskite-type $LaFePdO_x$ automotive catalysts. *Appl. Catal. A* 2005, *296*, 114–119.

74. Tanaka, H.; Taniguchi, M.; Kajita, N.; Uenishi, M.; Tan, I.; Sato, N.; Narita, K.; Kimura, M. Design of the intelligent catalyst for Japan ULEV standard. *Top. Catal.* 2004, *30/31*, 389–396.

75. Tanaka, H.; Taniguchi, M.; Uenishi, M.; Kajita, N.; Tan, I.; Nishihata, Y.; Mizuki, J.; Narita, K.; Kimura, M.; Kaneko, K. Self-Regenerating Rh- and Pt-Based Perovskite Catalysts for Automotive-Emissions Control. *Angew. Chem. Int. Ed.*2006, *45*, 5998–6002.

76. Dacquin, J.P.; Cabié, M.; Henry, C.R.; Lancelot, C.; Dujardin, C.; Raouf, S.R.; Granger, P. Structural changes of nano-Pt particles during thermal ageing: Support-induced effect and related impact on the catalytic performances. *J. Catal.*2010, *270*, 299–309.

77. Dacquin, J.P.; Lancelot, C.; Dujardin, C.; Cordier-Robert, C.; Granger, P. Support-Induced Effects of $LaFeO_3$ Perovskite on the Catalytic Performances of Supported Pt Catalysts in $DeNO_x$ Applications. *J. Phys. Chem. C* 2011, *115*, 1911–1921.

78. Katz, M.B.; Graham, G.W.; Duan, Y.; Liu, H.; Adamo, C.; Schlom, D.G.; Pan, X. Self-Regeneration of Pd-$LaFeO_3$Catalysts: New Insight from Atomic-Resolution Electron Microscopy. *J. Am. Chem. Soc.* 2011, *133*, 18090–18093.

79. Katz, M.B.; Zhang, S.; Duan, Y.; Wang, H.; Fang, M.; Zhang, K.; Li, B.; Graham, G.W.; Pan, X. Reversible precipitation/dissolution of precious-metal clusters in perovskite-based catalyst materials: Bulk *versus* surface re-dispersion. *J. Catal.* 2012, *293*, 145–148.

80. Zhou, K.; Chen, H.; Tian, Q.; Hao, Z.; Shen, D.; Xu, X. Pd-containing perovskite-type oxides used for three-way catalysts. *J. Mol. Catal. A* 2002, *189*, 225–232.

81. Lu, Y.; Michalow, K.A.; Matam, S.K.; Winkler, A.; Maegli, A.E.; Yoon, S.; Heel, A.; Weidenkaff, A.; Ferri, D. Methane abatement under stoichiometric conditions on perovskite-supported palladium catalysts prepared by flame spray synthesis. *Appl. Catal. B* 2014, *144*, 631–643.

82. Nijhuis, T.A.; Beers, A.E.W.; Vergunst, T.; Hoek, I.; Kapteijn, F.; Moulijn, J.A. Preparation of monolithic catalysts. *Catal. Rev. Sci. Eng.* 2001, *43*, 345–380.

83. Tomašić, V.; Jović, F. State-of-the-art in the monolithic catalysts/reactors. *Appl. Catal. A* 2006, *311*, 112–121.

84. Meille, V. Review on methods to deposit catalysts on structured surfaces. *Appl. Catal. A* 2006, *315*, 1–17.

85. Avila, P.; Montes, M.; Miró, E.E. Monolithic reactors for environmental applications: A review on preparation technologies. *Chem. Eng. J.* 2005, *109*, 11–36.

86. Lachman, I.M.; Williams, J.L. Extruded Monolithic Catalysts Supports. *Catal. Today* 1992, *14*, 317–329.

87. Bardhan, P. Ceramic honeycomb filters and catalysts. *Curr. Opin. Solid State Mater. Sci.* 1997, *2*, 577–583.

88. Isupova, L.A.; Sadykov, V.A.; Tikhov, S.F.; Kimkhai, O.N.; Kovalenko, O.N.; Kustova, G.N.; Ovsyannikova, I.A.; Dovbii, Z.A.; Kryukova, G.N.; Rozovskii, A.Y.; *et al.* Monolith perovskite catalysts for environmentally benign fuels combustion and toxic wastes incineration. *Catal. Today* 1996, *27*, 249–256.

89. Isupova, L.A.; Sadykov, V.A.; Solovyova, L.P.; Andrianova, M.P.; Ivanov, V.P.; Kryukova, G.N.; Kolomiichuk, V.N.; Avvakumov, E.G.; Pauli, I.A.; Andryushkova, O.V.; *et al.* Monolith perovskite catalysts of honeycomb structure for fuel combustion. *Stud. Surf. Sci. Catal.* 1995, *91*, 637–645.

90. Sadykov, V.A.; Isupova, L.A.; Tikhov, S.F.; Kimkhai, O.N. Perovskite catalysts: High surface area powders synthesis, monolith shaping and high-temperature applications. *Mat. Res. Soc. Symp. Proc.* 1995, *368*, 293–298.

91. Isupova, L.A.; Alikina, G.M.; Snegurenko, O.I.; Sadykov, V.A.; Tsybulya, S.V. Monolith honeycomb mixed oxide catalysts for methane oxidation. *Appl. Catal. B* 1999, *21*, 171–181.

92. Isupova, L.A.; Sadykov, V.A.; Alikina, G.M.; Snegurenko, O.I.; Tsybulya, S.V.; Salanov, A.N.; Rogov, V.A. Monolith honeycomb mixed oxide catalysts for methane oxidation. *Stud. Surf. Sci. Catal.* 2000, *130*, 3783–3788.

93. Ciambelli, P.; Palma, V.; Tikhov, S.F.; Sadykov, V.A.; Isupova, L.A.; Lisi, L. Catalytic activity of powder and monolith perovskites in methane combustion. *Catal. Today* 1999, *47*, 199–207.

94. Schneider, R.; Kiessling, D.; Wendt, G.; Burckhardt, W.; Winterstein, G. Perovskite-type oxide monolithic catalysts for combustion of chlorinated hydrocarbons. *Catal. Today* 1999, *47*, 429–435.

95. Khan, H.R.; Frey, H. R.f. plasma spray deposition of $LaMO_x$ (M = Co, Mn, Ni) films and the investigations of structure, morphology and the catalytic oxidation of CO and C_3H_8. *J. Alloys Compd.* 1993, *190*, 209–217.

96. Bahlawane, N.; Ngamou, P.H.T.; Kohse-Höinghaus, K. Structure, Electrical Properties, and Surface Reactivity of CVD-Made Functional Complex Oxides. *J. Electrochem. Soc.* 2010, *157*, D16–D20.

97. Sansernnivet, M.; Laosiripojana, N.; Assabumrungrat, S.; Charojrochkul, S. Fabrication of $La_{0.8}Sr_{0.2}CrO_3$-based Perovskite Film via Flame-Assisted Vapor Deposition for H_2 Production by Reforming. *Chem. Vap. Depos.* 2010, *16*, 311–321.

98. Jarligo, M.O.; Mauer, G.; Bram, M.; Baumann, S.; Vaßen, R. Plasma Spray Physical Vapor Deposition of $La_{1-x}Sr_xCo_yFe_{1-y}O_{3-\delta}$ Thin-Film Oxygen Transport Membrane on Porous Metallic Supports. *J. Therm. Spray Technol.*2014, *23*, 213–219.

99. Uusi-Esko, K.; Karppinen, M. Extensive Series of Hexagonal and Orthorhombic $RMnO_3$ (R = Y, La, Sm, Tb, Yb, Lu) Thin Films by Atomic Layer Deposition. *Chem. Mater.* 2011, *23*, 1835–1840.

100. Schneider, R.; Kiessling, D.; Wendt, G. Cordierite monolith supported perovskite-type oxides catalysts for the total oxidation of chlorinated hydrocarbons. *Appl. Catal. B* 2000, *28*, 187–195.

101. Isupova, L.A.; Alikina, G.M.; Tsybulya, S.V.; Salanov, A.N.; Blodyreva, N.N.; Rusina, E.S.; Ovsyannikova, I.A.; Rogov, V.A.; Bunina, R.V.; Sadykov, V.A. Honeycomb-supported perovskite catalysts for high-temperature processes. *Catal. Today* 2002, *75*, 305–315.

102. Li, L.; Shen, X.; Wang, P.; Meng, X.; Song, F. Soot capture and combustion for perovskite La-Mn-O based catalysts coated on honeycomb ceramic in practical diesel exhaust. *Appl. Surf. Sci.* 2011, *257*, 9519–9524.

103. Isupova, L.A.; Sutormina, E.F.; Kulilovskaya, N.A.; Plyasova, L.M.; Rudina, N.A.; Ovsyannikova, I.A.; Zootarskii, I.A.; Sadykov, V.A. Honeycomb supported perovskite catalysts for ammonia oxidation processes. *Catal. Today* 2005, *105*, 429–435.

104. Sannino, D.; Vaiano, V.; Ciambelli, P.; Isupova, L.A. Structured catalysts for photo-Fenton oxidation of acetic acid.*Catal. Today* 2011, *161*, 255–259.

105. Pechini, M.P. Method of preparing lead and alkaline earth titanates and niobates and coating method using the same to form a capacitor. U.S. Patent 3,330,697, 11 July 1967.

106. Qi, A.; Wang, S.; Fu, G.; Ni, C.; Wu, D. La-Ce-Ni-O monolithic perovskite catalysts potential for gasoline autothermal reforming system. *Appl. Catal. A* 2005, *281*, 233–246.

107. Vergunst, T.; Kapteijn, F.; Moulijn, J.A. Monolithic catalysts—Non-uniform active phase distribution by impregnation.*Appl. Catal. A* 2001, *213*, 179–187.

108. Cauda, E.; Fino, D.; Saracco, G.; Specchia, V. Nanosized Pt-perovskite catalyst for the regeneration of a wall-flow filter for soot removal from diesel exhaust gases. *Top. Catal.* 2004, *30/31*, 299–303.

109. Russo, N.; Furfori, S.; Fino, D.; Saracco, G.; Specchia, V. Lanthanum cobaltite catalysts for diesel soot combustion.*Appl. Catal. B* 2008, *83*, 85–95.

110. Russo, N.; Fino, D.; Saracco, G.; Specchia, V. Promotion effect of Au on perovskite catalysts for the regeneration of diesel particulate filters. *Catal. Today* 2008, *137*, 306–311.

111. Mescia, D.; Caroca, J.C.; Russo, N.; Labhsetwar, N.; Fino, D.; Saracco, G.; Specchia, V. Towards a single brick solution for the abatement of NO_x and soot from diesel engine exhausts. *Catal. Today* 2008, *137*, 300–305.

112. Fino, D.; Russo, N.; Cauda, E.; Saracco, G.; Specchia, V. La-Li-Cr perovskite catalysts for diesel particulate combustion.*Catal. Today* 2006, *114*, 31–39.

113. Furfori, S.; Russo, N.; Fino, D.; Saracco, G.; Specchia, V. NO SCR reduction by hydrogen generated in line on perovskite-type catalysts for automotive diesel exhaust gas treatment. *Chem. Eng. Sci.* 2010, *65*, 120–127.

114. Cauda, E.; Fino, D.; Saracco, G.; Specchia, V. Preparation and regeneration of a catalytic diesel particulate filter. *Chem. Eng. Sci.* 2007, *62*, 5182–5185.

115. Mescia, D.; Cauda, E.; Russo, N.; Fino, D.; Saracco, G.; Specchia, V. Towards practical application of lanthanum chromite catalysts for diesel particulate combustion. *Catal. Today* 2006, *117*, 369–375.

116. Fino, D.; Fino, P.; Saracco, G.; Specchia, V. Innovative means for the catalytic regeneration of particulate traps for diesel exhaust cleaning. *Chem. Eng. Sci.* 2003, *58*, 951–958.

117. Arendt, E.; Maione, A.; Klisinska, A.; Sanz, O.; Montes, M.; Suarez, S.; Blanco, J.; Ruiz, P. Structuration of $LaMnO_3$perovskite catalysts on ceramic and metallic monoliths: Physico-chemical characterisation and catalytic activity in methane combustion. *Appl. Catal. A* 2008, *339*, 1–14.

118. Fabbrini, L.; Rossetti, I.; Forni, L. Effect of primer on honeycomb-supported $La_{0.9}Ce_{0.1}CoO_{3\pm\delta}$ perovskite for methane catalytic flameless combustion. *Appl. Catal. B* 2003, *44*, 107–116.

119. Podyacheva, O.Y.; Ketov, A.A.; Ismagilov, Z.R.; Ushakov, V.A.; Bos, A.; Veringa, H.J. Metal foam supported perovskite catalysts. *React. Kinet. Catal. Lett.* 1997, *60*, 243–250.

120. Forni, L.; Rossetti, I. Catalytic combustion of hydrocarbons over perovskites. *Appl. Catal. B* 2002, *38*, 29–37.

121. Fabbrini, L.; Rossetti, I.; Forni, L. La_2O_3 as primer for supporting $La_{0.9}Ce_{0.1}CoO_{3\pm\delta}$ on cordieritic honeycombs. *Appl. Catal. B* 2005, *56*, 221–227.

122. Fabbrini, L.; Rossetti, I.; Forni, L. Effect of honeycomb supporting on activity of $LaBO_{3\pm\delta}$ perovskite-like catalysts for methane flameless combustion. *Appl. Catal. B* 2006, *63*, 131–136.

123. Arai, H.; Machida, M. Thermal stabilization of catalyst supports and their application to high-temperature catalytic combustion. *Appl. Catal. A* 1996, *138*, 161–176.

124. Zhang-Steenwinkel, Y.; van der Zande, L.M.; Castricum, H.L.; Bliek, A.; van den Brink, R.W.; Elzinga, G.D. Microwave-assisted *in-situ* regeneration of a perovskite coated diesel soot filter. *Chem. Eng. Sci.* 2005, *60*, 797–

804.

125. Kucharczyk, B.; Tylus, W. Effect of Pd or Ag additive on the activity and stability of monolithic $LaCoO_3$ perovskites for catalytic combustion of methane. *Catal. Today* 2004, *90*, 121–126.

126. Kucharczyk, B.; Tylus, W. Partial substitution of lanthanum with silver in the $LaMnO_3$ perovskite: Effect of the modification on the activity of monolithic catalysts in the reactions of methane and carbon oxide oxidation. *Appl. Catal. A* 2008, *335*, 28–36.

127. Cimino, S.; Pirone, R.; Russo, G. Thermal Stability of Perovskite-Based Monolithic Reactors in the Catalytic Combustion of Methane. *Ind. Eng. Chem. Res.* 2001, *40*, 80–85.

128. Cimino, S.; Pirone, R.; Lisi, L. Zirconia supported $LaMnO_3$ monoliths for the catalytic combustion of methane. *Appl. Catal. B* 2002, *35*, 243–254.

129. Cimino, S.; Lisi, L.; Pirone, R.; Russo, G.; Turco, M. Methane combustion on perovskites-based structured catalysts.*Catal. Today* 2000, *59*, 19–31.

130. Jiratova, K.; Moravkova, L.; Malecha, J.; Koutsky, B. Ceramic foam-supported perovskites as catalysts for combustion of methane. *Collect. Czechoslov. Chem. Commun.* 1997, *62*, 875–883.

131. Zwinkels, M.F.M.; Haussner, O.; Govind Menon, P.; Järås, S.G. Preparation and characterization of $LaCrO_3$ and Cr_2O_3methane combustion catalysts supported on $LaAl_{11}O_{18}$- and Al_2O_3-coated monoliths. *Catal. Today* 1999, *47*, 73–82.

132. Cimino, S.; Di Benedetto, A.; Pirone, R.; Russo, G. Transient behaviour of perovskite-based monolithic reactors in the catalytic combustion of methane. *Catal. Today* 2001, *69*, 95–103.

133. Muroi, T.; Nojiri, N.; Deguchi, T. Recent progress in catalytic technology in Japan—II (1994–2009). *Appl. Catal. A* 2010,*389*, 27–45.

134. Dimopoulos Eggenschwiler, P.; Tsinoglou, D.N.; Seyfert, J.; Bach, C.; Vogt, U.; Gorbar, M. Ceramic foam substrates for automotive catalyst applications: Fluid mechanic analysis. *Exp. Fluids* 2009, *47*, 209–222.

135. Patcas, F.C.; Garrido, G.I.; Kraushaar-Czanetzki, B. CO oxidation over structured carriers: A comparison of ceramic foams, honeycombs and beads. *Chem. Eng. Sci.* 2007, *62*, 3984–3990.

136. Populoh, S.; Trottmann, M.; Brunko, O.C.; Thiel, P.; Weidenkaff, A. Construction of a high temperature TEG measurement system for the evaluation of thermoelectric oxide modules. *Funct. Mater. Lett.* 2013, *6*.

CHAPTER 4

CHARACTERIZATION OF CATALYTIC CORRELATIONS IN COPPER-CERIA-BASED CATALYSTS FOR PREFERENTIAL OXIDATION OF CO IN H2-RICH STREAMS

Arturo Martínez-Arias, Daniel Gamarra, Ana B. Hungría, Marcos Fernández-García, Guillermo Munuera, Aitor Hornés, Parthasarathi Bera, José C. Conesa and Antonio López Cámara

Keywords: Copper-ceria catalysts; CO preferential oxidation (CO-PROX); diffuse reflectance infrared Fourier transform spectroscopy (DRIFTS); X-ray absorption near edge spectroscopy (XANES); Raman; X-ray diffraction (XRD); X-ray photoelectron spectroscopy (XPS)

INTRODUCTION

Production of H_2 for polymer fuel cells (PEMFC) is usually accomplished by a multi-step process that includes catalytic reforming of hydrocarbons or oxygenated hydrocarbons followed by water-gas shift (WGS) [1,2]. The gas stream obtained after these processes presents, in most cases, a relatively high CO concentration that disallows efficient handling of the fuel by the Pt alloy anode usually employed in the PEMFC. Preferential (or selective) oxidation of CO in the H_2-rich stream resulting from such processes (CO-PROX) has been recognized as one of the most straightforward and cost-effective methods to achieve acceptable CO concentrations (below *ca.* 100 ppm) [3,4,5,6,7].

Different types of catalysts have shown their efficiency for the CO-PROX process. These can be classified into three general groups as a function of their nature and/or respective catalytic properties. The first group involves supported noble metal catalysts (mainly Pt ones) and follows from the first developments done by Engelhard researchers in the context of processes related to production of clean hydrogen for ammonia synthesis [8]. These are mostly employed and commercially available catalysts, although they exhibit a relatively low selectivity for the reaction of interest (CO oxidation) at practical operating temperature (between 373 and 473 K), which can make necessary interstage cooling operations to avoid extensive heating as a consequence of the exothermicity of the (H_2 and CO) oxidation reactions involved [9]. A second group of active catalysts involves supported gold catalysts, well known for their outstanding performance for CO oxidation [10,11,12,13,14]. These show a high CO-PROX activity with a good match between their activity window and the PEMFC anode operating temperature (353–403 K). However, they can have the drawback of their poor resistance to the presence of CO_2 in the reactant mixture [3,11,12,15]. The third group is constituted by catalysts based on closely interacting copper oxide and ceria, which have shown promising properties in terms of activity, selectivity and resistance to CO_2 and H_2O, while their lower cost (particularly in comparison to catalysts based on supported platinum, taking also into account the relatively high platinum loading required to optimize CO-PROX performance) could make them strongly competitive [3 ,4,6,7,16,17,18,19,20,21,22,23,24].

The particular ability of copper-ceria catalysts for the CO-PROX or related processes has been essentially attributed to the synergistic redox properties exhibited upon formation of copper oxide-ceria interfacial sites [4,6,17,18,25,26,27,2 8,29,30,31,32,33,34,35]. In this sense, generally speaking,

Citation: Arturo Martínez-Arias, Daniel Gamarra , Ana B. Hungría, Marcos Fernández-García , Guillermo Munuera , Aitor Hornés , Parthasarathi Bera , José C. Conesa and Antonio López Cámara, Characterization of Catalytic Correlations in Copper-Ceria-Based Catalysts for Preferential Oxidation of CO in H2-Rich Streams, doi:10.3390/catal3020378

the properties of copper oxide entities for CO oxidation promotion depend strongly on their dispersion degree and/or related degree of interaction with ceria [18,25,26,36,37]. Nevertheless, although reaction models for CO oxidation (competing or not with H_2 oxidation, as it occurs under CO-PROX conditions) have been proposed on the basis of indirect analysis of redox or catalytic properties for this type of catalyst [38,39]; direct evidence on redox changes taking place in the catalyst under the reactant atmosphere, which could provide details on the nature of active sites for the processes taking place under CO-PROX conditions in this type of catalyst, has, to the best of our knowledge, only recently been reported in a series of contributions from our laboratory [18,40,41]. In turn, such redox characteristics have been shown recently to depend strongly on the type of face exposed by the ceria support, which has important consequences for the CO-PROX performance of this type of catalyst [42].

The present contribution reviews such studies and also analyzes issues related to catalytic/redox correlations on catalysts of this type in order to attempt to establish the most relevant aspects of the state of the art in this field.

RESULTS AND DISCUSSION

Redox and Catalytic Properties as a Function of Changes in the Ceria-Based Support Nature

The ability of catalysts based on combinations between copper oxide and ceria for CO oxidation or the CO-PROX process has been related to the promotion of redox properties, which is achieved upon combination of both oxides, taking into account that they appear to operate under a redox-type catalytic mechanism [6]. The existence of a redox promotion has been mainly based on intensive investigation done with different spectroscopic techniques, as well as temperature-programmed reduction studies in which it has generally been shown that the CO oxidation rate can be correlated with the degree of ceria-promoted reducibility attained on the dispersed copper oxide entities [6,26,29,39,43,44]. In turn, the level of promotion of such reduction can be also affected by modifying the nature of the support within structurally related doped ceria materials [18,27]. In this sense, the first approach of the authors' laboratory to the catalytic properties of this type of system was done in a work in which a set of catalysts prepared by impregnation with the same copper loading (1 wt.%)—but differing in the nature of the ceria-related support that was employed—were tested for the CO-PROX process [18]. Supports were chosen on the basis of previous experience in the field of three-way catalysts (TWC), as well as considering differences in oxygen transport properties [45]. Thus, in addition to the simple pure ceria support, Ce-Zr and Ce-Tb mixed oxide supports were employed, taking into account that they could provide enhanced physicochemical and catalytic characteristics based on previous experience in the TWC field, particularly in terms of enhanced redox (oxygen transport) properties [46]. Preparation and characterization details for these systems can be found elsewhere [18]; it must be noted that S_{BET} values around 100 m^2/g were exhibited by all these systems. As displayed in Figure 1, all catalysts are essentially constituted by fluorite nanocrystals (stable structure for the corresponding supports; also in agreement with HREM investigation displaying the presence of more or less rounded *ca.* 5–8 nm nanocrystals [46,47]) onto which copper oxide entities have been dispersed. The chemical state of copper has been shown to be similar for all samples and, in turn, similar to that observed for a CuO reference, with small differences being attributed to interactions with the ceria-related support [18,37], as evidenced by XANES (Figure 1). The existence of a fully oxidized state, Cu(II), of copper in the initial calcined catalysts is also in agreement with XPS investigation, as shown in 2. In turn values of XPS atomic ratios are in agreement with achievement of a relatively high dispersion degree in the copper oxide entities in all cases). However, in spite of the fact that mixed oxide supports could provide enhanced redox properties to the catalysts [46], overall CO-PROX performance is shown to be optimized (both in terms of CO conversion and CO_2 selectivity—*i.e.*, O_2 selectivity towards oxidizing CO in competition with H_2) for the system supported on pure ceria, as displayed in 3. In turn such behavior cannot be attributed to an enhanced copper oxide dispersion in the case of CuC according to analysis of Ar^+-sputtering XPS experiments (Figure 2), which point to an increase in the dispersion as CuC ≈ CuCT4 < CuCT1 < CuCZ [18]. In order to attempt to rationalize this behavior, the catalysts of the Ce-Tb series have been examined by means of CO-TPR tests [48]. In this sense, recent work by Luo *et al.* [32] provides an interesting rationalization of redox/catalytic correlations in catalysts combining copper and cerium oxides; on the basis of classical CO-TPR results, they could differentiate between the reducibilities of the three types of oxidized copper entities generically proposed to be present in this type of catalyst (in decreasing order of reducibility or CO oxidation activity): finely dispersed CuO; bulk-like CuO; and, finally, Cu^{2+} in the ceria lattice, likely substitutionally. Nevertheless, such an investigation has been undertaken on a series of catalysts combining in all cases copper oxide and ceria while, as previously mentioned, changes in the support nature can modify such redox or catalytic properties [46]. In this sense, as shown in Figure 4, the reduction profile (under CO) of CuC is characterized by a low intensity reduction peak and a somewhat stronger peak at about 170 °C and 225 °C, respectively, which are referred to as α and β peaks, according to the usual nomenclature employed in the literature [32,33,34]. In addition, there is a high intensity peak at 298 °C (γ peak). These peaks below 300 °C can be assigned to the reduction of Cu^{2+} ions belonging to different types of oxidized copper entities with differing degrees of interaction with the underlying ceria support. The presence of more than one reduction peak in the CuC catalyst, as opposed to one peak in the case of pure CuO (Figure 4), is consistent with the existence of more than one type of copper oxide species in CuC, which is in agreement with the irregular shape detected in the Cu/Ce atomic ratio evolution during Ar^+-sputtering experiments (Figure 2) [18]. The reduction of copper species responsible for α and β peaks occurs at lower temperatures compared to pure CuO, while the γ peak maximum appears fairly close to that of bulk CuO. This justifies attribution of the former two peaks to well-dispersed copper oxide species whose reduction would be appreciably promoted by ceria, while the latter accounts for the presence of less dispersed (even though still diffraction silent; see Figure 1) copper oxide entities. Such an interpretation is also in agreement with results achieved by EPR in previous studies [29]. Apparent downward shifts of the reduction of copper species are observed in the presence of Tb in the support (Figure 4). Thus, the reduction profile of CuCT4 contains a low temperature peak at 120 °C, whereas two peaks appear at higher temperatures (190 °C and 250 °C). The presence of multiple peaks must reflect, as in the case of CuC, the heterogeneity of copper oxide entities in these samples, which is in agreement also with Ar^+-sputtering experiments (Figure 2). Concerning CuCT1, the reduction starts around 50 °C and an intense peak at 155 °C is observed. There is also a hump at 250 °C in this case, which could in part be related to the existence of residual WGS activity (as a consequence of interaction of CO with surface hydroxyls) [49]. Similar results can be found in reports by Wang *et al.* [38,50], in which the formation of a low temperature α peak in case of CuO on samarium-doped ceria has been attributed to the effect of the presence of surface oxygen vacancies within the oxygen ion conducting support. They have inferred that an interfacial metal oxide-support interaction mechanism is involved at the inception of the reduction, followed by the induced successive reduction of bulk copper oxides, somewhat in accordance with a proposed model for the reduction of components of this type of system upon interaction with CO [29]. Similarly, our CO-TPR results (Figure 4) suggest that the low temperature reducibility of finely dispersed CuO on CuCT4 and CuCT1 is enhanced in comparison to CuC.

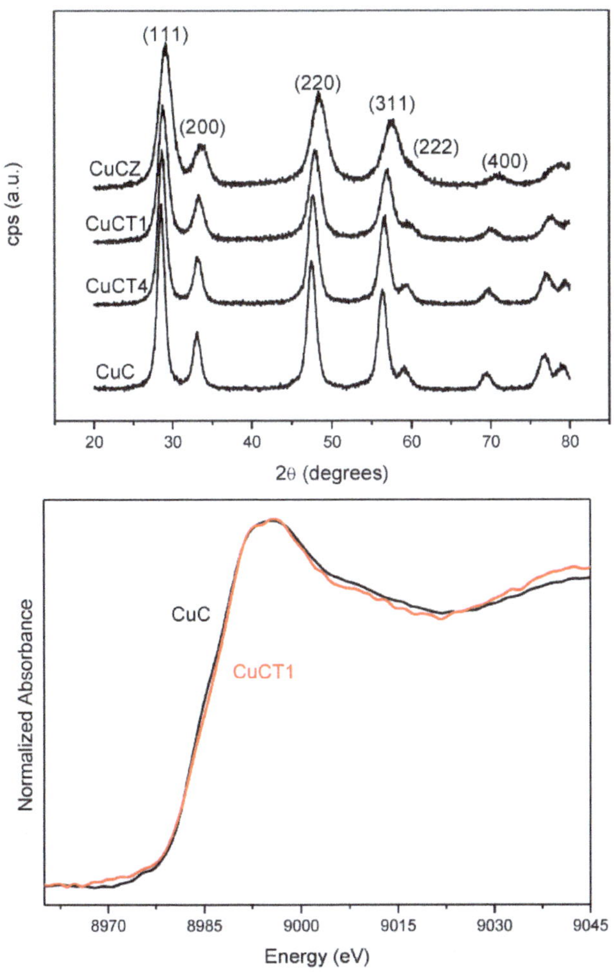

Figure 1: Top: X-ray diffractograms of catalysts with 1 wt.% copper oxide dispersed on ceria (CuC), Ce-Zr mixed oxide (CuCZ) and Ce-Tb mixed oxides with 4:1 (CuCT4) and 1:1 (CuCT1) Ce:Tb atomic ratios. Bottom: XANES spectra for the indicated catalysts. Note: This figure is adapted with permission from [18]. Copyright © 2005, Elsevier B.V.

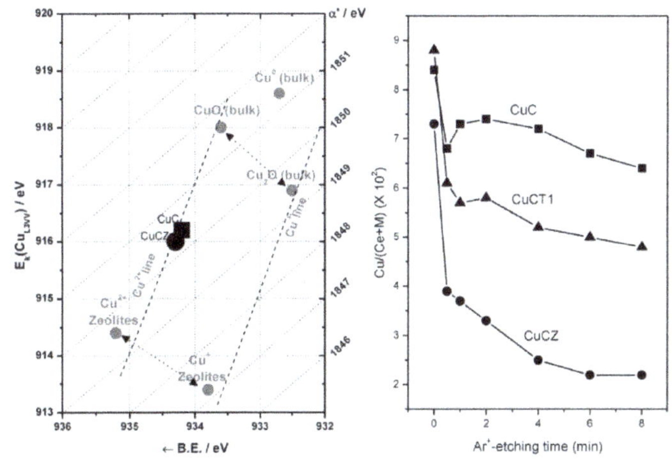

Figure 2. Left: Wagner diagram showing Cu (2p and AES) XP parameters observed for CuC and CuCZ samples. Right: Profile of the Cu/(Ce + M), M = Zr or Tb, atomic ratio as a function of the Ar^+-sputtering time for the indicated samples. Note: This figure is adapted with permission from [18]. Copyright © 2005, Elsevier B.V.

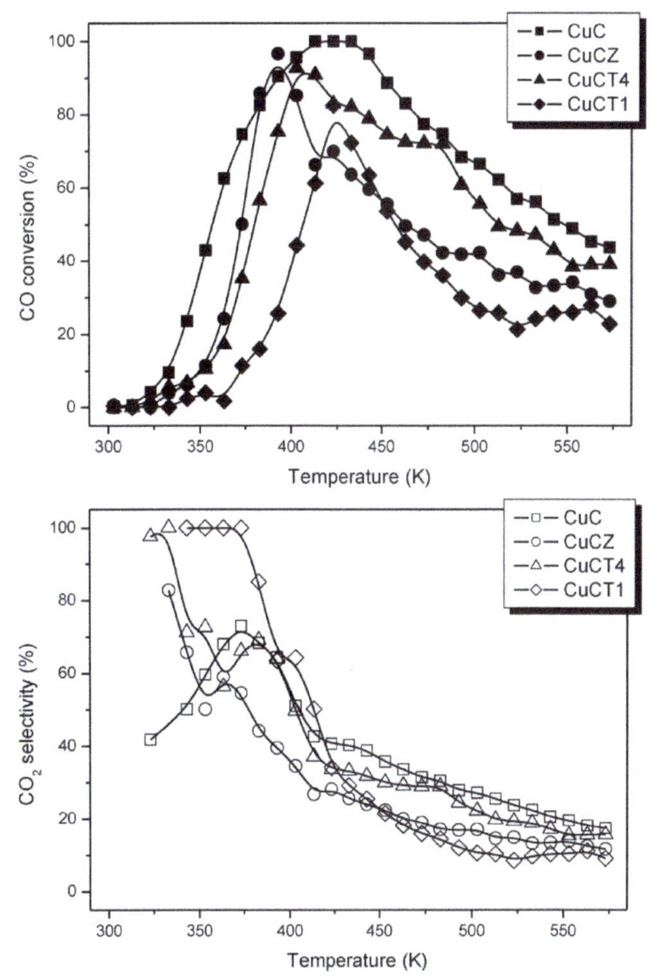

Figure 3. Profiles of CO conversion (top) and selectivity for the CO-PROX process (bottom) obtained during catalytic tests under 1% CO + 1.25% O_2 + 50% H_2 (Ar balance) for the indicated catalysts. Note: This figure is adapted with permission from [18]. Copyright © 2005, Elsevier B.V.

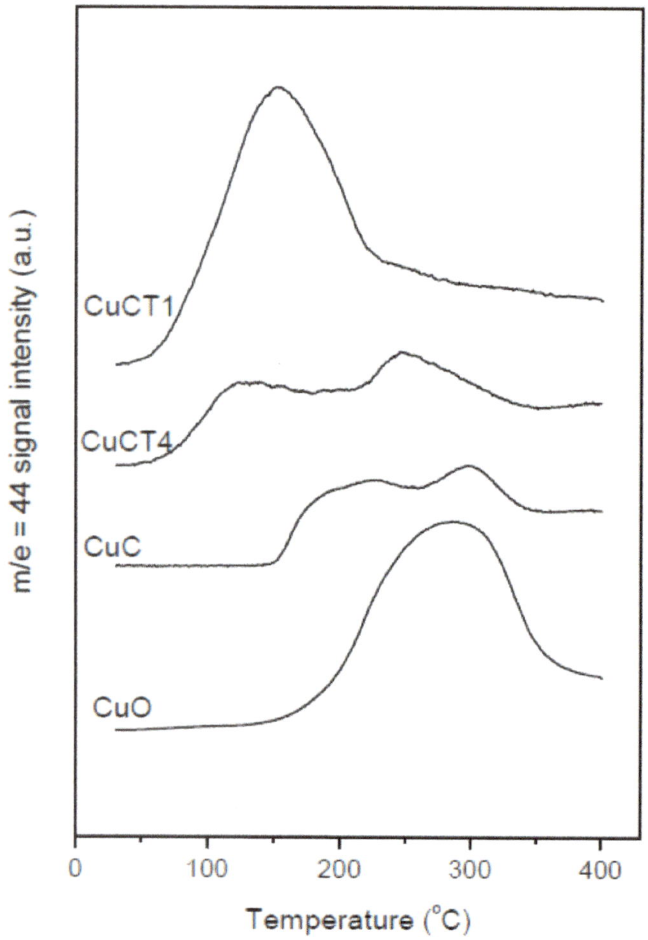

Figure 4. CO_2 production during the CO-TPR runs performed with a tubular reactor over the indicated samples. A CuO reference sample prepared by microemulsion has been included for comparative purpose. Note: This figure is reproduced with permission from [48]. Copyright © 2009, Elsevier Inc.

However, such apparently enhanced reduction of dispersed copper oxide entities upon Tb-doping of ceria (as inferred from interpretation of the CO-TPR profiles; Figure 4) does not correlate with the gradual decrease in CO oxidation activity under CO-PROX conditions (Figure 3). It has to be noted that such redox/catalytic correlations appear to be well established on the basis of both CO-TPR and H_2-TPR experiments [26,32,33,34]. In conclusion, these results evidence that such a type of correlation is not fully valid when the support nature changes. This being the case, what could then be the reason for such discrepancies, taking also into account that correlation between copper oxide reducibility and CO oxidation activity has been widely demonstrated not only based on mentioned TPR experiments but also on the basis of various spectroscopic techniques [40,41]? The origin of the discrepancy can be related to the redox phenomena taking place at relatively low temperatures (even subambient) during the gas equilibration period carried out at room temperature prior to starting the recording of the CO-TPR data typically done after such gas equilibration when the heating ramp under diluted CO is launched. Indeed, a process of CO consumption, with no CO_2 evolution, is detected during such equilibration period, in agreement also with previous experiments [48,51,52]. In this respect, amounts of CO consumed at 20 °C prior to launching the ramp are estimated as 230, 210 and 175 µmol/g for CuC, CuCT4 and CuCT1 samples, in good qualitative agreement with the aforementioned decrease of CO oxidation activity upon increasing the amount of terbium in the support (Figure 3). The respective CO oxidation activity is, in turn, reflected by the respective intensity of a Cu^+-carbonyl species (attribution in accordance with previous works, including ^{12}CO-^{13}CO isotopic substitution experiments and analysis of carbonyls thermal stability, where full details can be found [27,29,48,49,51]) formed upon interaction of the calcined samples with CO at room temperature, Figure 5. Since the catalysts were shown to display a fully oxidized state of copper in the initial calcined sample, formation of such carbonyl species upon room temperature interaction evidences the existence of a low temperature copper oxide reduction process, also in agreement with concomitant formation of carbonate-type species upon such room temperature interaction with CO [49,51]. Such a room temperature copper oxide reduction process has been also demonstrated to take place upon interaction with the CO-PROX reactant mixture [18]. Therefore, the extent of support promotion of such a low temperature reduction process, rather than reduction processes occurring at higher temperatures that are the ones typically reflected by conventional TPR tests, appears to correlate with respective CO oxidation activity in this type of catalyst, as will be discussed in the following. It may be noted that additional evidence for this room temperature reduction process was obtained in previous work by XPS and EPR [29]. In turn, further demonstration of the involvement of a partially reduced state of copper on the CO oxidation activity was also recently achieved for this type of catalyst by *operando*-DRIFTS showing in turn the existence of an induction process related to the aforementioned reduction of copper oxide [53].

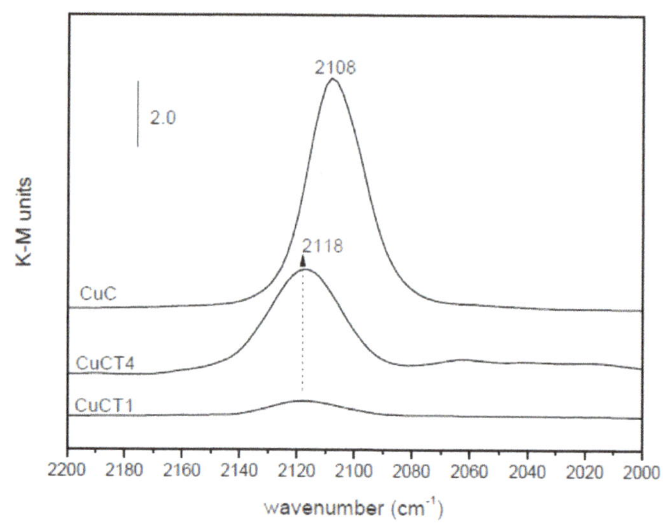

Figure 5: DRIFTS spectra in the carbonyl stretching region following interaction of the indicated catalysts with a diluted CO flow at room temperature.

Nature of Active Sites/Entities and Further Hints on Catalytic/Redox Correlations

Considering the results exposed in the former section, a set of catalysts with most active configuration involving the combination of copper oxide and ceria were explored with regards to their CO-PROX properties by means of spectroscopic techniques under reaction conditions (the so-called *operando* techniques). Two different preparation methods (impregnation and microemulsion-coprecipitation) and different copper loadings have been used in order to attempt to cover the widest possible range of physicochemical characteristics in the oxide entities constituting the catalysts. The main characteristics of the initial calcined catalysts at their structural, morphological and electronic levels, on the basis of XRD, Raman, HRTEM-XEDS, S_{BET} measurements, XANES, and XPS were reported in a former contribution, in which full details of catalytic activity results taken with a tubular reactor can also be found [40]. A brief summary of the characterization results is given in the following and some relevant data are collected in Table 1.

Table 1: Main textural and structural characteristics of the indicated copper-ceria catalysts. The initial number in the samples prepared by impregnation reflects the respective Cu wt.%, while atomic amounts of copper and cerium are indicated in the names employed for the samples prepared by microemulsion-coprecipitation [40].

Sample	Synthesis method	S_{BET} (m^2/g)	Lattice parameter [a] (Å)	Crystal size [a] (nm)	Phases detected [b]
$0.5CuO/CeO_2$	impregnation	116	5.410	7.6	Fluorite CeO_2
$1CuO/CeO_2$	impregnation	107	5.410	7.8	Fluorite CeO_2
$5CuO/CeO_2$	impregnation	101	5.413	8.1	Fluorite CeO_2, tenorite CuO
$Ce_{0.95}Cu_{0.05}O_2$	microemulsion coprecipitation	130	5.410	7.0	Fluorite $Ce_{1-x}Cu_xO_2$
$Ce_{0.8}Cu_{0.2}O_2$	microemulsion coprecipitation	151	5.413	6.6	Fluorite$Ce_{1-x}Cu_xO_2$

[a] For the fluorite phase; [b] Based on XRD and Raman [40].

Thus, XRD displayed only peaks corresponding to the fluorite structure of ceria, except for $5CuO/CeO_2$ for which weak and narrow peaks of tenorite CuO were detected, in agreement also with HRTEM, XANES and XPS results. As shown inTable 1, lattice parameters estimated from XRD analysis of the fluorite peaks are close to those expected for pure ceria for all the samples. It must be taken into account, however, that copper introduction into the ceria fluorite lattice is not expected to induce significant changes in this parameter [29,37]. Indeed, on the basis of analysis of lattice microstrain (determined from analysis of XRD results) and Raman results, significant differences between samples prepared by impregnation and microemulsion-coprecipitation were revealed [37,40]. These have been related to the fact that, as expected, all the copper remains essentially at the sample surface in the samples prepared by copper impregnation, while at least a part of the copper appears to be incorporated into the ceria fluorite lattice for those prepared by coprecipitation within microemulsion. In any case, it must be noted that the latter type of sample cannot be fully considered as pure Ce-Cu mixed oxides, since a certain copper surface segregation, increasing with the copper loading, is evident on the basis of mainly XEDS and Ar^+-sputtering XPS analyses [40]. Accordingly, catalysts of the $Ce_{1-x}Cu_xO_2$ series (Table 1) can probably be better described as $CuO/Ce_{1-z}Cu_zO_{2-y}$ (with undetermined *z* and *y*, and in any case with the amount of segregated CuO increasing with *x* [40]), since single solid solution of the copper is not fully achieved in any case (most particularly for Cu content >10 at.% [40]). In turn, although CuO-type clusters dispersed on the ceria support predominate for the catalysts prepared by impregnation, differences between catalysts of this series as a function of copper loading have been shown to be related, as noted in Table 1, to the presence of large crystalline CuO particles, in accordance also with electron microscopy investigation [40], in $5CuO/CeO_2$. This must be the result of having exceeded the copper oxide dispersion capacity of the CeO_2 support above a certain copper oxide surface loading (in our case, detected for the sample with 5 wt.% Cu) [40]. On the other hand, the presence of copper induces some surface area decrease in the samples prepared by impregnation, probably due to some copper covering of interparticle pores (Table 1). In contrast, the surface area appreciably increases with the copper amount for the samples prepared by microemulsion–coprecipitation, in correlation with the aforementioned introduction of copper into the ceria lattice, which basically induces a decrease in primary particle size (Table 1). On the other hand, concerning the electronic state of copper, XANES (as also shown below) and XPS analyses revealed that the starting sample calcined under dilute O_2 at 500 °C presents copper in a fully oxidized Cu^{2+} chemical state with relatively few differences (except for the crystalline CuO detected in $5CuO/CeO_2$) between the samples concerning copper electronic characteristics [37,40].

Concerning the catalytic/redox behavior of the series of samples collected in Table 1, DRIFTS experiments under CO-PROX conditions reveal the formation of bands of a similar nature for all catalysts upon contact with the reactant mixture at reaction temperatures between 303 and 523 K. These basically appear in three distinct spectral zones, as illustrated in Figure 6 for the $1CuO/CeO_2$ catalyst. The first zone displays bands corresponding mainly to hydroxyl species (isolated ones of various types giving sharp bands in the 3720–3600 cm^{-1} range and associated species giving a broad band extending from*ca.* 3800 to 3000 cm^{-1}) [54,55]. A second spectral zone below 1700 cm^{-1} exhibits most intense bands of carbonate or related species [56,57,58]. Stretching vibrations of these species appearing in this region are ascribed to bidentate carbonates (bands at *ca.* 1583 and 1297; note this attribution was recently revisited and tridentate coordination was suggested for this species on the basis of combined IR-DFT analysis [59]); a combination band at *ca.* 2880 cm^{-1} (see the highest wavenumber zone at the left of Figure 6), particularly apparent in lower temperature spectra, is also attributed to these bidentate carbonate species. Polydentate carbonates showing the symmetric and antisymmetric stretching vibrations of the terminal CO bonds at*ca.* 1478 and 1356 cm^{-1} are also detected. The band at 1216 cm^{-1}, along with that at 1399 cm^{-1} and a shoulder at *ca.* 1600 cm^{-1}, are attributed to hydrogen carbonate species [57,58]; this is confirmed by the presence of a sharp OH stretching vibration at *ca.* 3618 cm^{-1} also belonging to these species; note that this type of carbonate must be formed upon interaction of CO with monodentate hydroxyls (giving rise to a band at *ca.* 3710 cm^{-1} in the spectrum of the original sample) [55,58], in accordance with observation by DRIFTS, Figure 6. The third spectral zone (at intermediate frequencies) shows the formation of CO_2 (g), evolving in accordance with CO oxidation activity, and a carbonyl species (as discussed earlier, a Cu^+-carbonyl giving rise to a band at *ca.* 2120–2110 cm^{-1} [27,29,49,51]. As discussed previously, the presence of these Cu^+-CO species already upon initial contact at 303 K with the reactant mixture is consistent with the easy reduction of copper in the catalysts, considering that copper is fully oxidized in the initial calcined catalysts [40]. As also noted earlier, in the case of $Cu/Ce_{1-x}Tb_xO_{2-y}$ catalysts, one of the main differences between the samples is related to the intensity of this Cu^+-carbonyl band. Furthermore, as displayed in Figure 7 a correlation can be established between the intensity of such carbonyl species and the CO oxidation rate observed for every sample. Since such carbonyl species must be formed as a consequence of a support-promoted reduction process of the copper and taking also into account that the relatively low frequency of this band, with respect to those expected for this type of carbonyls [60], has been related to the interaction between the partially reduced copper centers and the underlying ceria [27,29,40,41,51], the correlation evidenced by Figure 7 reveals that the active species for CO oxidation under CO-PROX conditions must be related to surface-dispersed partially reduced copper oxide species interacting with the support, *i.e.*, at CuO-support interfacial positions. This is consistent with the aforementioned results collected in previous reports [27,29,40,49], which show that these copper species are the most reducible upon contact with CO at low temperatures within a process whereby the ceria support that is in contact with the copper species can also become reduced [27,29,61,62]. Thus, differences between the CO oxidation activities under CO-PROX conditions for this type of catalyst is related to the extent of support promotion of such partial (to Cu^+) CuO reduction at interfacial sites which is attained in each case. It is worth noting that the correlation between the intensity of the Cu^+-carbonyl and CO oxidation activity over samples of this type was also independently demonstrated later by Baertsch *et al.* [63].

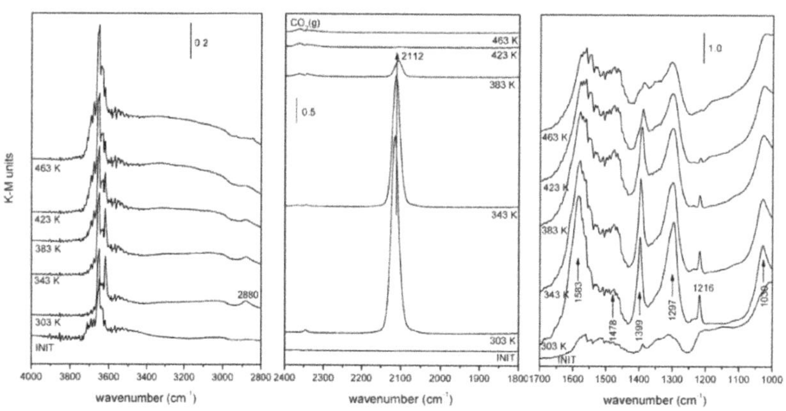

Figure 6: DRIFTS spectra under CO-H_2-O_2 reactant mixture flow at the indicated temperatures for $1CuO/CeO_2$. The spectrum at the bottom corresponds to that recorded at 303 K after pretreatment under diluted O_2 at 773 K, prior to contact with the reactant mixture. Note: This figure is adapted with permission from [40]. Copyright © 2007, American Chemical Society.

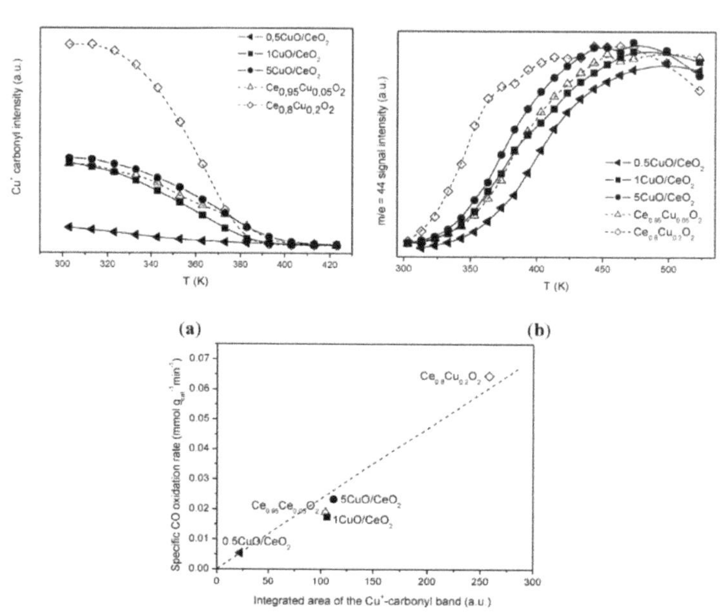

Figure 7: (**a**) intensity of the Cu^+-carbonyl as a function of the reaction temperature under CO-H_2-O_2 mixture for the indicated catalysts. (**b**) Intensity of the QMS m/e = 44 signal, corresponding to CO_2, during the tests performed with the DRIFTS cell under CO-H_2-O_2 mixture. (**c**) correlation between the intensity of the Cu^+-carbonyl just prior to CO oxidation onset and the initial CO oxidation activity. Note: This figure is adapted with permission from [40]. Copyright © 2007, American Chemical Society.

On the other hand, the reaction mechanism under which this type of catalyst operates for CO oxidation has been proposed to be of a redox, generally speaking, of the Mars-van Krevelen [64,65], type [6]. This has been, indeed, demonstrated for the $1CuO/CeO_2$ catalyst for which kinetic data are consistent with Mars-van Krevelen expressions [66]:

$$r_{CO} = \frac{k_{CO} k_{O_2} p_{CO} p_{O_2}^{n}}{0.5 k_{CO} p_{CO} + k_{O_2} p_{O_2}^{n}}$$

$$k_{CO} = A_{CO} \exp\left(\frac{-E_{a,CO}}{RT}\right)$$

$$k_{O_2} = A_{O_2} \exp\left(\frac{-E_{a,O_2}}{RT}\right)$$

with $A_{CO} = 1.61 \times 10^5$ mol/g·s·bar, $E_{a,CO} = 4.57 \times 10^4$ J/mol, $A_{O2} = 2.87 \times 10^3$ mol/g·s·bar^n, $E_{a,O2} = 5.51 \times 10^4$ J/mol and $n = 0.09$ [66]. In this respect, it becomes pertinent to enquire as to the nature of the active oxygen species involved in the reaction mechanism. It must be recalled in this sense that different oxygen species (superoxide, peroxide, oxide) have been proposed to be involved as active sites in this type of system, which can certainly depend on the specific degree of hydroxylation present in each case [67,68]. To help gain perspective, Raman spectra were collected under reaction conditions for $Cu_{0.2}Ce_{0.8}O_2$. They do not show the formation of peroxide or superoxide species under CO-PROX conditions; bands are expected at *ca.* 840 and 1125 cm^{-1}, respectively [64,69] (Figure 8). The shift of the fluorite F_{2g} mode band at *ca.* 460 cm^{-1} with increasing reaction temperature is in agreement with thermal expansion of the fluorite lattice (in accordance with spectra simultaneously recorded under the same conditions for CeO_2) and reduction at the highest temperatures [64,70], while the decrease of the band at *ca.* 600 cm^{-1} (related to oxygen vacancies [37,64]) at the highest reaction temperatures is consistent with copper migration to the surface as a consequence of reduction to metallic copper [37], in agreement with XANES results described below. The partially reduced oxygen species (superoxide, peroxide) were proposed to be involved during CO-PROX processes in hydroxylated ceria-supported gold catalysts [70]. Their absence under CO-PROX conditions in $Cu_{0.2}Ce_{0.8}O_2$ suggests that oxygen species involved in the redox processes can be related to oxide anions, in agreement with a recent investigation in which redox changes under CO/O_2 in $1CuO/CeO_2$ were observed to occur without involvement of superoxides or peroxides even at 303 K [64], at which they can be stable [29].

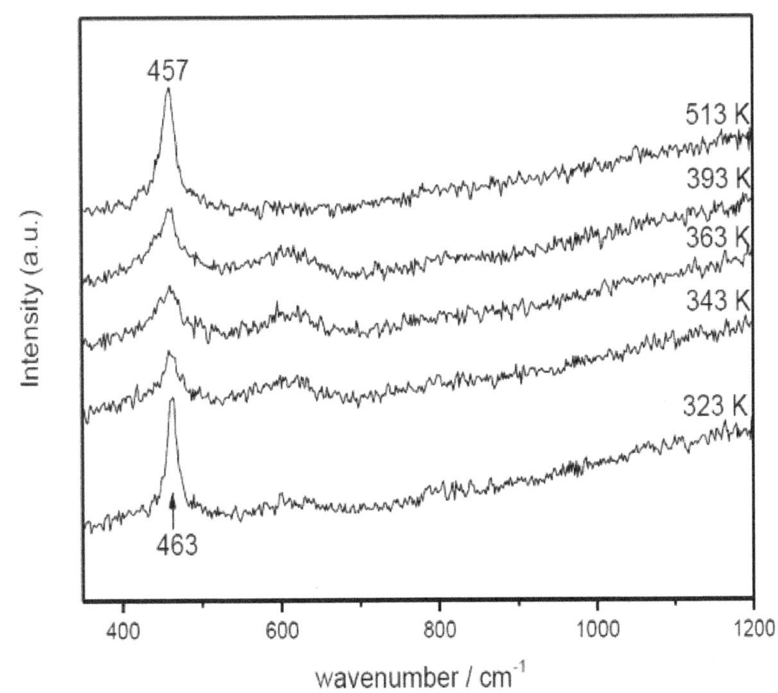

Figure 8: Raman spectra obtained for $Ce_{0.8}Cu_{0.2}O_2$ under CO-PROX conditions at the indicated temperatures. Note: This figure is adapted with permission from [41]. Copyright © 2007, American Chemical Society.

The analysis of the Cu K-edge XANES spectra of $Cu_{0.2}Ce_{0.8}O_2$ and $5CuO/CeO_2$ under CO-PROX conditions indicates the presence of three different chemical species during the course of runs. The first one corresponds to a Cu^{2+} chemical state displaying geometry similar to that found in CuO, though it also displays some particularities as a consequence of interactions with the support, as discussed in more detail elsewhere [37]. The Cu^{2+} component predominates at low reaction temperature, as illustrated by Figure 9. Since DRIFTS results described earlier revealed that Cu^+ species are formed at low temperature, it must be noted that the amount of Cu^+-carbonyls formed must correspond to a relatively low portion of the whole copper present in the sample (note in this sense a maximal limit of *ca.* 10% as intrinsic error of the XANES technique/analysis). They must be exclusively related to interfacial sites in close interaction with the support which present the highest redox activity, in accordance with existence of a support promoting effect on the CuO reduction, as discussed previously [29]. A component corresponding to zero-valent Cu^0, as identified from a comparison with a Cu foil reference, predominates at the end of the runs (Figure 9). Additionally, an intermediate species is detected during spectra analysis and attributed to a Cu^+ state on the basis of its 1s → 4p/3d transition energy and spectral shape. Joint analysis of the evolutions of the various copper species and the gases evolving during the CO-PROX tests allow separating different relevant zones (Figure 9). The first one (zone I) at lowest reaction temperature involves exclusively activity for the CO oxidation reaction and has been discussed above on the basis of DRIFTS experiments. The second zone (II) displays a correlation between the onset of H_2 oxidation and, at slightly lower temperature, the onset of massive copper reduction to Cu^+. This correlation suggests the involvement of the latter species in H_2 oxidation, in agreement with the high reactivity shown by partially reduced copper oxide towards hydrogen [71]. Note that as a difference from CO oxidation, H_2 oxidation takes place only when the reduction is propagated to zones of the copper oxide nanoparticles far from the interface between the two oxide components. In this respect, H_2 oxidation can be most dependent on the specific properties (size, shape [72]) of the dispersed copper oxide nanoparticles, as pointed out previously [18]. In contrast, CO oxidation properties are most likely governed by the characteristics of the CuO-support contacts, *i.e.*, the interfacial properties [41]. A third zone (III) is detected at higher temperatures whereby the H_2 oxidation reaction rate changes in coincidence with a sharp increase of the Cu^+contribution. This can be related to the formation of less active Cu_2O and/or to sintering of the copper prior to generation of metallic copper [40,73]. This is detected at the highest reaction temperature and its formation fairly coincides with the formation of Ce^{3+} states (not shown; see [41] for details). It may be pointed out that the copper segregation produced by this reduction process can contribute to the deactivation observed in this type of system when maintained under the reactant mixture at relatively high temperature [7,20]. Validation of the correlations observed for $Cu_{0.2}Ce_{0.8}O_2$ is provided by observation of similar ones for $5CuO/CeO_2$ [41].

The existence of different active sites (interfacial ones and on top of the reduced copper oxide particles) for each of the two reactions basically involved in the CO-PROX process over this type of catalyst (CO and H_2 oxidation) is in agreement with kinetic analysis showing the absence of interference between CO and H_2 during the two reactions (*i.e.*, the rates of CO and H_2 oxidation do not depend on partial pressure of H_2 and CO, respectively) [39]. This has, however, been refuted recently by Kydd *et al.* [73], who suggest that some degree of competency between H_2 and CO for the active sites may arise at relatively high reaction temperatures, in the non-selective region, when CO desorption from active reduced copper sites become favored, even though, to the best of our knowledge, no direct evidence for this is available yet [73].

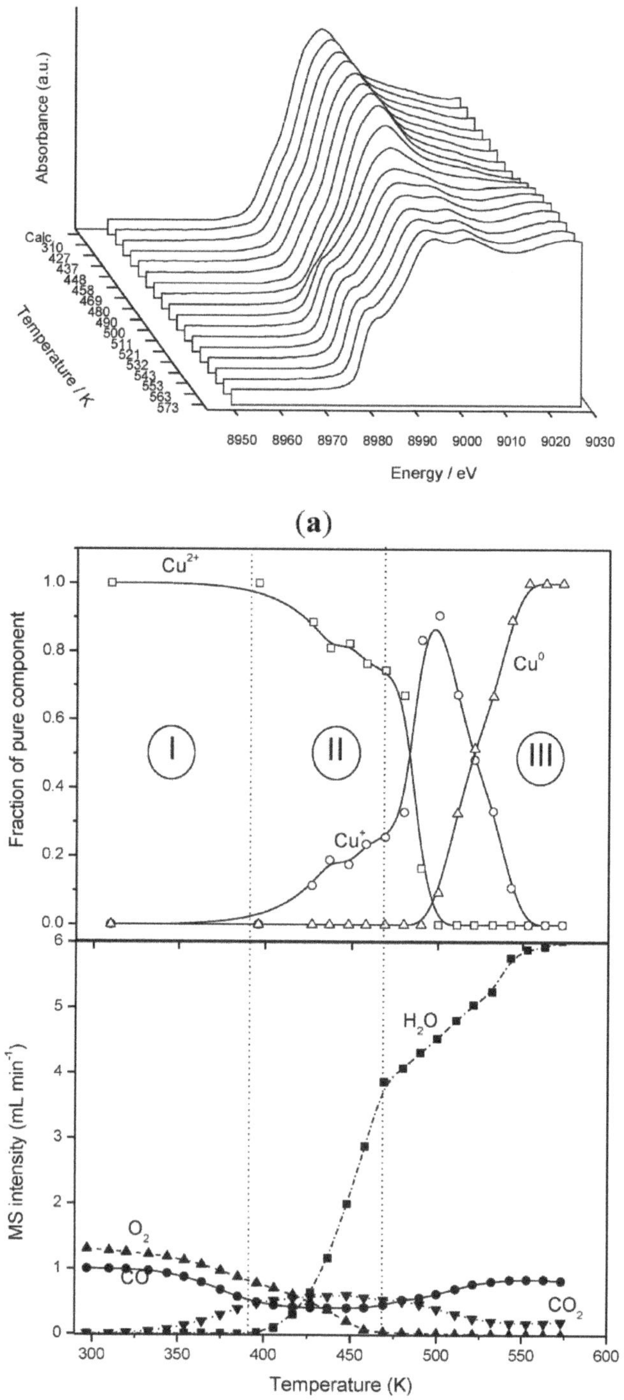

Figure 9: Cu-K edge XANES spectra under CO-H_2-O_2 mixture over $Ce_{0.8}Cu_{0.2}O_2$ (**a**). Evolution of chemical species extracted from spectra analysis and of the various gases detected in the course of the *Operando* test with the XAFS cell (**b**), quantified in accordance with calibration and appropriate gas interference corrections. Note: This figure is adapted with permission from [41]. Copyright © 2007, American Chemical Society.

Another recent finding is related to a very recent study in which three samples of copper oxide supported by ceria with different surface morphological characteristics (nanocubes exposing (100) faces at their surface, nanorods exposing (110) and (100) faces and nanospheres with polyhedral shape) have been examined with respect to their CO-PROX performance [42]. An absence of correlation between CO oxidation activity and intensity of Cu^+-carbonyls formed under CO-PROX reaction conditions (as explored by *operando*-DRIFTS) is observed in that case, which contrasts with the results exposed above (obtained with series of different copper loading samples supported on a ceria support with the same morphology or with samples prepared by copper-cerium microemulsion/coprecipitation.

This reveals differences in the catalytic properties of the dispersed copper oxide entities as a function of the surface characteristics of the ceria support with which they interact, *i.e.*, a structural dependence of the CO oxidation reaction involved in the CO-PROX process. On the whole, the main conclusion of such a study can be that optimized CO-PROX catalysts can be obtained by using ceria nanocubes as support. This is based on the fact that such morphology could allow achieving maximum dispersion (*i.e.*, avoiding copper oxide segregation into separated big copper oxide particles not interacting with the underlying support, as detected, for instance, when using ceria nanospheres as support; Table 1) in this case, and partly as a consequence of the relatively low specific surface area of ceria nanocubes [42] and larger particle size within homogeneously distributed dispersed copper oxide particles. Interaction of such dispersed copper oxide particles with the underlying (100) face of ceria can also be most relevant to explain the enhanced CO-PROX catalytic properties. The latter is related to the enhanced CO_2 selectivity, which favorably balances the decrease of CO oxidation activity produced as a consequence of the stronger interaction with the support. The higher CO_2 selectivity observed over the specimen supported on ceria nanocubes is shown to be related to difficulties in achieving full reduction of the dispersed copper oxide particles under CO-PROX conditions, on the basis of mainly *operando*-DRIFTS experiments, and also in agreement with arguments exposed above. On the other hand, other morphological details which could be relevant to this type of catalyst/process is related to control of the porous microstructure of the system, as shown in a recent study [74].

SUMMARY AND MAIN CONCLUSIONS

Catalysts constituted by combinations between copper oxide and ceria (or structurally related doped ceria) are amongst the most active catalysts within operative temperature ranges for the process of preferential oxidation of CO in H_2-rich stream (CO-PROX). This is an economically and practically interesting procedure for the purification of hydrogen extracted from hydrocarbon sources, particularly focusing on its use as fuel of proton exchange membrane fuel cells in mobile or small-scale applications. Although general agreement exists concerning the necessity of establishing contacts between both components' oxides, simply achievable upon dispersing at highest possible level copper oxide on high surface ceria, it is required to go a step further in the sense of understanding basic aspects of the CO-PROX process over this type of system in order to get optimized configurations, particularly considering that the main competitors within catalysts active for the process are very robust platinum-based catalysts. In this context, the present contribution basically revises work done in the authors' laboratories focused on gaining basic insights in two interrelated aspects: (1) determining the nature of the sites active for the various reactions involved in the process; and (2) establishing catalytic/redox correlations. Concerning the first point, it must be noted that basically two reactions take place below about 200 °C (before formation of WGS-active metallic copper under CO-PROX reaction conditions) in the presence of a CO-PROX mixture over this type of catalyst: CO and H_2oxidation. In turn, concerning the second point, when starting from a fully oxidized catalyst, changes can be produced in the oxide components during the course of the interaction with the reactant mixture, *i.e.*, considering the reducing nature of the mixture, reduction of the oxides components becomes thermodynamically favored. On this basis, *operando*-spectroscopic analysis of the catalysts has been carried out and allows concluding on the existence of a low temperature support promoting effect on the partial reduction (to Cu^+) of interfacial sites of the dispersed CuO entities. Such interfacial partially reduced copper sites (Cu^+ species) apparently constitute the most active sites for the CO oxidation reaction on the basis of the results obtained. This is mainly based on the finding of an interesting correlation between the intensity of Cu^+-carbonyls formed under reaction conditions and the CO oxidation rate respectively observed. This has been recently shown to depend also on the specific surface morphology of the ceria support which evidently determines the CO oxidation activity of the active reduced copper entities. On the other hand, the extension of the reduction over the copper oxide entities (*i.e.*, to sites out of the interface and accordingly most likely not so strongly affected by contact with the ceria support) upon interaction with the reactant mixture can provide the most active sites for the H_2 oxidation reaction. The possible separation between the two types of the most active sites that can be involved during the two (H_2 and CO) oxidation reactions taking place under CO-PROX conditions over this type of system (previous macroscopic kinetics experiments have also pointed out towards the presence of independent active sites for the two reactions [39]) can open the possibility to control their catalytic properties and design optimized systems for this important application. Indeed, this hypothesis has allowed the design of newly optimized catalysts of this type by using inverse CeO_2/CuO configurations, as recently reported [75]. Nevertheless, it must be noted that other authors have argued for different active sites for each one of the two oxidation reactions; *i.e.*, it has been proposed, although no direct demonstration is available to the best of our knowledge, that H_2 and CO can compete, at least above a certain reaction temperature, for the same active sites. This hypothesis proposes that H_2 oxidation starts as soon as the temperature is sufficiently high and CO desorption from copper-active sites becomes favored. We are presently working in our laboratory on understanding better how this functions.

On the other hand, the results point out some discrepancies as to the use of classical TPR runs for getting catalytic/redox correlations since redox processes of relevancy to explain CO oxidation can take place at relatively low temperature, typically during gases equilibration prior to data collection. In addition, as pointed out in previous works, the reaction mechanism under which the catalysts operate for CO oxidation can be of a redox Mars-van Krevelen type with oxide anions apparently constituting the active centers, according to Raman investigation. Other aspects of relevancy which have not been treated in detail in this contribution, are related to understanding the catalysts' deactivation mechanisms [76,77], particularly in the presence of CO_2 and H_2O in the reactant mixture, which, as it occurs for nanogold systems, could limit the practical application of this type of catalyst. It must, however, be mentioned in this sense that monolithic forms of CO-PROX catalysts combining copper and ceria have displayed high performance for the process, even in the presence of important amounts of CO_2 and H_2O in the reactant stream [78]. On the other hand, copper segregation and formation of metallic copper is shown to occur at $T > ca.$ 473 K under CO-H_2-O_2 mixture and can also contribute, in addition to formation of interfacial carbonate species and molecular water blocking effects [76,77], to deactivation observed under a simple CO-PROX mixture.

ACKNOWLEDGMENTS

This work was funded by Ministerio de Ciencia e Innovación and Ministerio de Economía y Competitividad (Plan Nacional Projects CTQ2009-14527 and CTQ2012-32928) and Comunidad de Madrid (Project DIVERCEL, Ref.: S2009/ENE-1475). Support from EU COST CM1104 action is also acknowledged. Antonio López Cámara acknowledges a PhD grant from the CSIC JAE program.

CONFLICT OF INTEREST

The authors declare no conflict of interest.

REFERENCES

1. Rostrup-Nielsen, J.R.; Sehested, J.; Norskov, J.K. Hydrogen and synthesis gas by steam- and CO_2 reforming. *Adv. Catal.* 2002, *47*, 65–139.
2. Fu, Q.; Saltsburg, H.; Flytzani-Stephanopoulos, M. Active nonmetallic Au and Pt species on ceria-based water-gas shift catalysts. *Science* 2003, *301*, 935–938.
3. Avgouropoulos, G.; Ioannides, T.; Papadopoulou, C.; Batista, J.; Hocevar, S.; Matralis, H.K. A comparative study of Pt/γ-Al_2O_3, Au/α-Fe_2O_3 and CuO-CeO_2 catalysts for the selective oxidation of carbon monoxide in excess hydrogen.*Catal. Today* 2002, *75*, 157–167.
4. Oh, S.H.; Sinkevitch, R.M. Carbon monoxide removal from hydrogen-rich fuel cell feedstreams by selective catalytic oxidation. *J. Catal.* 1993, *142*, 254–262.
5. Wang, J.B.; Lin, S.; Huang, T. Selective CO oxidation in rich hydrogen over CuO/samaria-doped ceria. *Appl. Catal. A*2002, *232*, 107–120.
6. Sedmak, G.; Hocevar, S.; Levec, J. Kinetics of selective CO oxidation in excess of H_2 over the nanostructured $Cu_{0.1}Ce_{0.9}O_{2-y}$ catalyst. *J. Catal.* 2003, *213*, 135–150.
7. Kim, D.H.; Cha, J.E. A CuO-CeO_2 Mixed-oxide catalyst for CO clean-up by selective oxidation in hydrogen-rich mixtures. *Catal. Lett.* 2003, *86*, 107–112.
8. Korotkikh, O.; Farrauto, R. Selective catalytic oxidation of CO in H_2: Fuel cell applications. *Catal. Today* 2000, *62*, 249–254.
9. Ghenciu, A.F. Review of fuel processing catalysts for hydrogen production in PEM fuel cell systems. *Curr. Opin. Solid State Mater. Sci.* 2002, *6*, 389–399.
10. Haruta, M.; Date, M. Advances in the catalysis of Au nanoparticles. *Appl. Catal. A* 2001, *222*, 427–437.
11. Schubert, M.M.; Venugopal, A.; Kahlich, M.J.; Plzak, V.; Behm, R.J. Influence of H_2O and CO_2 on the selective CO oxidation in H_2-rich gases over Au/α-Fe_2O_3. *J. Catal.* 2004, *222*, 32–40.
12. Luengnaruemitchai, A.; Osuwan, S.; Gulari, E. Selective catalytic oxidation of CO in the presence of H_2 over gold catalyst. *Int. J. Hydrog. Energy* 2004, *29*, 429–435.
13. Deng, W.; de Jesus, J.; Saltsburg, H.; Flytzani-Stephanopoulos, M. Low-content gold-ceria catalysts for the water-gas shift and preferential CO oxidation reactions. *Appl. Catal. A* 2005, *291*, 126–135.
14. Carrettin, S.; Concepcion, P.; Corma, A.; López, Nieto, J.M.; Puentes, V.F. Nanocrystalline CeO_2 increases the activity of Au for CO oxidation by two orders of magnitude. *Angew. Chem. Int. Ed.* 2004, *43*, 2538–2540.
15. Avgouropoulos, G.; Papavasiliou, J.; Tabakova, T.; Idakiev, V.; Ioannides, T. A comparative study of ceria-supported gold and copper oxide catalysts for preferential CO oxidation reaction. *Chem. Eng. J.* 2006, *124*, 41–45.
16. Liu, Y.; Fu, Q.; Flytzani-Stephanopoulos, M. Preferential oxidation of CO in H_2 over CuO-CeO_2 catalysts. *Catal. Today*2004, *93–95*, 241–246.
17. Marbán, G.; Fuertes, A.B. Highly active and selective CuO_x/CeO_2 catalyst prepared by a single-step citrate method for preferential oxidation of carbon monoxide. *Appl. Catal. B* 2005, *57*, 43–53.
18. Martínez-Arias, A.; Hungría, A.B.; Fernández-García, M.; Conesa, J.C.; Munuera, G. Preferential oxidation of CO in a H_2-rich stream over CuO/CeO_2 and CuO/(Ce,M)O_x (M = Zr, Tb) catalysts. *J. Power Sour.* 2005, *151*, 32–42.
19. Mariño, F.; Descorme, C.; Duprez, D. Supported base metal catalysts for the preferential oxidation of carbon monoxide in the presence of excess hydrogen (PROX). *Appl. Catal. B* 2005, *58*, 175–183.
20. Martínez-Arias, A.; Hungría, A.B.; Munuera, G.; Gamarra, D. Preferential oxidation of CO in rich H_2 over CuO/CeO_2: Details of selectivity and deactivation under the reactant stream. *Appl. Catal. B* 2006, *65*, 207–216.
21. Park, J.-W.; Jeong, J.-H.; Yoon, W.-L.; Jung, H.; Lee, H.-T.; Lee, D.-K.; Park, Y.-K.; Rhee, Y.W. Activity and characterization of the Co-promoted CuO-CeO_2/γ-Al_2O_3 catalyst for the selective oxidation of CO in excess hydrogen.*Appl. Catal. A* 2004, *274*, 25–32.
22. Jobbagy, M.; Mariño, F.; Schönbrod, B.; Baronetti, G.; Laborde, M. Synthesis of copper-promoted CeO_2 catalysts. *Chem. Mater.* 2006, *18*, 1945–1950.
23. Moretti, E.; Lenarda, M.; Storaro, L.; Talon, A.; Montanari, T.; Busca, G.; Rodríguez-Castellón, E.; Jiménez-López, A.; Turco, M.; Bagnasco, G.; *et al.* One-

step synthesis of a structurally organized mesoporous $CuO-CeO_2-Al_2O_3$ system for the preferential CO oxidation. *Appl. Catal. A* 2008, *335*, 46–55.

24. Gurbani, A.; Ayastuy, J.L.; González-Marcos, M.P.; Herrero, J.E.; Guil, J.M.; Gutiérrez-Ortiz, M.A. Comparative study of $CuO-CeO_2$ catalysts prepared by wet impregnation and deposition-precipitation. *Int. J. Hydrog. Energy* 2009, *34*, 547–553.
25. Martínez-Arias, A.; Soria, J.; Cataluña, R.; Conesa, J.C.; Cortés Corberán, V. Influence of ceria dispersión on the catalytic performance of $Cu/CeO_2/Al_2O_3$ catalysts for the CO oxidation reaction. *Stud. Surf. Sci. Catal.* 1998, *116*, 591–600.
26. Liu, W.; Sarofim, A.F.; Flytzani-Stephanopoulos, M. Complete oxidation of carbon monoxide and methane over metal-promoted fluorite oxide catalysts. *Chem. Eng. Sci.* 1995, *49*, 4871–4888.
27. Martínez-Arias, A.; Fernández-García, M.; Gálvez, O.; Coronado, J.M.; Anderson, J.A.; Conesa, J.C.; Soria, J.; Munuera, G. Comparative study on redox properties and catalytic behavior for CO Oxidation of CuO/CeO_2 and $CuO/ZrCeO_4$catalysts. *J. Catal.* 2000, *195*, 207–216.
28. Skårman, B.; Grandjean, D.; Benfield, R.E.; Hinz, A.; Andersson, A.; Wallenberg, L.R. Carbon monoxide oxidation on nanostructured CuO_x/CeO_2 composite particles characterized by HREM, XPS, XAS, and high-energy diffraction. *J. Catal.* 2002, *211*, 119–133.
29. Martínez-Arias, A.; Hungría, A.B.; Fernández-García, M.; Conesa, J.C.; Munuera, G. Interfacial redox processes under CO/O_2 in a nanoceria-supported copper oxide catalyst. *J. Phys. Chem. B* 2004, *108*, 17983–17991.
30. Ilichev, A.N.; Firsova, A.A.; Korchak, V.N. Mechanism of CO oxidation in excess H_2 over CuO/CeO_2 catalysts: ESR and TPD studies. *Kinet. Catal.* 2006, *47*, 585–592.
31. Gamarra, D.; Hornés, A.; Koppány, Z.; Schay, Z.; Munuera, G.; Soria, J.; Martínez-Arias, A. Catalytic processes during preferential oxidation of CO in H_2-rich streams over catalysts based on copper-ceria. *J. Power Sour.* 2007, *169*, 110–116.
32. Luo, M.-F.; Song, Y.-P.; Lu, J.Q.; Wang, X.-Y.; Pu, Z.-Y. Identification of CuO species in high surface area $CuO-CeO_2$catalysts and their catalytic activities for CO oxidation. *J. Phys. Chem. C* 2007, *111*, 12686–12692.
33. Luo, M.-F.; Ma, J.-M.; Lu, J.-Q.; Song, Y.-P.; Wang, Y.-J. High-surface area $CuO-CeO_2$ catalysts prepared by a surfactant-templated method for low-temperature CO oxidation. *J. Catal.* 2007, *246*, 52–59.
34. Pintar, A.; Batista, J.; Hocevar, S. TPR, TPO, and TPD examinations of $Cu_{0.15}Ce_{0.85}O_{2-y}$ mixed oxides prepared by co-precipitation, by the sol-gel peroxide route, and by citric acid-assisted synthesis. *J. Coll. Interf. Sci.* 2005, *285*, 218–231.
35. Manzoli, M.; di Monte, R.; Boccuzzi, F.; Coluccia, S.; Kaspar, J. CO oxidation over $CuO_x-CeO_2-ZrO_2$ catalysts: Transient behaviour and role of copper clusters in contact with ceria. *Appl. Catal. B* 2005, *61*, 192–205.
36. Avgouropoulos, G.; Ioannides, T.; Matralis, H. Influence of the preparation method on the performance of $CuO-CeO_2$catalysts for the selective oxidation of CO. *Appl. Catal. B* 2005, *56*, 87–93.
37. Wang, X.Q.; Rodriguez, J.A.; Hanson, J.C.; Gamarra, D.; Martínez-Arias, A.; Fernández-García, M. Unusual physical and chemical properties of Cu in $Ce_{1-x}Cu_xO_2$ oxides. *J. Phys. Chem. B* 2005, *109*, 19595–19603.
38. Wang, J.B.; Tsai, D.-H.; Huang, T.-J. Synergistic Catalysis of carbon monoxide oxidation over copper oxide supported on samaria-doped ceria. *J. Catal.* 2002, *208*, 370–380.
39. Lee, H.C.; Kim, D.H. Kinetics of CO and H_2 oxidation over $CuO-CeO_2$ catalyst in H_2 mixtures with CO_2 and H_2O.*Catal. Today* 2008, *132*, 109–116.
40. Gamarra, D.; Munuera, G.; Hungría, A.B.; Fernández-García, M.; Conesa, J.C.; Midgley, P.A.; Wang, X.Q.; Hanson, J.C.; Rodriguez, J.A.; Martínez-Arias, A. Structure-activity relationship in nanostructured copper-ceria-based preferential CO oxidation catalysts. *J. Phys. Chem. C* 2007, *111*, 11026–11038.
41. Gamarra, D.; Belver, C.; Fernández-García, M.; Martínez-Arias, A. Selective CO oxidation in excess H2 over copper-ceria catalysts: Identification of active entities/species. *J. Am. Chem. Soc.* 2007, *129*, 12064–12065.
42. Gamarra, D.; Lopez Camara, A.; Monte, M.; Rasmussen, S.B.; Chinchilla, L.E.; Hungria, A.B.; Munuera, G.; Gyorffy, N.; Schay, Z.; Corberan, V.C.; *et al.* Preferential oxidation of CO in excess H_2 over CuO/CeO_2 catalysts: Characterization and performance as a function of the exposed face present in the CeO_2 support. *Appl. Catal. B* 2013,*130–131*, 224–238.
43. Bion, N.; Epron, F.; Moreno, M.; Mariño, F.; Duprez, D. Preferential oxidation of carbon monoxide in the presence of hydrogen (PROX) over noble metals and transition metal oxides: Advantages and drawbacks. *Top. Catal.* 2008, *51*, 76–88.
44. López, I.; Valdés-Solís, T.; Marbán, G. An attempt to rank copper-based catalysts used in the CO-PROX reaction. *Int. J. Hydrog. Energy* 2008, *33*, 197–205.
45. Trovarelli, A. *Catalysis by Ceria and Related Materials*; Imperial College Press: London, UK, 2002.
46. Hungría, A.B.; Martínez-Arias, A.; Fernández-García, M.; Iglesias-Juez, A.; Guerrero-Ruiz, A.; Calvino, J.J.; Conesa, J.C.; Soria, J. Structural, morphological, and oxygen handling properties of nanosized cerium-terbium mixed oxides prepared by microemulsion. *Chem. Mater.* 2003, *15*, 4309–4316.
47. Martínez-Arias, A.; Fernández-García, M.; Hungría, A.B.; Conesa, J.C.; Munuera, G. Spectroscopic characterization of heterogeneity and redox effects in zirconium-cerium (1:1) mixed oxides prepared by microemulsion methods. *J. Phys. Chem. B* 2003, *107*, 2667–2677.
48. Hornés, A.; Bera, P.; López Cámara, A.; Gamarra, D.; Munuera, G.; Martínez-Arias, A. CO-TPR-DRIFTS-MS *in situ*study of $CuO/Ce_{1-x}Tb_2O_{2-y}$ (x = 0, 0.2 axnd

0.5) catalysts: Support effects on redox properties and CO oxidation catalysis. *J. Catal.* 2009, *268*, 367–375.

49. Bera, P.; López Cámara, A.; Hornés, A.; Martínez-Arias, A. Comparative *in situ* DRIFTS-MS study of ^{12}CO- and ^{13}CO-TPR on CuO/CeO_2 catalyst. *J. Phys. Chem. C* 2009, *113*, 10689–10695.

50. Wang, J.B.; Shih, W.-H.; Huang, T.-J. Study of Sm_2O_3-doped CeO_2/Al_2O_3-supported copper catalyst for CO oxidation.*Appl. Catal. A* 2000, *203*, 191–199.

51. Martínez-Arias, A.; Fernández-García, M.; Soria, J.; Conesa, J.C. Spectroscopic study of a Cu/CeO_2 catalyst subjected to redox treatments in carbon monoxide and oxygen. *J. Catal.* 1999, *182*, 367–377.

52. Caputo, T.; Lisi, L.; Pirone, R.; Russo, G. On the role of redox properties of CuO/CeO_2 catalysts in the preferential oxidation of CO in H_2-rich gases. *Appl. Catal. A* 2008, *348*, 42–53.

53. Martínez-Arias, A.; Hungría, A.B.; Fernández-García, M.; Iglesias-Juez, A.; Soria, J.; Conesa, J.C.; Anderson, J.A.; Munuera, G. Operando DRIFTS study of the redox and catalytic properties of $CuO/Ce_{1-x}Tb_xO_{2-d}$ ($x = 0$–0.5) catalysts: Evidence of an induction step during CO oxidation. *Phys. Chem. Chem. Phys.* 2012, *14*, 2144–2151.

54. Badri, A.; Binet, C.; Lavalley, J.-C. An FTIR study of surface ceria hydroxy groups during a redox process with H_2. *J. Chem. Soc. Faraday Trans.* 1996, *92*, 4669–4673.

55. Binet, C.; Daturi, M.; Lavalley, J.-C. IR study of polycrystalline ceria properties in oxidised and reduced states. *Catal. Today* 1999, *50*, 207–225.

56. Li, C.; Sakata, Y.; Arai, T.; Domen, K.; Maruya, K.-I.; Onishi, T. Carbon monoxide and carbon dioxide adsorption on cerium oxide studied by fourier-transform infrared spectroscopy. Part 1.—Formation of carbonate species on dehydroxylated CeO_2, at room temperature. *J. Chem. Soc. Faraday Trans. 1* 1989, *85*, 929–943.

57. Li, C.; Sakata, Y.; Arai, T.; Domen, K.; Maruya, K.-I.; Onishi, T. Adsorption of carbon monoxide and carbon dioxide on cerium oxide studied by Fourier-transform infrared spectroscopy. Part 2.—Formation of formate species on partially reduced CeO_2 at room temperature. *J. Chem. Soc. Faraday Trans. 1* 1989, *85*, 1451–1461.

58. Pozdnyakova, O.; Teschner, D.; Wootsch, A.; Kröhnert, J.; Steinhauer, B.; Sauer, H.; Toth, L.; Jentoft, F.C.; Knop-Gericke, A.; Paál, Z.; *et al.* Preferential CO oxidation in hydrogen (PROX) on ceria-supported catalysts, part I: Oxidation state and surface species on Pt/CeO_2 under reaction conditions. *J. Catal.* 2006, *237*, 1–16.

59. Vayssilov, N.; Mihaylov, M.; Petkov, P.S.; Hadjiivanov, K.I.; Neyman, K.M. reassignment of the vibrational spectra of carbonates, formates, and related surface species on ceria: A combined density functional and infrared spectroscopy investigation. *J. Phys. Chem. C* 2011, *115*, 23435–23454.

60. Scarano, D.; Bordiga, S.; Lamberti, C.; Spoto, G.; Ricchiardi, G.; Zecchina, A.; Otero Areán, C. FTIR study of the interaction of CO with pure and silica-supported copper(I) oxide. *Surf. Sci.* 1998, *411*, 272–285.

61. Martínez-Arias, A.; Fernández-García, M.; Hungría, A.B.; Iglesias-Juez, A.; Gálvez, O.; Anderson, J.A.; Conesa, J.C.; Soria, J.; Munuera, G. Redox interplay at copper oxide-$(Ce,Zr)O_x$ interfaces: Influence of the presence of NO on the catalytic activity for CO oxidation over $CuO/CeZrO_4$. *J. Catal.* 2003, *214*, 261–272.

62. Martínez-Arias, A.; Gamarra, D.; Fernández-García, M.; Wang, X.Q.; Hanson, J.C.; Rodriguez, J.A. Comparative study on redox properties of nanosized CeO_2 and CuO/CeO_2 under CO/O_2. *J. Catal.* 2006, *240*, 1–7.

63. Polster, C.S.; Nair, H.; Baertsch, C.D. Study of active sites and mechanism responsible for highly selective CO oxidation in H2 rich atmospheres on a mixed Cu and Ce oxide catalysts. *J. Catal.* 2009, *266*, 308–319.

64. Mars, P.; van Krevelen, D.W. Oxidations carried out by means of vanadium oxide catalysts. *Chem. Eng. Sci.* 1954, *3*, 41–59.

65. Vannice, M.A. An analysis of the Mars-van Krevelen rate expression. *Catal. Today* 2007, *123*, 18–22.

66. Martínez-Arias, A.; Gamarra, D.; Fernández-García, M.; Hornés, A.; Belver, C. Spectroscopic study on the nature of active entities in copper-ceria CO-PROX catalysts. *Top. Catal.* 2009, *52*, 1425–1432.

67. Martínez-Arias, A.; Cataluña, R.; Conesa, J.C.; Soria, J. Effect of copper-ceria interactions on copper reduction in a $Cu/CeO_2/Al_2O_3$ catalyst subjected to thermal treatments in CO. *J. Phys. Chem. B* 1998, *102*, 809–817.

68. Guzman, J.; Carrettin, S.; Corma, A. spectroscopic evidence for the supply of reactive oxygen during CO oxidation catalyzed by gold supported on nanocrystalline CeO_2. *J. Am. Chem. Soc.* 2005, *127*, 3286–3287.

69. Pushkarev, V.V.; Kovalchuk, V.I.; d'Itri, J.L. probing defect sites on the ceo_2 surface with dioxygen. *J. Phys. Chem. B*2004, *108*, 5341–5348.

70. Spanier, J.E.; Robinson, R.D.; Zhang, F.; Chan, S.-W.; Herman, I.P. Size-dependent properties of CeO_{2-y} nanoparticles as studied by Raman scattering. *Phys. Rev. B* 2001, *64*, 245407.

71. Kim, J.Y.; Rodriguez, J.A.; Hanson, J.C.; Frenkel, A.I.; Lee, P.L. Reduction of CuO and Cu_2O with H_2: H embedding and kinetic effects in the formation of suboxides. *J. Am. Chem. Soc.* 2003, *125*, 10684–10692.

72. Fernández-García, M.; Martínez-Arias, A.; Hanson, J.C.; Rodriguez, J.A. Nanostructured oxides in chemistry: Characterization and properties. *Chem. Rev.* 2004, *104*, 4063–4104.

73. Kydd, R.; Ferri, D.; Hug, P.; Scott, J.; Teoh, W.Y.; Amal, R. Temperature-induced evolution of reaction sites and mechanisms during preferential oxidation of CO. *J. Catal.* 2011, *277*, 64–71.

74. Yen, H.; Seo, Y.; Kaliaguine, S.; Kleitz, F. Tailored mesostructured copper/ceria catalysts with enhanced performance for preferential oxidation of CO at low temperature. *Angew. Chem. Int. Ed.* 2012, *51*, 12032–12035.

75. Hornés, A.; Hungría, A.B.; Bera, P.; López Cámara, A.; Fernández-García, M.; Martínez-Arias, A.; Barrio, L.; Estrella, M.; Zhou, G.; Fonseca, J.J.; *et al.* Inverse

CeO2/CuO catalyst as an alternative to classical direct configurations for preferential oxidation of CO in hydrogen-rich stream. *J. Am. Chem. Soc.* 2010, *132*, 34–35.

76. Gamarra, D.; Martínez-Arias, A. Preferential oxidation of CO in rich H_2 over CuO/CeO_2: *Operando*-DRIFTS analysis of deactivating effect of CO_2 and H_2O. *J. Catal.* 2009, *263*, 189–195.

77. Gamarra, D.; Fernández-García, M.; Belver, C.; Martínez-Arias, A. Operando DRIFTS and XANES study of deactivating effect of CO_2 on a $Ce_{0.8}Cu_{0.2}O_2$ CO-PROX catalyst. *J. Phys. Chem. C* 2010, *114*, 18576–18582.

78. Zeng, S.H.; Liu, Y.; Wang, Y.Q. CuO-CeO_2/Al_2O_3/FeCrAl monolithic catalysts prepared by sol-pyrolysis method for preferential oxidation of carbon monoxide. *Catal. Lett.* 2007, *117*, 119–125.

CHAPTER 5

NEW TRENDS IN OXIDATIVE FUNCTIONALIZATION OF CARBON

Georgiy B. Shul'pin

Keywords: Alkanes; alcohols; alkyl hydroperoxides; aromatics; hydrogen peroxide; mechanisms of oxidation; metal-complex catalysis; oxygen; oxygenation; peroxyacids

INTRODUCTION

This review is devoted to functionalization of carbon–hydrogen bonds [1,2,3,4,5,6,7,8,9,10,11,12,13] in various organic compounds, which is a very important process both from academic and practical point of view. Replacing hydrogen atoms in carbon–hydrogen compounds by functional groups X leads to valuable functionalized C–X compounds (Scheme 1a).

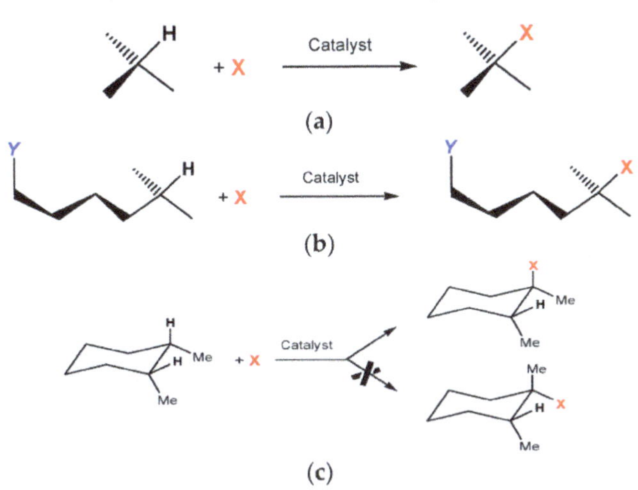

Scheme 1: Functionalization of various C–H bonds.

Citation: Georgiy B. Shul'pin, New Trends in Oxidative Functionalization of Carbon, doi:10.3390/catal6040050

Remote functionalization of carbon–hydrogen compounds bearing functional groups *Y* under the action of metal complex catalysts can give valuable difunctionalized alkanes (Scheme 1b).

Some catalytic systems are capable of functionalizing carbon–hydrogen bonds stereospecifically (Scheme 1c).

Functionalization of the carbon–hydrogen bond primarily proceeds via its activation when the reactivity of this σ-bond increases toward a reagent. As a result, the bond is capable of splitting, and in many cases this rupture of a saturated carbon–hydrogen bond is implied in the literature when the term "activation" is used. However, it is necessary to emphasize that splitting the bond is actually *a consequence* of its activation [1]. Alternative definitions have been proposed by Periana and colleagues who

"*define CH activation as a facile CH cleavage reaction with an 'MX' species that proceeds by coordination of an alkane to the inner-sphere of 'M' (either via an intermediate 'alkane complex' or a transition state) leading to a M–C intermediate*" [14]

and wrote about

"*the CH activation reaction as a coordination reaction with a reactive species 'M', that proceeds without the involvement of free radicals, carbocations, or carbanions to generate discrete M–R intermediates*". [14,15,16,17]

Labinger and Bercaw gave the definition of

"*reactions that have clearly been shown to involve bond formation between a metal and an alkane carbon*" [18]

and Crabtree wrote:

"*Activation is considered to be the binding of a substrate to a metal center. This can be followed by a functionalization step in which the substrate is transformed...The term 'CH activation' emphasizes the distinct reactivity pattern of low valent metal complexes from that of classical organic reagents which break CH bonds by electrophilic or radical routes. In a classical route, radicals such as •OH readily abstract an H• atom from alkanes, RH, to give the alkyl radical R•. Included in this group are some metal catalyzed oxidations, such as the Gif reaction and Fenton chemistry*".

[2]

Finally, the term "C–H activation" refers to a metal-mediated pathway involving the formation of metal-carbon bonds [19].

In recent years and decades, many reactions of inert saturated hydrocarbons with various metal compounds under mild conditions have been described [1]. Such processes with the participation of metal complexes occur at relatively low temperatures and can be selective. The oxidations of hydrocarbons with molecular oxygen and donors of an oxygen atom (hydrogen peroxide, alkyl hydroperoxides, and peroxy acids) constitute an important field of contemporary catalytic chemistry. We divide all carbon–hydrogen functionalization processes into three types taking into account the mechanistic point of view [1]. The first group includes reactions involving "true" or "organometallic" functionalization of the carbon–hydrogen bond. In these reactions compounds containing metal-carbon σ-bonds (organometallic derivatives) are formed as intermediates or as final products. We include in the second group reactions in which the contact between the complex and the carbon–hydrogen bond is only via a ligand of the complex during the whole process of the carbon–hydrogen bond cleavage. The σ-C–M bond is not generated directly at any stage. In these reactions the function of the metal complex usually consists in abstracting an electron or a hydrogen atom from the carbon–hydrogen compound. In the processes that belong to the third type, a complex activates initially not the hydrocarbon but it reacts with another reactant (for example, with hydrogen peroxide or molecular oxygen). The formed reactive species (for example, hydroxyl radical) then attacks the hydrocarbon molecule without any participation in the latter process of the metal complex. The "direct activation" in this case does not involve the participation of the metal catalyst [1]. The three categories of metal-mediated C–H functionalization processes are shown below Scheme 2.

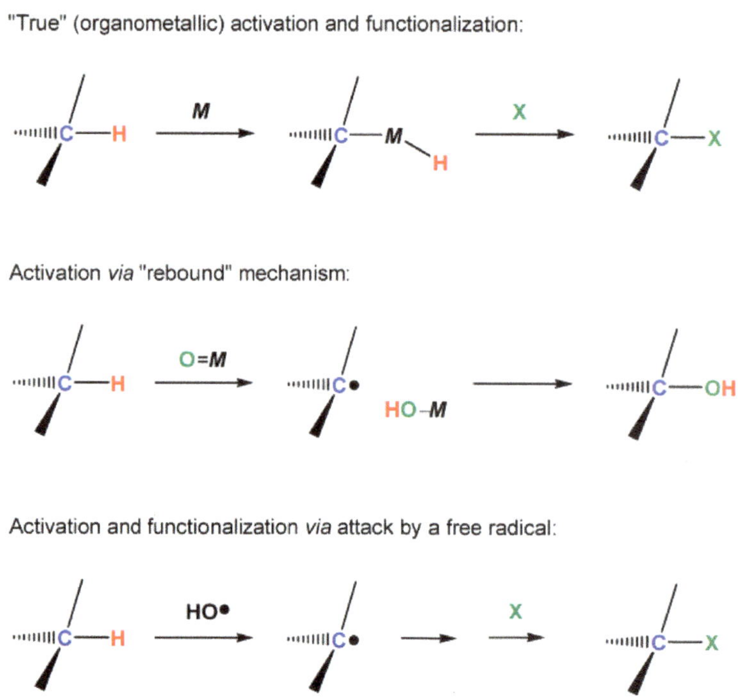

Scheme 2: Three types of metal-mediated C–H functionalization.

In living organisms under the action of certain metal-containing enzymes, the carbon–hydrogen oxidations proceed as reactions of the second or third type [1]. Although these oxidations occur via the formation of reactive radicals, they are rather selective and produce what is necessary for microorganisms compounds and energy. Biodegradation of hydrocarbons requires also metal-containing enzymes.

The Shilov reaction [1] was the first example of the organometallic (first type) activation which allows us to oxidize alkanes in aqueous solutions under catalysis by platinum(II) complexes. Curiously, the mechanism of this first reaction is very complex and is not completely clear. It will be discussed in more detail later. Numerous publications have appeared since Shilov's discovery which described unambiguous formation of σ-alkyl derivatives in alkane activation by metal complexes. In this review, however, we will discuss mainly the processes of oxygenation of carbon–hydrogen bonds to afford compounds with C–O bonds. All hydrocarbon oxidations occurring in living organisms [1,20,21,22,23,24,25,26,27,28,2

9] from the mechanistic point of view are also of the second or third types. Some non-organometallic complexes, for example, metal derivatives of porphyrins can play the role of models of certain metal-containing enzyme centers. In this review, we will consider recent publications devoted to metal-catalyzed liquid-phase reactions of carbon–hydrogen compounds. Many reviews on some aspects of carbon–hydrogen functionalization have been published in recent years [30,31,32,33,34,35,36,37,38,39,40,41,42,43,44,45,46,47,48,49,50,51,52].

MAIN CHARACTERISTICS OF CATALYSTS AND CATALYZED REACTIONS

Definitions

A reaction can be characterized [53] by the yield (%) (Equation (1)):

(amount of a product/amount of either starting substrate or reagent) × 100

- Important parameters [54,55] characterizing the catalyst are turnover number (TON) (Equation (2))
- TON = moles of product/moles of catalyst
- and turnover frequency (TOF) (Equation (3))
- TOF = TON/time.

In functionalization of inert hydrocarbons, these substrate compete with a solvent. In these cases, usually acetic and trifluoroacetic acid [56], acetonitrile [57], and water [58,59] are employed. Alcohols are more easily oxidizable compounds in comparison with alkanes. It is noteworthy, however, that, in the aerobic oxidation of alkanes catalyzed by $FeCl_3$ under visible light irradiation, alcohols can be used as solvents [60]. In the initial stage, photohomolysis of the Fe^{III}–Cl bonds leads to the generation of reactive Cl• radicals [61] which attack both the alkane and the alcohols used as a solvent with comparable rates. This leads to the formation of oxygenation products from the alkanes in quite high yield.

"To strengthen the relationship between catalysis and green chemistry, methods making use of neoteric solvents or alternative media rather than hazardous solvents, are more and more considered and represent unique challenges and opportunities for the future of catalysis".

[62]

Ionic liquids can be presented as examples [63,64,65,66].

In recent decades, traditional methods of stimulating catalyzed reactions (heating) new methods were added: photochemical initiation [67,68,69,70,71,72], microwave heating [73], electrocatalytic [74], and sonochemical [75] transformations.

Selectivity

Typically alkanes are transformed into derivatives which are more reactive in comparison with parent hydrocarbons, and this makes the selectivity one of the most important problems in functionalizations of saturated hydrocarbons as well as aromatic and other carbon–hydrogen compounds.

We will begin our discussion here from the term "*substrate selectivity*." It is especially important to discuss the substrate selectivity in the oxidations in living systems because the cell contains myriad of various compounds that might be potential substrates.

The term "*product selectivity*" is a concept more important for the reactions made *in vitro*. Indeed, the same interaction of a few reactants can lead depending on conditions to different distribution of products. By optimizing the conditions, a chemist can obtain predominantly one desirable product (isomer).

Usually, reactions give rise to the formation of a few possible positional isomers, such as monosubstituted benzenes. Electrophilic substitution in monosubstituted benzenes C_6H_5X gives comparable amounts of the three isomers. The distribution of isomers is *ortho* ≈ *para* > *meta* if X is an electron-releasing group and *meta* > *ortho* ≈ *para* if X is an electron-withdrawing group. In reactions of homolytic arylation, both withdrawing and releasing groups X direct the substituent mainly into *ortho*- and *para*-positions. Normal alkanes contain carbon–hydrogen bonds of two types: primary carbon–hydrogen bonds of two terminal methyl groups and secondary carbon–hydrogen bonds of internal methylene groups. Reactivities of methylene carbon–hydrogen bonds relative reactivity of the methyl group often depend on the nature of an oxidizing species and thus can give valuable information on the mechanism of the oxidation. Parameter C(1):C(2):C(3):C(4) is relative normalized (calculated taking into account the number of hydrogen atoms at each carbon) reactivities of hydrogen atoms at carbons 1, 2, 3, and 4 of the chain of normal alkane. This parameter is often discussed in mechanistic studies. Substitution of hydrogen atoms in normal alkanes with the participation of very reactive radicals like Cl• or HO•proceeds non-selectively, which is reflected by low selectivity parameter: C(1):C(2):C(3):C(4) ≈ 1:5:5:5. The stochiometric uncatalyzed oxygenations of normal alkanes with peracids occur without the participation of free radicals and exhibit high selectivity; more inert methyl groups are not oxidized at all. Some catalytic systems direct the substituent predominantly to only one position (region). We say that such reactions are "*regioselective*" or regiospecific. Regioselectivity is the preferential formation of one constitutional isomer over another (constitutional isomers of the same molecular formula differ in how their atoms are connected).

In reactions of branched alkanes, the bond selectivity parameter 1°:2°:3° is widely used. This value is relative normalized reactivities of hydrogen atoms at primary, secondary, and tertiary carbons of a branched alkane. Reactions that involve free radicals are usually less selective compared to reactions that proceed via concerted non-radical mechanism. In some reactions, the nature of a solvent can influence dramatically changing preferable carbon–hydrogen bonds in the same molecule [76] Scheme 3:

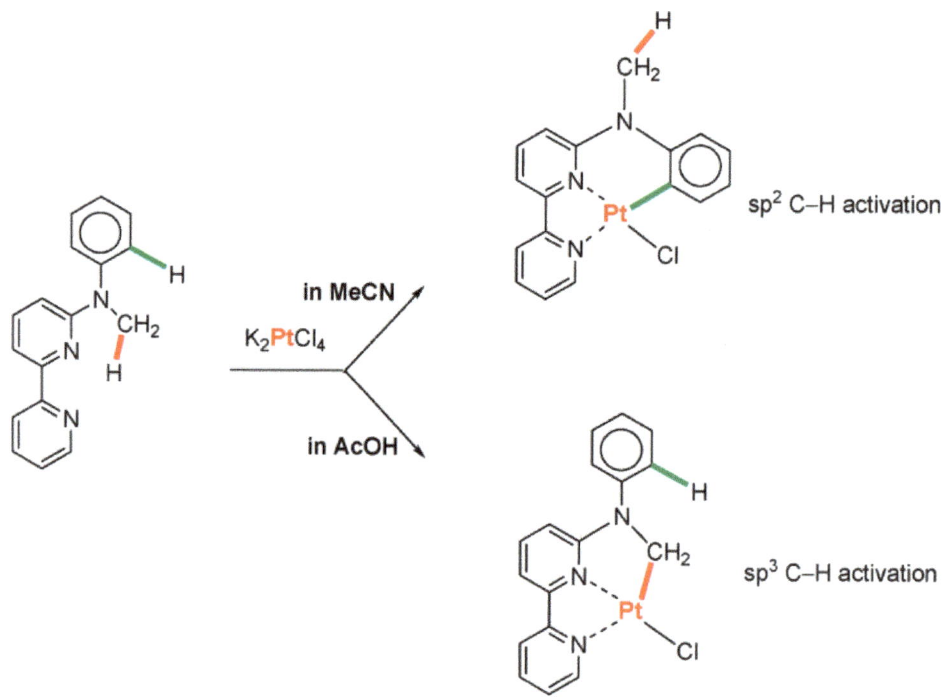

Scheme 3: Activation of aryl or methyl C–H groups by the Pt(II) complex.

An investigation of "*stereoselectivity*" helps to understand the mechanism of the interaction between the active species and carbon–hydrogen bonds. The transformation of a test compound *cis*-1,2-dimethylcyclohexane with or without retention of configuration is shown in Scheme 4.

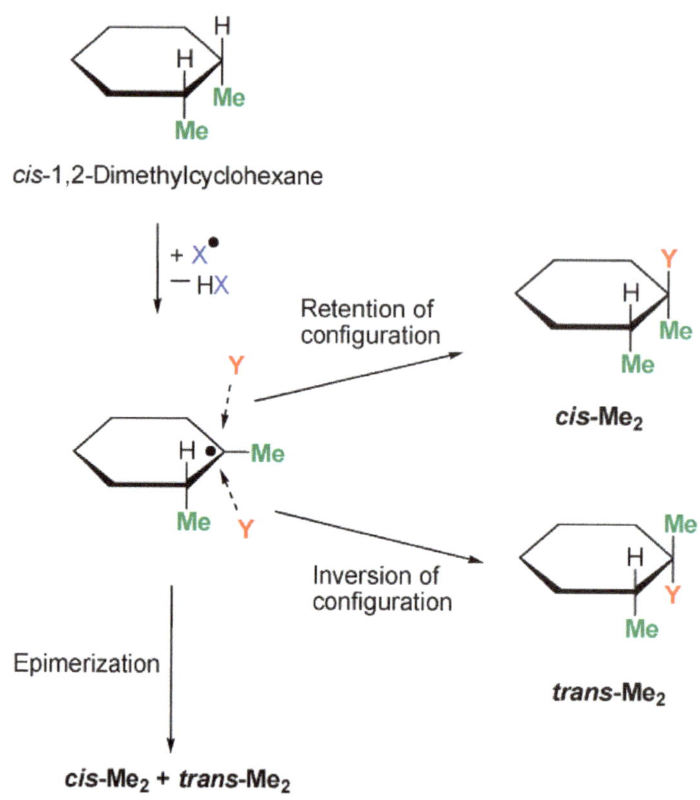

Scheme 4: Activation of *cis*-1,2-dimethylcyclohexane.

If two methyl groups in the product (in the case Y = OH) are in *cis*-orientation, we say that the reaction proceeds with retention of configuration. In the second case, if methyls are in mutual *trans*-orientation, we say about inversion of configuration. Usually, both reactions proceed with comparable rates and give ***cis*-Me$_2$** and ***trans*-Me$_2$** isomers in approximately equal amounts (epimerization). If the reaction proceeds without the participation of reactive radicals but via a concerted mechanism (as in the cases of peracids), the products are generated with partial or complete retention of configuration.

Effect of Additives: Pyrazine-2-Carboxylic Acid (PCA)

It is known that the addition of some compounds can dramatically change the reaction rate (increasing or decreasing it) and also affect the selectivity of the reaction. Thus, pyridine present in the solution of Gif systems induces the transformation of initially formed alkyl hydroperoxide into the corresponding ketone ("ketonization" of the alkane occurs). Light irradiation of the emulsion of cyclohexane with an aqueous solution of $Fe(ClO_4)_3$ in air exclusively gives cyclohexanone [77]. One can propose that water, like pyridine, dehydrates the cyclohexyl hydroperoxide. This is supported by finding that irradiation of the solution of cyclohexyl hydroperoxide and $Fe(ClO_4)_3$ in water selectively gave cyclohexanone.

Some time ago, we found [78,79,80] that pyrazine-2-carboxylic acid (PCA) and analogous compounds (2,3-pyrazinedicarboxylic, picolinic, and dipicolinic acids) can be used as highly active co-catalysts in the vanadate-catalyzed oxidations of alkanes, arenes, alcohols, and other substrates, under mild conditions by peroxides in air. The $[VO_3]^-$/PCA/H_2O_2 system has been studied in detail [81,82,83,84,85,86,87,88,89,90,91,92]. We then continued studies of alkane oxidations in the presence of PCA under catalysis by complexes of some other metals: iron [93,94,95,96], manganese [97,98], and rhenium [99].

The complex $(OC)_3Fe(\mu\text{-}PhS)_2Fe(CO)_3$ (compound **1**; see Figure 1) catalyzes the oxygenation of alkanes and benzene with hydrogen peroxide in air in an acetonitrile solution [100]. Alkyl hydroperoxides are formed as the main reaction products in the alkane oxidation (turnover numbers of 2300 were attained); benzene is transformed into phenol. The addition of PCA improves the oxidation. The duration of the lag period in the oxygenate accumulation can be substantially reduced if pyridine is added. Some dependencies are shown in Figure 1. Interestingly, the addition of picolinic acid (PIC) instead of PCA leads to the complete oxidation inhibition (Figure 1A).

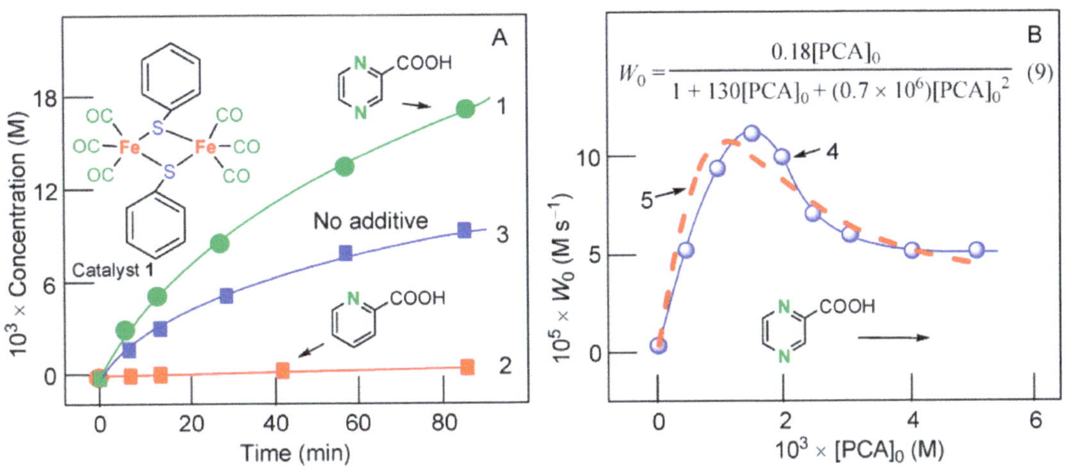

Figure 1: Graph (**A**). Accumulation of oxygenates (cyclohexyl hydroperoxide + cyclohexanol + cyclohexanone) with time in the cyclohexane oxidation with H_2O_2catalyzed by compound **1** in acetonitrile in the presence of two different co-catalysts: PCA (3.0×10^{-3} M; curve 1) and PIC (3.0×10^{-3} M; curve 2) or in the absence of additives (curve 3). Conditions: $[\mathbf{1}]_0 = 2.0 \times 10^{-4}$ M, $[pyridine]_0 = 0.2$ M, $[H_2O_2]_0 = 0.64$ M, $[cyclohexane]_0 = 0.37$ M, 50 °C. Graph (**B**). Curve 4: dependence of the initial rateW_0 of oxygenate accumulation in the cyclohexane oxidation with H_2O_2 catalyzed by **1**in acetonitrile on the initial concentrations of PCA. Curve 5: simulated dependence of the initial rate W_0 on $[PCA]_0$ corresponding to Equation (6). Conditions: $[\mathbf{1}]_0 = 2.0 \times 10^{-4}$ M, $[pyridine]_0 = 0.1$ M, $[H_2O_2]_0 = 0.64$ M, $[cyclohexane]_0 = 0.37$ M, $[H_2O]_{total} = 1.21$ M, 50 °C. Adapted from Ref. [100].

The dependence of the initial oxidation rate W_0 on the initial concentration of added co-catalyst PCA in the presence of pyridine is presented by curve 1 in Figure 1B. For the kinetic Equations (4)–(7).

1+n py⇌1·(py)n (4)

1·(py)n+PCA⇌K1 1·(py)n−1 ·PCA+py (5)

1·(py)n−1·PCA+PCA⇌K2 1·(py)n−2·(PCA)2+py (6)

1·(py)n−1·PCA+H2O2→k X (7)

where X is a species active in the cyclohexane oxidation, we have Equation (8)

WX=kK1[1·(py)n]0[PCA]0[H2O2]01+K1[PCA]0+K1K2[PCA]02 (8)

W0=0.18[PCA]01+130[PCA]0+(0.7×106)[PCA]02 (9)

It is easy to obtain Equation (9) shown in Figure 1B for conditions of the discussed experiments.

More recently, further research of other chemists has resulted in new interesting catalytic systems based on the combination of PCA. For example, Pombeiro and colleagues described catalytic oxidations with complexes of iron, copper, vanadium as well as gold nanoparticles [101,102,103,104] (see also reviews [105,106]).

NEW CATALYSTS AND LIGANDS

New types of catalysts for efficient oxidation of carbon–hydrogen compounds have been reported in recent years: heteropolycompounds [107,108,109,110], supported complex catalysts [111,112], and porous materials [113,114,115,116,117,118,119,120,121,122].

Metal Ions Most Active in Oxidation Catalysis

In the early stages of development of homogeneous carbon–hydrogen oxidations only a few transition metals were more or less carefully studied as catalysts [1]. These ions were iron, copper, manganese, *etc.* During the past several decades, many metals have been studied in catalytic activation and functionalization of carbon–hydrogen bonds. The reader is directed to selected recent examples with gold [123], cobalt [124,125,126] chromium [127], copper [128,129,130,131] iron [132,133,134,135], iridium [136], manganese [137,138,139], molybdenum [140], nickel [141], osmium [142,143,144,145,146,147,148], palladium [149,150,151], rhenium [152], rhodium [153,154], ruthenium [155,156], and vanadium [157,158].

A new method of synthesis of heterometallic coordination compounds by spontaneous self-assembly (SSA) [159] led recently to complexes which turned out to be very efficient catalysts in oxidations with peroxides. For example, Nesterov *et al.* prepared the heterobimetallic complex with a Schiff-base ligand [$Co_4Fe_2OSae_8$] (compound **2**; where Sae = L^4 and H_2Sae is salicylidene-2-ethanolamine (Figure 2)) [160]. In oxidation of cyclohexane into cyclohexyl hydroperoxide under the action of H_2O_2, this compounds exhibits a very high TON of 3570, which is the highest value ever observed for an iron-containing catalyst. It has been indicated [160] via mass spectral and comprehensive kinetic studies that hexanuclear **2** dissociates when dissolved in CH_3CN to form a trinuclear particle $[Co_2Fe(L^4)_4]^+$. This particle, observed in concentrated (1×10^{-4} M) solutions of **2**, is the catalytically active species responsible for the fast reaction rates greater than 3×10^4 M·s^{-1} (maximum turnover frequency TOF = 11,200 h^{-1} = 3.1 s^{-1}). The catalytic system is believed to proceed by free-radical pathways with the hydroxyl radical as the main attacking species.

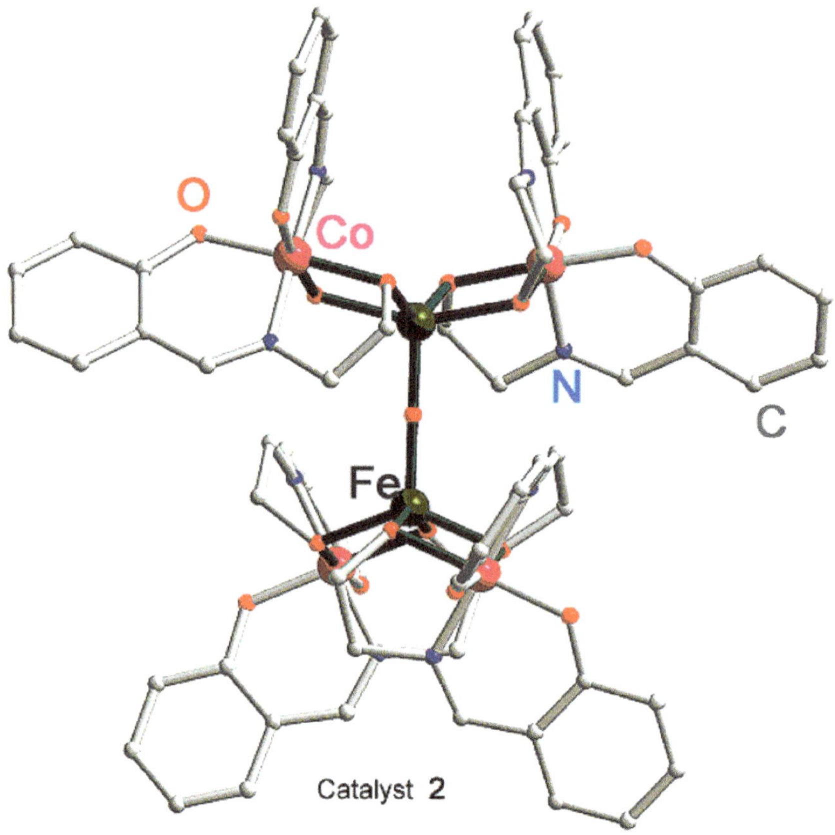

Figure 2: Molecular structure of $[Co^{III}_4Fe^{III}_2O(L^4)_8]$·4DMF·$H_2O$ (complex **2**). Solvated DMF and water molecules have been omitted for clarity. Adapted from Ref. [160].

A few years ago, we discovered that the salt of non-transition metal aluminum catalyzes the alkane hydroperoxidation with hydrogen peroxide [161]. It has been proposed that interacting with an aluminum derivative hydrogen peroxide is decomposed to afford hydroxyl radicals that attack the alkane molecule. This reaction is by no means a Fenton-like process because, in a Fenton reaction with Fe(II) or a Fenton-like reaction (for example with Fe^{III}), a transition metal can change its oxidation state ($Fe^{II} \rightleftharpoons Fe^{III}$) in the course of the hydrogen peroxide decomposition. Aluminum cation is unable to change its valency. To explain the catalytic action of aluminum cation, our co-authors Kozlov and Kuznetsov [162] assumed that, in the case of Al^{III}, not one but simultaneously *two* H_2O_2 molecules interact (entering into cation's coordination sphere). The first H_2O_2 molecule—as a model of the electron-rich Fe^{II} ion—delivers one electron (like in the Fenton mechanism) to the system, and this leads to the splitting the O–O bond in the second H_2O_2 molecule. Figure 3 presents a catalytic cycle generating two free radicals, HO$^•$ and HOO$^•$. According to DFT calculations, H_2O_2 molecule in complex **4** is highly activated toward the homolysis in the free hydrogen peroxide (6.1 *vs.* 39.4 kcal/mol). This approach has been used for some other non-transition metals [163,164,165,166,167]. The overall activation barriers (kcal/mol) of the generation of hydroxyl radicals were calculated: Ga (19.5) < Sc (22.2) < In (23.5) ≈ Y (23.7) < La (25.2) ≈ Al (25.6).

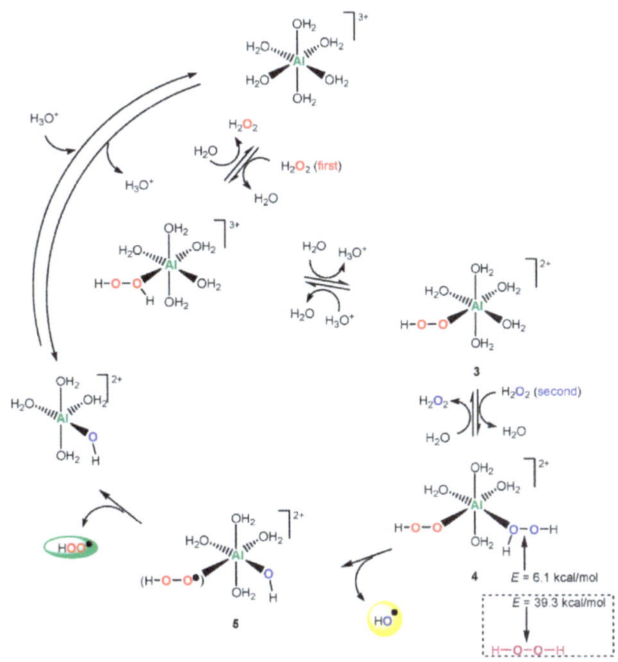

Figure 3: The Al-catalyzed decomposition of hydrogen peroxide. Adapted from Ref. [162].

Very recently, Meyerstein and colleagues used our approach involving two H_2O_2 molecules in DFT calculations of Fenton-like reactions with the participation of transition metals [168]. It was also shown that the Fenton-like reaction of $Co(H_2O)_6^{2+}$proceeds via the binding of three H_2O_2 molecules to the cobalt(II) ion forming the transient $(H_2O)_3Co^{II}(OOH)_2(H_2O_2)$ [169]. Harvey studied the mechanism of the Fenton reaction via computational methods [170].

New Types of Ligands and Catalysts

In this section, we will briefly consider work that has been published in recent years describing new types of ligands and catalysts. Some examples have been discussed in previous sections (for example, see Figure 3).

Heme and nonheme oxidation catalysts have been intensively used in oxidations of alkanes aromatics, as well as in epoxidations [171,172,173,174]. Metalloporphyrins [175,176,177,178] play a very important role in such studies; they are models of enzyme reaction centers.

Sorokin and colleagues used phthalocyanines as catalysts in oxidations with peroxides [179,180]. Thus, mild oxidation of ethane to acetic acid by H_2O_2 was catalyzed by supported mu-nitrido diiron phthalocyanes [181], methane was oxidized to methanol, formaldehyde, and formic acid [182]. Maksimov, Karakhanov, and colleagues used analogous catalysts containing ions of iron and copper [183].

Talsi, Bryliakov, and colleagues published a series of papers on the C–H oxidations with H_2O_2 in the presence of aminopyridine complexes of manganese(II) of the types **6**–**10**. It was shown that complexes **6**–**8** efficiently (with TON up to 970) induce the oxidation of substrates containing aliphatic carbon–hydrogen groups, such as cyclohexane, adamantane, and (−)-acetoxy-*p*-menthane (Figure 4), with high regioselectivity (preferential oxidation of tertiary C–H bonds) [184]. Along with a few more catalysts, Mn-aminopyridines have been regarded as demonstrating good promise for practical use [185]. The introduction of electron-donating groups (complexes **9** and **10**) improved the oxidation selectivity; furthermore, catalysts **9** and **10** surprisingly demonstrated a much better efficiency in the benzylic oxidation of cumenes and ethylbenzene compared to the parent catalyst **6** [186]. All Mn-aminopyridine catalysts were active in acetonitrile solutions with the addition of co-catalytic additive (acetic acid), and typically required only 0.1 mol. % catalyst loadings and 1.3 equiv. of H_2O_2(*vs*. substrate). Some examples of preparative oxidations on catalysts of this type are presented in Figure 4.

The mechanism of carbon–hydrogen bond oxidation in the presence of Mn-aminopyridine complexes was considered. The authors confirmed the electrophilic nature of the active oxidizing species. The latter demonstrated high 3°/2° ratios (40–49), alcohol/ketone ratios > 5, and high stereospecificity in the oxidation of *cis*-1,2-dimethylcyclohexane [187]; kinetic studies revealed a sizable kinetic isotopic effect (k_H/k_D 4.5–4.7), indicating that that the H atom lies on the reaction coordinate and participates in the C–H bond breaking step. Altogether, the experimental data provided indirect evidence in favor of the participation of a high-valent (presumably oxometal(V) [188]) species, ruling out the contribution of a free-radical-driven oxidation pathway. The key findings pointing to the oxomanganese(V) active species were (i) the incorporation of ^{18}O from added $H_2^{18}O$ into the oxidizing product, corroborating the "water-assisted" mechanism and (ii) partial inversion of *cis*configuration during the oxidation of *cis*-1,2-dimethylcyclohexane, apparently proceeding via formation of relatively short-lived carbon-centered tertiary radicals typical of the classic "rebound mechanism," postulated for the oxoferryl-driven cytochrome P450 reaction cycle. So far, no direct spectroscopic evidence for the oxomanganese(V) reactive species in Mn-aminopyridine-based systems has been reported. However, in catalyst systems relying on isostructural Fe catalysts, such evidence was gained by EPR spectroscopy taking advantage of the low-spin ($S = {}^1/_2$) structure of known oxoiron(V) intermediates [189].

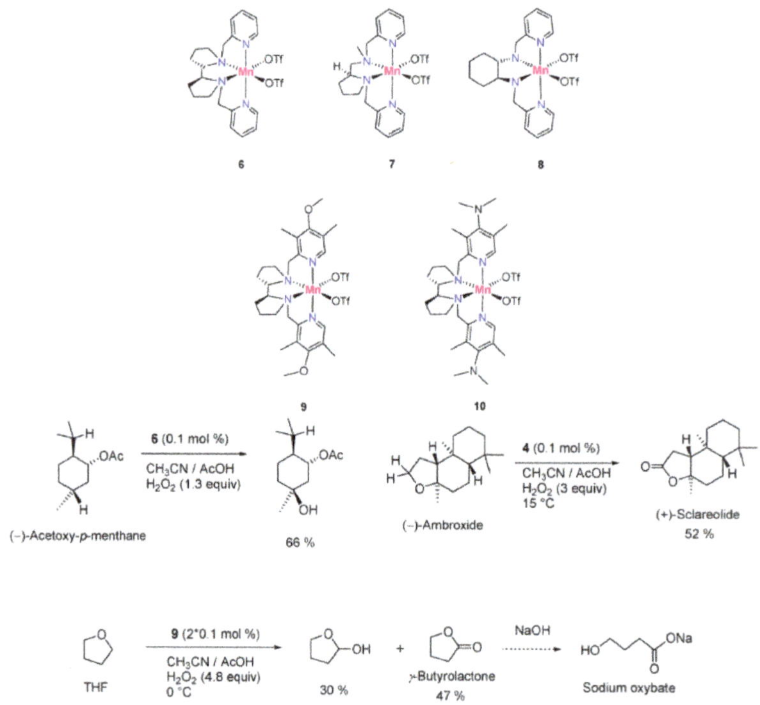

Figure 4: Examples of preparative-scale selective C–H oxidations catalyzed by Mn-aminopyridine complexes.

Costas in recent years has carried out many valuable works on functionalization of carbon–hydrogen bond catalyzed by nonheme complexes with polyamino ligands [190,191,192,193]. Additionally, very important works on complexes with polydentate ligands as catalysts have been reported by Nam and colleagues [194].

Complexes with ligands of new types have been recently synthesized, and in some cases these complexes turned out to be efficient catalysts for C–H functionalization. We have to mention here Tp^xM complexes which were used by Pérez and other authors for alkane functionalization by carbene insertion [195,196,197,198,199,200].

The nature of ligand and ligand robustness play a very important role in metal-complex catalysis [201]. New types of ligands have been proposed in recent decades which form efficient catalysts with transition metal ions. These ligands are: N-heterocyclic carbene (NHC) [202,203], pincer-like [204,205,206,207,208], redox-active ligands [209,210,211,212], and scorpionates [213,214,215,216,217,218].

Polynuclear Metal Complexes As Catalysts

Multi-metallic metal complexes often have higher catalytic activity in comparison with mononuclear compounds [219,220].

Recently, Bilyachenko *et al.* have proposed a new convenient route to metallasilsesquioxanes and synthesized a series of complexes containing various transition metal ions. The most simple compound is the binuclear cage-like copper silsesquioxane $[(PhSiO_{1.5})_{10}(CuO)_2(NaO_{0.5})_2$ 4EtOH] (complex **11**) with a "Cooling Tower" structure (Figure 5) [221]. In addition, isomers containing four [222] and six [223] copper ions have been isolated. Their structures are presented in Figure 6 and Figure 7, respectively. Al these compounds exhibited high catalytic activity in oxidation of benzene, alkanes, and alcohols with peroxides [224].

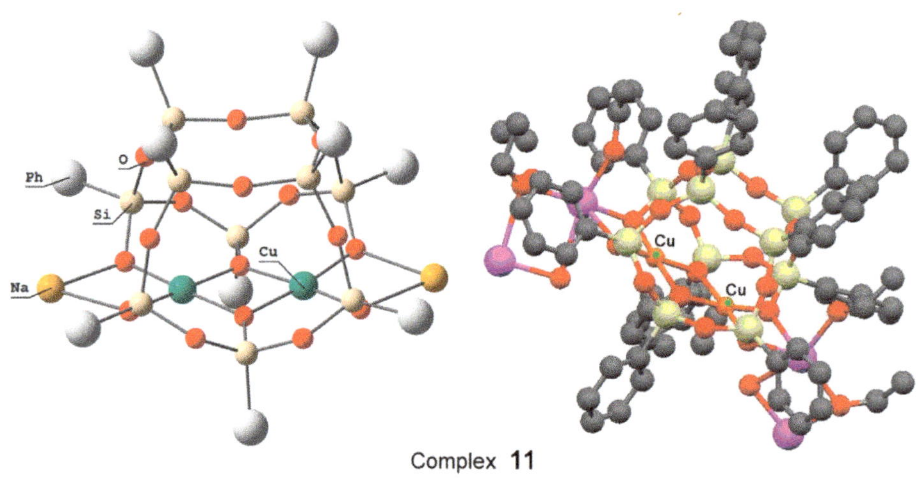

Figure 5: The structure of compound $[(PhSiO_{1.5})_{10}(CuO)_2(NaO_{0.5})_2$ 4EtOH] (**11**) (left) and a fragment of polymeric structure (right). Adapted from Ref. [221].

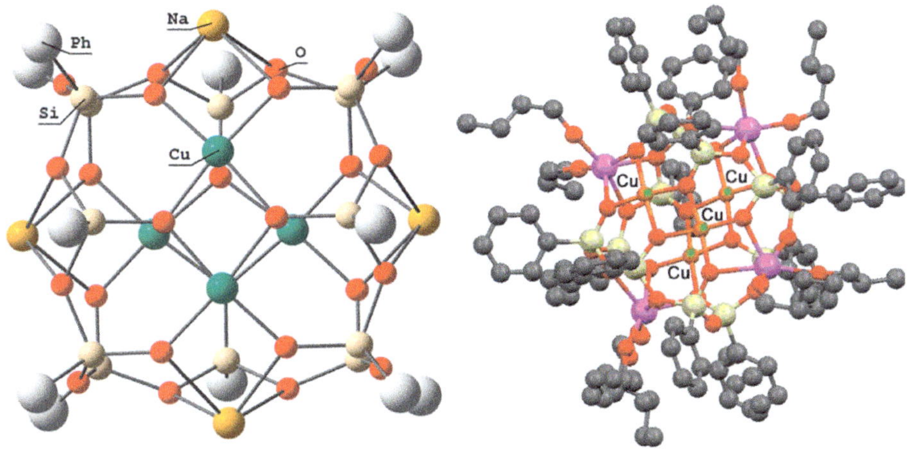

Complex **12**

Figure 6: The structure of "Globule"-like compound $[(PhSiO_{1.5})_{12}(CuO)_4(NaO_{0.5})_4]$ (**12**). Adapted from Ref. [222].

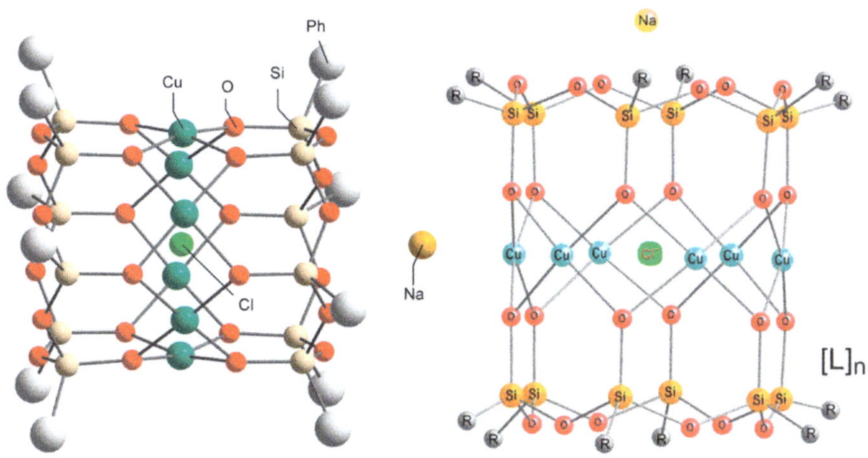

Complex **13**

Figure 7: The structure of hexanuclear cylinder-like copper silsesquioxane $[(PhSiO_{1.5})_{12}(CuO)_6(NaCl)(C_4H_8O_2)_{12}(H_2O)_2]$ (**13**), L = 1,4-dioxane/H_2O. Adapted from Ref. [223].

The hexanuclear oxo cluster $[(PhSiO_{1.5})_{20}(FeO_{1.5})_6(NaO_{0.5})_8(C_7H_8)(C_4H_9OH)_9]$ **14** (for the strucure, see Figure 8) containing iron and sodium cations efficiently catalyzed oxidation of cyclohexane and benzene in acetonitrile solution to the corresponding alkyl hydroperoxides and phenol, respectively, by hydrogen peroxide in air in the presence of nitric acid [225]. Another hexairon complex **15** [226] (for the structure, see Figure 9) catalyzed the oxidation of cyclohexane affording the products in 46% yield (Figure 10), which is very high for oxidation of alkanes.

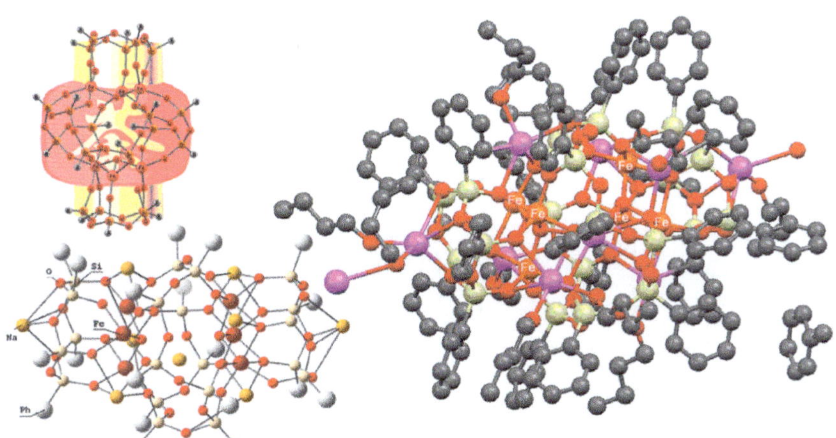

Complex **14**

Figure 8: The structure of heteronuclear Na_8Fe_6-oxo cluster bearing silsesquioxane ligands $[(PhSiO_{1.5})_{20}(FeO_{1.5})_6(NaO_{0.5})_8(C_7H_8)(C_4H_9OH)_9]$ (**14**). Adapted from Ref. [225].

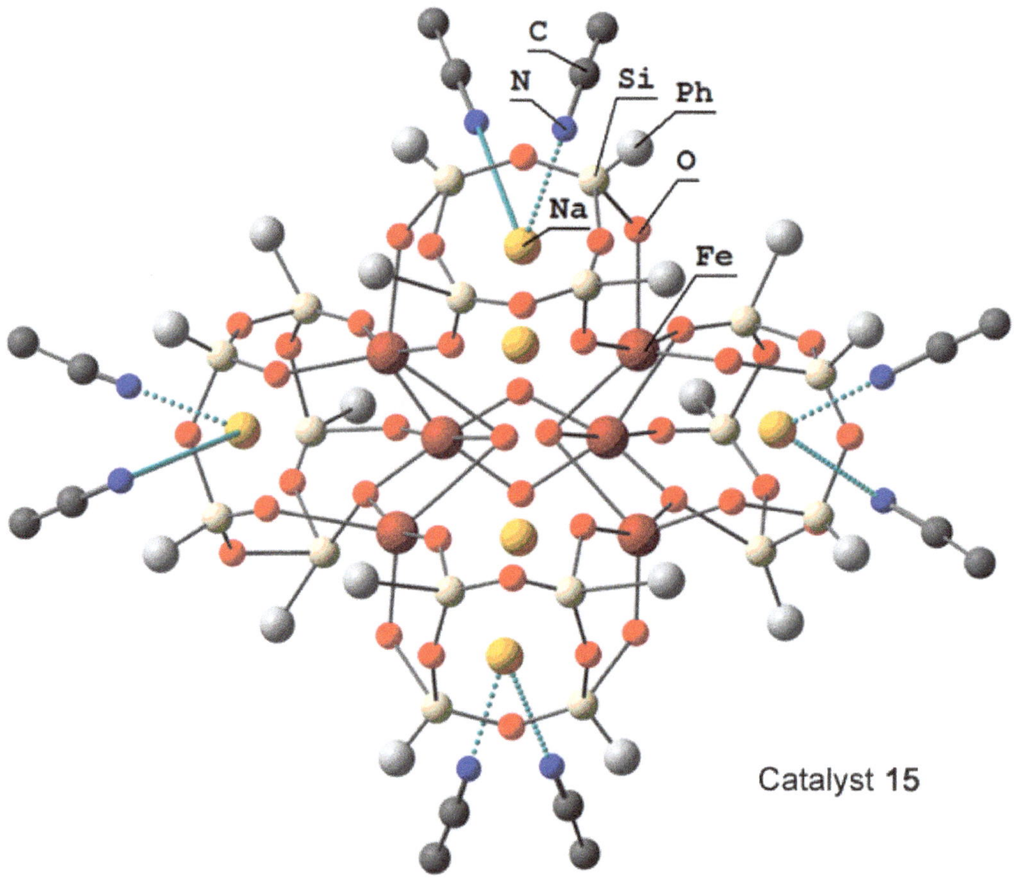

Figure 9: Molecular structure of hexairon compound **15**. Adapted from Ref. [226].

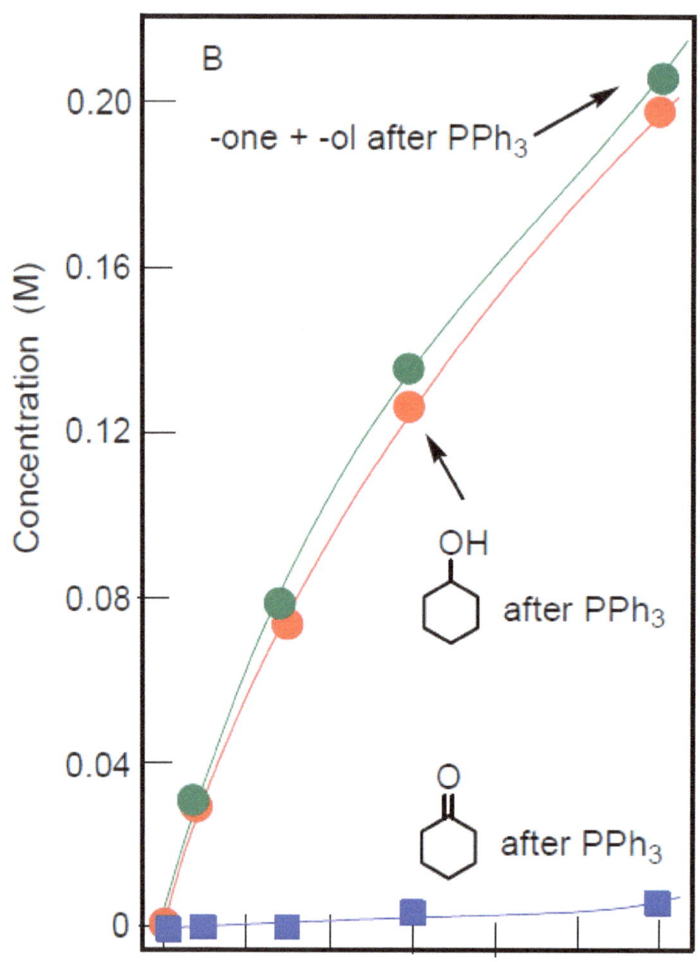

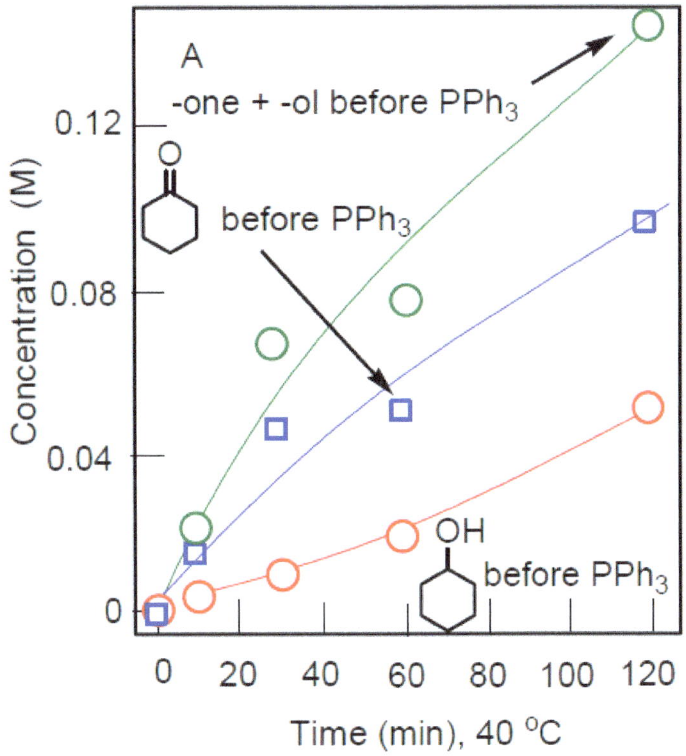

Figure 10: Accumulation of cyclohexanol and cyclohexanone with time in the cyclohexane (0.45 M) oxidation with H_2O_2 (50%, aqueous, 1.5 M) catalyzed by compound **15** (5 × 10^{-4} M) in the presence of CF_3COOH (0.05 M). Solvent was acetonitrile (total volume of the reaction solution was 5 mL); temperature was 40 °C. Concentrations measured before (Graph **A**) and after (Graph **B**) reduction of the samples with PPh_3 are shown.

Cations of cobalt and nickel gave cylinder-type complexes phenylsilsesquioxane $[(PhSiO_{1,5})_{10}(CoO)_5(NaOH)]$ (compound**16**) [227] and $(PhSiO_{1.5})_{12}(NiO)_6(NaCl)$ (compound **17**) [228]. Their "drum-like" structures are presented in Figure 11 andFigure 12, respectively. Compound **16** exhibited activity in catalytic highly stereoselective oxidation of 1,2-cyclohexane with*m*-CPBA (Table 1).

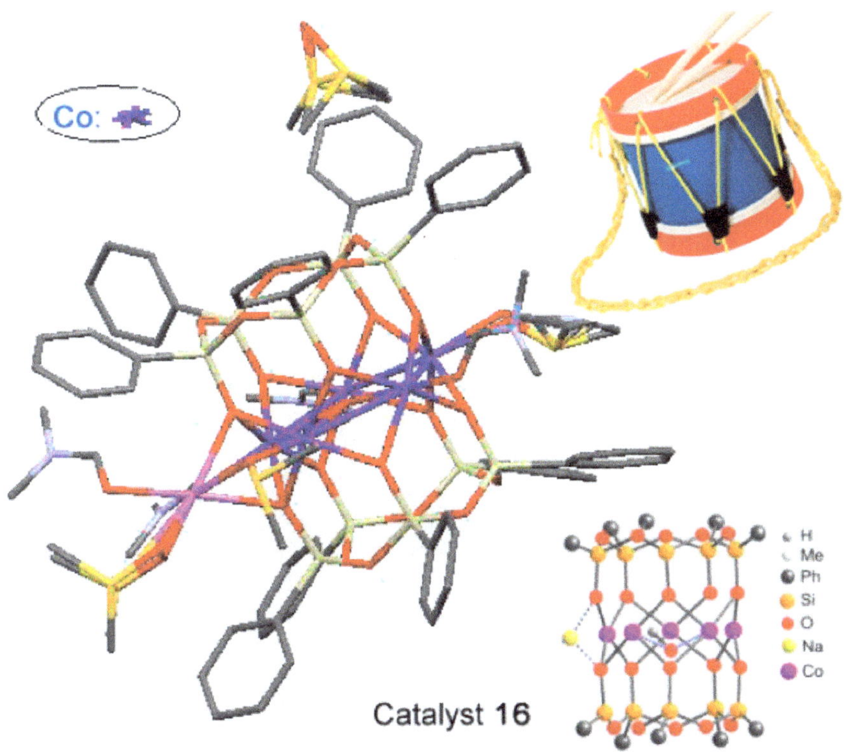

Figure 11: Molecular structure of compound $[(PhSiO_{1,5})_{10}(CoO)_5(NaOH)]$, **16**. Adapted from Ref. [227].

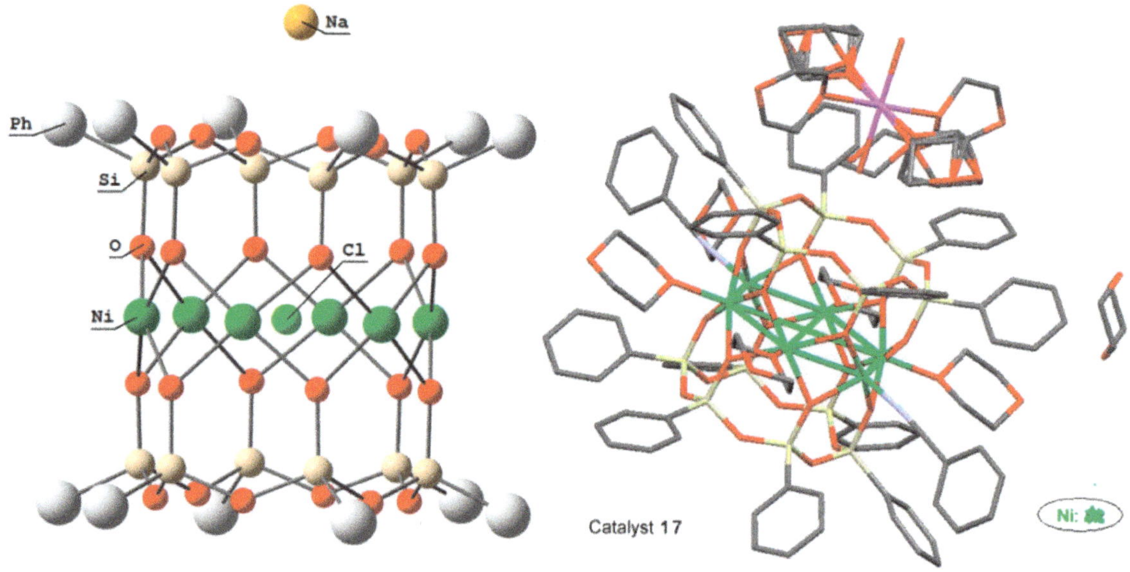

Figure 12: Molecular structure of compound $(PhSiO_{1.5})_{12}(NiO)_6(NaCl)$, **17**. Adapted from Ref. [228].

Table 1: Oxygenation of isomeric dimethylcyclohexanes with *meta*-chloroperoxybenzoic acid (*m*-CPBA) catalyzed by hexanuclear cobalt complex **16** [227][a].

Entry	Substrate	T (°C)	$[HNO_3]$ (M)	Time (min)	Reduction with PPh_3	*t*-Dimethyl-Alcohol (mM)	c-Dimethyl-Alcohol (mM)	Yield (%)	Ratio t/c
1[b]	*cis*-1,2-DMCH	40	0	120	yes	4	5.7	7	0.70
2[b]	*cis*-1,2-DMCH	50	0	30	yes	4	5	6	0.77
3[b]	*cis*-1,2-DMCH	50	0.05	30	yes	3	21	17	0.13
4[b]	*cis*-1,2-DMCH	40	0.05	300	yes	3	15	13	0.20
5	*cis*-1,2-DMCH	40	0.05	30	yes	4	83	62	0.05
6	*cis*-1,2-DMCH	40	0.05	15	no	3	55	41	0.05
7	*cis*-1,2-DMCH	40	0.05	15	yes	3	55	41	0.05
8	*trans*-1,2-DMCH	40	0.05	15	yes	33	2	25	16
9	*cis*-1,4-DMCH	40	0.05	15	yes	19	59	56	0.32
10	*trans*-1,4-DMCH	40	0.05	15	yes	33	2	25	16

Notes: [a] Reactions were performed in MeCN. Concentrations: $[Substrate]_0$ = 0.14 M, $[m\text{-CPBA}]_0$ = 0.26 M, $[catalyst\ 16]_0 = 5 \times 10^{-4}$ M. [b] In the absence of catalyst **16**.

Metal-organic frameworks (MOFs), or porous coordination polymers, have attracted the attention of chemists [229,230,231,232,233]. Pombeiro, Martins, Guedes da Silva [234,235,236,237,238], Kirillov [234,235,239,240,241], and other scientists [242] synthesized various polynuclear complexes, coordination polymers, and derived metal integrated mesoporous matrices and demonstrated high catalytic activity of these materials in oxidations with peroxides. Figure 13,Figure 14 and Figure 15 show the examples of such catalysts.

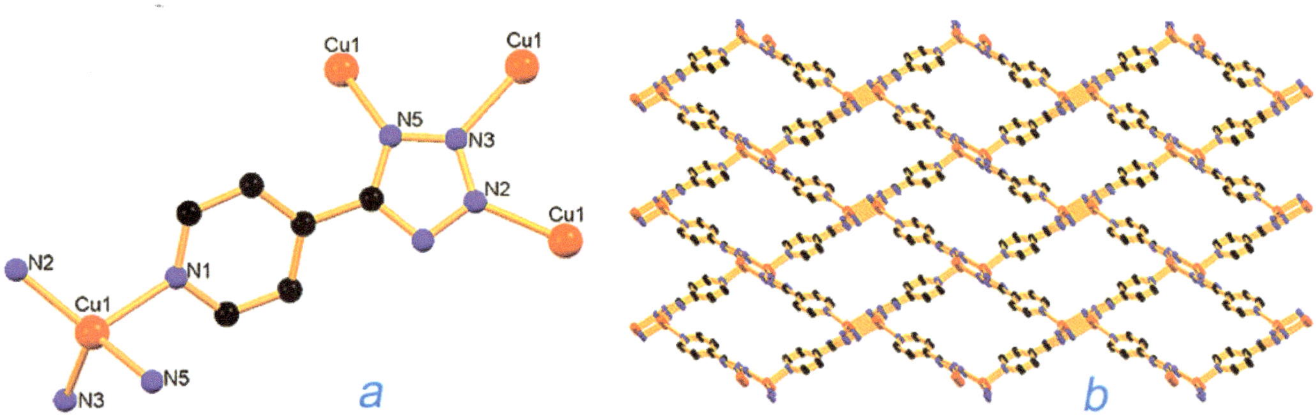

Figure 13: Structural fragments of the Cu(I) catalyst $[Cu(mu_4\text{-}4\text{-ptz})]_n$ [4-ptz = 5-(4-pyridyl)tetrazolate], **18**, representing (**a**) the basic unit; (**b**) a view of rhomboid voids along the *a*-axis. Adapted from Ref. [238].

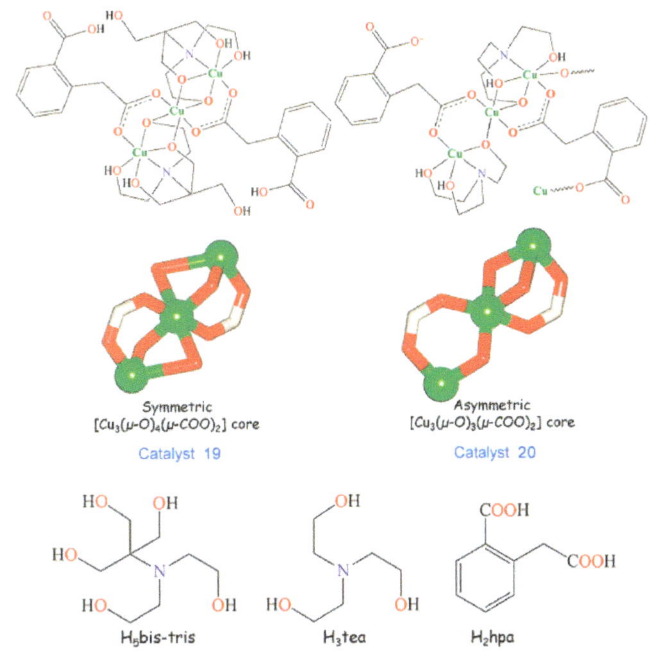

Figure 14: Catalysts $[Cu_3(\mu_2\text{-}H_3\text{bis-tris})_2(\mu_2\text{-}Hhpa)_2]\cdot H_2O$ (**19**) and $[Cu_3(\mu_2\text{-}H_2tea)_2(\mu_2\text{-}hpa)(\mu_3\text{-}hpa)]_n$ (**20**) bearing distinct tricopper(II) cores where H_3tea is triethanolamine, H_2hpa is homophthalic acid, H_5bistris is bis(2-hydroxyethyl)amino-tris(hydroxymethyl)methane. Adapted from Ref. [241].

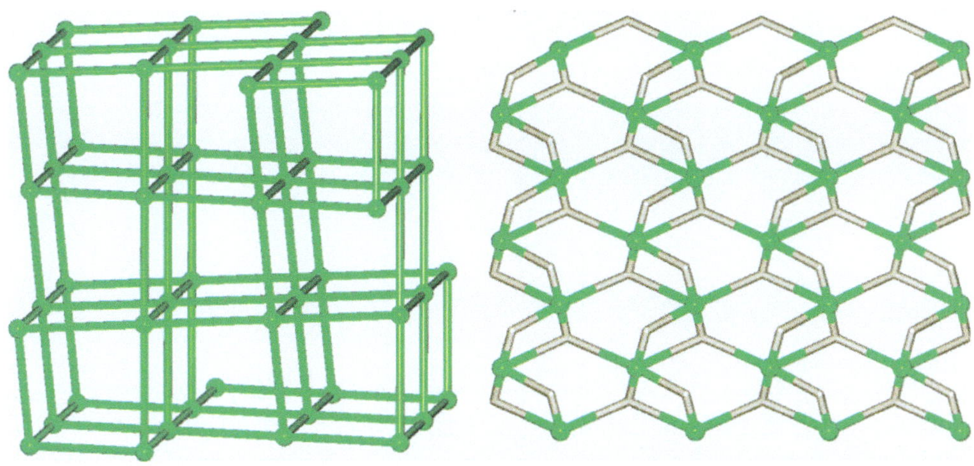

Figure 15: Topological representations of the H-bonded networks in catalysts **19** (left) and **20** (right). Green balls correspond to centroids of 6-connected or 5-connected molecular nodes. Adapted from Ref. [241].

MECHANISM OF OXIDATION BY PEROXIDES: RADICAL, OR NOT RADICAL, THAT IS THE QUESTION

In many cases, organic reactions occur with the formation of free carbon- and oxygen-centered radicals [243,244,245,246], and alkyl hydroperoxides are formed. A few tests can be used to detect the formation of ROOH from RH.

Detection and Determination of Alkyl Hydroperoxides Comparing GC before and after Treating Samples with PPh_3

We proposed and developed a simple method for analysis of alkyl hydroperoxides generated in various alkane oxidations in solutions [11,12,13,247,248,249,250,251,252]). A chemist can find that, if the direct injection of a reaction sample into the chromatograph gave comparable amounts of cyclohexanol and cyclohexanone (Figure 16, Figure 17a), the reduction of the sample with PPh_3 prior to GC analysis led to the noticeable predominance of the alcohol (Figure 17b). The comparison of the results obtained before and after the reduction clearly indicates that cyclohexyl hydroperoxide was formed as the main primary product. We can estimate the *real* concentrations of the alkyl hydroperoxide, alcohol, and ketone present in the reaction solution comparing the intensities of peaks attributed to the alcohol and ketone *before* and *after* the reduction. This method gives the very valuable information on the existence or non-existence of alkyl peroxides in the reaction mixture and allows us to estimate concentrations of the three products (ketone, alcohol, and alkyl hydroperoxide) formed in the reaction. In some cases, using non-polar columns and/or very clean detector, we can see in chromatogram a peak of alkyl hydroperoxide (Figure 16, Figure 17a), which disappears under the action of PPh_3 (Figure 17b).

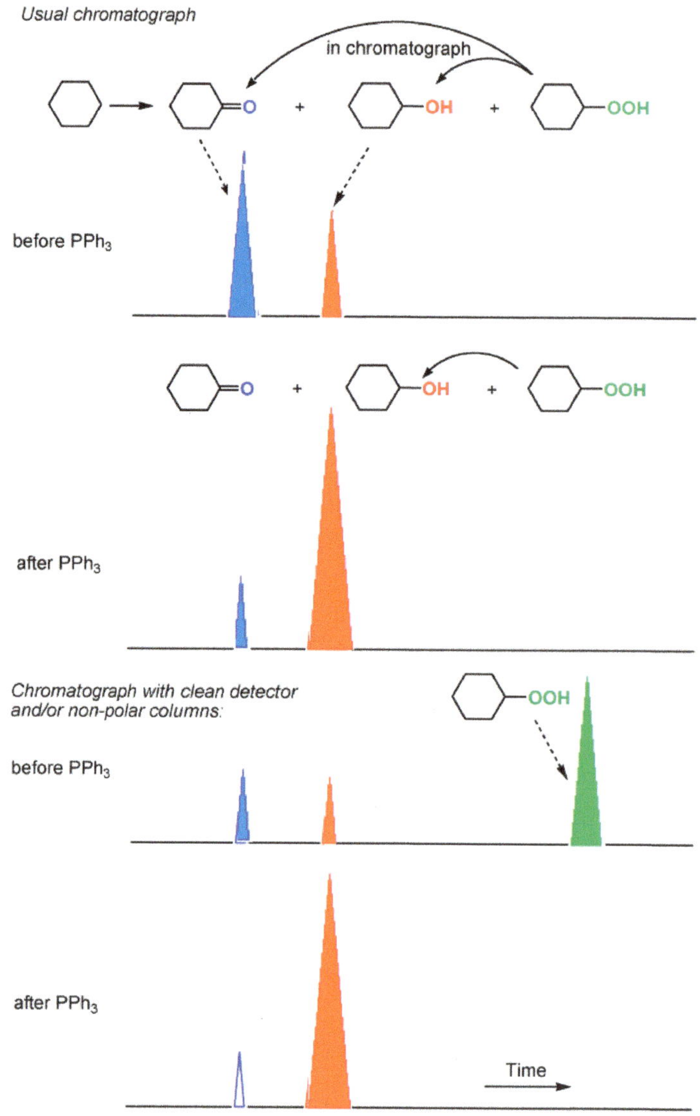

Figure 16: Detection of alkyl hydroperoxide ROOH by GC.

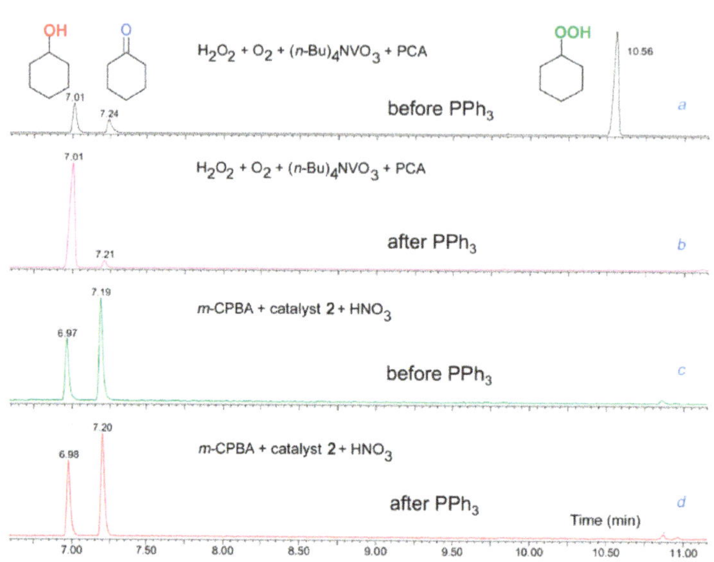

Figure 17: Chromatograms of reaction products obtained in the cyclohexane oxidation in acetonitrile by the H_2O_2/O_2/vanadate anion/PCA reagent [78,79,80,81,82,83,84,85,86,87,88,89,90,91,92] and the *m*-CPBA/catalyst **2**/HNO_3 system (see below [160]). For both systems the chromatograms were recorded before and after the addition of PPh_3. Conditions: $[\text{cyclohexane}]_0 = 0.46$ M; (***a*** and ***b***): $[H_2O_2]_0 = 0.75$ M, $[(n\text{-Bu})_4NVO_3]_0 = 1 \times 10^{-4}$ M, $[\text{PCA}]_0 = 4 \times 10^{-4}$ M, 40 °C; (***c*** and ***d***): $[m\text{-CPBA}]_0 = 0.028$ M, [complex **2**]$_0 = 2.7 \times 10^{-6}$ M, $[HNO_3]_0 = 5.5 \times 10^{-3}$ M, 50 °C. A Perkin-Elmer Clarus 600 gas chromatograph, equipped with non-polar capillary columns (SGE BPX5; 30 m × 0.32 mm × 25 μm) was used for analyses of the reaction mixtures. Adapted from Ref. [160].

Generally speaking, the [alcohol]/[ketone] ratio (presented in many publication as the *A*/*K* ratio) has no meaning and gives valuable information only if [alkyl hydroperoxide] = 0 [253]. Thus, it is more correct to write "the *A*/*K* ratio determined after reduction of the sample with PPh_3" [254,255]. Evidently, in this case, [A] = [alcohol] + [alkyl hydroperoxide].

When we have only the chromatogram obtained *after* the addition of PPh_3, we cannot definitively say that ROOH is present in the reaction solution (for example, [256]). In cases when the difference between chromatograms obtained *before*and *after* reduction of samples with PPh_3 indicates the formation of cyclohexyl hydroperoxide (that is when the investigator knows definitely that ROOH is present in the reaction mixture), it is convenient to characterize the reaction by the following ratio: ([alkyl hydroperoxide] + [alcohol])/[ketone]. This ratio is measured by GC for the samples *after* treating with PPh_3, and ([alkyl hydroperoxide] + [alcohol]) = $[\text{alcohol}]_{\text{after PPh3}}$. Although this rough parameter (see, for example, a recent paper by Kühn [257]) does not give the information on the amount of alkyl hydroperoxide that is present in the reaction mixture, it approximately describes the selectivity relative to the three products (alkyl hydroperoxide, alcohol and ketone).

The so-called Shul'pin method [257,258,259] of analysis of oxygenates discussed above in this section and based on GC and PPh_3 has been used by other authors (selected examples are given in Refs. [260,261,262,263,264,265,266,267,268,269 ,270,271,272,273,274,275,276,277,278]).

Finally, if the GC analysis shows that there is no difference between chromatograms obtained before and after reduction with PPh_3, it indicates that the samples do not contain alkyl hydroperoxides. Consequently, the reaction proceeds without the participation of free radicals, unless it was proposed, for example, for an oxidation with hydrogen peroxide via hydroxyl radicals (Fenton-like scheme): $H_2O_2 \rightarrow HO^\bullet$; $HO^\bullet + RH \rightarrow R^\bullet$; $R^\bullet + O_2 \rightarrow ROO^\bullet$; $ROO^\bullet \rightarrow \rightarrow ROOH$. Possibly, we have a concerted mechanism in this case. Examples of such reactions are given below in Table 1 and Figure 17c,d for the catalyzed oxidation of 1,2-dimethylcyclohexane with *m*-CPBA.

Competitive Oxidation of Alkane and Solvent

Kozlov demonstrated that the mode of dependence of the initial rate of oxygenate formation W_0 on initial concentration of cyclohexane (W_0 is approaching to a plateau at relatively high concentrations of cyclohexane; an example is shown in Figure 18A) can lead to the conclusion on the participation of hydroxyl radicals in the reaction (see, for, example, Ref. [160]) by the competitive interaction of hydroxyl radicals both with cyclohexane and acetonitrile. The following kinetic scheme describes the dependence of W_0 on $[\text{cyclohexane}]_0$ with the plateau:

$$H_2O_2 + \text{catalyst} \rightarrow HO^\bullet \quad (\text{rate } W_5) \tag{10}$$

$$HO^\bullet + CyH \rightarrow Cy^\bullet \rightarrow \rightarrow CyOOH \quad (\text{constant } k_6) \tag{11}$$

$$HO^\bullet + CH_3CN \rightarrow \rightarrow \text{products} \ (\text{constant } k_7) \tag{12}$$

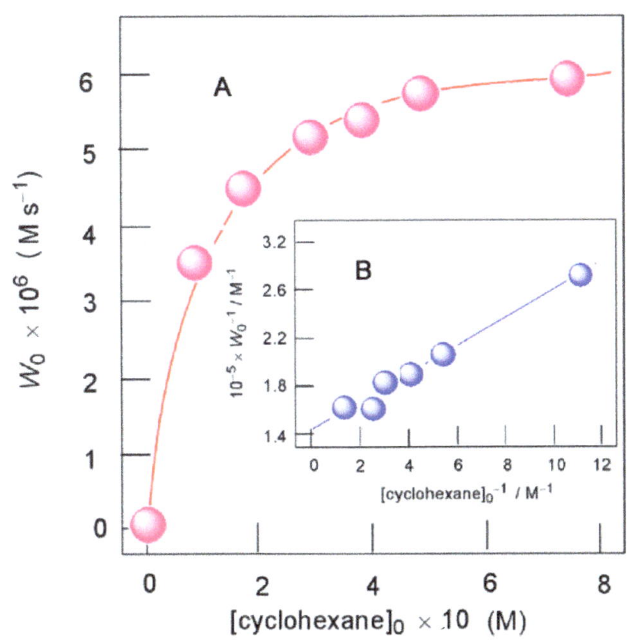

Figure 18: Oxidation of cyclohexane with H_2O_2 (1.0 M) catalyzed by compound **2** (3.2×10^{-5} M) in the presence of HNO_3 (0.04 M) in MeCN at 50 °C. Graph **A**: dependence of the initial rate of oxygenate formation W_0 on initial concentration of cyclohexane. Graph **B**: anamorphosis of curve shown in Graph A in coordinates $[\text{cyclohexane}]_0^{-1} - W_0^{-1}$. Adapted from Ref. [160].

Here, the initiation stage (10) corresponds to the catalytic generation of hydroxyl radicals with rate W_5, stage (11) corresponds to the sequence of transformations of CyH into CyOOH (with kinetically rate-limiting interaction between $HO^\bullet$and CyH), and stage (12) is the reaction of $HO^\bullet$ with acetonitrile. The analysis of this kinetic scheme in the quasi-stationary approximation relative concentration of hydroxyl radicals leads to the following equation:

$$W_0 = -\frac{d[CyH]}{dt} = \frac{d[CyOOH]}{dt} = \frac{W_i}{1 + \frac{k_7[MeCN]}{k_6[CyH]}} \tag{13}$$

The term

$$\frac{1}{\frac{d[CyOOH]}{dt}} \quad (14)$$

should be a linear function of $1/[CyH]_0$ (see Figure 18B). The following ratio can be obtained:

$$\frac{k_7[MeCN]}{k_6} = 8 \times 10^{-2}\,M \quad (15)$$

If we assume [MeCN] = 18 M in the reaction solution, we obtain $k_7/k_6 = 4.4 \times 10^{-3}$. Values $k_7[MeCN]/k_6$ (0.08 M) and k_7/k_6(0.0044) are close to the corresponding parameters obtained for oxygenation of cyclohexane in acetonitrile with H_2O_2. The experimental data are in agreement with this dependence, which is shown in Figure 18A.

On the Rebound and Concerted Mechanisms

Radicals can be generated in the solvent cage, as in the so-called "oxygen rebound mechanism" [279,280,281,282]: $M{=}O + RH \rightarrow M{-}OH + R^{\bullet} \rightarrow M + R{-}OH$. The regioselectivity of such reactions is relatively high, e.g., C(1):C(2):C(3):C(4) ≈ 1:40:40:40 (compare with the parameter typical for free-radical oxidation: 1:5:5:5). Reactions of such type are in many cases stereoselective and even stereospecific, *i.e.*, for the oxidation of *cis*-1,2-dimethylcyclohexane parameter trans/cis <<1. However, if radicals $R^{\bullet}$ partly leave the solvent cage and react with O_2, the formation of ROOH is possible. The $H_2O_2/[Mn_2(TMTACN)_2O_3]^{2+}$ (**21**)/RCOOH system has been assumed to oxidize via a "crypto-radical" ("hydroperoxo-rebound") mechanism (Figure 19) [283].

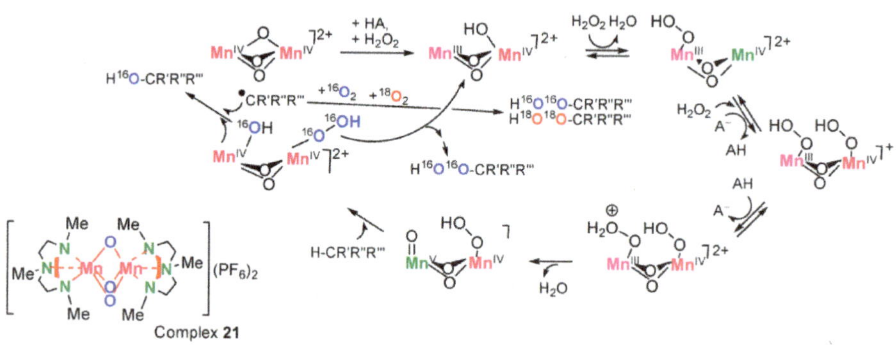

Figure 19: A catalytic cycle proposed for the alkane hydroperoxidation catalyzed by complex **21** in the presence of $^{18}O_2$ (in an atmosphere $^{16}O_2$+$^{18}O_2$). AH is a carboxylic acid where A^- is its anion. Adapted from Ref. [283].

m-Chloroperbenzoic acid in the presence of certain metal compounds efficiently oxidizes alkanes at low temperature with a very high degree of stereoselectivity (*trans*/*cis* << 1) [160,227,284,285]. Apparently, radicals are not produced intermediately in the course of such oxidation, which occurs via a "concerted" mechanism.

SOME NEW REACTIONS

Shilov Chemistry

So-called Shilov chemistry [1,286,287,288,289,290,291] includes two reactions (discovered in 1969 and 1972 [1]). It was demonstrated that $Pt^{II}Cl_4^{2-}$ ion catalyzes H/D exchange in methane in a D_2O/CD_3COOD solution and, if $Pt^{IV}Cl_6^{2-}$ is added, the latter oxidizes methane to methanol. The catalytic cycle in which σ-methyl complexes of platinum(II) and platinum(IV) are involved is shown in Figure 20.

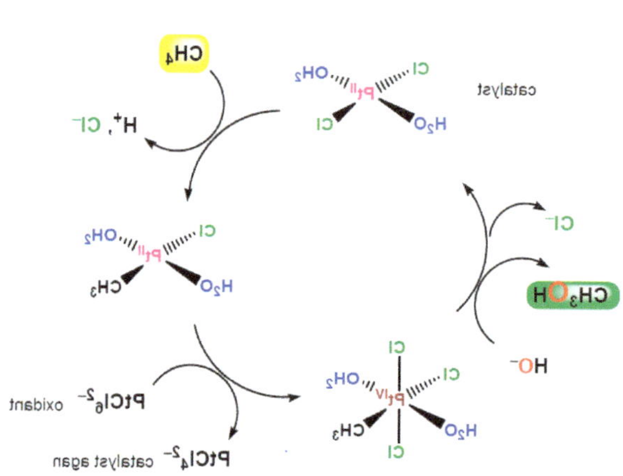

Figure 20: A schematically depicted catalytic cycle proposed for Shilov chemistry (the methane oxidation to methanol by $Pt^{IV}Cl_6^{2-}$ catalyzed by $Pt^{II}Cl_4^{2-}$).

The initial step of the reaction between alkane and ion $Pt^{II}Cl_4^{2-}$ is the formation of a σ-alkyl platinum(II) derivative, which in the second step is oxidized by platinum(IV) present in the solution to give alkanol (and also alkyl chloride):

$$Alk–H + Cl–Pt^{II}–Cl \rightarrow Alk–Pt^{II}–Cl + HCl$$

$$Alk–Pt^{II}–Cl + Pt(IV) + H_2O \rightarrow Alk–OH + Pt^{II} + Pt(II) + HCl$$

Recently, Rudakov [287] proposed an intimate mechanism for the first step of organometallic alkane activation by the Pt^{II} species (Figure 21). An active form of the aquachloride species $Cl_2(H_2O)_2Pt^{II}$ (**A**) is an adduct of the hydride $^{\bullet}Pt^{III}$ derivative with hydroxyl radical in the form of unseparated radical pair $Cl_2(H_2O)H(^{\bullet}Pt^{III})\cdots O^{\bullet}H$ (**B**). This coordinatively unsaturated complex has a vacancy to add an R fragment from the reacting alkane, RH. Indeed, coordinated hydroxyl radical can abstract a hydrogen atom from RH to produce radical $R^{\bullet}$. A collapse of radical pair ($^{\bullet}Pt^{III}$) and $R^{\bullet}$ affords an alkylhydride derivative of platinum(IV) (**C**) postulated for one stage of the catalytic cycle [1,287]. Thus, considering such a hypothesis, we can say that one of the main peculiarities of Shilov chemistry is a crypto-radical character of its intimate mechanism ("rebound mechanism"; see Section 4.3). It is possible to compare the platinum complex $Cl_2Pt^{II}(OH_2)_2$ (**A**) with a wasp, which "pricks" (activates) alkanes using its sting (coordinated hydroxyl radical). It is useful to remember that Bach and Dmitrenko [292] proposed a mechanism of alkane oxidation catalyzed by cytochrome P450. It involves the rearrangement (somersault motion) of the metal hydroperoxide to its inverted isomeric form with a hydroxyl radical that is hydrogen bonded to the metal oxide: MO–OH → M=O···HO•. The coordinated hydroxyl radical abstracts a hydrogen atom from the alkane carbon–hydrogen bond. This mechanism has similarities with the hypothetical platinum "wasp's sting mechanism" described in this paragraph (Figure 21).

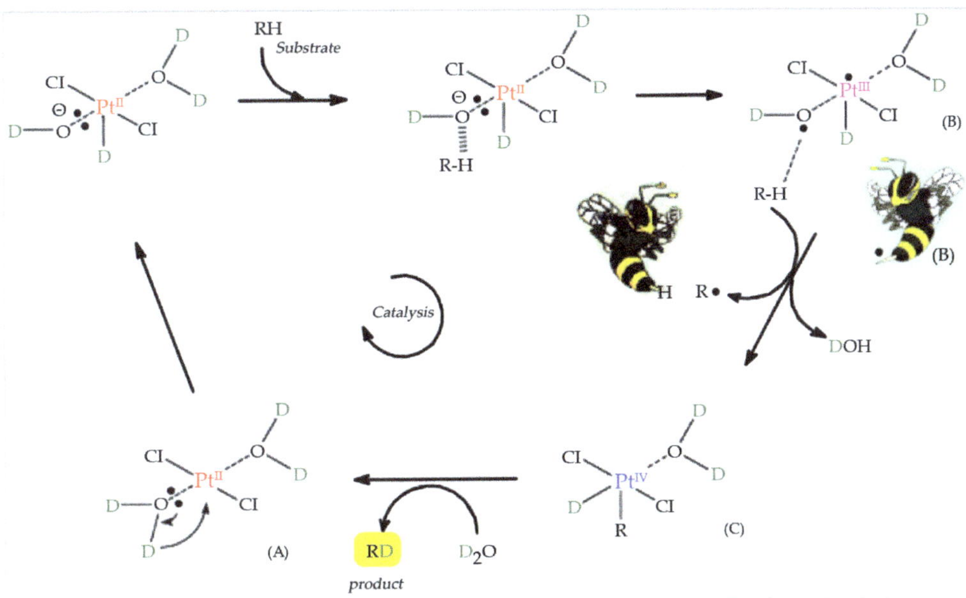

Figure 21: A hypothetical "wasp's sting mechanism" of the first step of Shilov chemistry (interaction between methane and $Pt^{II}Cl_4^{2-}$).

Hexachloroplatinate used originally as the stoichiometric oxidant in alkane reactions is a very expensive reagent and is not convenient to use in the synthesis. The Pt(II)–Pt(IV)–H_2O_2 system used by Thorn and colleagues to hydroxylate *n*-propanol selectively to 1,3-propanediol [293] exhibited very low efficiency: The amount of 1,3-propandiol corresponded to 1.3 turnovers of the entire platinum content (about 0.09 h^{-1}).

Sigma-aryl complexes of platinum(IV) obtained via the electrophilic substitution reaction between H_2PtCl_6 and arenes in CF_3COOH or CH_3COOH (so-called Shul'pin reaction as noted by Matsumoto [294], Sanford [295], and Jones [296]) (Scheme 5) turned out to be more stable than sigma-alkyl complexes and have been isolated in the form stabilized by NH_3 ligands. These complexes were demonstrated to play the role of models of sigma-alkyl platinum complexes generated as intermediates in the reactions with alkanes [1,287,297,298,299]. For selected recent original publications on organoplatinum chemistry including carbon–hydrogen bond functionalizations, see [300,301,302,303,304].

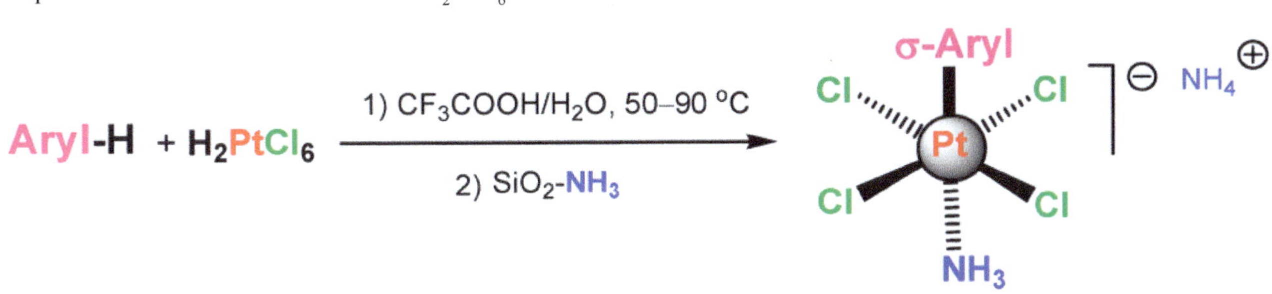

Scheme 5: The reaction of H_2PtCl_6 with arenes to afford σ-aryl derivatives of Pt(IV).

Periana System

Platinum complexes derived from the bidiazine ligand family, discovered by Periana and colleagues [305,306,307], catalyze oxidation of methane by sulfuric acid to a methanol derivative in >70% one-pass yield based on methane. These complexes are very stable in concentrated sulfuric acid at relatively high temperature and are among the most effective catalysts for methane conversion. A simplified scheme of the catalytic cycle is shown in Figure 22.

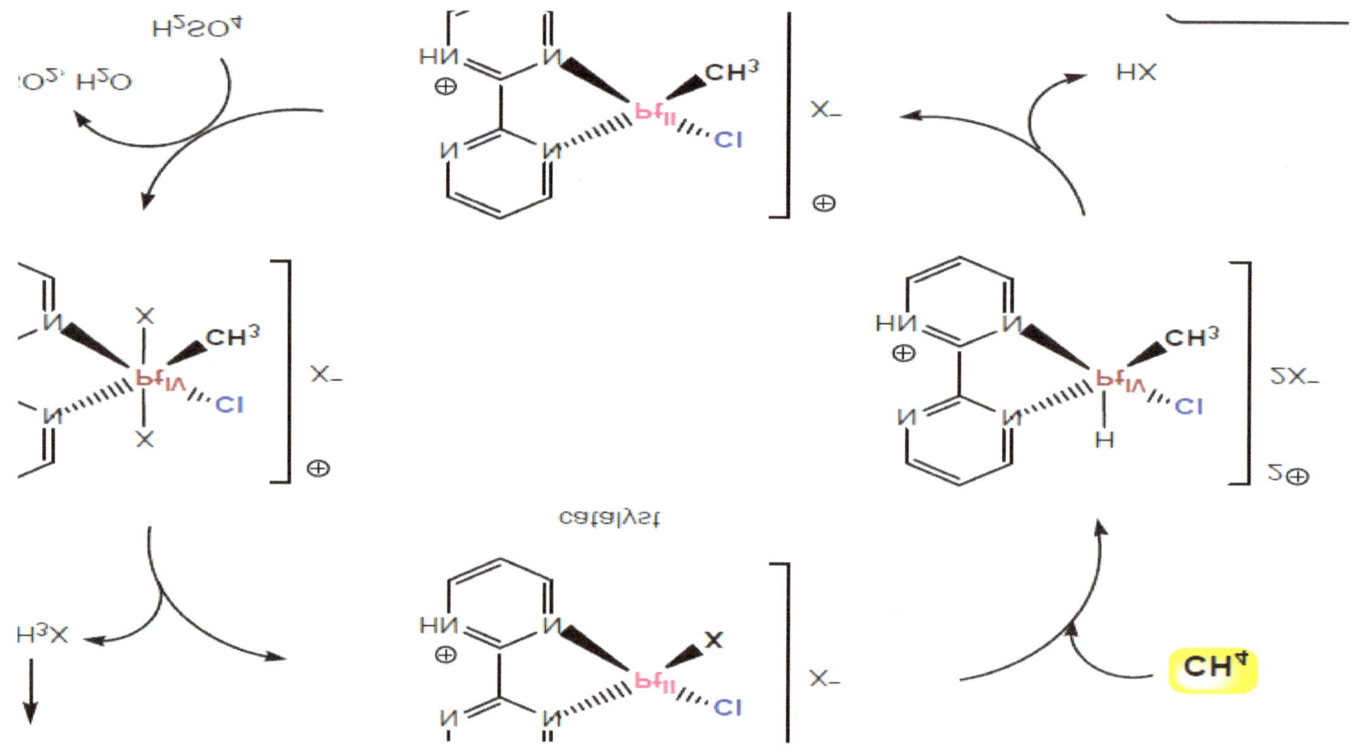

Figure 22: A catalytic cycle for the methane oxidation by Periana system.

Carboxylation of C–H Compounds (Sen-Fujiwara-Pombeiro Reaction)

In 1992, Sen reported that, in aqueous medium anion-radical $SO_4^{-\bullet}$ (generated from $S_2O_8^{2-}$) abstracts,

"*a hydrogen atom from methane and ethane to form the corresponding alkyl radicals which could be trapped efficiently by carbon monoxide, the resultant acyl radicals being ultimately converted into the homologous carboxylic acids*".

[308]

The reaction can be presented by Figure 23.

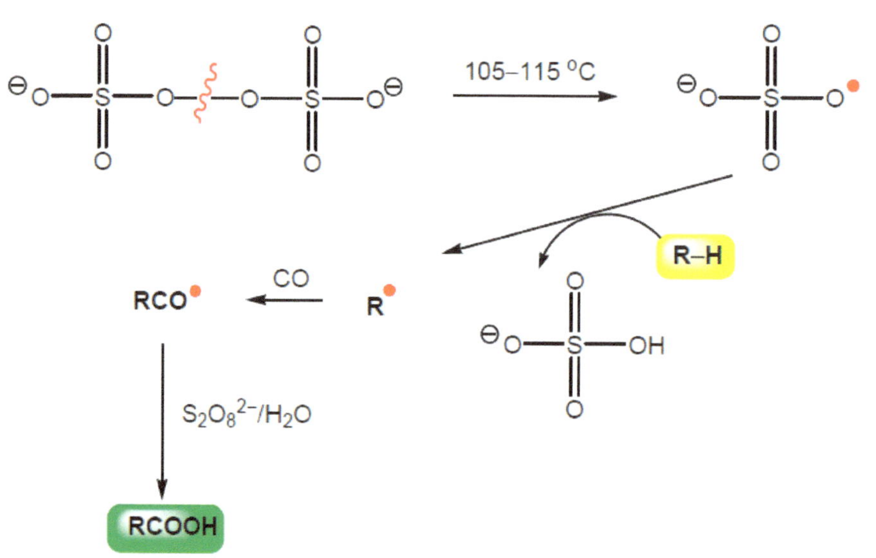

Figure 23: Hydrocarbonylation of alkane, RH, with hexaoxo-μ-peroxodisulfat, $S_2O_8^{2-}$, and CO. Adapted from Ref. [308].

Later, Fujiwara [309,310,311] and Pombeiro [312,313,314,315] demonstrated that this reaction can be sufficiently improved if various catalysts are used. Derivatives of copper, vanadium, and other metals served as catalysts. A simplified scheme of this reaction is shown in Figure 24. A number of works have been published by other authors in recent years [316].

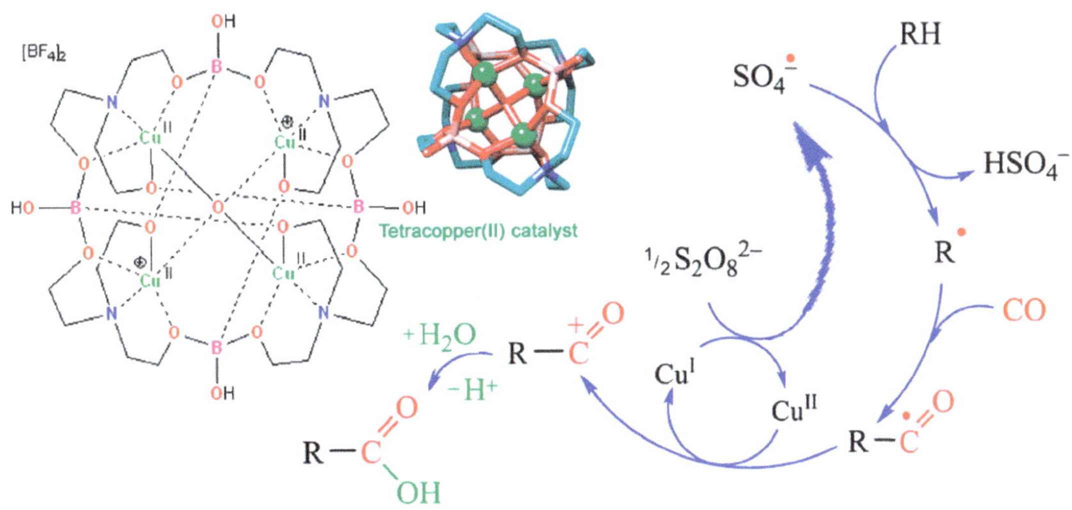

Figure 24: Alkane (RH) carboxylation with $S_2O_8^{2-}$/CO system catalyzed by a copper salt. Adapted from Refer. [313].

OUTLOOK

We can see a few routes to further develop the field of metal-catalyzed functionalization of C–H compounds. One of these ways is the search of complexes of as yet unexplored metals. It has been found recently that osmium complexes catalyze the alkane oxidation with huge TONs [144,145,146,148]. For example, the oxidation of cyclohexane with hydrogen peroxide at 60 °C and low concentration of Os-catalyst (10^{-7} M) gave a turnover number (TON) of 200,000 after 24 h. [146].

Polynuclear metal complexes containing voluminous organic [63,66,118,119,129,130,131,186,213,214,215,216,217,218,234,235,236,237,238,239,240,241] or siloxane [221,222,223,224,225,226,227,228] ligands are also very efficient in oxidations of organic compounds including saturated hydrocarbons with peroxides. For example, dinuclear copper complex **11** [221,224] and hexairon complex **15** (for the structure, see Figure 9) affording the products of cyclohexane oxidation in a very high yield of 46% (see Figure 10 [226]) can be considered as "inorganic models of methane monooxygenases." These enzymes contain dinuclear copper [317,318] and iron [319] reaction centers, respectively. A new method of synthesis of heterometallic coordination compounds by spontaneous self-assembly (SSA) gave complexes which are very efficient catalysts in oxidations with peroxides (turnover frequency TOF was up to 11200 h^{-1} = 3.1 s^{-1}) [160]. This approach modeling natural enzymes can be called biomimetic route. Various polynuclear complexes [320,321,322] including siloxane derivatives [221,222,223,224,225,226,227,228] containing bulky ligands are especially attractive as catalysts in oxidations with peroxides. Bulky ligands around the reaction centers give rise in these cases to the selectivity enhancement [12,323].

Another route that uses as oxidants dioxygen or even concentrated sulfuric acid [305,306,307,308] that require high temperatures seems to be more attractive from the point of view of industrial applications. It is noteworthy that the Periana reaction occurring under severe conditions employs a metal complex catalyst (but not a simple salt). We may hope that new practically useful systems will be discovered in the near future.

ACKNOWLEDGMENTS

This work was supported by the Russian Foundation for Basic Research (grant No. 16-03-00254) and the "Science without Borders Program, Brazil-Russia," CAPES (grant A017-2013).

CONFLICTS OF INTEREST

The author declares no conflict of interest.

REFERENCES

1. Shilov, A.E.; Shul'pin, G.B. *Activation and Catalytic Reactions of Saturated Hydrocarbons in the Presence of Metal Complexes*; Kluwer Academic Publishers: New York, NY, USA; Boston, MA, USA; Dordrecht, Holland; London, UK; Moscow, Russia, 2002.
2. Crabtree, R.H. Organomeallic alkane CH activation. *J. Organomet. Chem.* 2004, *689*, 4083–4091.
3. Denisov, E.T.; Afanas'ev, I.B. *Oxidation and Antioxidants in Organic Chemistry and Biochemistry*; Taylor & Francis: Boca Raton, FL, USA; London, UK; New York, NY, USA; Singapore, 2005.
4. Hartwig, J.F. Evolution of C–H bond functionalization from methane to methodokogy (Review). *J. Am. Chem. Soc.*2016, *138*, 2–24.
5. Davies, H.M.L.; Morton, D. Recent advances in C–H functionalization. *J. Org. Chem.* 2016, *81*, 343–350.
6. Newhouse, T.; Baran, P.S. If C–H bonds could talk: Selective C–H bond oxidation. *Angew. Chem. Int. Ed.* 2011, *50*, 3362–3374.

7. Zhou, M.; Crabtree, R.H. C–H Oxidation by platinum group metal oxo or peroxo species. *Chem. Soc. Rev.* 2011, *40*, 1875–1884.
8. Webb, J.R.; Bolaño, T.; Gunnoe, B. Catalytic oxy-functionalization of methane and other hydrocarbons: Fundamental advancement and new strategies. *ChemSusChem* 2011, *4*, 37–49.
9. Chepaikin, E.G. Homogeneous catalysis in the oxidative functionalization of alkanes in protic media. *Russ. Chem. Rev.*2011, *80*, 363–396.
10. Labinger, J.A.; Bercaw, J.E. The role of higher oxidation state species in platinum-mediated C–H bond activation and functionalization. *Top. Organomet. Chem.* 2011, *35*, 29–60.
11. Shul'pin, G.B. Hydrocarbon oxygenations with peroxides catalyzed by metal compounds. *Mini-Rev. Org. Chem.* 2009,*6*, 95–104.
12. Shul'pin, G.B. Selectivity enhancement in functionalization of C–H bonds: A review. *Org. Biomol. Chem.* 2010, *8*, 4217–4228.
13. Shul'pin, G.B. C–H functionalization: Thoroughly tuning ligands at a metal ion, a chemist can greatly enhance catalyst's activity and selectivity. *Dalton Trans.* 2013, *42*, 12794–12818.
14. Periana, R.A.; Bhalla, G.; Tenn, W.J., III; Young, K.J.H.; Liu, X.Y.; Mironov, O.; Jones, C.J.; Ziatdinov, V.R. Perspectives on some challenges and approaches for developing the next generation of selective, low temperature, oxidation catalysts for alkane hydroxylation based on the CH activation reaction. *J. Mol. Catal. A* 2004, *220*, 7–25.
15. Conley, B.L.; Ganesh, S.K.; Gonzales, J.M.; Ess, D.H.; Nielsen, R.J.; Ziatdinov, V.R.; Oxgaard, J.; Goddard, W.A., III; Periana, R.A. Facile oxy-functionalization of a nucleophilic metal alkyl with a *cis*- dioxo metal species via a (2+3) transition state. *Angew. Chem. Int. Ed.* 2008, *47*, 7849–7852.
16. Bischof, S.M.; Ess, D.H.; Meier, S.K.; Oxgaard, J.; Nielsen, R.J.; Bhalla, G.; Goddard, W.A., III; Periana, R.A. Benzene C–H bond activation in carboxylic acids catalyzed by O-donor iridium(III) complexes: An experimental and density functional study. *Organometallics* 2010, *29*, 742–756.
17. Hashiguchi, B.G.; Young, K.J.H.; Yousufuddin, M.; Goddard, W.A., III; Periana, R.A. Acceleration of nucleophilic CH activation by strongly basic solvents. *J. Am. Chem. Soc.* 2010, *132*, 12542–12545.
18. Labinger, J.A.; Bercaw, J.E. Understanding and exploiting C–H bond activation. *Nature* 2002, *417*, 507–514.
19. Munz, D.; Webster-Gardiner, M.; Fu, R.; Strassner, T.; Goddard, W.A., III; Gunnoe, T.B. Proton or metal? The H/D exchange of arenes in acidic solvents. *ACS Catal.* 2015, *5*, 769–775.
20. Burton, S.G. Oxidizing enzymes as biocatalysts. *Trends Biotechnol.* 2003, *21*, 543–549.
21. Coon, M.J. Omega oxygenases: Nonheme-iron enzymes and P450 cytochromes. *Biochem. Biophys. Res. Commun.* 2005,*338*, 378–385.
22. Stone, K.L.; Borovik, A.S. Lessons from nature: Unraveling biological C–H bond activation. *Curr. Opin. Chem. Biol.*2009, *13*, 114–118.
23. Austin, R.N.; Groves, J.T. Alkane-oxidizing metalloenzymes in the carbon cycle. *Metallomics* 2011, *3*, 775–787.
24. Yang, G.; Ding, Y. Recent advances in biocatalyst discovery, development and applications. *Bioorg. Med. Chem.* 2014,*22*, 5604–5612.
25. Matsuo, T.; Hirota, S. Artificial enzymes with protein scaffolds: Structural design and modification. *Bioorg. Med. Chem.*2014, *22*, 5638–5656.
26. Doble, M.V.; Ward, A.C.C.; Deuss, P.J.; Jarvis, A.G.; Kamer, P.C.J. Catalyst design in oxidation chemistry; from $KMnO_4$to artificial metalloenzymes. *Bioorg. Med. Chem.* 2014, *22*, 5657–5677.
27. Blomberg, M.R.A.; Borowsky, T.; Himo, F.; Liao, R.-Z.; Siegbahn, P.E.M. Quantum chemical studies of mechanisms for metalloenzymes. *Chem. Rev.* 2014, *114*, 3601–3658.
28. Poulos, T.L. Heme enzyme structure and function. *Chem. Rev.* 2014, *114*, 3919–3962.
29. Zhuk, T.S.; Goldman, M.; Hofmann, J.; Pohl, J.C.S.; Zorn, H. Preparative aerobic oxidations with basidiomycetous enzymes: CH-functionalization of adamantane. *J. Mol. Catal. B* 2015, *122*, 87–92.
30. Ishii, Y.; Sakaguchi, S. Recent progress in aerobic oxidation of hydrocarbons by *N*-hydroxyimides. *Catal. Today* 2006,*117*, 105–113.
31. Vastin, B.A.; Hall, M.B. The molecular and electronic structure of carbon–hydrogen bond activation and transition metal assisited hydrogen transfer. *Coord. Chem. Rev.* 2009, *253*, 1202–1218.
32. Gunay, A.; Theopold, H.H. C–H bond activations by metal oxo compounds. *Chem. Rev.* 2010, *110*, 1060–1081.
33. Chepaikin, E.G. Oxidative functionalization of alkanes under dioxygen in the presenceof homogeneous noble metal catalysts. *J. Mol. Catal. A* 2014, *385*, 160–174.
34. *Alkane C–H Activation by Single-Site Metal Catalysis*; Pérez, P.J., Ed.; Springer: Heidelberg, Germany, 2012.
35. Cavaliere, V.N.; Mindiola, D.J. Methane: A new frontier in organometallic chemistry. *Chem. Sci.* 2012, *3*, 3356–3365.
36. Campbell, A.N.; Stahl, S.S. Overcoming the "Oxidant Problem": Strategies to use O_2 as the oxidant in organometallic C–H oxidation reactions catalysed by Pd (and Cu). *Acc. Chem. Res.* 2012, *45*, 851–863.
37. Hartwig, J.F. Borylation and silylation of C–H bonds: A platform for diverse C–H bond functionalizations. *Acc. Chem. Res.* 2012, *45*, 864–873.
38. Hashiguchi, B.G.; Bischof, S.M.; Konnick, M.M.; Periana, R.A. Designing catalysts for functionalization

of unactivated C–H bonds based on the CH activation reaction. *Acc. Chem. Res.* 2012, *45*, 885–898.

39. Roizen, J.L.; Harvey, M.E.; Du Bois, J. Metal-catalyzed nitrogen-atom transfer methods for the oxidation of aliphatic C–H bonds. *Acc. Chem. Res.* 2012, *45*, 911–922.
40. Sivaramakrishna, A.; Suman, P.; Goud, E.V.; Janardan, S.; Sravani, C.; Sandep, T.; Vijayakrishna, K.; Clayton, H.S. Review: Active homogeneous reagents and catalysts in *n*-alkane activation reactions. *J. Coord. Chem.* 2013, *66*, 2091–2109.
41. Caballero, A.; Pérez, P.J. Methane as raw material in synthetic chemistry: The final frontier. *Chem. Soc. Rev.* 2013, *42*, 8809–8820.
42. Sun, M.; Zhang, J.; Putaj, P.; Caps, V.; Lefebre, F.; Pelletier, J.; Basset, J.-M. Catalytic oxidation of light alkanes (C_1-C_4) by heteropoly compounds. *Chem. Rev.* 2014, *114*, 981–1019.
43. Hussain, I.; Singh, T. Synthesis of biaryls through aromatic C–H bond activation: A review of recent developments.*Adv. Synth. Catal.* 2014, *356*, 1661–1696.
44. Guan, W.; Sayyed, F.B.; Zeng, G.; Sakaki, S. σ-Bond activation of small molecules and reactions catalyzed by transition-metal complexes: Theoretical understanding of electronic processes. *Inorg. Chem.* 2014, *53*, 6444–6457.
45. Landaeta, V.R.; Rodríguez-Lugo, R.E. Catalytic oxygenation of organic substrates: Toward greener ways for incorporating oxygen. *Inorg. Chim. Acta* 2015, *431*, 21–47.
46. Horn, R.; Schlogl, R. Methane activation by heterogeneous catalysis. *Catal. Lett.* 2015, *145*, 23–39.
47. Souillart, L.; Cramer, N. Catalytic C–C bond activation via oxidative addition to transition metals. *Chem. Rev.* 2015,*115*, 9410–9464.
48. Kuninobu, Y.; Sueki, S. C–H Bond transformations leading to the synthesis of organic functional materials. *Synthesis*2015, *47*, 3823–3845.
49. Reay, A.J.; Fairlamb, J.S. Catalytic C–H bond functionalisation chemistry: The case for quasi-heterogeneous catalysis.*Chem. Commun.* 2015, *51*, 16289–16307.
50. Chepaikin, E.G.; Borsch, V.N. Rhodium complexes in homogeneous catalytic systems for oxidative functionalization of alkanes: Experiment and quantum-chemical calculations. *J. Organomet. Chem.* 2015, *793*, 78–82.
51. Munz, D.; Strassner, T. Alkane C–H functionalization and oxidation with molecular oxygen. *Inorg. Chem.* 2015, *54*, 5043–5052.
52. Khan, M.S.; Haque, A.; Al-Suti, M.K.; Raithby, P.R. Recent advances in the application of group-10 transition metal based catalysts in C–H activation and functionalization. *J. Organomet. Chem.* 2015, *793*, 114–133.
53. van Leeuwen, P.W.N.M.; Chadwick, J.C. *Homogeneous Catalysis: Activity, Stability and Deactivation*; Wiley: Weinheim, Germany, 2011.
54. Kozuch, S.; Shaik, S. How to conceptualize catalytic cycles? The energetic span model. *Acc. Chem. Res.* 2011, *44*, 101–110.
55. Kozuch, S.; Martin, J.M.L. "Turning over" definitions in catalytic cycles. *ACS Catal.* 2012, *2*, 2787–2794.
56. Kirillova, M.V.; Da Silva, J.A.L.; Fraústo da Silva, J.J.R.; Pombeiro, A.J.L. Direct and efficient transformation of gaseous alkanes into carboxylic acids catalyzed by vanadium containing heteropolyacids. *Appl. Catal. A* 2007, *332*, 159–165.
57. Shul'pin, G.B.; Kirillova, M.V.; Kozlov, Y.N.; Shul'pina, L.S.; Kudinov, A.R.; Pombeiro, A.J.L. Decamethylosmocene-catalyzed efficient oxidation of saturated and aromatic hydrocarbons and alcohols with hydrogen peroxide in the presence of pyridine. Part 3 from the series "Oxidations Catalyzed by Osmium Compounds". *J. Catal.* 2011, *277*, 164–172.
58. Shul'pin, G.B.; Nizova, G.V.; Kozlov, Y.N.; Arutyunov, V.S.; dos Santos, A.C.M.; Ferreira, A.C.T.; Mandelli, D. Oxidations by the system "hydrogen peroxide–$[Mn_2L_2O_3][PF_6]_2$ (L = 1,4,7-trimethyl-1,4,7-triazacyclononane)–oxalic acid." Part 6. Oxidation of methane and other alkanes and olefins in water. *J. Organomet. Chem.* 2005, *690*, 4498–4504.
59. Sorokin, A.B.; Kudrik, E.V.; Alvarez, L.X.; Afanasiev, P.; Millet, J.M.M.; Bouchu, D. Oxidation of methane and ethylene in water at ambient conditions. *Catal. Today* 2010, *157*, 149–154.
60. Shul'pin, G.B.; Druzhinina, A.N. Iron(III) chloride catalysed photooxygenation of alcohol solutions of alkanes by atmospheric oxygen. *Mendeleev Commun.* 1992, *2*, 36–37.
61. Shul'pin, G.B.; Nizova, G.V.; Kozlov, Y.N. Photochemical aerobic oxidation of alkanes promoted by iron complexes. Part 32 of the series "Photoinduced Reactions of Organic Compounds with Metal Complexes". *New J. Chem.* 1996, *20*, 1243–1256.
62. Lamaty, F.; Dupont, J. Preface to the special issue on "Catalysis in neoteric solvent or alternative media". *Catal. Commun.* 2015, *63*, 1.
63. Jlassi, R.; Ribeiro, A.P.C.; Guedes da Silva, M.F.C.; Mahmudov, K.T.; Kopylovich, M.N.; Anisimova, T.B.; Pombeiro, A.J.L. Polynuclear copper(II) complexes as catalysts for the peroxidative oxidation of cyclohexane in a room-temperature ionic liquid. *Eur. J. Inorg. Chem.* 2014, 4541–4550.
64. Chepaikin, E.G.; Bezruchenko, A.P.; Menchikova, G.N.; Moiseeva, N.I.; Gekhman, A.E. Direct catalytic oxidation of lower alkanes in ionic liquid media. *Petrol. Chem.* 2014, *54*, 374–381.
65. Lentini, S.; Galloni, P.; Garcia-Bosch, I.; Costas, M.; Conte, V. Ionic liquids as reaction media in catalytic oxidations with manganese and iron pyridyl triazacyclononane complexes. *Inorg. Chim. Acta* 2014, *410*, 60–64.
66. Ribeiro, A.P.C.; Martins, L.M.D.R.S.; Hazra, S.; Pombeiro, A.J.L. Catalytic oxidation of cyclohexane with hydrogen peroxide and a tetracopper(II) complex in an ionic liquid. *C. R. Chim.* 2015, *18*, 758–765.

67. Zhang, Y.; Zhang, N.; Tang, Z.-R.; Xu, Y.-J. Transforming CdS into an efficient visible light photocatalyst for selective oxidation of saturated primary C–H bonds under ambient conditions. *Chem. Sci.* 2012, *3*, 2812–2822.

68. Ohtani, B. Titania photocatalysis beyond recombination: A critical review. *Catalysts* 2013, *3*, 942–953.

69. Vaiano, V.; Sannino, D.; Almeida, A.R.; Mul, G.; Ciambelli, P. Investigation of the deactivation phenomena occurring in the cyclohexane photocatalytic oxidative dehydrogenation on MoO_x/TiO_2 through gas phase and *in situ* DRIFTS analyses. *Catalysts* 2013, *3*, 978–997.

70. Hoffmann, N. Combining photoredox and metal catalysis. *ChemCatChem* 2015, *7*, 393–394.

71. Protti, S.; Fagnoni, M.; Ravelli, D. Photocatalytic C–H activation by hydrogen-atom transfer in synthesis. *ChemCatChem* 2015, *7*, 1516–1523.

72. Liu, Y.; Chen, L.; Yuan, Q.; He, J.; Au, C.-T.; Yin, S.-F. A green and efficient photocatalytic route for the highly-selective oxidation of saturated alpha-carbon C–H bonds in aromatic alkanes over flower-like Bi_2WO_6. *Chem. Commun.* 2016,*52*, 1274–1277.

73. Bermudez, J.M.; Menendez, J.A.; Arenillas, A.; Martínez-Palou, R.; Antonio, A.; Romero, A.A.; Luque, R. Graphene oxide-catalysed oxidation reaction of unsaturated compounds under microwave irradiation. *Catal. Commun.* 2015, *72*, 133–137.

74. Joglekar, M.; Nguyen, V.; Pylypenko, S.; Ngo, C.; Li, Q.; O'Reilly, M.E.; Gray, T.S.; Hubbard, W.A.; Gunnoe, T.B.; Herring, A.M.; Trewyn, B.G. Organometallic complexes anchored to conductive carbon for electrocatalytic oxidation of methane at low temperature. *J. Am. Chem. Soc.* 2016, *216*, 116–125.

75. Cravotto, G.; Borretto, E.; Oliverio, M.; Procopio, A.; Penoni, A. Organic reactions in water or biphasic aqueous systems under sonochemical conditions. A review on catalytic effects. *Catal. Commun.* 2015, *63*, 2–9.

76. Garner, A.W.; Harris, C.F.; Vezzu, D.A.K.; Pike, R.D.; Huo, S. Solvent-controlled switch of selectivity between sp^2 and sp^3 C–H bond activation by platinum(II). *Chem. Commun.* 2011, *47*, 1902–1904.

77. Shul'pin, G.B.; Nizova, G.V. Selective photochemical ketonization of cyclohexane by air in aqueous emulsion in the presence of iron ions. *Mendeleev Commun.* 1995, 143–145.

78. Shul'pin, G.B.; Attanasio, D.; Suber, L. Efficient H_2O_2 oxidation of alkanes and arenes to alkyl peroxides and phenols catalyzed by the system vanadate-pyrazine-2-carboxylic acid. *J. Catal.* 1993, *142*, 147–152.

79. Shul'pin, G.B.; Attanasio, D.; Suber, L. Oxidations by a H_2O_2-VO_3^--pyrazine-2-carboxylic acid reagent. 1. Oxidations of alkanes in CH_3CN to produce alkyl peroxides. *Russ. Chem. Bull.* 1993, *42*, 55–59.

80. Nizova, G.V.; Shul'pin, G.B. Oxidation by a H_2O_2-vanadium complex-2-pyrazinecarboxylic acid reagent. 3. Evidence for hydroxyl radical formation. *Russ. Chem. Bull.* 1994, *43*, 1146–1148.

81. Shul'pin, G.B.; Drago, R.S.; Gonzalez, M. Oxidations by a "H_2O_2-vanadium complex–pyrazine-2-carboxylic acid" reagent. 5. Oxidation of lower alkanes with the formation of carbonyl compounds. *Russ. Chem. Bull.* 1996, *45*, 2386–2388.

82. Shul'pin, G.B.; Guerreiro, M.C.; Schuchardt, U. Oxidations by the reagent O_2-H_2O_2-vanadium complex-pyrazine-2-carboxylic acid. Part 7. Hydroperoxidation of higher alkanes. *Tetrahedron* 1996, *52*, 13051–13062.

83. Nizova, G.V.; Süss-Fink, G.; Shul'pin, G.B. Oxidations by the reagent «O_2-H_2O_2-vanadium complex-pyrazine-2-carboxylic acid»-8. Efficient oxygenation of methane and other lower alkanes in acetonitrile. *Tetrahedron* 1997, *53*, 3603–3614.

84. Shul'pin, G.B.; Ishii, Y.; Sakaguchi, S.; Iwahama, T. Oxidations with the "O_2-H_2O_2-vanadium complex-pyrazine-2-carboxylic acid" reagent. 11. Oxidation of styrene, phenylacetylene, and their derivatives with the formation of benzaldehyde and benzoic acid. *Russ. Chem. Bull.* 1999, *48*, 887–890.

85. Shul'pin, G.B.; Kozlov, Y.N.; Nizova, G.V.; Süss-Fink, G.; Stanislas, S.; Kitaygorodskiy, A.; Kulikova, V.S. Oxidations by the reagent "O_2-H_2O_2-vanadium derivative–pyrazine-2-carboxylic acid" Part 12. Main features, kinetics and mechanism of alkane hydroperoxidation. *J. Chem. Soc. Perkin Trans.* 2001, *2*, 1351–1371.

86. De la Cruz, M.H.C.; Kozlov, Y.N.; Lachter, E.R.; Shul'pin, G.B. Oxidations by the reagent "O_2-H_2O_2-vanadium derivative-pyrazine-2-carboxylic acid." Part 13. Kinetics and mechanism of the benzene hydroxylation. *New J. Chem.*2003, *27*, 634–638.

87. Khaliullin, R.Z.; Bell, A.T.; Head-Gordon, M. A density functional theory study of the mechanism of free radical generation in the system vanadate/PCA/H_2O_2. *J. Phys. Chem. B* 2005, *109*, 17984–17992.

88. Kozlov, Y.N.; Romakh, V.B.; Kitaygorodskiy, A.; Buglyó, P.; Süss-Fink, G.; Shul'pin, G.B. Oxidation of 2-propanol and cyclohexane by the reagent "Hydrogen peroxide-vanadate anion-pyrazine-2-carboxylic acid": Kinetics and mechanism. *J. Phys. Chem. A* 2007, *111*, 7736–7752.

89. Romakh, V.B.; Süss-Fink, G.; Shul'pin, G.B. Vanadate ion-catalyzed oxidation of methane with hydrogen peroxide in an aqueous solution. *Petrol. Chem.* 2008, *48*, 440–443.

90. Kirillova, M.V.; Kuznetsov, M.L.; Romakh, V.B.; Shul'pina, L.S.; Fraústo da Silva, J.J.R.; Pombeiro, A.J.L.; Shul'pin, G.B. Mechanism of oxidations with H_2O_2 catalyzed by vanadate anion or oxovanadium(V) triethanolaminate (vanadatrane) in combination with pyrazine-2-carboxylic acid (PCA): Kinetic and DFT studies. *J. Catal.* 2009, *267*, 140–157.

91. Gusevskaya, E.V.; Menini, L.; Parreira, L.A.; Mesquita, R.A.; Kozlov, Y.N.; Shul'pin, G.B. Oxidation of isoeugenol to vanillin by the "H_2O_2-vanadate-pyrazine-2-carboxylic acid" reagent. *J. Mol. Catal. A* 2012, *363–364*, 140–147.

92. Sutradhar, M.; Shvydkiy, N.V.; da Silva, M.F.C.G.; Kirillova, M.V.; Kozlov, Y.N.; Pombeiro, A.J.L.; Shul'pin, G.B. New binuclear oxovanadium(V) complex as a catalyst in combination with pyrazinecarboxylic acid (PCA) for efficient alkane oxygenation by H_2O_2. *Dalton Trans.* 2013, *42*, 11791–11803.

93. Nizova, G.V.; Krebs, B.; Süss-Fink, G.; Schindler, S.; Westerheide, L.; Gonzalez Cuervo, L.; Shul'pin, G.B. Hydroperoxidation of methane and other alkanes with H_2O_2 catalysed by a dinuclear iron complex and an amino acid. *Tetrahedron* 2002, *58*, 9231–9237.

94. Shul'pin, G.B.; Nizova, G.V.; Kozlov, Y.N.; Gonzalez Cuervo, L.; Süss-Fink, G. Hydrogen peroxide oxygenation of alkanes including methane and ethane catalyzed by iron complexes in acetonitrile. *Adv. Synth. Catal.* 2004, *346*, 317–332.

95. Romakh, V.B.; Therrien, B.; Süss-Fink, G.; Shul'pin, G.B. Synthesis, molecular structure and catalytic potential of the tetrairon complex $[Fe_4(N_3O_2\text{-}L)_4(\mu\text{-}O)_2]^{4+}$ (L = 1-carboxymethyl-4,7-dimethyl-1,4,7-triazacyclononane). *Inorg. Chem.*2007, *46*, 3166–3175.

96. Shul'pina, L.S.; Kudinov, A.R.; Mandelli, D.; Carvalho, W.A.; Kozlov, Y.N.; Vinogradov, M.M.; Ikonnikov, N.S.; Shul'pin, G.B. Oxidation of alkanes and benzene with hydrogen peroxide catalyzed by ferrocene in the presence of acids. *J. Organomet. Chem.* 2015, *793*, 217–231.

97. Nizova, G.V.; Shul'pin, G.B. A unique rate-accelerating effect of certain amino acids in the H_2O_2 oxidation of alkanes catalyzed by a dinuclear manganese complex containing 1,4,7-trimethyl-1,4,7-triazacyclononane. Part 9 from the series "Oxidations by the system 'hydrogen peroxide–$[Mn_2L_2O_3][PF_6]_2$ (L =1,4,7-trimethyl-1,4,7-triazacyclononane)–carboxylic acid". *Tetrahedron* 2007, *63*, 7997–8001.

98. Shul'pin, G.B.; Matthes, M.G.; Romakh, V.B.; Barbosa, M.I.F.; Aoyagi, J.L.T.; Mandelli, D. Oxidations by the system "hydrogen peroxide–$[Mn_2L_2O_3][PF_6]_2$ (L = 1,4,7-trimethyl-1,4,7-triazacyclononane)–carboxylic acid." Part 10: Co-catalytic effect of different carboxylic acids in the oxidation of cyclohexane, cyclohexanol, and acetone. *Tetrahedron*2008, *64*, 2143–2152.

99. Schuchardt, U.; Mandelli, D.; Shul'pin, G.B. Methyltrioxorhenium catalyzed oxidation of saturated and aromatic hydrocarbons by H_2O_2 in air. *Tetrahedron Lett.* 1996, *37*, 6487–6490.

100. Karslyan, E.E.; Shul'pina, L.S.; Kozlov, Y.N.; Pombeiro, A.J.L.; Shul'pin, G.B. Oxygenation of saturated and aromatic hydrocarbons with H_2O_2 catalyzed by the carbonyl thiophenolate iron complex $(OC)_3Fe(PhS)_2Fe(CO)_3$. *Catal. Today*2013, *218–219*, 93–98.

101. Fernandes, R.R.; Kirillova, M.V.; da Silva, J.A.L.; Fraústo da Silva, J.J.R.; Pombeiro, A.J.L. Oxidations of cycloalkanes and benzene by hydrogen peroxide catalyzed by an $\{Fe^{III}N_2S_2\}$ centre. *Appl. Catal. A* 2009, *353*, 107–112.

102. Fernandes, R.R.; Lasri, J.; Guedes da Silva, M.F.C.; da Silva, J.A.L.; Fraústo da Silva, J.J.R.; Pombeiro, A.J.L. Mild alkane C–H and O-H oxidations catalysed by mixed-N,S copper, iron and vanadium systems. *Appl. Catal. A* 2011, *402*, 110–120.

103. Milunovic, M.N.M.; Martins, L.M.D.R.S.; Alegria, E.C.B.A.; Pombeiro, A.J.L.; Krachler, R.; Trettenhahn, G.; Turta, C.; Shova, S.; Arion, V.B. Hexanuclear and undecanuclear iron(III) carboxylates as catalyst precursors for cyclohexane oxidation. *Dalton Trans.* 2013, *40*, 14388–14401.

104. Carabineiro, S.A.C.; Martins, L.M.D.R.S.; Avalos-Borja, M.; Buijnsterse, J.G.; Pombeiro, A.J.L.; Figueiredo, J.L. Gold nanoparticles supported on carbon materials for cyclohexane oxidation with hydrogen peroxide. *Appl. Catal. A* 2013,*467*, 279–290.

105. Kirillov, A.M.; Shul'pin, G.B. Pyrazinecarboxylic acid and analogs: Highly efficient co-catalysts in the metal-complex-catalyzed oxidation of organic compounds. *Coord. Chem. Rev.* 2013, *257*, 732–754.

106. Kirillova, M.V.; Kirillov, A.M. 2-Pyrazinecarboxylic Acid. In *Encyclopedia of Reagents for Organic Synthesis*; Fuchs, P.L., Ed.; Wiley: Chichester, UK.

107. Piquemal, J.-Y.; Briot, E.; Brégeault, J.-M. Preparation of materials in the presence of hydrogen peroxide: From discrete or "zero-dimensional" objects to bulk materials. *Dalton Trans.* 2013, *42*, 29–45.

108. Khenkin, A.M.; Neumann, R. Carbon–hydrogen bond activation of arenes and alkylarenes by electron transfer followed by oxygen transfer catalyzed by vanadium substituted polyoxometalates—A comparative study of the reactivity of different polyoxometalate compounds. *J. Organomet. Chem.* 2015, *793*, 134–138.

109. Zalomaeva, O.V.; Evtushok, V.Y.; Maksimov, G.M.; Kholdeeva, O.A. Selective oxidation of pseudocumene and 2-methylnaphthalene with aqueous hydrogen peroxide catalyzed by Keggin divanadium-substituted polyoxotungstate.*J. Organomet. Chem.* 2015, *793*, 210–216.

110. Jing, F.; Katryniok, B.; Bordes-Richard, E.; Dumeignil, F.; Paul, S. Structural evolution under reaction conditions of supported $(NH_4)_3HPMo_{11}VO_{40}$ catalysts for the selective oxidation of isobutane. *Catalysts* 2015, *5*, 460–477.

111. Alberto, J.; Garcia, H. Doped graphenes in catalysis. *J. Mol. Catal. A* 2015, *408*, 296–309.

112. Gruselle, M. Apatites: A new family of catalysts in organic synthesis. *J. Organomet. Chem.* 2015, *793*, 93–101.

113. Rahman, A.K.M.L.; Kumashiro, M.; Ishihara, T. Direct synthesis of formic acid by partial oxidation of methane on H-ZSM-5 solid acid catalyst. *Catal. Commun.* 2011, *12*, 1198–1200.

114. Wannakao, S.; Warakulwit, C.; Kongpatpanich, K.; Probst, M.; Limtrakul, J. Methane activation in gold cation-exchanged zeolites: A DFT study. *ACS Catal.* 2012, *2*, 986–992.

115. Forde, M.M.; Armstrong, R.D.; Hammond, C.; He, Q.; Jenkins, R.L.; Kondrat, S.A.; Dimitratos, N.; Lopez-Sanchez, J.A.; Tylor, S.H.; Willock, D.; *et al.* Partial

oxidation of ethane to oxygenates using Fe- and Cu-containing ZSM-5. *J. Am. Chem. Soc.* 2013, *135*, 11087–11099.

116. Liquid Phase Oxidation via Heterogeneous Catalysis; Clerici, M.G., Kholdeeva, O.A., Eds.; Wiley: Hoboken, NJ, USA, 2013.

117. Kholdeeva, O.A.; Skobelev, I.Y.; Ivanchikova, I.D.; Kovalenko, K.A.; Fedin., V.P.; Sorokin, A.B. Hydrocarbon oxidation over Fe- and Cr-containing metal-organic frameworks MIL-100 and MIL-101—A comparative study. *Catal. Today* 2014,*238*, 54–61.

118. Kholdeeva, O.A. Recent developments in liquid-phase selective oxidation using environmentally benign oxidants and mesoporous metal silicates. *Catal. Sci. Technol.* 2014, *4*, 1869–1889.

119. Alayon, E.M.C.; Nachtegaal, M.; Bodi, A.; Rabocchiari, M.; van Bokhoven, J.A. Bis(mu-oxo) *versus* mono(mu-oxo)dicopper cores in a zeolite for converting methane to methanol: An *in situ* XAS and DFT investigation. *Phys. Chem. Chem. Phys.* 2015, *17*, 7681–7693.

120. Zou, H.; Sun, Q.; Fan, D.; Fu, W.; Liu, L.; Wang, R. Facile synthesis of Yolk/Core-Shell structured TS-1@ Mesosilica composites for enhanced hydroxylation of phenol. *Catalysts* 2015, *5*, 2134–2146.

121. Dragomirova, R.; Wohlrab, S. Zeolite membranes in catalysis—From separate units to particle coatings. *Catalysts* 2015,*5*, 2161–2222.

122. Hammond, C.; Hermans, I.; Dimitratos, N. Biomimetic oxidation with Fe-ZSM-5 and H_2O_2? Identification of an active, extra-framework binuclear core and an Fe^{III}-OOH intermediate with resonance-enhanced Raman spectroscopy.*ChemCatChem* 2015, *7*, 434–440.

123. Xie, J.; Pan, C.; Abdukader, A.; Zhu, C. Gold-catalyzed C(sp^3)–H bond functionalization. *Chem. Soc. Rev.* 2014, *43*, 5245–5256.

124. Strassner, T.; Ahrens, S.; Muehlhofer, M.; Munz, D.; Zeller, A. Cobalt-catalyzed oxidation of methane to methyl trifluoroacetate by dioxygen. *Eur. J. Inorg. Chem.* 2013, 3659–3663.

125. Bellows, S.M.; Thomas, R.; Cundari, T.R.; Jones, W.D. Methane is the best substrate for C(sp^3)–H activation with Cp*(PMe_3)Co(Me)(OTf): A density functional theory study. *Organometallics* 2015, *34*, 4032–4038.

126. Moselage, M.; Li, J.; Ackermann, L. Cobalt-catalyzed C–H activation. *ACS Catal.* 2016, *6*, 498–525.

127. Goo, Y.R.; Maity, A.C.; Cho, K.-B.; Lee, Y.-M.; Seo, M.S.; Park, Y.J.; Cho, J.; Nam, W. Tuning the reactivity of chromium(III)-superoxo species by coordinating axial ligands. *Inorg. Chem.* 2015, *54*, 10513–10520.

128. Punniyamurthy, T.; Rout, L. Recent advances in copper-catalyzed oxidation of organic compounds. *Coord. Chem. Rev.*2008, *252*, 134–154.

129. Kopylovich, M.N.; Nunes, A.C.C.; Mahmudov, K.T.; Haukka, M.; Mac Leod, T.C.O.; Martins, L.M.D.R.S.; Kuznetsov, M.L.; Pombeiro, A.J.L. Complexes of copper(II) with 3-(*ortho*-substituted phenylhydrazo) pentane-2,4-diones: Syntheses, properties and catalytic activity for cyclohexane oxidation. *Dalton Trans.* 2011, *40*, 2822–2836.

130. Perraud, O.; Sorokin, A.B.; Dutasta, J.P.; Martinez, A. Oxidation of cycloalkanes by H_2O_2 using a copper-hemicryptophane complex as a catalyst. *Chem. Commun.* 2013, *49*, 1288–1290.

131. Truong, T.; Nguyen, K.D.; Doan, S.H.; Phan, N.T.S. Efficient and recyclable $Cu_2(BPDC)_2(DABCO)$-catalyzed directamination of activated sp^3 C–H bonds by N–H heterocycles. *Appl. Catal. A* 2016, *510*, 27–33.

132. Bauer, E.B. Iron catalysis: Historic overview and current tends. *Top. Organomet. Chem.* 2015, *50*, 1–18.

133. Martínez-Ferraté, O.; Britovsek, G.J.P.; Carmen Claver, C.; van Leeuwen, PW.N.M. C–H benzylic oxidation promoted by dinuclear iron DBDOC iminopyridine complexes. *Inorg. Chim. Acta* 2015, *431*, 156–160.

134. Nesterov, D.S.; Nesterova, O.V.; Guedes da Silva, M.F.C.; Pombeiro, A.J.L. Catalytic behaviour of a novel Fe(III) Schiff base complex in the mild oxidation of cyclohexane. *Catal. Sci. Technol.* 2015, *5*, 1801–1812.

135. Summerscales, O.T.; Scott, B.L.; Viswanathan, H.S.; Sutton, A.D. Synthesis and reactivity of *cis*-$FeH_2(dcpe)_2$ (dcpe = 1,2-bis(dicyclohexylphosphino) ethane). *Inorg. Chem. Commun.* 2016, *63*, 57–60.

136. D'Amato, E.M.; Neumann, C.N.; Ritter, T. Selective aromatic C–H hydroxylation enabled by η^6-coordination to iridium(III). *Organometallics* 2015, *34*, 4626–4631.

137. Wegermann, C.A.; Ribeiro, R.R.; Ucoski, G.M.; Nakagaki, S.; Nunes, F.S.; Drechsel, S.M. Study of the catalytic activity of non-heme manganese complexes toward oxidation of cyclooctene and cyclohexene. *Appl. Catal. A* 2014, *471*, 56–62.

138. De Paula, R.; Santos, I.C.M.S.; Simões, M.M.Q.; Graça, M.; Neves, P.M.S.; Cavaleiro, J.A.S. An immobilized imidazolyl manganese porphyrin for the oxidation of olefins. *J. Mol. Catal. A* 2015, *404–405*, 156–166.

139. Sankaralingam, M.; Jeon, S.H.; Lee, Y.-M.; Seo, M.S.; Ohkubo, K.; Fukuzumi, S.; Nam, W. An amphoteric reactivity of a mixed-valent bis(μ-oxo) dimanganese(III,IV) complex acting as an electrophile and a nucleophile. *Dalton Trans.* 2016,*45*, 376–383.

140. Bailey, B.C.; Richard, R.; Schrock, R.R.; Kundu, S.; Goldman, A.S.; Huang, Z.; Brookhart, M. Evaluation of molybdenum and tungsten metathesis catalysts for homogeneous tandem alkane metathesis. *Organometallics* 2009, *28*, 355–360.

141. Nagataki, T.; Ishii, K.; Tachi, Y.; Itoh, S. Ligand effects on Ni^{II}-catalysed alkane-hydroxylation with *m*-CPBA. *Dalton Trans.* 2007, *11*, 1120–1128.

142. Shul'pin, G.B.; Kudinov, A.R.; Shul'pina, L.S.; Petrovskaya, E.A. Oxidations catalyzed by osmium compounds. Part 1: Efficient alkane oxidation with peroxides catalyzed by an olefin carbonyl osmium(0) complex. *J. Organomet. Chem.*2006, *691*, 837–845.

143. Yuan, Q.; Deng, W.; Zhang, Q.; Wang, Y. Osmium-catalyzed selective oxidations of methane and ethane with hydrogen peroxide in aqueous medium. *Adv. Synth. Catal.* 2007, *349*, 1199–1209.

144. Shul'pin, G.B.; Süss-Fink, G.; Shul'pina, L.S. Oxygenation of alkanes with hydrogen peroxide catalysed by osmium complexes. *Chem. Commun.* 2000, 1131–1132.

145. Shul'pin, G.B.; Kozlov, Y.N.; Shul'pina, L.S.; Carvalho, W.; Mandelli, D. Oxidation reactions catalyzed by osmium compounds. Part 4. Highly efficient oxidation of hydrocarbons and alcohols including glycerol by the $H_2O_2/Os_3(CO)_{12}$/pyridine reagent. *RSC Adv.* 2013, *3*, 15065–15074.

146. Vinogradov, M.M.; Kozlov, Y.N.; Nesterov, D.S.; Shul'pina, L.S.; Pombeiro, A.J.L.; Shul'pin, G.B. Oxidation of hydrocarbons with H_2O_2/O_2 catalyzed by osmium complexes containing π-cymene ligands in acetonitrile. *Catal. Sci. Technol.* 2014, *4*, 3214–3226.

147. Chen, M.; Yi, P.; Kwong, H.-K.; Zeng, R.J.; Lau, K.-C.; Lau, T.-C. Catalytic oxidation of alkanes by a (salen) osmium(VI) nitrido complex using H_2O_2 as the terminal oxidant. *Chem. Commun.* 2015, *51*, 13686–13689.

148. Vinogradov, M.M.; Shul'pina, L.S.; Kozlov, Y.N.; Kudinov, A.R.; Ikonnikov, N.S.; Shul'pin, G.B. Oxidation of hydrocarbons and alcohols with peroxides catalyzed by new π-cymene osmium complexes. *J. Organometal. Chem.*2015, *784*, 52–61.

149. Lyons, T.W.; Sanford, M.S. Palladium-catalyzed ligand-directed C–H functionalization reactions. *Chem. Rev.* 2010, *110*, 1147–1169.

150. Musaev, D.G.; Figg, T.M.; Kaledin, A.L. Versatile reactivity of Pd-catalysts: Mechanistic features of the mono-*N*-protected amino acid ligand and cesium-halide base in Pd-catalyzed C–H bond functionalization. *Chem. Soc. Rev.*2014, *43*, 5009–5031.

151. Zultanski, S.L.; Stahl, S.S. Palladium-catalyzed aerobic acetoxylation of benzene using NO_x-based redox mediators. *J. Organomet. Chem.* 2015, *793*, 263–268.

152. Gryca, I.; Machura, B.; Małecki, J.G.; Kusz, J.; Shul'pina, L.S.; Ikonnikov, N.S.; Shul'pin, G.B. *p*-Tolylimido rhenium(V) complexes with phenolate-based ligands: Synthesis, X-ray studies and catalytic activity in oxidation with *tert*-butylhydroperoxide. *Dalton Trans.* 2016, *45*, 334–351.

153. Shul'pin, G.B.; Muratov, D.V.; Shul'pina, L.S.; Kudinov, A.R.; Strelkova, T.V.; Petrovskiy, P.V. Oxygenation of aromatic hydrocarbons with hydrogen peroxide catalyzed by rhodium carbonyl complexes. *Appl. Organomet. Chem.* 2008, *22*, 684–688.

154. Lin, Y.; Zhu, L.; Lan, Y.; Rao, Y. Development of a rhodium(II)-catalyzed chemoselective C(sp^3)-H oxygenation. *Chem. Eur. J.* 2015, *21*, 14937–14942.

155. Li, B.; Darcel, C.; Roisnel, T.; Dixneuf, P.H. Cycloruthenation of aryl imines and N-heteroaryl benzenes via C–H bond activation with Ru(II) and acetate partners. *J. Organomet. Chem.* 2015, *793*, 200–209.

156. Ackermann, L. Robust ruthenium(II)-catalyzed C–H arylations: Carboxylate assistance for the efficient synthesis of angiotensin-II-Receptor blockers. *Org. Process Res. Dev.* 2015, *19*, 260–269.

157. da Silva, J.A.L.; Fraústo da Silva, J.J.R.; Pombeiro, A.J.L. Oxovanadium complexes in catalytic oxidations. *Coord. Chem. Rev.* 2011, *255*, 2232–2248.

158. Ma, X.T.; Xing, N.; Yan, Z.; Zhang, X.; Wu, Q.; Xing, Y.H. Peroxo- and oxovanadium(IV) complexes with tridentate *N*-heterocycle ligands: Synthesis, structure, and catalytic performance. *New J. Chem.* 2015, *39*, 1067–1074.

159. Nesterov, D.S.; Nesterova, O.V.; Kokozay, V.N.; Pombeiro, A.J.L. Polynuclear Heterometallic Complexes from Metal Powders: The "Direct Synthesis" Approach. *Eur. J. Inorg. Chem.* 2014, 4496–4517.

160. Nesterov, D.S.; Chygorin, E.N.; Kokozay, V.N.; Bon, V.V.; Boča, R.; Kozlov, Y.N.; Shul'pina, L.S.; Jezierska, J.; Ozarowski, A.; Pombeiro, A.J.L.; *et al.* Heterometallic $Co^{III}_4Fe^{III}_2$ schiff base complex: Structure, electron paramagnetic resonance, and alkane oxidation catalytic activity. *Inorg. Chem.* 2012, *51*, 9110–9122.

161. Mandelli, D.; Chiacchio, K.C.; Kozlov, Y.N.; Shul'pin, G.B. Hydroperoxidation of alkanes with hydrogen peroxide catalyzed by aluminium nitrate in acetonitrile. *Tetrahedron Lett.* 2008, *49*, 6693–6697.

162. Kuznetsov, M.L.; Kozlov, Y.N.; Mandelli, D.; Pombeiro, A.J.L.; Shul'pin, G.B. Mechanism of Al^{3+}-catalyzed oxidations of hydrocarbons: Dramatic activation of H_2O_2 toward O–O homolysis in complex $[Al(H_2O)_4(OOH)(H_2O_2)]^{2+}$ explains the formation of HO• radicals. *Inorg. Chem.* 2011, *50*, 3996–4005.

163. Novikov, A.S.; Kuznetsov, M.L.; Pombeiro, A.J.L.; Bokach, N.A.; Shul'pin, G.B. Generation of HO radical from hydrogen peroxide catalyzed by aqua complexes of the group III metals $[M(H_2O)n]^{3+}$ (M = Ga, In, Sc, Y, or La): A theoretical study. *ACS Catal.* 2013, *3*, 1195–1208.

164. Kuznetsov, M.L.; Teixeira, F.A.; Bokach, N.A.; Pombeiro, A.J.L.; Shul'pin, G.B. Radical decomposition of hydrogen peroxide catalyzed by aqua complexes $[M(H_2O)n]^{2+}$ (M = Be, Zn, Cd). *J. Catal.* 2014, *313*, 135–148.

165. Rocha, B.G.M.; Kuznetsov, M.L.; Kozlov, Y.N.; Pombeiro, A.J.L.; Shul'pin, G.B. Simple soluble Bi(III) salts as efficient catalysts for the oxidation of alkanes with H_2O_2. *Catal. Sci. Technol.* 2015, *5*, 2174–2187.

166. Kuznetsov, M.L.; Rocha, B.G.M.; Pombeiro, A.J.L.; Shul'pin, G.B. Oxidation of olefins with hydrogen peroxide catalyzed by bismuth salts: A mechanistic study. *ACS Catal.* 2015, *5*, 3823–3835.

167. Novikov, A.S.; Kuznetsov, M.L.; Rocha, B.G.M.; Pombeiro, A.J.L.; Shul'pin, G.B. Oxidation of olefins with H_2O_2catalysed by salts of group III metals (Ga, In, Sc, Y and La): Epoxidation *versus* hydroperoxidation. *Catal. Sci. Technol.*2016, *6*, 1343–1356.

168. Popivker, I.; Zilbermann, I.; Maimon, E.; Cohen, H.; Meyerstein, D. The "Fenton like" reaction of MoO_4^{3-} involves two H_2O_2 molecules. *Dalton Trans.* 2013, *42*, 16666–16668.

169. Burg, A.; Shusterman, I.; Kornweitz, H.; Meyerstein, D. Three H_2O_2 molecules are involved in the "Fenton-like" reaction between $Co(H_2O)_6^{2+}$ and H_2O_2. *Dalton Trans.* 2014, *43*, 9111–9115.

170. Petit, A.S.; Pennifold, R.C.R.; Harvey, J.N. Electronic structure and formation of simple ferryloxo complexes: Mechanism of the Fenton reaction. *Inorg. Chem.* 2014, *53*, 6473–6481.

171. Grau, M.; Britovsek, G.J.P. High-valent iron in biomimetic alkane oxidation catalysis. *Top. Organomet. Chem.* 2015, *50*, 145–172.

172. Liu, W.; Groves, J.T. Manganese catalyzed C–H halogenation. *Acc. Chem. Res.* 2015, *48*, 1727–1735.

173. Oloo, W.N.; Que, L., Jr. Bioinspired nonheme iron catalysts for C–H and C=C bond oxidation: Insights into the nature of the metal-based oxidants. *Acc. Chem. Res.* 2015, *48*, 2612–2621.

174. Lindhorst, A.C.; Haslinger, S.; Kühn, F.E. Molecular iron complexes as catalysts for selective C–H bond oxygenation reactions. *Chem. Commun.* 2015, *51*, 17193–17212.

175. Costas, M. Selective C–H oxidation catalyzed by metalloporphyrins. *Coord. Chem. Rev.* 2011, *255*, 2912–2932.

176. Che, C.-M.; Lo, V.K.-Y.; Ahou, C.-Y.; Huang, J.-S. Selective functionalisation of saturated C–H bonds with metalloporphyrin catalysts. *Chem. Soc. Rev.* 2011, *40*, 1950–1975.

177. Kudrik, E.V.; Afanasiev, P.; Alvarez, L.X.; Dubourdeaux, P.; Clémancey, M.; Latour, J.-M.; Blondin, G.; Bouchu, D.; Albrieux, F.; Nefedov, S.E.; *et al.* An N-bridged high-valent diiron–oxo species on a porphyrin platform that can oxidize methane. *Nat. Chem.* 2012, *4*, 1024–1029.

178. da Silva, V.S.; Meireles, A.M.; da Silva Martins, D.C.; Rebouças, J.S.; DeFreitas-Silva, G.; Idemori, Y.M. Effect of imidazole on biomimetic cyclohexane oxidation by first-, second-, and third-generation manganese porphyrins using PhIO and$PhI(OAc)_2$ as oxidants. *Appl. Catal. A* 2015, *491*, 17–27.

179. Kudrik, E.V.; Sorokin, A.B. μ-Nitrido bridged diiron phthalocyanines: Old complexes for new catalytic applications.*Macroheterocycles* 2011, *4*, 154–160.

180. Sorokin, A.B. Phthalocyanine metal complexes in catalysis. *Chem. Rev.* 2013, *113*, 8152–8191.

181. Alvarez, L.X.; Sorokin, A.B. Mild oxidation of ethane to acetic acid by H_2O_2 catalyzed by supported μ-nitrido diiron phthalocyanines. *J. Organomet. Chem.* 2015, *793*, 139–144.

182. Sorokin, A.B.; Kudrik, E.V.; Bouchu, D. Bio-inspired oxidation of methane in water catalyzed by N-bridged diiron phthalocyanine complex. *Chem. Commun.* 2008, 2562–2564.

183. Maksimov, A.L.; Kardasheva, Y.S.; Predeina, V.V.; Kluev, M.V.; Ramazanov, D.N.; Talanova, M.Y.; Karakhanov, E.A. Iron and copper complexes with nitrogen_containing ligands as catalysts for cyclohexane oxidation with hydrogen peroxide under mild reaction conditions. *Petrol. Chem.* 2012, *52*, 318–326.

184. Ottenbacher, R.V.; Samsonenko, D.G.; Talsi, E.P.; Bryliakov, K.P. Highly efficient, regioselective, and stereospecific oxidation of aliphatic C–H groups with H_2O_2, catalyzed by aminopyridine manganese complexes. *Org. Lett.* 2012, *14*, 4310–4313.

185. Adams, A.M.; Du Bois, J.; Malik, H.A. Comparative study of the limitations and challenges in atom-transfer C–H oxidations. *Org. Lett.* 2015, *17*, 6066–6069.

186. Ottenbacher, R.V.; Talsi, E.P.; Bryliakov, K.P. Mechanism of Selective C–H hydroxylation mediated by manganese-aminopyridine enzyme models. *ACS Catal.* 2015, *5*, 39–44.

187. Talsi, E.P.; Ottenbacher, R.V.; Bryliakov, K.P. Bioinspired oxidations of aliphatic C–H groups with H_2O_2 in the presence of manganese complexes. *J. Organomet. Chem.* 2015, *793*, 102–107.

188. Bryliakov, K.P.; Talsi, E.P. Active sites and mechanisms of bioinspired oxidation with H_2O_2, catalyzed by non-heme Fe and related Mn complexes. *Coord. Chem. Rev.* 2014, *276*, 73–96.

189. Lyakin, O.Y.; Zima, A.M.; Samsonenko, D.G.; Bryliakov, K.P.; Talsi, E.P. EPR Spectroscopic detection of the elusive Fe^{V}=O intermediates in selective catalytic oxofunctionalizations of hydrocarbons mediated by biomimetic ferric complexes. *ACS Catal.* 2015, *5*, 2702–2707.

190. Gómez, L.; Canta, M.; Font, D.; Prat, I.; Ribas, X.; Costas, M. Regioselective oxidation of nonactivated alkyl C–H groups using highly structured non-heme iron catalysts. *J. Org. Chem.* 2013, *78*, 1421–1433.

191. Prat, I.; Company, A.; Corona, T.; Parella, T.; Ribas, X.; Costas, M. Assessing the impact of electronic and steric tuning of the ligand in the spin state and catalytic oxidation ability of the Fe^{II}(Pytacn) family of complexes. *Inorg. Chem.* 2013,*52*, 9229–9244.

192. Canta, M.; Font, D.; Gómez, L.; Ribas, X.; Costas, M. The iron(II) complex $[Fe(CF_3SO_3)_2(mcp)]$ as a convenient, readily available catalyst for the selective oxidation of methylenic sites in alkanes. *Adv. Synth. Catal.* 2014, *356*, 818–830.

193. Codola, Z.; Lloret-Fillol, J.; Costas, M. Aminopyridine iron and manganese complexes as molecular catalysts for challenging oxidative transformations. *Prog. Inorg. Chem.* 2014, *59*, 447–531.

194. Nam, W. Synthetic mononuclear nonheme iron–oxygen intermediates. *Acc. Chem. Res.* 2015, *48*, 2415–2423.

195. Fuentes, M.A.; Munoz, B.K.; Jacob, K.; Vendier, L.; Caballero, A.; Etienne, M.; Pérez, P.J. Functionalization of non-activated C–H bonds of alkanes: An effective and recyclable catalytic system based on fluorinated silver catalysts and solvents. *Chem. Eur. J.* 2013, *19*, 1327–1334.

196. Caballero, A.; Pérez, P.J. Catalyst design in the alkane C–H bond functionalization of alkanes by carbene insertion with Tp^xM complexes (Tp^x = hydrotrispyrazolylborate ligand; M = Cu, Ag). *J. Organomet. Chem.* 2015, *793*, 108–113.

197. Gava, R.; Olmos, A.; Noverges, B.; Varea, T.; Alvarez, E.; Belderrain, T.R.; Caballero, A.; Asensio, G.; Pérez, P.J. Discovering copper for methane C–H bond functionalization. *ACS Catal.* 2015, *5*, 3726–3730.

198. Caballero, A.; Díaz-Requejo, M.M.; Fructos, M.R.; Olmos, A.; Urbano, J.; Pérez, P.J. Catalytic functionalization of low reactive C(sp^3)-H and C(sp^2)-H bonds of alkanes and arenes by carbene transfer from diazo compounds. *Dalton Trans.*2015, *44*, 20295–20307.

199. Gava, R.; Olmos, A.; Noverges, B.; Varea, T.; Funes-Ardoiz, I.; Belderrain, T.R.; Caballero, A.; Maseras, F.; Asensio, G.; Pérez, P.J. Functionalization of C_nH_{2n+2} alkanes: Supercritical carbon dioxide enhances the reactivity towards primary carbon–hydrogen bonds. *ChemCatChem* 2015, *7*, 3254–3260.

200. Kulkarni, N.V.; Dash, C.; Jayaratna, N.B.; Ridlen, S.G.; Khani, S.K.; Das, A.; Kou, X.; Yousufuddin, M.; Cundari, T.R.; Dias, H.V.R. Zinc(II)-Mediated carbene insertion into C–H bonds in alkanes. *Inorg. Chem.* 2015, *54*, 11043–11045.

201. Crabtree, R.H. The stability of organometallic ligands in oxidation catalysis. *J. Organomet. Chem.* 2014, *751*, 174–180.

202. Davies, H.M.L.; Manning, J.R. Catalytic C–H functionalization by metal carbenoid and nitrenoid insertion. *Nature*2008, *451*, 417–424.

203. Gaillard, S.; Cazin, C.S.J.; Nolan, S.P. N-Heterocyclic carbene gold(I) and copper(I) complexes in C–H bond activation.*Acc. Chem Res.* 2012, *45*, 778–787.

204. Haibach, M.C.; Kundu, S.; Brookhart, M.; Goldman, A.S. Alkane metathesis by tandem alkane dehydrogenation-olefin metathesis catalysis and related chemistry. *Acc. Chem. Res.* 2012, *45*, 947–958.

205. Organometallic Pincer Chemistry; van Koten, G., Milstein, D., Eds.; Springer: Heidelberg, Germany, 2013.

206. Pincer and Pincer-Type Complexes: Application in Organic Synthesis and Catalysis; Szabo, K.J., Wendt, O.F., Eds.; Wiley-VCH: Weinheim, Germany, 2014.

207. Schaefer, B.A.; Margulieux, G.W.; Small, B.L.; Chirik, P.J. Evaluation of cobalt complexes bearing tridentate pincer ligands for catalytic C–H borylation. *Organometallics* 2015, *34*, 1307–1320.

208. Andrew, R.E.; González-Sebastián, L.; Chaplin, A.B. NHC-Based pincer ligands: Carbenes with a bite. *Dalton Trans.*2016, *45*, 1299–1305.

209. Kaim, W. Manifestations of noninnocent ligand behavior. *Inorg. Chem.* 2011, *50*, 9752–9765.

210. Blanchard, S.; Derat, E.; Desage-El Murr, M.; Fensterbank, L.; Malacria, M.; Desage-El Murr, M.; Mouriès-Mansuy, V. Non-innocent ligands: New opportunities in iron catalysis. *Eur. J. Inorg. Chem.* 2012, *2012*, 376–389.

211. Gambarotta, S.; Budzelaar, P.H.M. Redox-active ligands and organic radical chemistry. *Inorg. Chem.* 2011, *50*, 9879–9887.

212. Olatunji-Ojo, O.A.; Cundari, T.R. C–H Activation by multiply bonded complexes with potentially noninnocent ligands: A computational study. *Inorg. Chem.* 2013, *52*, 8106–8113.

213. Martins, L.M.D.R.; De Almeida, M.P.; Carabineiro, S.A.C.; Figueiredo, J.L.; Pombeiro, A.J.L. Heterogenisation of a C-scorpionate Fe^{II} complex on carbon materials for cyclohexane oxidation with hydrogen peroxide. *ChemCatChem* 2013,*5*, 3847–3856.

214. Silva, T.F.S.; Mac Leod, T.C.O.; Martins, L.M.D.R.S.; Guedes da Silva, M.F.C.; Schiavon, M.A.; Pombeiro, A.J.L. Pyrazole or tris(pyrazolyl)ethanol oxo-vanadium(IV) complexes as homogeneous or supported catalysts for oxidation of cyclohexane under mild conditions. *J. Mol. Catal. A* 2013, *367*, 52–60.

215. Martins, L.M.D.R.S.; Martins, A.; Alegria, E.C.B.A.; Carvalho, A.P.; Pombeiro, A.J.L. Efficient cyclohexane oxidation with hydrogen peroxide catalysed by a C-scorpionate iron(II) complex immobilized on desilicated MOR zeolite. *Appl. Catal. A* 2013, *464–465*, 43–50.

216. De Almeida, M.P.; Martins, L.M.D.R.S.; Carabineiro, S.A.C.; Lauterbach, T.; Rominger, F.; Hashmi, A.S.K.; Pombeiro, A.J.L. Homogeneous and heterogenised new gold C-scorpionate complexes as catalysts for cyclohexane oxidation.*Catal. Sci. Technol.* 2013, *3*, 3056–3069.

217. Martins, L.M.D.R.S.; Pombeiro, A.J.L. Tris(pyrazol-1-yl)methane metal complexes for catalytic mild oxidative functionalizations of alkanes, alkenes and ketones. *Coord. Chem. Rev.* 2014, *265*, 74–88.

218. Silva, T.F.S.; Martins, L.M.D.R.S.; Guedes da Silva, M.F.C.; Kuznetov, M.L.; Fernandes, A.R.; Silva, A.; Pan, C.-J.; Lee, J.-F.; Hwang, B.-J.; Pombeiro, A.J.L. Cobalt complexes with pyrazole ligands as catalyst precursors for the peroxidative oxidation of cyclohexane: X-ray absorption spectroscopy studies and biological applications. *Chem. Asian J.* 2014, *9*, 1132–1143.

219. Powers, D.C.; Ritter, T. Bimetallic redox synergy in oxidative palladium catalysis. *Acc. Chem. Res.* 2012, *45*, 840–850.

220. Adams, R.D.; Rassolov, V.; Wong, Y.O. Binuclear aromatic C–H bond activation at a Dirhenium Site. *Angew. Chem. Int. Ed.* 2016, *55*, 1324–1327.

221. Bilyachenko, A.N.; Dronova, M.S.; Yalymov, A.I.; Korlyukov, A.A.; Shul'pina, L.S.; Arkhipov, D.E.; Shubina, E.S.; Levitsky, M.M.; Kirilin, A.D.; Shul'pin, G.B. New binuclear cage-like copper(II) silsesquioxane ("Cooling Tower"); Its high catalytic activity in oxidation of benzene and alcohols. *Eur. J. Inorg. Chem.* 2013, *2013*, 5240–5246.

222. Dronova, M.S.; Bilyachenko, A.N.; Yalymov, A.I.; Kozlov, Y.N.; Shul'pina, L.S.; Korlyukov, A.A.; Arkhipov, D.E.; Levitsky, M.M.; Shubina, E.S.; Shul'pin, G.B. Solvent-controlled synthesis of tetranuclear cage-like copper(II) silsesquioxanes. Remarkable features of the cage structures and their high catalytic activity in oxidation with peroxides. *Dalton Trans.* 2014, *43*, 872–882.

223. Bilyachenko, A.N.; Dronova, M.S.; Yalymov, A.I.; Lamaty, F.; Bantreil, X.; Martinez, J.; Bizet, C.; Shul'pina, L.S.; Korlyukov, A.A.; Arkhipov, D.E.; *et al.* Cage-like Copper(II) silsesquioxanes: Transmetalation reactions, structural, quantum chemical and catalytic studies. *Chem. Eur. J.* 2015, *21*, 8758–8770.

224. Vinogradov, M.M.; Kozlov, Y.N.; Bilyachenko, A.N.; Nesterov, D.S.; Shul'pina, L.S.; Zubavichus, Y.V.; Pombeiro, A.J.L.; Levitsky, M.M.; Yalymov, A.I.; Shul'pin, G.B. Alkane oxidation with peroxides catalyzed by cage-like copper(II) silsesquioxanes. *New J. Chem.* 2015, *39*, 187–199.

225. Bilyachenko, A.N.; Levitsky, M.M.; Yalymov, A.I.; Korlyukov, A.A.; Vologzhanina, A.V.; Kozlov, Y.N.; Shul'pina, L.S.; Nesterov, D.N.; Pombeiro, A.J.L.; Lamaty, F.; *et al.* A heteronuclear Na_8Fe_6-oxo cluster of unusual "Asian Lantern"-like type: Its synthesis, structure, spin glass behavior and high catalytic activity. 2016. in preparation.

226. Bilyachenko, A.N.; Yalymov, A.I.; Levitsky, M.M.; Korlyukov, A.A.; Shul'pina, L.S.; Es'kova, M.A.; Khrustalev, V.N.; Dorovatovsky, P.V.; Lamaty, F.; Bantreil, X.; *et al.* Unusual heterometallic (Fe_6Na_7 and Fe_6Na_6) cage silsesquioxanes: Synthesis, structure and high catalytic activity. 2016. in preparation.

227. Bilyachenko, A.N.; Yalymov, A.I.; Levitsky, M.M.; Korlyukov, A.A.; Es'kova, M.A.; Long, J.; Larionova, J.; Guari, Y.; Shul'pina, L.S.; Ikonnikov, N.I.; *et al.* First pentanuclear Co(II) silsesquioxane: Synthesis, structure, slow dynamic behaviour of the magnetization and high activity in stereoselective alkane oxidations with peroxides. 2016. in preparation.

228. Bilyachenko, A.N.; Yalymov, A.I.; Shul'pina, L.S.; Korlyukov, A.A.; Vologzhanina, A.V.; Es'kova, M.A.; Shubina, E.S.; Levitsky, M.M.; Shul'pin, G.B. Novel cage-like hexanuclear nickel(II) silsesquioxane. Synthesis, structure and catalytic activity. 2016. in preparation.

229. Gascon, J.; Corma, A.; Kapteijn, F.; Llabrés i Xamena, F.X. Metal organic framework catalysis: Quo vadis? *ACS Catal.*2014, *4*, 361–378.

230. Van Der Voort, P.; Leus, K.; Liu, Y.-Y.; Vandichel, M.; Van Spebroeck, V.; Waroquier, M.; Biswas, S. Vanadium metal–organic frameworks: Structures and applications. *New J. Chem.* 2014, *38*, 1853–1867.

231. Ivanchikova, I.D.; Skobelev, I.Y.; Kholdeeva, O.A. Kinetics and mechanism of anthracene oxidation with *tert*-butyl hydroperoxide over metal-organic frameworks Cr-MIL-101 and Cr-MIL-100. *J. Organomet. Chem.* 2015, *793*, 175–181.

232. Ruano, D.; Díaz-García, M.; Alfayate, A.; Sánchez-Sánchez, M. Nanocrystalline M–MOF-74 as heterogeneous catalysts in the oxidation of cyclohexene: Correlation of the activity and redox potential. *ChemCatChem* 2015, *7*, 674–681.

233. Yao, R.-X.; Quiao, X.-H.; Cui, X.; Jia, X.-X.; Zhang, X.-M. Hexagonal Co_6 and zigzag Co_4 cluster based magnetic MOFs with a pcu net for selective catalysis. *Inorg. Chem. Front.* 2016, *3*, 78–85.

234. Kirillov, A.M.; Kirillova, M.V.; Pombeiro, A.J.L. Homogeneous multicopper catalysts for oxidation and hydrocarboxylation of alkanes. *Adv. Inorg. Chem.* 2013, *65*, 1–31.

235. Gupta, S.; Kirillova, M.V.; Guedes da Silva, M.F.C.; Pombeiro, A.J.L.; Kirillov, A.M. Alkali metal directed assembly of heterometallic V^V/M (M = Na, K, Cs) coordination polymers: Structures, topological analysis, and oxidation catalytic properties. *Inorg. Chem.* 2013, *52*, 8601–8611.

236. Das, S.K.; Mukherjee, S.; Lopes, L.M.; Ilharco, L.M.; Ferraria, A.M.; do Rego, A.M.B.; Pombeiro, A.J.L. Synthesis, characterization and heterogeneous catalytic application of copper integrated mesoporous matrices. *Dalton Trans.*2014, *43*, 3215–3226.

237. Timokhin, I.; Pettinari, C.; Marchetti, F.; Pettinari, R.; Condello, F.; Galli, S.; Alegria, E.C.B.A.; Martins, L.M.D.R.S.; Pombeiro, A.J.L. Novel coordination polymers with (Pyrazolato)-based tectons: Catalytic activity in the peroxidative oxidation of alcohols and cyclohexane. *Cryst. Growth Des.* 2015, *15*, 2303–2317.

238. Martins, L.; Nasani, R.; Saha, M.; Mobin, S.; Mukhopadhyay, S.; Pombeiro, A. Greener selective cycloalkane oxidations with hydrogen peroxide catalyzed by copper-5-(4-pyridyl)tetrazolate metal-organic frameworks. *Molecules* 2015, *20*, 19203–19220.

239. Fernandes, T.A.; Santos, C.I.M.; André, V.; Kłak, J.; Kirillova, M.V.; Kirillov, A.M. Copper(II) coordination polymers self-assembled from aminoalcohols and pyromellitic acid: Highly active precatalysts for the mild water-promoted oxidation of alkanes. *Inorg. Chem.* 2016, *55*, 125–135.

240. Dias, S.S.P.; Kirillova, M.V.; André, V.; Kłak, J.; Kirillov, A.M. New tetracopper(II) cubane cores driven by a diamino alcohol: Self-assembly synthesis, structural and topological features, and magnetic and catalytic oxidation properties.*Inorg. Chem.* 2015, *54*, 5204–5212.

241. Dias, S.S.P.; Kirillova, M.V.; André, V.; Kłak, J.; Kirillov, A.M. New tricopper(II) cores self-assembled from aminoalcohol biobuffers and homophthalic acid: Synthesis, structural and topological features, magnetic properties and mild catalytic oxidation of cyclic and linear C_5-C_8 alkanes. *Inorg. Chem. Front.* 2015, *2*, 525–537.

242. Palomas, D.; Kalamaras, C.; Haycock, P.; White, A.J.P.; Hellgardt, K.; Horton, A.; Crimmin, M.R. Re-evaluating selectivity as a determining factor in peroxidative methane oxidation by multimetallic copper complexes. *Catal. Sci. Technol.* 2015, *5*, 4108–4115.

243. Schreiner, P.R.; Fokin, A.A. Selective alkane C–H-bond functionalizations utilizing oxidative single-electron transfer and organocatalysis. *Chem. Rec.* 2004, *3*, 247–257.

244. Limberg, C. The role of radicals in metal-assisted oxygenationreactions. *Angew. Chem. Int. Ed.* 2003, *42*, 5932–5954.

245. Ding, X.-L.; Wu, X.-N.; Zhao, Y.-X.; He, S.-G. C–H bond activation by oxygen-centered radicals over atomic clusters.*Acc. Chem. Res.* 2012, *45*, 382–390.

246. Studer, A.; Curran, D.P. Catalysis of radical reactions: A radical chemistry perspective. *Angew. Chem. Int. Ed.* 2016, *55*, 58–102.

247. Shul'pin, G.B.; Druzhinina, A.N. Hydroperoxidation of alkanes by atmospheric oxygen in the presence of hydroquinone or quinone catalyzed by copper(II) acetate

under visible light irradiation. *React. Kinet. Catal. Lett.* 1992,*47*, 207–211.

248. Shul'pin, G.B.; Nizova, G.V. Formation of alkyl peroxides in oxidation of alkanes by H_2O_2 catalyzed by transition metal complexes. *React. Kinet. Catal. Lett.* 1992, *48*, 333–338.
249. Shul'pin, G.B. Metal-catalysed hydrocarbon oxygenations in solutions: The dramatic role of additives: A review. *J. Mol. Catal A* 2002, *189*, 39–66.
250. Shul'pin, G.B. Metal-catalysed hydrocarbon oxidations. *C. R. Chim.* 2003, *6*, 163–178.
251. Shul'pin, G.B.; Kozlov, Y.N.; Shul'pina, L.S.; Kudinov, A.R.; Mandelli, D. Extremely efficient alkane oxidation by a new catalytic reagent $H_2O_2/Os_3(CO)_{12}/$pyridine. *Inorg. Chem.* 2009, *48*, 10480–10482.
252. Shul'pin, G.B.; Kozlov, Y.N.; Shul'pina, L.S.; Petrovskiy, P.V. Oxidation of alkanes and alcohols with hydrogen peroxide catalyzed by complex $Os_3(CO)_{10}(\mu\text{-}H)_2$. *Appl. Organometal. Chem.* 2010, *24*, 464–472.
253. Das, B.; Al-Hunaiti, A.; Haukka, M.; Demeshko, S.; Meyer, S.; Shteinman, A.A.; Meyer, F.; Repo, T.; Norlander, E. Catalytic oxidation of alkanes and alkenes by H_2O_2 with a μ-oxido diiron(III) complex as catalyst/catalyst precursor.*Eur. J. Inorg. Chem.* 2015, 3590–3601.
254. Contaldi, S.; Di Nicola, C.; Garau, F.; Karabach, Y.Y.; Martins, L.M.D.R.S.; Monari, M.; Pandolfo, L.; Pettinari, C.; Pombeiro, A.J.L. New coordination polymers based on the triangular $[Cu_3(\mu_3\text{-}OH)(\mu\text{-}pz)_3]^{2+}$ unit and unsaturated carboxylates. *Dalton Trans.* 2009, 4928–4941.
255. Kirillova, M.V.; Kirillov, A.M.; Mandelli, D.; Carvalho, W.A.; Pombeiro, A.J.L.; Shul'pin, G.B. Mild homogeneous oxidation of alkanes and alcohols including glycerol with *tert*-butyl hydroperoxide catalyzed by a tetracopper(II) complex. *J. Catal.* 2010, *272*, 9–17.
256. Xue, S.; Chen, G.; Long, Z.; Zhou, Y.; Wang, J. Efficient and recyclable multi-cationic polyoxometalate-based hybrid catalyst for heterogeneous cyclohexane oxidation with H_2O_2. *RSC Adv.* 2015, *5*, 19306–19314.
257. Haslinger, S.; Raba, A.; Cokoja, M.; Pöthig, A.; Kühn, F.E. Iron-catalyzed oxidation of unreactive C–H bonds: Utilizing bio-inspired axial ligand modification to increase catalyst stability. *J. Catal.* 2015, *331*, 147–153.
258. Olivo, G.; Nardi, M.; Vidal, D.; Barbieri, A.; Lapi, A.; Gomez, L.; Lanzalunga, O.; Costas, M.; Di Stefano, S. C–H bond oxidation catalyzed by an imine-based iron complex: A mechanistic insight. *Inorg. Chem.* 2015, *54*, 10141–10152.
259. Sutradhar, M.; Martins, L.M.D.R.S.; Guedes da Silva, M.F.C.; Mahmudov, K.T.; Liu, C.-M.; Pombeiro, A.J.L. Trinuclear Cu^{II} structural isomers: Coordination, magnetism, electrochemistry and catalytic activity towards the oxidation of alkanes. *Eur. J. Inorg. Chem.* 2015, 3959–3969.
260. Vanoppen, D.L.; de Vos, D.E.; Genet, M.J.; Rouxhet, P.G.; Jacobs, P.A. Cobalt-containing molecular sieves as catalysts for the low conversion autoxidation of pure cyclohexane. *Angew. Chem. Int. Ed. Engl.* 1995, *34*, 560–563.
261. Knops-Gerrits, P.P.; Trujillo, C.A.; Zhan, B.Z.; Li, X.Y.; Rouxhet, P.; Jacobs, P.A. Oxidation catalysis with well-characterised vanadyl bis-bipyridine complexes encapsulated in NaY zeolite. *Top. Catal.* 1996, *3*, 437–449.
262. Bonchio, M.; Carraro, M.; Scorrano, G.; Kortz, U. Microwave-assisted fast cyclohexane oxygenation catalyzed by iron-substituted polyoxotungstates. *Adv. Synth. Catal.* 2005, *347*, 1909–1912.
263. De Castries, A.; Magnier, E.; Momotton, S.; Fensterbank, H.; Larpent, C. An expedient synthesis of perfluorinated tetraazamacrocycles: New ligands for copper-catalyzed oxidation under fluorous biphasic conditions. *Eur. J. Org. Chem.* 2006, 4685–4692.
264. Fornal, E.; Giannotti, C. Photocatalyzed oxidation of cyclohexane with heterogenized decatungstate. *Photochem. Photobiol. A* 2007, *188*, 279–286.
265. Zhou, L.; Chen, Y.; Yang, X.; Su, Y.; Zhang, W.; Xu, J. Electronic effect of substituent of quinones on their catalytic performance in hydrocarbons oxidation. *Catal. Lett.* 2008, *125*, 154–159.
266. Aprile, C.; Corma, A.; Domine, M.E.; Garcia, H.; Mitchell, C. A cascade aerobic epoxidation of alkenes over Au/$CeO_2$and Ti-mesoporous material by "*in situ*" formed peroxides. *J. Catal.* 2009, *264*, 44–53.
267. Jiang, D.; Mallat, T.; Meier, D.M.; Urakawa, A.; Baiker, A. Copper metal–organic framework: Structure and activity in the allylic oxidation of cyclohexene with molecular oxygen. *J. Catal.* 2010, *270*, 26–33.
268. Lloyd, R.; Jenkins, R.L.; Piccinini, M.; He, Q.; Kiely, C.J.; Carley, A.F.; Golunski, S.E.; Bethell, D.; Bartley, J.K.; Hutchings, G.J. Low-temperature aerobic oxidation of decane using an oxygen-free radical initiator. *J. Catal.* 2011, *283*, 161–167.
269. Maksimchuk, N.V.; Kovalenko, K.A.; Fedin, V.P.; Kholdeeva, O.A. Cyclohexane selective oxidation over metal-organic frameworks of MIL-101 family: Superior catalytic activity and selectivity. *Chem. Commun.* 2012, *48*, 6812–6814.
270. Skobelev, I.Y.; Kovalenko, K.A.; Fedin, V.P.; Sorokin, A.B.; Kholdeeva, O.A. Allylic oxidation of alkenes with molecular oxygen catalyzed by porous coordination polymers Fe-MIL-101 and Cr-MIL-101[1]. *Kinet. Catal.* 2013, *54*, 607–614.
271. Skobelev, I.V.; Sorokin, A.B.; Kovalenko, K.A.; Fedin, V.P.; Kholdeeva, O.A. Solvent-free allylic oxidation of alkenes with O_2 mediated by Fe- and Cr-MIL-101. *J. Catal.* 2013, *298*, 61–69.
272. Cao, Y.; Yu, H.; Peng, F.; Wang, H. Selective allylic oxidation of cyclohexene catalyzed by nitrogen-doped carbon nanotubes. *ACS Catal.* 2014, *4*, 1617–1625.
273. Mavrogiorgou, A.; Papastergiou, M.; Deligiannakis, Y.; Louloudi, M.J. Activated carbon functionalized with Mn(II) schiff base complexes asefficient alkene oxidation catalysts: Solid support matters. *J. Mol. Catal. A* 2014, *393*, 8–17.

274. Makgwane, P.R.; Ray, S.S. Efficient room temperature oxidation of cyclohexane over highly active hetero-mixed WO_3/V_2O_5 oxide catalyst. *Catal. Commun.* 2014, *54*, 118–123.

275. Cao, Y.; Li, Y.; Yu, H.; Peng, F.; Wang, H. Aerobic oxidation of α-pinene catalyzed by carbon nanotubes. *Catal. Sci. Technol.* 2015, *5*, 3935–3944.

276. Liu, X.; Xing, N.; Song, J.; Wu, Q.; Yan, Z.; Zhang, Y.; Xing, Y. Two novel oxomolybdenum(V)-trispyrazolylborate complexes: Synthesis, structure, and catalytic performance in the cyclohexane oxidation. *Polyhedron* 2015, *102*, 386–393.

277. Yan, Z.D.; Xing, N.; Zhang, Y.; Ma, X.T.; Song, J.; Liu, X.; Xing, Y.H. Oxidovanadium(IV) complexes with tridentate N-heterocycle ligands: Synthesis, structure, and efficient catalyst for cyclohexane oxidation to cyclohexanone. *Polyhedron*2015, *102*, 600–608.

278. Zhao, Q.; Chen, K.; Zhang, W.; Yao, J.; Li, H. Efficient metal-free oxidation of ethylbenzene with molecular oxygen utilizing the synergistic combination of NHPI analogues. *J. Mol. Catal. A* 2015, *402*, 79–82.

279. Kojima, T.; Nakayama, K.; Ikemura, K.; Ogura, T.; Fukuzumi, S. Formation of a ruthenium(iv)-oxo complex by electron-transfer oxidation of a coordinatively saturated ruthenium(II) complex and detection of oxygen-rebound intermediates in C–H bond oxygenation. *J. Am. Chem. Soc.* 2011, *133*, 11692–11700.

280. Cho, K.-B.; Wu, X.; Lee, Y.-M.; Kwon, Y.H.; Shaik, S.; Nam, W. Evidence for an alternative to the oxygen rebound mechanism in C–H bond activation by non-heme $Fe^{IV}O$ complexes. *J. Am. Chem. Soc.* 2012, *134*, 20222–20225.

281. Postils, V.; Company, A.; Solà, M.; Costas, M.; Luis, J.M. Computational insight into the mechanism of alkane hydroxylation by non-heme Fe(PyTACN) iron complexes. *Effects of the substrate and solvent. Inorg. Chem.* 2015, *54*, 8223–8236.

282. Cho, K.-B.; Nam, W. A theoretical study into a trans-dioxo Mn^V porphyrin complex that does not follow the oxygen rebound mechanism in C–H bond activation reactions. *Chem. Commun.* 2016, *52*, 904–907.

283. Shul'pin, G.B.; Nesterov, D.S.; Shul'pina, L.S.; Pombeiro, A.J.L. A hydroperoxo-rebound mechanism of alkane oxidation by the system "H_2O_2–O_2–$[Mn_2L_2O_3][PF_6]_2$ (L=1,4,7-trimethyl-1,4,7-triazacyclononane)–acid" with involvement of atmospheric dioxygen. 2016. in preparation.

284. Shul'pin, G.B.; Stoeckli-Evans, H.; Mandelli, D.; Kozlov, Y.N.; Tesouro Vallina, A.; Woitiski, C.B.; Jimenez, R.S.; Carvalho, W.A. Oxidation of alkanes with *m*-chloroperbenzoic acid catalyzed by iron(III) chloride and a polydentate amine. *J. Mol. Catal. A* 2004, *219*, 255–264.

285. Nesterova, O.V.; Nesterov, D.S.; Kokozay, V.N.; Shul'pin, G.B.; Pombeiro, A.J.L. Highly efficient stereospecific alkane hydroxylation with *m*-chloroperbenzoic acid catalyzed by a heterometallic Co/Fe complex. 2016. in preparation.

286. Shul'pin, G.B.; Adams, R.D. Biosketch—Shilov special issue. *J. Organomet. Chem.* 2015, *793*, 2–3.

287. Rudakov, E.S.; Shul'pin, G.B. Stable organoplatinum complexes as intermediates and models in hydrocarbon functionalization (review). *J. Organomet. Chem.* 2015, *793*, 4–16.

288. Shestakov, A.F.; Goldshleger, N.F. Catalysts for C–H functionalization: Platinum, gold and even nanodiamonds. *J. Organomet. Chem.* 2015, *793*, 17–33.

289. Shteinman, A.A. Shilov alkane platinum chemistry: 45 years. *J. Organomet. Chem.* 2015, *793*, 34–40.

290. Crabtree, R.H. A.E. Shilov's influence on early work in organometallic CH activation and functionalization. *J. Organomet. Chem.* 2015, *793*, 41–46.

291. Labinger, J.A.; Bercaw, J.E. Mechanistic studies on the Shilov system: A retrospective. *J. Organomet. Chem.* 2015, *793*, 47–53.

292. Bach, R.D.; Dmitrenko, O. The "Somersault" mechanism for the P-450 hydroxylation of hydrocarbons. The intervention of transient inverted metastable hydroperoxides. *J. Am. Chem. Soc.* 2006, *128*, 1474–1488.

293. DeVries, N.; Roe, D.C.; Thorn, D.L. Catalytic hydroxylation using chloroplatinum compounds. *J. Mol. Catal. A* 2002,*189*, 17–22.

294. Ochiai, M.; Fukui, K.; Iwatsuki, S.; Ishihara, K.; Matsumoto, K. Synthesis of aryl-platinum dinuclear complexes via*ortho* C–H bond activation of phenol and transmetalation of arylboronic acid. *Organometallics* 2005, *24*, 5528–5536.

295. Wagner, A.M.; Hickman, A.J.; Sanford, M.S. Platinum-catalyzed C–H arylation of simple arenes. *J. Am. Chem. Soc.*2013, *135*, 15710–15713.

296. Kovach, J.; Brennessel, W.W.; Jones, W.D. Electrophilic C–H activation of benzene with a Shilov-inspired rhodium(III) diimine complex. *J. Organomet. Chem.* 2015, *793*, 192–199.

297. Shul'pin, G.B. The reaction of H_2PtCl_6 with aromatic compounds affording the sigma-aryl complexes of Pt(IV). III. The synthesis of Pt(IV) complexes of benzenes containing electron-withdrawing substituents. *J. Organometal. Chem.*1981, *212*, 267–274.

298. Shul'pin, G.B. Reaction of aromatic compounds with H_2PtCl_6 in CF_3COOH-H_2O system leading to the preparation of anionic sigma-aryl complexes of Pt(IV). *J. Gen. Chem. USSR* 1981, *51*, 1808–1818.

299. Shul'pin, G.B.; Nizova, G.V.; Nikitaev, A.T. The reaction of $PtCl_6^{2-}$ with aromatic compounds to afford anionic sigma-aryl complexes of Pt(IV). VIII. Kinetics and mechanisms of thermal, photochemical and gamma-induced reactions with arenes and arylmercury compounds (electrophilic substitution involving electron transfer). *J. Organomet. Chem.*1984, *276*, 115–153.

300. Albert, J.; Bosque, R.; Crespo, M.; Granell, J.; Rodriguez, J.; Zafrilla, J. Platinum-mediated C–H bond activation of arene solvents and subsequent C–C bond formation. *Organometallics* 2010, *29*, 4619–4627.

301. Prince, B.M.; Cundari, T.R. C–H Bond Activation of Methane by Pt^{II}–N-Heterocyclic Carbene Complexes. The importance of having the ligands in the right place at the right time. *Organometallics* 2012, *31*, 1042–1048.
302. Pal, S.; Zavalij, P.Y.; Vedernikov, A.N. Concurrent B to-Pt methyl migration and B-center retention in aerobic oxidation of methylborato platinum(II) complexes. *Organometallics* 2015, *34*, 5183–5190.
303. Furukawa, T.; Tobisu, M.; Chatani, N. C–H Functionalization at sterically congested positions by the platinum-catalyzed borylation of arenes. *J. Am. Chem. Soc.* 2015, *137*, 12211–12214.
304. Lee, M.; Sanford, M.S. Platinum-catalyzed, terminal-selective C(sp^3)-H oxidation of aliphatic amines. *J. Am. Chem. Soc.*2015, *137*, 12796–12799.
305. Gunsalus, N.J.; Konnick, M.M.; Hashiguchi, B.G.; Periana, R.A. Discrete molecular catalysis for methane functionalization. *Isr. J. Chem.* 2014, *54*, 1467–1480.
306. Mironov, O.A.; Bischof, S.M.; Konnick, M.M.; Hashiguchi, B.G.; Ziatdinov, V.R.; Goddard, W.A., III; Ahlquist, M.; Periana, R.A. Using reduced catalysts for oxidation reactions: Mechanistic studies of the "Periana-Catalytica" system for CH4 oxidation. *J. Am. Chem. Soc.* 2013, *135*, 14644–14658.
307. Konnick, M.M.; Bischhof, S.M.; Yousufuddin, M.; Hashiguchi, B.G.; Ess, D.H.; Periana, R.A. A mechanistic change results in 100 times faster CH functionalization for ethane *versus* methane by a homogeneous Pt catalyst. *J. Am. Chem. Soc.* 2014, *136*, 10085–10094.
308. Lin, M.; Sen, A. Oxidation and oxidative carbonylation of methane and ethane by hexaoxo-μ-peroxodisulfate(2−) ion in aqueous medium. A model for alkane oxidation through the hydrogen-atom abstraction pathway. *J. Chem. Soc. Chem. Commun.* 1992, 892–893.
309. Nishiguchi, T.; Nakata, K.; Takaki, K.; Fujiwara, Y. Transition metal catalyzed acetic acid synthesis from methane and CO. *Chem. Lett.* 1992, *21*, 1141–1142.
310. Nakata, K.; Yamaoka, Y.; Miyata, T.; Taniguchi, Y.; Takaki, K.; Fujiwara, Y. Palladium and/or copper-catalyzed carboxylation of small alkanes such as methane and ethane with carbon monoxide. *J. Organomet. Chem.* 1994, *473*, 329–334.
311. Jia, C.; Kitamura, T.; Fujiwara, Y. Catalytic functionalization of arenes and alkanes via C–H bond activation. *Acc. Chem. Res.* 2001, *34*, 633–639.
312. Reis, P.M.; Silva, J.A.L.; Palavra, A.F.; Frausto da Silva, J.J.R.; Pombeiro, A.J.L. Vanadium-catalyzed carboxylation of linear and cyclic C_5 and C_6 alkanes. *J. Catal.* 2005, *235*, 333–340.
313. Kirillova, M.V.; Kirillov, A.M.; Pombeiro, A.J.L. Mild, single-pot hydrocarboxylation of linear C_5–C_9 alkanes into branched monocarboxylic C_6–C_{10} acids in copper-catalyzed aqueous systems. *Appl. Catal. A* 2011, *401*, 106–113.
314. Sutradhar, M.; Kirillova, M.V.; Guedes da Silva, M.F.C.; Liu, C.-M.; Pombeiro, A.J.L. Tautomeric effect of hydrazone Schiff bases in tetranuclear Cu(II) complexes: Magnetism and catalytic activity towards mild hydrocarboxylation of alkanes. *Dalton Trans.* 2013, *42*, 16578–16587.
315. Nesterova, O.V.; Kirillova, M.V.; Guedes da Silva, M.F.C.; Boča, R.; Pombeiro, A.J.L. How to force a classical chelating ligand to a metal non-chelating bridge: The observation of a rare coordination mode of diethanolamine in the 1D complex $\{[Cu_2(Piv)_4(H_3tBuDea)](Piv)\}_n$. *CrystEngComm* 2014, *16*, 775–783.
316. Phan, A.; Czaja, A.U.; Gándara, F.; Knobler, C.B.; Yaghi, O.M. Metal-organic frameworks of vanadium as catalysts for conversion of methane to acetic acid. *Inorg. Chem.* 2011, *50*, 7388–7390.
317. Itoyama, S.; Doitomi, K.; Kamachi, T.; Shiota, Y.; Yoshizawa, K. Possible peroxo state of the dicopper site of particulate methane monooxygenase from combined quantum mechanics and molecular mechanics calculations. *Inorg. Chem.*2016.
318. Culpepper, M.A.; Cutsail III, G.E.; Gunderson, W.A.; Hoffman, B.M.; Rosenzweig, A.C. Identification of the valence and coordination environment of the particulate methane monooxygenase copper centers by advanced EPR characterization. *J. Am. Chem. Soc.* 2014, *136*, 11767–11775.
319. Wang, W.; Lippard, S.J. Diiron oxidation state control of substrate access to the active site of soluble methane monooxygenase mediated by the regulatory component. *J. Am. Chem. Soc.* 2014, *136*, 2244–2247.
320. Kirillova, M.V.; Kirillov, A.M.; Martins, A.N.C.; Graiff, C.; Tiripicchio, A.; Pombeiro, A.J.L. Topologically unique heterometallic CuII/Li coordination polymers self-assembled from N,N-bis(2-Hydroxyethyl)-2-aminoethanesulfonic acid biobuffer: Versatile catalyst precursors for mild hydrocarboxylation of alkanes to carboxylic acids. *Inorg. Chem.*2012, *51*, 5224–5234.
321. Kirillov, A.M.; Karabach, Y.Y.; Kirillova, M.V.; Haukka, M.; Pombeiro, A.J.L. Topologically unique 2D heterometallic CuII/Mg coordination polymer: Synthesis, structural features, and catalytic use in alkane hydrocarboxylation. *Cryst. Growth Des.* 2012, *12*, 1069–1074.
322. Fernandes, T.A.; Santos, C.I.M.; André, V.; Dias, S.S.P.; Kirillova, M.V.; Kirillov, A.M. New aqua-soluble dicopper(II) aminoalcoholate cores for mild and water-assisted catalytic oxidation of alkanes. *Catal. Sci. Technol.* 2016.
323. Shul'pin, G.B. Chapter 6.04 Selectivity in C–H functionalizations. In *Comprehensive Inorganic Chemistry II*, 2nd ed.; Reedijk, K.J., Poeppelmeier, L.C., Eds.; Elsevier: Amsterdam, The Netherlands, 2013; Volume 6, pp. 79–104.

CHAPTER 6

UPGRADING OF HEAVY CRUDE OIL WITH W-ZR CATALYST

P. Schacht, L. Díaz-García, J. Aguilar, S. Ramírez, P. Pérez-Romo, M. Espinosa

Keywords: Heavy Crude Oil, Upgrading, Tungsten-Zirconium, Catalyst

INTRODUCTION

The International Energy Agency reported in 2009 that the demand for crude oil has increased over the last 40 years by 150% and, by 20% in the past two decades, to reach a current demand of 80 million barrels per day. It is projected to grow 1.4% per year for the next 30 years and reach 118 million barrels per day by 2030 [1] [2] . This oil demand comes at time when there is a gradual reduction in the supply of relatively cheap conventional crude and new discoveries of light field oil do not cover this demand [2] [3] . It is estimated that there are 8 - 9 trillion barrels of heavy crude oil and bitumen around the world, of which potentially 900 billion barrels of crude oil are commercially exploitable with modern technology [3] -[5] . In 2015, the growth of oil sand exploitation will exceed 2 to 3 times than the conventional crude oil production. Moreover, better methods are needed to upgrade partially the heavy crude oil before further processing in refineries in order to convert it into light crude oil [4] -[7] .

In the catalytic hydrocracking of heavy crude oil, it is necessary to snap the asphaltenes, because they are extremely large molecules. These molecules consist of highly condensed material, heterocyclic and aromatic rings containing sulfur, nitrogen, oxygen, and metals (mainly vanadium and nickel). Therefore, some of the catalysts for hydrogenation contain nickel, tungsten, molybdenum, cobalt or zirconium [8] [9] . For example, molybdenum, tungsten and nickel are widely used as catalysts in refining processes such as hydrocracking, hydrodesulphurization and hydrodenitrogenation. These catalysts are supported on porous materials like crystalline alumina [10] .

Moreover, the acid-base properties of these catalysts may help to increase the yield of gasoline and other light hydrocarbons in crude oil processing and also to produce environmentally friendly products [9] -[14] .

The purpose of this work is to study the behavior of an acid catalyst formulated with tungsten and zirconium, in the upgrading of the properties of a heavy crude oil. The study of viscosity decreasing and the changes in structural composition of heavy crude oil after the reaction were analyzed [15] -[17] . The paper also examines the decreasing of asphaltenes, resins, sulfur and nitrogen compounds and the increase of API gravity.

EXPERIMENTAL SECTION

Preparation of Catalyst

The active elements of this catalyst are zirconium, tungsten and oxygen. 5 grams of $ZrOCl_2·8H_2O$ (Aldrich 99%) were dissolved in 300 mL of water using a beaker; the content was stirred at room temperature. A solution of 15 grams of $Na_2WO_4·2H_2O$

Citation: Schacht, P. , Díaz-García, L. , Aguilar, J. and Ramírez, S. (2014) Upgrading of Heavy Crude Oil with W-Zr Catalyst.Advances in Chemical Engineering and Science, 4, 250-257. doi: 10.4236/aces.2014.42028.

(FLUKA 98%) in 100 mL of water was also prepared and added to the solution of $ZrOCl_2 \cdot H_2O$. The resulting solution immediately turned turbid. After stirring this solution for about 30 minutes, 200 ml 1 N of HCl (MERICHEM 37%) was also added and the mixture was stirred for one more hour and allowing it to stand still overnight at room temperature. The precipitate formed was separated by filtration.

Activity Tests

The activity tests were carried out in a Parr batch reactor. During the experiments, 200 g of heavy oil were mixed with 5 g of the W-Zr catalyst. Before each experimental test, the reactor was purged with H_2 and stabilized at the required reaction pressure, temperature and stirring rate. The hydrogen pressure was increased to 10.8 MPa, and the reactor was heated to the reaction temperature of 653 K, at 1000 rpm. Three reaction times were evaluated 30, 45 and 60 minutes, the product samples were recovered at the end of each test. The physical and chemical properties of the products from heavy oil upgrading were characterized in accord to the ASTM methods.

Characterization

Thermogravimetric analyses (TGA-DTA) were obtained with a Perkin-Elmer 1700. The nature of the acid sites of catalyst was analyzed by Pyridine thermo desorption-FTIR, using a spectrophotometer Nicolet FTIR Magna 560. A quartz cell was used for infrared spectrometry with 4 cm^{-1} of resolution, using KBr as blank. Temperature programmed reduction was carried out in an equipment AMI-200 ZETON-Altamira coupled to thermal conductivity detector (TCD) using a certified gases mixture, 10 vol-% of Hydrogen in Argon from 303 K to 1123 K with heating rate of 10K/min and recording the changes in sample with thermal conductivity detector. The detector calibration was carried out at 313 K using a 57 uL loop. In order to characterize the metallic sites of this W-Zr liquid phase catalyst, this was supported on inert alumina, dried at 383 K in air static atmosphere during 4 hours before H_2-TPR. Physical and chemical properties of heavy oil and products were characterized. API gravity was measured according ASTM-D-287 method. The viscosity was determined using a rotary viscometer; heavy crude oil was put into a sample cup and kept at 298, 310 and 327 K over 20 min, then measurement was conducted according to ASTM-D445 method. SARA analysis (saturates, aromatics, resins and asphaltenes) were determined by ASTM-D-4124 method. Sulfur content in the crude oil was measured by ASTMD4294 method.

RESULTS AND DISCUSSION

The catalyst W-Zr was characterized using the solid precursor's mechanical mixture in order to obtain the thermal transformations by TGA and DTA.

Thermogravimetric Analysis

The DTA and TGA plots of catalyst are shown in Figure 1 and Figure 2 respectively. The obtained results showed that the catalyst weight is reduced which depends, in part of the composition of the sample [18] [19] .

In a first zone, the catalyst showed changes at temperatures in the range of 298 - 373 K. The weight loss was about 3.6 wt% and it was attributed to the removal of fisisorbed water. These are stable octahedral structures of H_2O molecules that form hydrogen bonds with the acid protons.

In a second stage the weight loss was 33.6 wt% and occurred between 388 and 585 K and corresponded to the removal of crystallization water that leads to the formation of anhydrous species such as Na_2WO_4 and $ZrOCl_2$. These two stages showed different mass loss; both are highly endothermic [20] .

The third stage is characterized by the decomposition of the tungstic acid which occurs with great difficulty. A weight loss of 6.2 wt%, between 585 and 613 K, attributed to the destruction of stable bonds in this volatile compound.

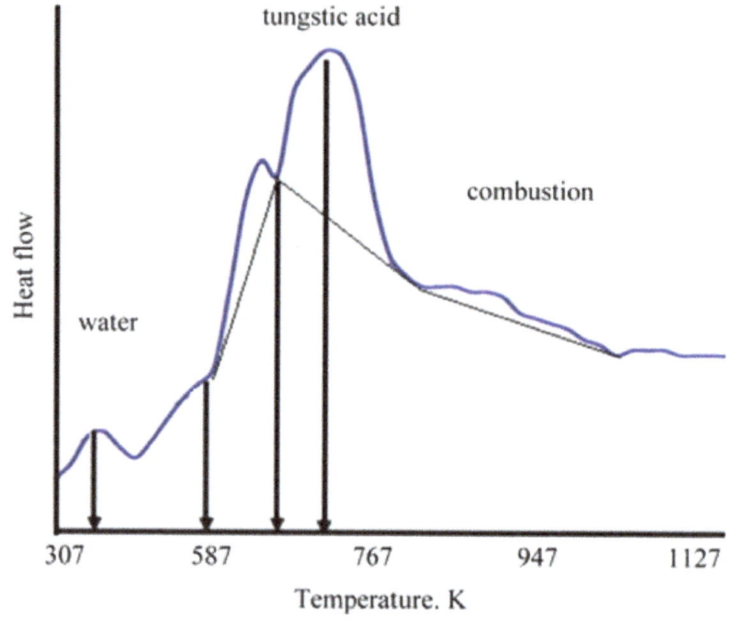

Figure 1: DTA curve for W-Zr catalyst.

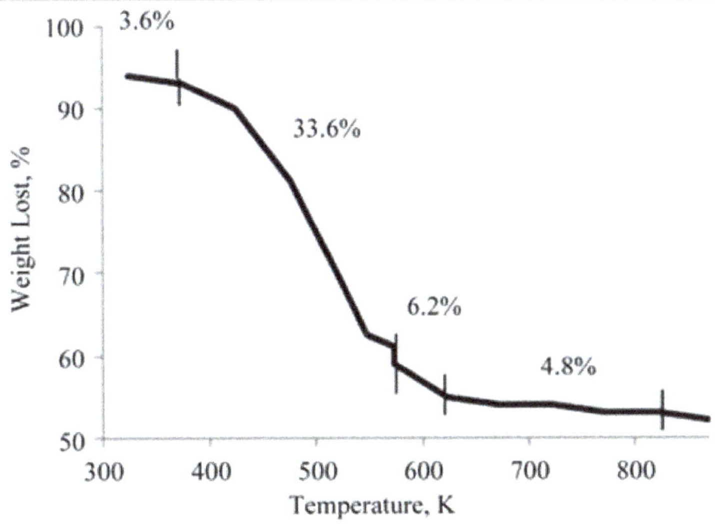

Figure 2: TGA curve for W-Zr catalyst.

The maximum rate of mass loss and DTA curve shifts to a positive zone, above the baseline, which confirmed an exothermic process, see Table 1.

The fourth stage was also characterized by small variations in rate and weight loss of 4.8 wt% between 613 and 803 K, which was attributed to the breaking of binding metal-chlorides. All reactions were highly exothermic, during the increment of the temperature in the analysis.

The mass loss was related to the following stages, according to the enthalpy changes observed by DTA: (1) desorption of fisisorbed water before 383 K, (2) at 388 K removal of crystallization water leading to anhydrous species such as Na_2WO_4 and $ZrOCl_2$, (3) decomposition of tungstic acid between 585 and 613 K and (4) oxidation of remaining Cl-Zr and Cl-W species, breaking the bindings of metal-chlorides between 613 and 803 K [21] .

Pyridine Thermo Programmed Desorption (TPD) by FTIR

The IR spectroscopy is widely used to detect adsorbed pyridine and to distinguish different acid sites. FTIR pyridine adsorption spectra at 298 K and the spectra of TPD at 373, 473 and 523K of W-Zr catalyst were shown in Figure 3. The spectra showed a sharp pyridine absorption-desorption bands at 1487, 1539, 1628 and 1643 cm^{-1}. Pyridine molecules bonded to Lewis acid sites absorb at 1604 and 1444 cm^{-1}, whereas those responsible for Brönsted acid sites (pyridinium ion) showed absorbance at 1539 and 1628 cm^{-1} [22] . The band at 1487 cm^{-1} is a combined band originated from pyridine bonds to both Brönsted and Lewis acid sites. Table 2 shows the Bronsted acid sites, while Figure 3 shows the FTIR pyridine TPD at different temperatures which could be confirmed elsewhere [23] [24] .

Table 1: DTA and TG results for W-Zr catalysts.

Total	First Peak	Second Peak
Hydrogen consumption	Zirconia reduction 22.75 milimoles H_2	Tungsten reduction 83.25 milimoles H_2
106.00 milimoles H_2		
	Maximum @837K	Maximum @1039K

Table 2: Brönsted Acid Sites at different temperatures for W-Zr catalyst.

Temperature, K	Brönsted (a.u.)
523	4.923
473	5.180
373	6.628
298	6.458

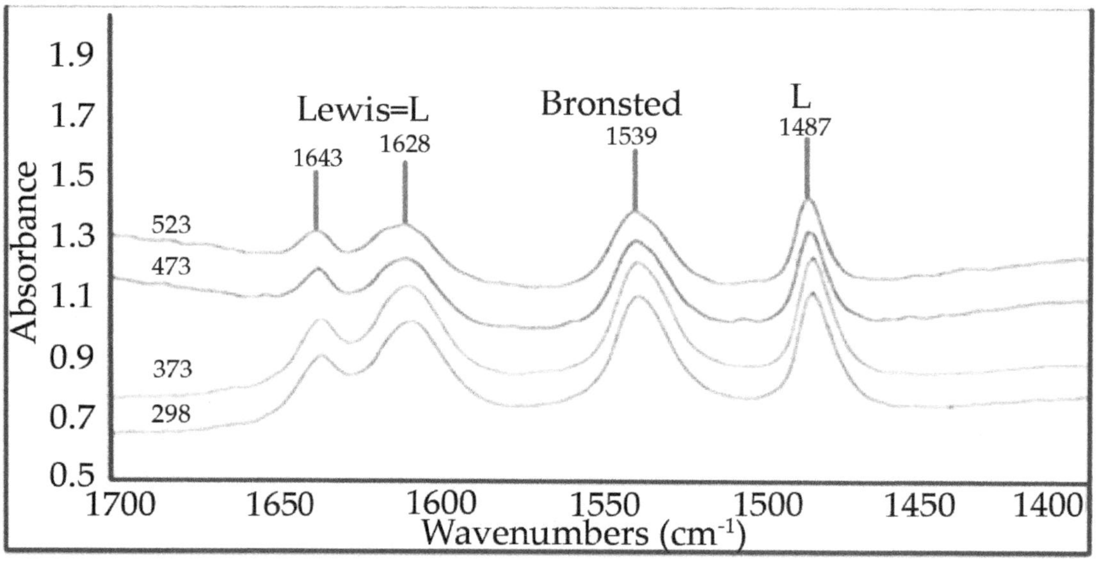

Figure 3: Acid sites distribution using Pyridine TPD-FTIR of W-Zr catalyst.

Temperature Programmed Reduction (H_2-TPR)

The W-Zr catalyst was analyzed by Hydrogen TPR, starting the experiment at 303K.calibrating at zero the signal of TCD. Two main reduction profiles are shown in Figure 4, one first peak observed with a maximum temperature at 837 K attributed to the tetragonal and monoclinic phases of zirconia reduction. And the second peak profile was shown with a maximum reduction temperature of 1039 K attributed to tungsten species reduction of higher hydrogen consumption, with 78.5% of total reduction obtained (see Table 3). In accord to these reduction results the active sites of W-Zr catalyst are able to hydrogenate heavy crude oil, considering the reaction temperature is possible that the influence of zirconia can be higher than tungsten in this catalyst.

Viscosity and API Gravity of Upgrading Products

After the reaction, the API gravity was increased from 12° to 16° and 18°, because of hydrocraking of resins and asphaltenes. Consequently, the kinematic viscosity was reduced from 15,416 to 391 cSt at 298 K, because of the production of smaller molecular weight structures. Increasing reaction time from 30 minutes to 45 and 60 minutes improved the reaction products quality and a similar behavior was reported elsewhere [25] . The Table 4 and Figure 5 show the API gravity increment, after the hydrocracking.

Effects on the Composition of the Heavy Crude Oil

SARA analyses (saturates, aromatics, resins and asphaltenes) of the heavy crude oil sample, shows that this sample contains 19, 22.5, 33.5 and 25 wt%, of each group of S, A, R and A respectively.

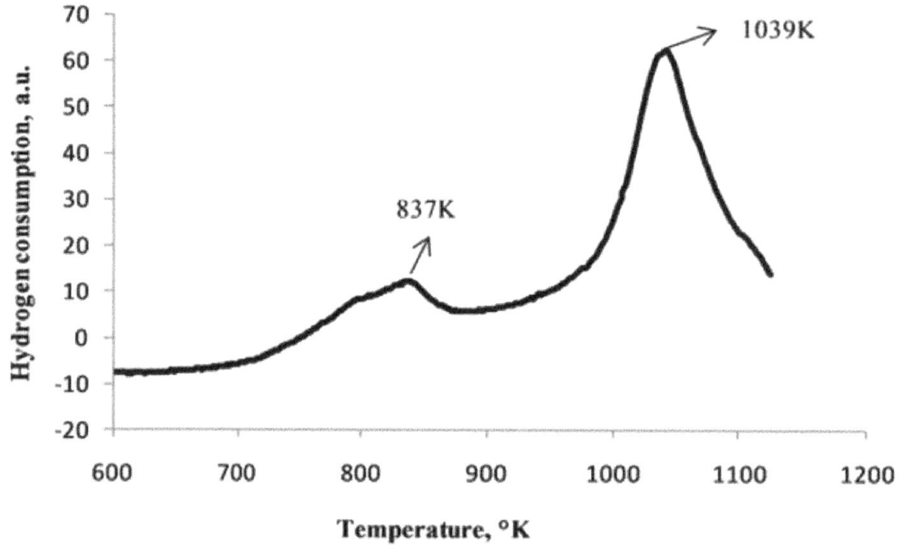

Figure 4: Temperature Programmed Reduction of the W-Zr catalyst.

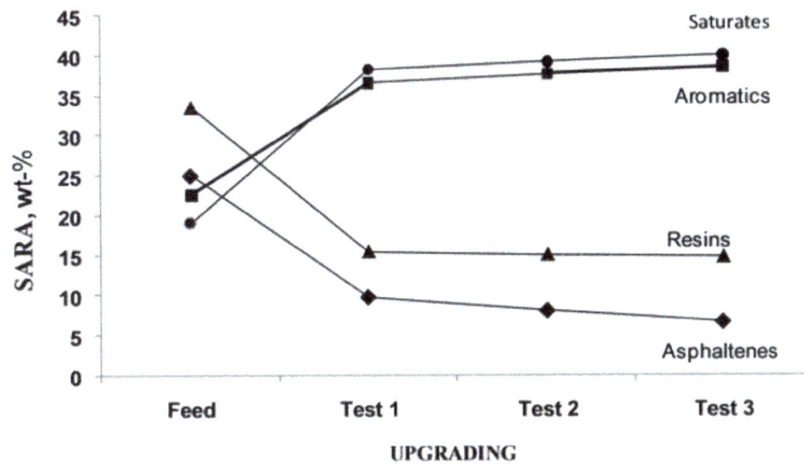

Figure 5: Effect of catalysts W-Zr on the hydrocarbon composition of heavy oil.

Table 3: Physical and chemical properties of the heavy crude oil and the products.

Properties		Heavy crude oil	Test 1 30 minutes	Test 2 45 minutes	Test 3 60 minutes
API		12.5	15.3	16.7	19.0
Viscosity, cSt (K)	@298.0	15.416	391	136	84
	@310.8	9897	278	100	60
	@327.4	5635	185	68	49
Sulfur, wt %		5.13	3.4	3.1	2.8
Nitrogen, wt %		0.78	0.69	0.64	0.62
SARA					
Saturates, wt%		19.0	38.2	39.1	40.0
Aromatics, wt%		22.5	36.5	37.7	38.5
Resins, wt%		33.5	15.5	15.1	14.8
Asphaltenes, wt%		25.0	9.8	8.1	6.7

Table 4: Hydrogen consumption of the reducible species for W-Zr catalyst.

Total	First Peak	Second Peak
Hydrogen consumption	Zirconia reduction 22.75 milimoles H_2	Tungsten reduction 83.25 milimoles H_2
106.00 milimoles H_2		
	Maximum @837K	Maximum @1039K

After reaction, it was observed that there is an increment in the saturated hydrocarbons and aromatic as a result of conversion of resins and asphaltenes, the asphaltenes content was also reduced. These results are in agreement with previous reports [25] .

The results obtained indicate several effects of the catalysts in the upgrading of the heavy oil. For example the analysis by families of hydrocarbons reveals that the asphaltene declined, approximately 67 wt% of these were converted to lighter hydrocarbons. On the other hand, the concentration of aromatics was increased from 22.5 in original oil to 36.5, 37.7 and 38.5 wt%, for the reaction times of 30, 45 and 60 minutes, respectively.

Effects on Content Sulfur and Nitrogen

The catalyst showed a high activity in the removal of sulfur of around 45 wt%. However, sulfur removal also depends on the type of the feed and the operating conditions, which were not severe in the present test. This high catalytic activity can be explained by the total acidity of the catalyst and the hydrocracking properties of the transition metals W and Zr [26] . The sulfur content in the heavy oil was removed up to 3.4, 3.1 and 2.8 wt%, at reaction times of 30, 45 and 60 minutes, respectively. The sulfur removal is caused by the scission of C–S bonds as explained by the removal of thiophenic sulfur using the transition metal salts [27] [28] . These last results showed that the distribution of compounds changed with the increment of reaction time, confirming that this change was due to the conversion of the molecules of resins and asphaltenes.

On the other hand, the content of nitrogen after reaction was reduced from 0.78% wt in the original oil to 0.69 wt% at 30 min, 0.64 wt% at 45 min, and 0.62 wt% at 60 minutes [27] [28] . The nitrogen removal was approximately 20 wt%. Table 4 summarizes the effect of the catalyst on the nitrogen content.

CONCLUSION

The hydrocracking and hydrogenation capability of the catalyst were responsible for the upgrading of heavy oil. The mild operating conditions in the present test were confirmed by the yields, and higher quality of the upgraded crude oil was obtained. The use of W/Zr catalysts is an interesting alternative for the upgrading of heavy crude oils. These catalysts showed high activity at three different residence times and achieved upgraded products with higher API gravity, lower kinematic viscosity, and lower content of sulfur, resins and asphaltenes.

REFERENCES

1. Kjarstad, J. and Johnsson, F. (2009) Resources and Future Supply of Oil. Energy Policy, 37, 441-464.
2. Laherrere, J. (2003) Forecast of Oil and Gas Supply to 2050. PETROTECH 2003, New Delhi.
3. Isaacs, E. (2006) Book Canadian Oil Sands: Development and Future Outlook. Managing Director, Alberta Energy Research Institute, Canada's National Energy Board, Calgary.http://www.asponews.org, http://www.energiekrise.de
4. Fumoto, E. Tago, T., Tsuji, T. and Takao, M. (2004) Recovery of Useful Hydrocarbons from Petroleum Residual Oil by Catalytic Cracking with Steam over Zirconia-Supporting Iron Oxide Catalyst. Energy Fuels, 18, 1770-1774.
5. Brian, C., Gordon, J. and Graves, W. (2007) Heavy Oil, Extra-Heavy Oil and Bitumen Unconventional Oil. National Petroleum Council Heavy Oil Subgroup of the Technology Task Group of the NPC Committee on Global Oil and Gas.http://www.npc.org/Study_Topic_Papers/22-TTG-Heavy-Oil.pdf
6. Greene, D.L., Hopson, L. and Li, J. (2006) Have We Run out of Oil Yet? Oil Peaking Analysis from an Optimist's Perspective. Energy Policy, 34, 515-531.
7. Söderbergh, B., Robelius, F. and Aleklett, K. (2007) A Crash Programme Scenario for the Canadian Oil Sands Industry. Energy Policy, 35, 1931-1947.
8. Mitsunori, S., Yoshlml, S. and Chlrato, T. (1986) Effect of Catalyst Pore Structure on Hydrotreating of Heavy Oil. Industrial & Engineering Chemistry Fundamentals, 25, 330-337.
9. Nor, A.L., Nilofar, A., Mohd, A.R. and Ali, A.B (2008) Investigation of Nanoscale Tungsten (VI) Oxide as a Catalyst for the Ozonolysis of Oleic Acid. European Journal of Inorganic Chemistry, 24, 463-467.
10. Díaz-Garcia, L., Cortez, M.T., et al. (2006) Influence of Alumina Crystal Size on the Hydrotreating Activity of Supported NiMo Catalysts Using Real Feedstock. Petroleum Science and Technology, 24, 485-506.
11. Supothina, S., Seeharaj, P., Yoriya, S. and Sriyudthsak, M. (2007) Synthesis of Tungsten Oxide Nanoparticles by Acid Precipitation Method. Ceramics International, 33, 931-936.http://dx.doi.org/10.1016/j.ceramint.2006.02.007
12. Benitez, V.M., Yori, J.C., Grau, J.M., Pieck, C.L. and Vera, C.R. (2006) Hydroisomerization and Cracking of n-Octane and n-Hexadecane over Zirconia Catalysts. Energy & Fuels, 20, 422-426. http://dx.doi.org/10.1021/ef050092j
13. Bianchini, C., Jiménez, M.V., Meli, A., Moneti, S., Patinec, V. and Vizza, F. (1997) Mimicking the HDS Activity of Promoted Tungsten Catalysts. A Homogeneous Modeling Study Using a Two-Component Tungsten/Rhodium System. Organometallics, 16, 5696-5705. http://dx.doi.org/10.1021/om970711e
14. Alhulmaidan, F., Haitham, M.S. and Al-Adwani, H. (2010) Hydrocracking of Atmospheric Residue Feedstock in Hydrotreating Processes. Kuwait Journal of Science, 37, 129-159.
15. Welz, B. and Sperling, M. (2007) Frontmatter. In: Atomic Absorption Spectrometry, 3rd Edition, Wiley-VCH Verlag GmbH, Weinheim.
16. Hsu, C.H., Chang, C.C., Tseng, C.M., Chan, C.C., Chao, W.H., Wu, Y.R., Wen, M.H., Hsieh, Y.T., Wang, Y.C., Chen, C.L., Wang, M.J. and Wu, M.K. (2013) An Ultra-Fast Response Gasochromic Device for Hydrogen Gas Detection. Sensors and Actuators B: Chemical, 186, 193-198. http://dx.doi.org/10.1016/j.snb.2013.06.004
17. Valigi, M., Gazzoli, D., Pettiti, I., Mattei, G., Colonna, S., De Rossi, S. and Ferraris, G. (2002) WO_x/ZrO_2 Catalysts ☆: Part 1. Preparation, Bulk and Surface Characterization. Applied Catalysis A: General, 231, 159-172. http://dx.doi.org/10.1016/S0926-860X(02)00050-9
18. Baker, J.E., Burch, R. and Golunski, S.E. (1989) Thermal analysis of catalyst surfaces. Acta, 329-342.
19. Haines, P.J. (2002) Principles of Thermal Analysis and Calorimetry. Royal Society of Chemistry, London, 220 Pages.
20. China, H., Campbell, S. and Kesler, O. (2007) Thermal and Electrochemical Stability of Tungsten Carbide Catalyst Supports. Journal of Power Sources, 164, 431-440.
21. Brostow, W., Menard, K.P. and Menard, N. (2009) Combustion Properties of Several Species of Wood. Chemistry & Chemical Technology, 3, 173-176.
22. Busca, G. (1998) Spectroscopic Characterization of the Acid Properties of Metal Oxide Catalysts. Catalysis Today, 41, 191-206. http://dx.doi.org/10.1016/S0920-5861(98)00049-2
23. Kim, T.Y., Sung, D., Choi, Y., Baek, J., Park, J.R. and Yi, J. (2012) Preparation and Characterization of Mesoporous Zr-WO_x/SiO_2 Catalysts for the Esterification of 1-Butanol with Acetic Acid. Journal of Materials Chemistry, 22, 10021-10028.http://dx.doi.org/10.1039/c2jm30904a
24. Lebarbier, V. and Clet, G. (2006) A Comparative Study of the Surface Structure, Acidity, and Catalytic Performance of Tungstated Zirconia Prepared from Crystalline Zirconia or Amorphous Zirconium Oxyhydroxide. Journal of Physical Chemistry B, 110, 13905-13911. http://dx.doi.org/10.1021/jp0571224
25. Wang, H., Wu, Y., He, L. and Liu, Z. (2012) Supporting Tungsten Oxide on Zirconia by Hydrothermal and

Impregnation Methods and Its Use as a Catalyst to Reduce the Viscosity of Heavy Crude Oil. Energy & Fuels, 26, 6518-6527. [Paper reference:2]

26. Joonaki, E., Ghanaatian, Sh. and Zargar, Gh. (2012) A New Approach to Simultaneously Enhancing Heavy Oil Recovery and Hindering Asphaltene Precipitation. Iranian Journal of Oil & Gas Science and Technology, 1, 37-42.

27. Rana, M., Samano, V., Ancheyta, J. and Diaz, J.A. (2007) A Review of Recent Advances on Process Technologies for Upgrading of Heavy Oils and Residua. Fuel, 86, 1216-1231.http://dx.doi.org/10.1016/j.fuel.2006.08.004

CHAPTER 7

NI-BASED CATALYSTS FOR LOW TEMPERATURE METHANE STEAM REFORMING

Hongjing Wu, Valeria La Parola, Giuseppe Pantaleo, Fabrizio Puleo, Anna M. Venezia and Leonarda F. Liotta

Keywords: Hydrocarbons steam reforming; nickel; gold; bimetallic; surface alloy; support modification

INTRODUCTION

Production of synthesis gas from natural gas and Fisher-Tropsch synthesis from the synthesis gas have become very important in the chemical industry for reasons related to the soaring petroleum price, depletion of oil reserves, and environmental problems with exhaust gases [1,2,3]. Steam reforming of methane and other hydrocarbons has been an extremely important process for the production of synthesis gas [4,5,6]. Steam reforming is the conventional method to produce synthesis gas from natural gas and is utilized on the industrial scale, which is highly endothermic reaction as indicated below (Equation (1)).

$$CH_4 + H_2O \rightarrow CO + 3H_2 \qquad \Delta H^0_{298\,K} = +206\ kJ{\cdot}mol^{-1} \qquad (1)$$

In industry, hydrogen is practically produced from natural gas and mainly used for the ammonia synthesis [7,8,9]. Recently, much attention has been paid to the hydrogen production relating to the fuel cell technology [10,11,12]. Moreover, many efforts have been made in developing dry (CO_2) reforming of methane as a practical way to utilize CO_2 as a building block in organic syntheses (Equation (2)) [13,14].

$$CH_4 + CO_2 \rightarrow 2CO + 2H_2 \qquad \Delta H^0_{298\,K} = +286\ kJ{\cdot}mol^{-1} \qquad (2)$$

It has been pointed out that one problem in methane reforming is carbon deposition on the catalyst surface, which causes the catalyst deactivation. There are two principal carbon deposition pathways for steam reforming of methane: the methane decomposition (Equation (3)) and the CO disproportion (Equation (4)).

$$CH_4 \rightarrow C + 2H_2 \qquad \Delta H^0_{298K} = +75\ kJ{\cdot}mol^{-1} \qquad (3)$$

$$2CO \rightarrow C + CO_2 \qquad \Delta H^0_{298K} = -172\ kJ{\cdot}mol^{-1} \qquad (4)$$

$$CH_4 + 2O_2 \rightarrow CO_2 + 2H_2O \qquad \Delta H^0_{298K} = -806\ kJ{\cdot}mol^{-1} \qquad (5)$$

The tendency to carbon deposition may depend on the atomic ratio O/C and H/C in the feed gas. Lower H_2O/CH_4 and H_2/CO ratios correspond to higher tendency toward coke formation [15]. This means that the tendency of carbon deposition in the dry reforming of methane is much higher than that in the steam reforming of methane. Coke formation is relatively severe over the catalysts used for CO_2 reforming at high temperatures (> 1073 K).

For steam and dry reforming of methane, the catalyst bed has to be heated from the outside of the reactor by the combustion of methane (Equation (5)). Therefore, the reaction rate is limited by the heat transfer and the scale merit of the syngas production by this method is low. Other combined technologies such as autothermal reforming, oxidative reforming, and so on, are used to realize the large-scale synthesis gas production. For example, oxidative steam reforming of methane, in which O_2 is introduced to the catalyst bed together with methane and steam, is a promising method for syngas production because the reaction conditions can be adjusted by balancing exothermic combustion and endothermic reforming to an autothermal system.

Citation: Wu, Valeria La Parola, Giuseppe Pantaleo, Fabrizio Puleo, Anna M. Venezia and Leonarda F. Liotta, Ni-Based Catalysts for Low Temperature Methane Steam Reforming, doi:10.3390/catal3020563

Reforming of methane with steam and CO_2 is catalyzed by various transition and noble metals such as Ni, Co, Ru, Rh, Pd, Pt, etc. The catalytic activity of metal catalysts supported on Al_2O_3-MgO was reported to follow the order: Ru > Rh > Ir > Ni > Pt [16]. Ni has a comparable activity to noble-metal catalysts, and that higher loading amount of Ni is feasible in terms of the catalyst cost in order to increase the activity per catalyst volume. Although Ni is inferior to noble metals regarding the resistance to carbon deposition and catalyst oxidation, the high resistance to coke formation and catalyst oxidation can be realized by the addition of very small amounts of noble metals. The modification of Ni catalysts with small amounts of noble metals such as Pd, Pt, Ru, Rh and Ir has been well reviewed [1,2,3] and it has been demonstrated as a promising approach to design catalysts with excellent performances for methane reforming.

The present review article is related to the effects of modification of Ni catalysts by other metals (such as Au, Ag, Sn, Cu, Co, Mo, Fe, Gd and B), with particular attention being devoted to bimetallic Ni-Au catalysts. It is well known that the preparation method strongly influences the structure of bimetallic systems. The first part of this work is, therefore, devoted to the structural differences of catalysts prepared by different methods. In the second part, the resistance to coke formation in the steam reforming of methane, as well as the relationship between the catalytic performance and the structure of bimetallic catalysts, are discussed.

PREPARATION AND STRUCTURE OF NI CATALYSTS MODIFIED WITH A SMALL AMOUNT OF SECOND METALS

Table 1 lists the chemical composition of the main Ni-M catalysts (M = Au, Ag, Sn, Cu, Co, Mo, Fe, Gd and B) for hydrocarbon reforming [17,18,19,20,21,22,23,24,25,26,27,28,29,30,31,32,33,34,35,36,37,38,39,40,41,42,43,44,45,46,47,48,49,50,51,52,53,54,55,56,57] recently reported in the literature. Some typical examples are discussed in detail in the present review.

Table 1: List of Ni catalysts modified with a second metal reported in this review: chemical composition and catalytic reactions.

Catalysts	Second metal	Ni content (wt%)	Molar ratio (M/Ni)	Reactions	Second metal effects	Refs.
Au/Ni/$MgAl_2O_4$	Au	16.5	0.0054	SR of n-butane	C	[17,18,19]
Au/Ni/$MgAl_2O_4$	Au	8.8	0.0034–0.034	SR of methane	C	[20]
Au/Ni/$MgAl_2O_4$	Au	8	0.019	SR of methane	C,A	[21]
Au-Ni/Al_2O_3	Au	8	0.038	SR of methane	C,A	[22,23,24]
Au-Ni/Al_2O_3	Au	8	0.075	PO of methane	C,A	[25]
Au-Ni/$CrAl_3O_6$	Au	5	0.0060–0.12	PO of methane	C,A	[26,27]
Au-Ni/MgO-Al_2O_3	Au	5	0.12	PO of methane	C	[28]
AuNiAl LDHs(Cp)	Au	79.5	0.011	SR of methanol	-	[29]
Au/Ni/GDC(Dp)	Au	65	0.01–0.04	SR of methane	C	[30]
AuNi/YSZ(Cs)	Au	50	0.01–0.05	SR of methane	C	[31]
Ag-Ni/Al_2O_3	Ag	8	0.038	SR of methane	C	[22,23,24]
Ag-Ni/Al_2O_3	Ag	15	0.0036–0.022	SR of methane	C	[32]
Ag-Ni/Al_2O_3	Ag	15	0.011	SR of methane	C	[33]
AgNi/YSZ(Cs)	Ag	50	0.01–0.05	SR of methane	C	[34]
Ag/Ni/$MgAl_2O_4$	Ag	7.93	0.5	SR of butane	C,S	[35]
Ag/Ni alloy	Ag	-	-	SR	C	[36]
Sn-Ni/Al_2O_3	Sn	7.0–71.8	0.014–0.17	SR	C	[37]
Sn/Ni/Al_2O_3	Sn	35	0.0049	SR of methane	C	[38]
SnNi/YSZ(Cp)	Sn	15	0.0049	SR	C	[39,40,41]
Sn-Ni/MgO-Al_2O_3	Sn	11.25	0.16	SR of methanol	C	[42]
SnNi3 alloy	Sn	-	0.33	SR	C	[43]
Cu-Ni/Al_2O_3	Cu	8.72	0.0092	SR of methane	A,S	[44]
Cu-Ni/Al_2O_3	Cu	3–7	0.40–2.16	SR of methanol	S,C	[45]
Cu-Ni/Al_2O_3	Cu	5	0.92	SR of methane	C	[46]
Cu-Ni/SiO_2	Cu	1.25–3.75	0.33–3	SR of ethanol	A,C,S	[47]
Cu-Ni/Al_2O_3	Cu	-	1	SR of methane	C	[48]
Cu-Ni-Mg-Al(Cp)	Cu	-	-	DR of methane	C	[49]
Cu/Ni(111) alloy	Cu	-	-	SR of methane	C	[50]
Co-Ni-Mg-Al(Cp)	Co	-	1.36	DR of methane	A,C	[49]
Co-Ni/YSZ	Co	25,34	0.44,0.92	SR of ethanol	S	[51]
Co-Ni/ZrO_2	Co	13,8.7	1.00,0.500	DR of methane	A,C,S	[52]
MoO_3/Ni/Al_2O_3	Mo	-	-	SR of n-butane	C	[53]
MoO_3/Ni/Al_2O_3	Mo	17	0.00048–0.048	SR of n-butane	A,C	[54]
Mo/Ni/Al_2O_3	Mo	60	0.00050–0.0020	SR of methane	A	[55]
Fe-Ni(Cp)	Fe	-	0.33–1	SR of ethanol	A,C	[56]
GdNi/Al_2O_3(Cs)	Gd	7	0.12	PO,SR,DR	C	[57]
B/Ni/Al_2O_3	B	15	0.18,0.36	SR of methane	C	[58]

M/Ni, sequential impregnation; M-Ni, co-impregnation; Cp, co-precipitation; Dp, deposition-precipitation; Cs, combustion synthesis; SR, steam reforming; PO, partial oxidation; DR, dry reforming; ATR, autothermal reforming; A, activity; C, coke resistant; S, selectivity.

Ni Metal Particles Modified with Au

Nørskov and co-workers [17,18,19] have reported that Au/Ni surface alloy on the Ni particles (on SiO_2 and $MgAl_2O_4$) was active for steam reforming and more resistant towards carbon formation than the pure Ni catalysts. The Ni/SiO_2 and Ni/$MgAl_2O_4$ with high surface area were prepared by precipitation-deposition and pore filling of the spinel support with Ni-nitrate, respectively. The bimetallic Au/Ni catalysts were prepared by carefully impregnating the reduced monometallic Ni catalysts with an aqueous solution containing $Au(NH_3)_4(NO_3)_3$. After filtering, the products were dried in air at 383 K. They described in details how Monte Carlo simulations and various physicochemical characterization tools such as transmission electron microscopy (TEM), in situ extended X-ray absorption fine structure measurements (EXAFS), and X-ray powder diffraction (XRD) were used to get information on the existence of the Au/Ni surface alloy in the small particle system encountered in the supported catalysts [18]. By fitting the back-transformed Au L_3 spectrum of silica supported Au/Ni catalyst, the nonzero value of the Au-Ni coordination number was observed and gave evidence for the presence of Ni in the nearest neighbor environment of Au. However, the Au-Au coordination number was larger than what could be expected for an Au surface alloy. Therefore, they took into account that XAFS probed all Au present in the catalyst; also Au presented as large Au particle as evidenced by TEM and XRD results. Linear combinations of calculated Au L_3 XAFS functions of bulk Au and Au/Ni surface alloys were compared to the experimental data by varying the amount of Au on Ni in the particles as well as the relative amount of bulk Au. The best fit results in 73% of the Au presented as bulk Au and the remainder of the Au presented as a surface alloy (molar ratio Au/Ni = 0.10) on a Ni particle with a diameter of 1.8 nm. No significant differences were observed between the EXAFS spectra and the Fourier transformed data of the freshly reduced catalyst and the catalyst after steam reforming in diluted n-butane.

King and co-workers [20] have reported that the addition of small quantities of gold to the surface of supported nickel catalysts has been proved to retard carbon formation during hydrocarbon steam reforming. After Ni/$MgAl_2O_4$being impregnated with Au, the sample was dried in ambient air, heated in flowing He at a rate of 1 K/min to 473 K, than held isothermally at 473 K for at least 8 h. Table 2 shows the results of fitting the back-transformed Au L_3-edge data with both the Au-Ni and Au-Au first coordination shells. Increasing Au loadings resulted in increased Au-Au interaction and a parallel decrease in Au-Ni interaction. For the 0.1 wt% Au/8.8 wt% Ni sample, an average coordination number was 5.3, with a negligible coordination number of Au-Au interaction. This means that at the lowest Au loading (0.1 wt%), only Au-Ni bonds are detected. Furthermore, the XANES analysis of the Au-Ni/$MgAl_2O_4$ catalysts revealed changes in both the Ni K-edge and Au L_3-edge white line intensities, indicative of transfer of electronic charge from Au to Ni as a result of surface alloy formation.

Table 2: Structural parameters derived from fitting Au L_3-edge EXAFS data for Au-8.8% Ni/$MgAl_2O_4$ after reduction at 500 °C [20].

u concentration	Au L_3-edge fitting Au-Ni				Au L_3-edge fitting Au-Au				R fac
(%)	C.N.	R (Å)	$\Delta\sigma^2 \times 10^3$	ΔE_0	C.N.	R (Å)	$\Delta\sigma^2 \times 10^3$	ΔE_0	(%
0.1	5.3 ± 1.1	2.58 ± 0.01	6.5 ± 1.6	0.2	-	-	-	-	4.7
0.2	4.9 ± 1.1	2.59 ± 0.02	7.4 ± 1.6	4.2	4.0 ± 1.3	2.97 ± 0.04	8.3 ± 4.3	5.4	4.9
0.4	3.7 ± 0.7	2.58 ± 0.01	5.5 ± 0.1	2.8	6.0 ± 1.5	2.82 ± 0.01	7.7 ± 1.8	4.9	2.1
1.0	2.8 ± 0.3	2.57 ± 0.01	4.3 ± 0.7	4.1	7.6 ± 1.0	2.84 ± 0.01	6.5 ± 0.6	6.3	4.7

Besides, the pretreatment also will influence the Au-Ni surface alloy formation. Table 3 shows a significant Au-Ni nearest-neighbor interaction when the Ni component is in the reduced state, but there appears to be no interaction between Au and oxidized Ni atoms.

Recently, Lazar et al. [22,23,24] have synthesized the 1.0 wt% Au-7.0 wt% Ni/γ-Al_2O_3 catalyst by co-impregnation method and found that the addition of Au improved the methane conversion, CO_2 selectivity and H_2 production at low temperatures (T < 600 °C). The impregnated catalysts were dried at room temperature overnight, calcined in Ar at 823 K and then, reduced in H_2 at 823 K. Taking all characterization results (XRD, H_2 chemisorption and TEM) into account, they concluded that in the Ni-Au/γ-Al_2O_3 catalysts prepared by co-impregnation and reduced at 823 K, no significant interaction of Au with Ni has been proved to appear and no formation of a superficial Au-Ni surface alloy was observed [24]. Maniecki et al. [25] have also prepared 2.0 wt% Au-10.0 wt% Ni/Al_2O_3 catalysts by co-impregnation method and found that the addition of gold improved catalyst stability and activity in POM reaction. After drying they were calcined for 4 h in air at 673 K. The temperature programmed in situ XRD patterns for NiO-Au mechanical mixture in (5% H_2 + 95% Ar) gas mixture is presented in Figure 1a. It can be observed that the reduction of free NiO takes place above 473 K and at the temperature of 563 K only metallic Ni and Au phases are detectable. Further increase of temperature up to 1153K leads to gradual shifts of 2θ values that characterize the Au-Ni alloy formation. Finally, at 1153 K only XRD lines assigned to Au-Ni alloy phase are visible. Figure 1b shows that Au-Ni alloy formation occurs in the mixture of 80 wt% NiO-20 wt% Au after partial oxidation of methane at 1153 K, after reduction in (5% H_2 + 95% Ar) stream, at 1153 K, and for 8 wt% Ni-2 wt% Au/Al_2O_3 catalyst after partial methane oxidation (POM) reaction at 1153 K. One can see that the Au-Ni alloy is formed in the reductive gas atmosphere and after reaction. The formation of Au-Ni alloy during high temperature POM reaction is experimentally proved on 8 wt% Ni-2 wt% Au/Al_2O_3 catalyst surface. The mechanism of the formation of the bimetallic particles mentioned above is described in Figure 2 on the basis of the above in situ XRD characterization results.

Table 3: Structural parameters derived from fitting Au L_3-edge EXAFS data for Au/$MgAl_2O_4$ following either oxidation or reduction, and for Au/Ni/$MgAl_2O_4$ after oxidation, all at 500 °C [20].

Catalysts	Pretreatment	*C.N.*	*R* (Å)	$\Delta\sigma^2 \times 10^3$	ΔE_0	*R* factor (%)
0.2% Au/MgO-Al_2O_3	Reduced at 500 °C	11.3 ± 0.6	2.86 ± 0.002	2.4 ± 0.3	5.62	0.4
0.2% Au/MgO-Al_2O_3	Oxidized at 500 °C	11.0 ± 0.6	2.86 ± 0.002	1.9 ± 0.3	5.62	0.4
0.2% Au-8.8% Ni/MgO-Al_2O_3	Oxidized at 500 °C	11.9 ± 0.9	2.87 ± 0.003	2.0 ± 0.4	4.89	1.0

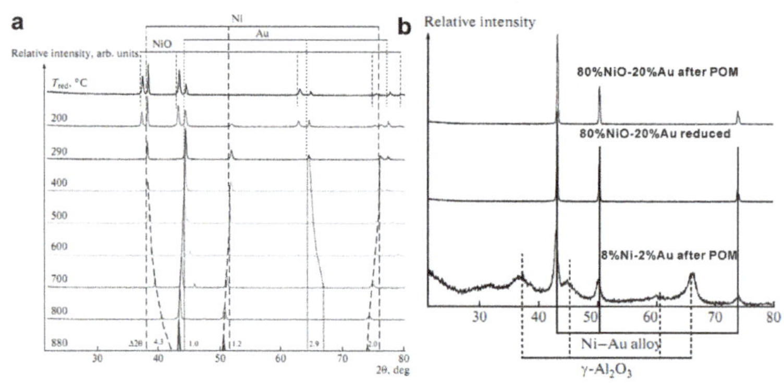

Figure 1: XRD patterns for (**a**) nickel oxide (80%)-gold (20%) mechanical mixture reduced in situ in (5% H_2 + 95% Ar) gas mixture at different temperatures, and (**b**) (80% NiO-20% Au) mixture after partial oxidation of methane at 1153 K, after reduction in (5% H_2 + 95% Ar) at 1153 K and for 8%Ni-2% Au/Al_2O_3 catalyst after reaction at 1153 K [25].

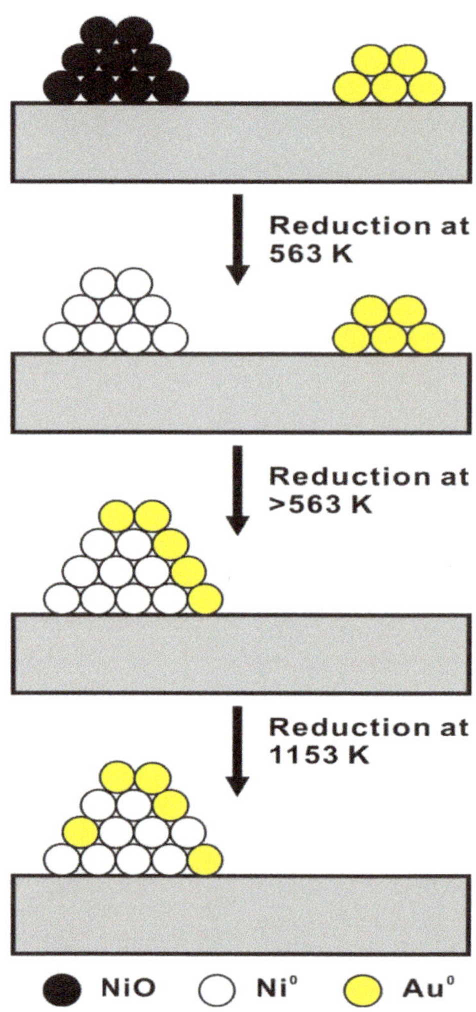

Figure 2: Model of the formation mechanism of Ni-Au bimetallic particles during the reduction pretreatment Al_2O_3 supported catalyst.

Ni Metal Particles Modified with Ag

Bueno and co-workers have prepared Ni/γ-Al_2O_3 modified with Ag by co-impregnation method [32,33]. The impregnated samples were dried at 353 K overnight and calcined under flowing air at 723 K for 2 h. Finally, the samples were reduced in H_2 flow, heated at 10 K/min and held at 1073 K for 2 h. Figure 3 shows the in situ Ni K-edge XANES-TPR-H_2 spectra of the oxidized 5 wt% Ni/γ-Al_2O_3 and that doped by 0.6 wt% Ag at different temperatures. The results indicate that two different species of Ni (II) are reduced in AgNi/γ-Al_2O_3. There is no obviously shift in the threshold for reduced catalysts with higher Ag loading, with low contribution of Ag-Ni bonds relative to that of bulk Ni-Ni bonds. However, the presence of Ag changes the surface structure of the Ni sites that may be related to the nucleation and growth of graphitic structure. The catalysts with Ag loading >0.3 wt% submitted to stability tests showed high resistance to coke deposition.

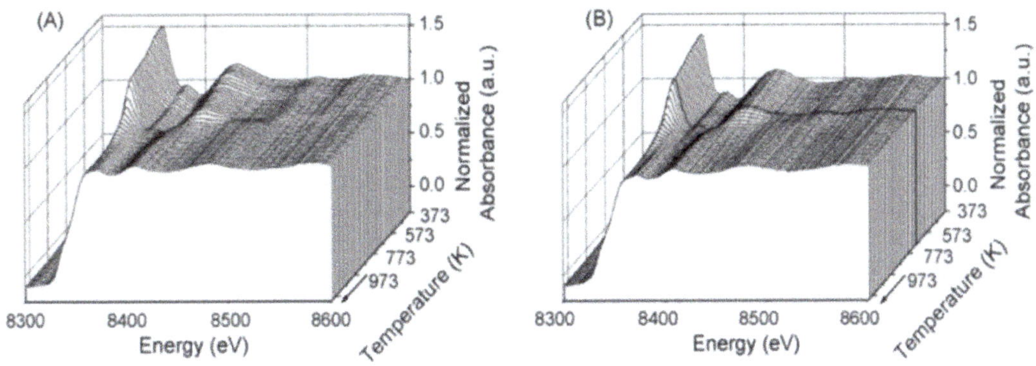

Figure 3: In situ XANES spectra of 0.6Ag5NiAl (**A**) and 5NiAl (**B**) samples under reduction conditions from 323 to 973 K [32].

Recently, Kang et al. [35] have prepared 15 mol% Ag/15 mol% Ni/$MgAl_2O_4$ and 15 mol% Ni/15 mol% Ag/ $MgAl_2O_4$catalysts by the sequential wetness impregnation method. After being impregnated with Ag or Ni, the sample was dried in ambient air, heated in flowing air at a rate of 10 K/min to 773 K, and then held isothermally at the final temperature for at least 1 h. The reduction of the catalysts were performed in H_2 at 973 K for 2 h, followed by cooling to room temperature under Ar. Figure 4 shows the catalytic phase change and n-butane reforming mechanism. The catalytic performances differ according to the order in which the metal precursors are added. Ag is added between Ni and Al in an attempt to decrease the catalytic deactivation induced by the strong sintering between them during n-butane steam reforming, while simultaneously improving the catalytic activity. Consequently, their synergistic effect is different from that between Au and Ni nanoparticles as mentioned above. It gives a new interpretation about the effect of Ag in the bimetallic catalysts.

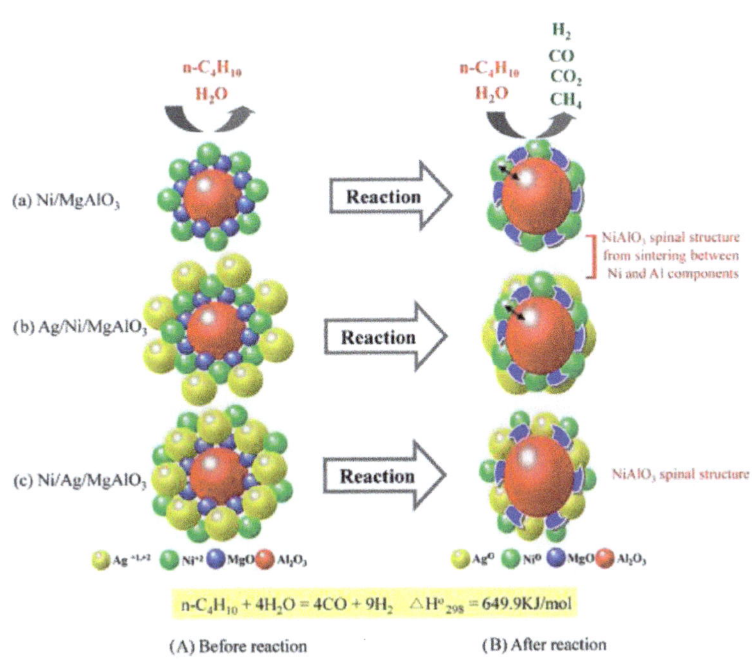

Figure 4: Expected phase transformation in the butane reforming before and after butane reforming reaction [35].

More recently, a density functional theory study of the effect Ag on the control of Ni-catalyzed carbon formation has been reported [36]. The calculated results indicated that the activation energy for methane dissociation was increased with the Ag coverage on Ni(211) surface than that on pure Ni(111), the active center was transferred from the stepped surface to the close-packed surface. The middle-step sites acted as the nucleation center for the growth of filamentous carbon and therefore had the potential to prevent catalyst nanoparticles from being destroyed. These findings provided a rational interpretation of the experimental observations that Ag/Ni surface alloy exhibited lower catalytic activity towards steam reforming of methane but high resistance to coke formation.

Ni Metal Particles Modified with Other Metals (Sn, Cu, Co, Mo, Fe, Gd and B)

Linic and co-workers have extensively investigated Ni and Sn-Ni alloy catalysts for hydrocarbons (methane, propane and isooctane) steam reforming [38,39,40,41]. By employing the Density Functional Theory (DFT), calculations on Ni and Sn/Ni catalysts supported over 8 mol% yttria-stabilized zirconia (YSZ) showed they identified Sn-Ni alloy as a potential carbon-tolerant reforming catalyst. Their DFT calculations showed that the formation energy of Sn/Ni surface alloy was lower than the formation energy of Sn bulk alloys or than the formation energy corresponding to pure Sn and Ni phases, therefore, suggesting that the formation of Sn/Ni surface alloy was thermodynamic favorable. It was deduced that Sn could displace Ni atoms from the step-edge sites, which effectively repelled C atoms from the low-coordinated step sites. This result indicated that Sn atoms lower the tendency of the surface alloy to form coke as confirmed in Figure 5b,c. Figure 5d shows the post-reaction STEM micrograph of Ni and Sn/Ni. Ni particles are completely covered with carbon deposits while Sn/Ni is carbon-free. Furthermore, elemental mapping of Sn/Ni particles via line scan (STEM/EDS) suggests surface alloying. Lower panel (Figure 5d) shows that Sn/Ni ratio diminishes as the probe moves from the particle bottom edge upward, indicating the Sn surface enrichment.

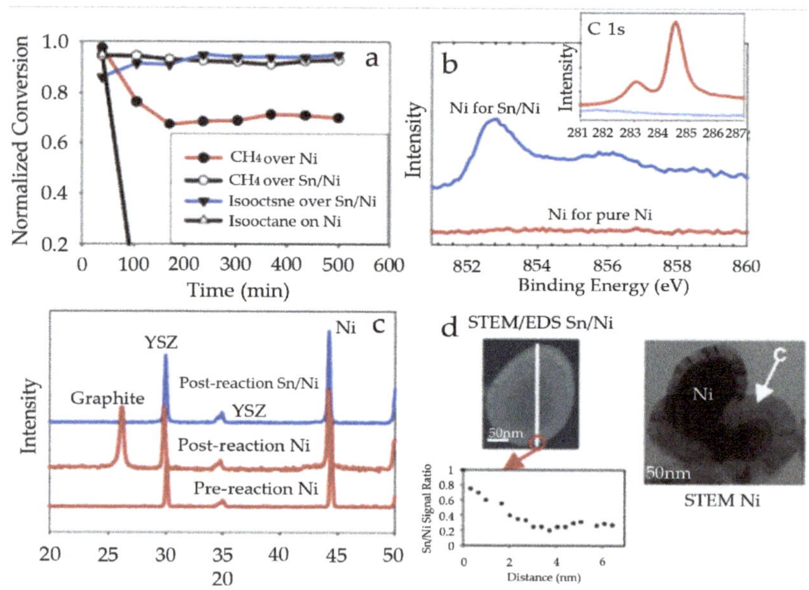

Figure 5: (**a**) normalized conversion; (**b**) Ni 2p XPS spectra after isooctane reforming on Ni (red) and on Sn/Ni (blue); (**c**) XRD spectra for post-isooctane Ni and Sn/Ni, and for fresh Ni catalysts, and (**d**) post-reaction STEM micrograph of Ni and Sn/Ni [38].

It has been reported that the introduction of Cu in the catalyst formulation suppresses coke deposition and the sintering of the active phase, in steam/dry reforming of hydrocarbons [44,45,46,47,48,49,50]. Fornasiero et al. [45] have comprehensively studied the methanol and ethanol steam reforming activity of a series of Ni_xCu_y/Al_2O_3 catalysts prepared by wet-impregnation method. Figure 6a shows a progressive shift of the metallic phase peaks. The XRD pattern shows diffraction peaks in an intermediate position, indicating the formation of a Ni-Cu surface alloy. The composition and the number of alloys depend on the Ni:Cu ratio. Figure 6b shows the comparison between the experimental cell parameter a of the metallic phases and the theoretical value calculated by Vegard's law for intermediate composition. The cell parameter of Ni_3Cu_7 and Ni_5Cu_5 is appreciably different with respect to the theoretical values. When the Ni:Cu ratio decreases the lattice parameter increases. There may be two metal phases: a Cu rich alloy and a Ni-rich alloy, existed in the Ni_3Cu_7. Whereas, only one cubic structure alloy is detectable for the metal phase in the Ni_5Cu_5. The discrepancy between the observed and theoretical cell parameters suggests that, also in this case, two metal phase alloys could be present.

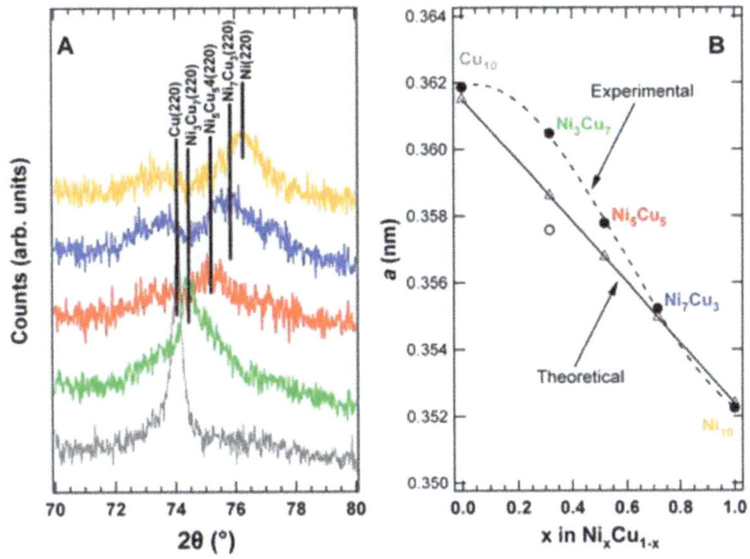

Figure 6: (**A**) Detail of the range 2θ = 70-80° of the XRD patterns of the reduced Cu, Ni and Ni_xCu_{1-x} samples, shows the shift of the (220) reflection with the composition, and (**B**) trend of the experimental cell parameter a with the composition of the metal phase, in comparison with the theoretical values [45].

The effect of the addition of Co on Ni-based systems was also studied [51,52]. Ni-Co catalysts with different composition $Ni_{(1-x)}Co_x$ (x = 0, 0.32, 0.5) were prepared on yttria-stabilised zirconia (YSZ) as catalysts for ethanol steam reforming. TEM micrographs of the catalysts, 50 wt%Ni/YSZ and Ni(25 wt%)-Co(25 wt%)/YSZ (labeled as $Ni_{25}Co_{25}$/YSZ) are shown inFigure 7a–c, respectively. The materials are composed of a powder which is heterogeneous in size. As detected by EDS analysis, 50 wt%Ni/YSZ is composed by relatively large Ni clusters (whose crystal size is tens of nm) deposited over small particles of YSZ support, better shown in Figure 7b (crystal size = 5nm). The TEM micrograph of $Ni_{25}Co_{25}$/YSZ catalyst is shown in Figure 7c. Isolated crystals of Ni-Co alloy, with particle size of 50–100 nm, were detected along with smaller particles mainly constituted by the YSZ support. XRD analysis of the Ni-Co catalysts confirmed the alloy formation, showing a progressive increase of the lattice parameter from 3.524 Å (for pure Ni) up to 3.529 Å determined for $Ni_{1-x}Co_x$alloys in agreement with Vegard's law (Figure 7d). Moreover, Gonzalez-delaCruz et al. [52] reported that bimetallic Co-Ni catalysts showed a better activity and stability than the nickel monometallic system. They proposed the formation in the bimetallic systems of a more reducible nickel-cobalt alloy phase, which remains completely metallic in contact with the CO_2/CH_4 reactants at any temperature. The strong synergic effect between Ni and Co sites accounts for the better performance of the bimetallic catalysts.

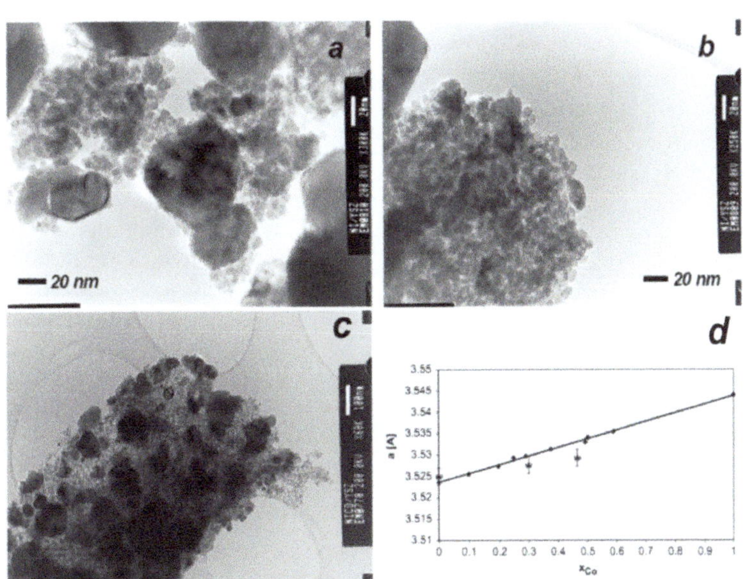

Figure 7: TEM micrographs of fresh catalysts: (**a** and **b**) 50 wt% Ni/YSZ sample; (**c**) Ni(25wt%)-Co(25wt%)/YSZ sample; (**d**) lattice parameter a (Å) vs. X_{Co}, Co loading, for Ni-Co alloys [51].

The promoting effect of small amounts of molybdenum oxide (0.1 wt%) on the resistance to coking of Ni/Al_2O_3 catalysts for methane and n-butane steam reforming were investigated by Borowiecki and Golebiowski [53,54]. The properties of Ni-Mo catalysts depend on the hydrogen-steam ratio in the reaction mixture which affects the mean degree of oxidation states of MoO_x. The addition of Mo to Ni/Al_2O_3 catalysts for methane steam reforming was recently investigated also by Maluf and Assaf [55], who found evidence of the transfer of electrons form MoO_x species to Ni, leading to an increase in the electron density of metallic Ni, and hence, the catalytic activity.

Interestingly, for the first time, a variety of Ni-Fe mixed oxides derived from Ni-Fe hydrotalcite has been reported by Abelló et al., for ethanol steam reforming [56]. They observed a positive role of iron in nickel-based catalysts in terms of enhanced catalytic activity, improved dispersion of nickel and lower carbon deposition.

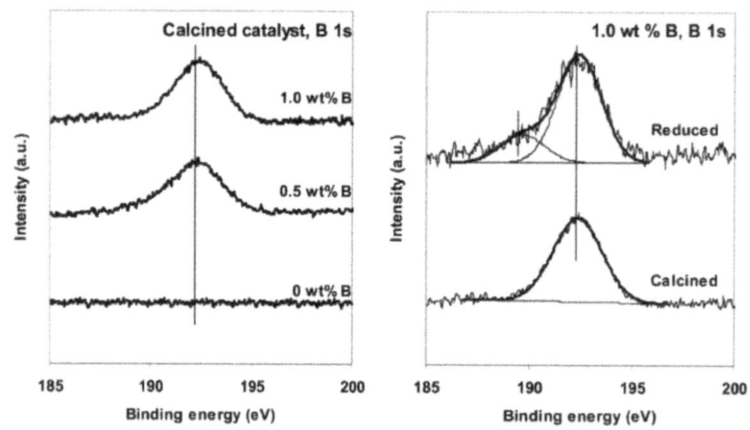

Figure 8: B 1s XPS spectra for calcined and reduced 15 wt% $Ni/\gamma-Al_2O_3$ catalysts with various B contents [58].

The effects of lanthanide promoters (La_2O_3, CeO_2, Pr_2O_3, Sm_2O_3 and Gd_2O_3) on a Ni/Al_2O_3 catalyst for methane partial oxidation, steam and CO_2 reforming were recently addressed by Shao and co-workers [57]. Raman spectroscopy indicated that the addition of lanthanide promoters can reduce the degree of graphitization of carbon deposited over nickel. Moreover, O_2-TPO results evidenced that among the investigated systems, $GdNi-Al_2O_3$ posses the best coke resistance.

Based on first principle calculations, Saeys and Borgna [58] proposed that a small amount of boron could possibly enhance the stability of Ni catalysts. Since B and C exhibit similar chemisorptions preferences on Ni catalysts, a small amount of B might selectively block the most stable binding sites. Blocking most active sites first by B, therefore, potentially reduces coke deposition. Supported Ni catalysts were prepared by aqueous slurry impregnation with a nickel nitrate solution to produce Ni content of 15 wt% on a commercial $\gamma-Al_2O_3$ support (BET area: 380 m^2/g). After drying at 353 K overnight and calcination at 673 K for 2 h, boric acid was sequentially introduced following the same preparation procedure to produce B content of 0.5 or 1.0 wt%. The XPS spectra (Figure 8) suggest that only B atoms interacting with Ni particles can be reduced, while a significant amount of boron oxide, probably interacting with the $\gamma-Al_2O_3$ support, cannot be reduced. Because $\gamma-Al_2O_3$ without Ni impregnated with 1 wt% B and subjected to the reduction procedure, no shift in the B 1s XPS peak could be found (not shown). The characterization studies indicate that boron might adsorb on both the Ni particles and the $\gamma-Al_2O_3$ support, and that 1 wt% B is sufficient to block the step and subsurface sites.

CATALYTIC ACTIVITY OF NI-AU CATALYSTS AND COMPARISON WITH OTHER NI-BASED BIMETALLIC SYSTEMS

Among the above mentioned bimetallic Ni-M systems, some examples have been selected for comments on catalytic activity.

Nørskov and co-workers have proved that Au-Ni surface alloy catalyst is a less reactive but more robust steam reforming catalyst than pure Ni catalyst by experimental tests and theoretical calculations by using CH_4 and n-butane as probe molecules [17,18,19]. The Au/Ni catalyst hardly deactivates, whereas the pure Ni catalyst rapidly forms carbon and deactivates. Thus, the Au/Ni catalyst exhibiting the surface alloy is more resistant towards carbon formation than the pure Ni catalyst. King and co-workers then have comprehensively studied the loss of activity as a result of Au promotion, as well as the effect of Au addition on methane steam reforming under typical conditions of industrial reformer operation [20]. They found that carbon formation was not totally suppressed, and with increasing temperature, the effect of Au addition decreased. Addition of 0.4 wt% Au to 8.8 wt% $Ni/MgAl_2O_4$ showed a small decrease in catalyst activity and a small increase in catalyst stability for methane steam reforming at 823 K. However, Guczi and co-workers [21] have recently reported that the addition of 0.5 wt% Au to $NiMgAl_2O_4$, although retards the methane dry reforming activity measured in temperature programmed reaction (Figure 9), the activity of the bimetallic Ni-Au system is improved during long time runs. In Figure 10the concentration and conversion of CO, CO_2 and CH_4 during overnight reaction at 948 K are displayed for $Ni/MgAl_2O_4$and $NiAu/MgAl_2O_4$ catalysts, respectively. Moreover, on the Au containing bimetallic catalyst the formation of carbon nanotubes was retarded or vanished, in the experimental conditions investigated.

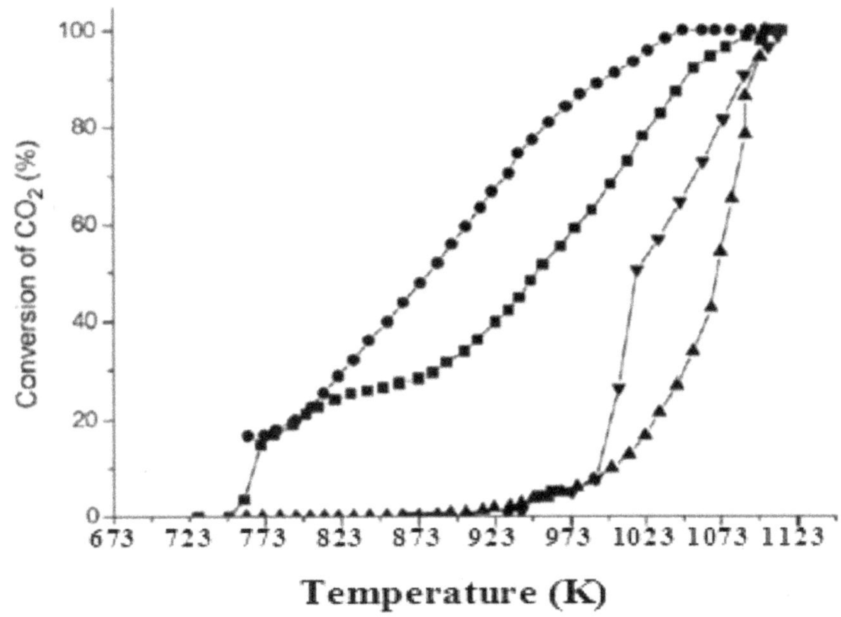

Figure 9: Comparison of $Ni/MgAl_2O_4$ and $NiAu/MgAl_2O_4$ catalysts in $CO_2 + CH_4$ in first reaction (● and ■, respectively) and after TPO (▼ and ▲, respectively) [21].

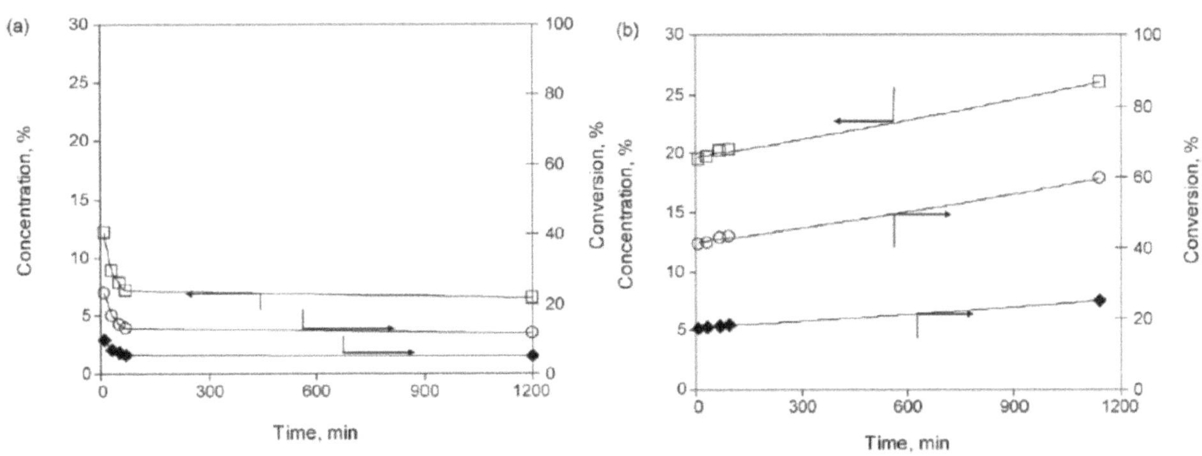

Figure 10: Concentration and conversion of CO (□), CO_2 (○) and (◆) in overnight reaction at 948 K: (**a**) on $Ni/MgAl_2O_4$; (**b**) on $NiAu/MgAl_2O_4$ [21].

Furthermore, Lazar et al. [22,23,24] have found that the addition of Au to the nickel catalyst improved the methane conversion, CO_2 selectivity and hydrogen production at low reaction temperatures (T < 873 K). At 973 K under their working conditions, the additives had no important effect in hydrogen production by methane steam reforming, and the catalysts deactivated. Therefore, the result is very consistent with the report by King et al [20]. On the other hand, Manieckiet al. have reported that the addition of Au to Ni/Al_2O_3 system improved catalyst stability and activity in partial oxidation of methane to synthesis gas [25,26,27,28]. Based on the results above mentioned, it can be found that there is discrepancy between results reported by different authors with respect to the effect of the addition of gold on catalytic activity of Ni-based catalysts. However, it seems to be confirmed that the addition of gold could improve the resistance of the Ni catalysts towards deactivation due to coke deposition.

Lazar et al. [22,23,24] have reported that the addition of Ag decreased the catalytic properties of Ni catalyst. The catalysts with Ag loading of 0.3 or 0.6 wt% submitted to stability tests showed high resistance to coke deposition reported by Buenoet al. [32]. They further found that at more oxidant conditions (autothermal reforming of methane, ATRM and POM) the redox process led to NiO formation and catalyst deactivation by $NiAl_2O_4$ formation [33]. At higher temperatures most probably due to Ag evaporation from Ag doped Ni/YSZ surface, severe anode catalytic and electro-catalytic activity degradation was observed by Neophytides et al. [34]. However, the addition of Ag could reduce the degree of carbon deposition and improve the H_2 product selectivity by eliminating the formation of C_2 hydrocarbons at temperature less than 1023 K in the butane steam reforming [35].

Catalytic steam reforming of methane and isooctane over pure Ni and 1% Sn:Ni catalysts are also shown in Figure 5a. The steam to carbon ratios are 0.5 and 1.5 for methane and isooctane, respectively. At 1073 K, the monometallic Ni catalyst rapidly deactivates. Unlike Ni, the Sn-Ni catalyst is active and stable for as long as it is kept on stream, approximately 12 h [38].

Nikolla et al. have shown that Sn/Ni alloy deposited over YSZ is much more carbon-tolerant than monometallic Ni in the steam reforming of methane, propane and isooctane at moderate steam to carbon ratios [39]. Figure 11 shows the normalized conversion of methane over the 1 wt% and 5 wt% Sn/Ni/YSZ and Ni/YSZ catalysts at a steam-to-carbon ratio of 0.5 and operating temperature of 1073 K. It is evident that Ni/YSZ lost about 45% of its activity after 2h of reaction, whereas the 1wt% Sn/Ni/YSZ was stable for the entire long-run time (15 h). By increasing the Sn loading from 1 to 5 wt%, methane conversion was reduced by ~ 25%, suggesting a negative effect on C-H bond activation.

The introduction of Cu in the Ni catalysts was also investigated for methanol and ethanol steam reforming [45]. The addition of Cu showed a positive effect inhibiting the formation of methane, an undesirable by-product. Although in the ethanol steam reforming the promoter effect of Cu was lesspromising, enhanced stability was observed in the two experiments, indicating a positive role was played by the Cu-Ni bimetallic catalysts.

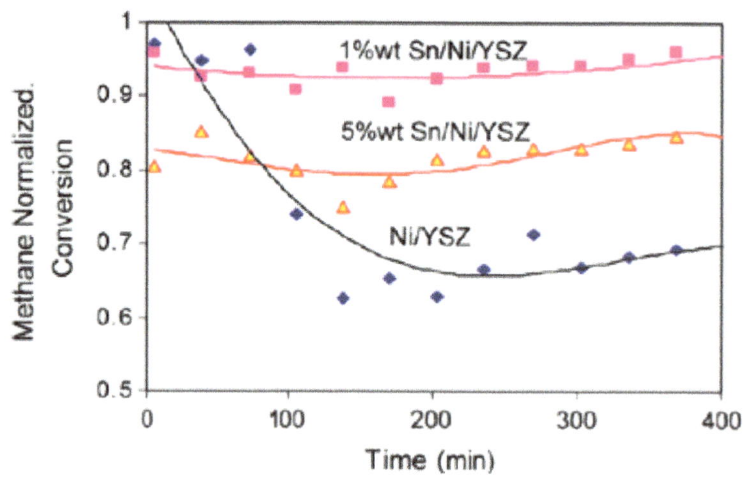

Figure 11: Normalized methane conversion (methane conversion divided by the highest obtained conversion) as a function of the time on stream for 1 wt% Sn/Ni/YSZ, 5 wt% Sn/Ni/YSZ and Ni/YSZ catalysts measured at the steam-to carbon ratio of 0.5 and 1073 K [39].

The effect of the addition of Co on the catalytic activity of Ni/YSZ catalyst for the ethanol steam reforming reaction was also studied by Resini et al. [51]. The addition of Co resulted in the inhibition of the dehydration reaction, as well as of methane production. Furthermore, Co also had an effect on the hydrogen production by increasing it, and thus, apparently favoring methane steam reforming.

Borowiecki et al. have reported that small amounts of MoO_3 (up to 0.1 wt%) caused an increase in activity of doped Ni catalysts [54]. By further increasing the amount of promoter (about 0.5 wt%), a drop occurs (Figure 12a). The catalytic properties of Ni-Mo catalysts depend also on the H_2 vol% concentration in the reaction stream. Figure 12b shows the dependence of the methane steam reforming rate on the hydrogen concentration at 873 K for Ni and different Ni-Mo catalysts. The activity of monometallic Ni gradually decreased by increasing hydrogen, while for Ni-Mo catalysts activity maxima were observed as a function of hydrogen content.

Shao and coworkers have, recently, compared the catalytic properties of lanthanide promoters Ni-Al_2O_3 catalysts, highlighting for GdNi-Al_2O_3 the best coke resistance and the good stability for methane partial oxidation at 1123 K for 300 h (Figure 13) [57].

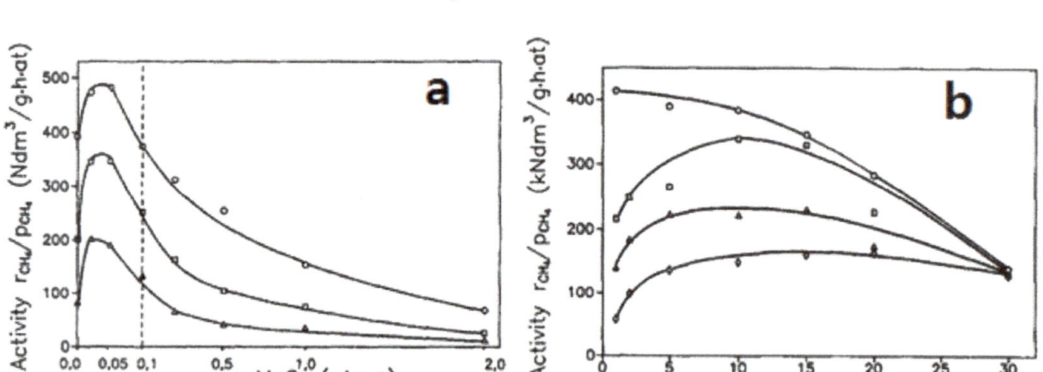

Figure 12: (**a**)Dependence of catalysts activity in steam reforming of methane at different temperatures as a function of the promoter amount (○ 873 K, □ 823 K, Δ 773 K); (**b**) Influence of hydrogen on the activity of catalysts in steam reforming of methane (○ Ni, □ Ni-Mo (0.5), Δ Ni-Mo (1.0), ◆ Ni-Mo (2.0)) [54].

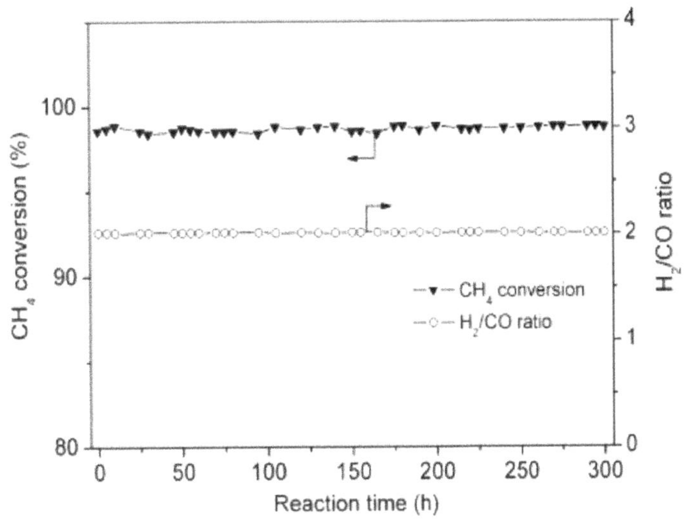

Figure 13: Time dependence of methane conversion and H_2 to CO ratio under CH_4:O_2= 2:1 conditions at 850 °C for GdNi-Al_2O_3 catalyst [57].

Interestingly, Borgna and Saeys [58] found that a small amount of boron could enhance the stability without compromising the catalytic activity. Catalytic tests showed that promotion with 1.0 wt% boron reduced the rate of deactivation by a factor of 3 and increased the initial methane conversion from 56% to 61%.

CONCLUSIONS

The addition of a small amount of second metals such as Au, Ag, Sn, Cu, Co, et al. to Ni catalysts significantly modifies the properties of metallic Ni particles by the formation of Ni-M bimetallic surface alloys. The structure of bimetallic surface alloys could be changed by choosing different preparation methods or pretreatments.

In particular, the addition of a small amount of Au to $MgAl_2O_4$-supported Ni catalysts by the sequential impregnation method leads to the formation of Au atoms, which segregated on the surface of the bimetallic particles when the Au loading is high, while well-mixed Au-Ni surface alloys are formed when the Au loading is low (0.1 wt%). The co-impregnation method results in different structures of Au-Ni bimetallic catalysts according to the different authors' reports. Some authors reported that Au-Ni surface alloys could be also formed on the Ni/Al_2O_3 catalysts prepared by co-impregnation method. While, some other authors found that no formation of a superficial Au-Ni surface alloy was observed.

Compared to monometallic Ni catalysts, Ni-M bimetallic catalysts exhibit superior catalytic performance in the methane steam reforming, i.e., high activity, high resistance to carbon formation and sintering of metal particles, and high selectivity to H_2 product. The improved catalytic performance of Ni catalysts modified with a small amount of second metals has been attributed to the synergic effect between a second metal and Ni by the formation of a superficial bimetallic alloy (such as Ni-Au, Ni-Ag, Ni-Sn, Ni-Cu, Ni-Co, etc.).

As a result, we expect that the following mentioned studies would lead to a further progress in the development of coke resistant bimetallic Ni-based catalysts for steam/dry reforming of methane:

The DFT studies of Ni containing bi/trimetallic alloys or Ni catalysts modified with a second inorganic element would lead to design and preparation of novel Ni-based catalysts;

The comprehensive understanding of the formation mechanisms of bimetallic surface alloys and their effects on catalytic performance and carbon formation would lead to design and preparation of new Ni-based catalysts with improved activity, stability and selectivity;

We feel that the development of novel Ni catalysts modified with a second inorganic element (such as Mo, Fe, Gd and B) should deserve further research studies for steam/dry reforming of methane;

The design of Ni catalysts with selected crystal planes of Ni and Ni-M alloyed particles could find applications in steam/dry reforming of methane.

ACKNOWLEDGMENTS

The authors acknowledge the financial support provided by EFOR-CNR Project (Energies by Renewable sources). The China Scholarship Council is sincerely acknowledged for supporting Wu's scholarship. Hongjing Wu thanks the Excellent Doctorate Foundation, Doctorate Foundation of Northwestern Polytechnical University, and the Scholarship Award for Excellent Doctoral Student granted by Ministry of Education.

REFERENCES

1. Tomishige, K. Oxidative steam reforming of methane over Ni catalysts modified with noble metals. J. Jpn. Pet. Inst.2007, 50, 287–298.
2. Liu, C.J.; Ye, J.Y.; Jiang, J.J.; Pan, Y.X. Progresses in the preparation of coke resistant Ni-based catalyst for steam and CO2 reforming of methane. Chemcatchem 2011, 3, 529–541.
3. Li, D.L.; Nakagawa, Y.; Tomishige, K. Methane reforming to synthesis gas over Ni catalysts modified with noble metals. Appl. Catal. A 2011, 408, 1–24.

4. Trimm, D.L. Coke formation and minimisation during steam reforming reactions. Catal. Today 1997, 37, 233–238.
5. Rafiqul, L.; Weber, C.; Lehmann, B.; Voss, A. Energy efficiency improvements in ammonia production-presperctives and uncertainties. Energy 2005, 30, 2487–2504.
6. Rostrup-Nielsen, J.; Trimm, D.L. Mechanisms of carbon formation on nickel-containing catalysts. J. Catal. 1977, 48, 155–165.
7. Pená, M.A.; Gómez, J.P.; Fierro, J.L.G. New catalytic routes for syngas and hydrogen production. Appl. Catal. A 1996,144, 7–57.
8. Armor, J.N. The multiple roles for catalysis in the production of H_2. Appl. Catal. A 1999, 176, 159–176.
9. Rostrup-Nielsen, J.R.; Rostrup-Nielsen, T. Large-scale hydrogen production. Cattech 2002, 6, 150–159.
10. Silva, L.C.; Murata, V.V.; Hori, C.E.; Assis, A.J. Hydrogen production from methane steam reforming: Parametric and gradient based optimization of a Pd-based membrane reactor. Optim. Eng. 2010, 11, 441–458.
11. Roh, H.S.; Jung, Y.; Koo, K.Y.; Jung, U.H.; Seo, Y.S.; Yoon, W.L. H_2 production over Ni/γ-Al_2O_3 catalyst prepared by a homogeneous precipitation method using urea for direct internal reforming (DIR) in a molten carbonate fuel cell (MCFC). Chem. Lett. 2009, 38, 1162–1163.
12. Rameshan, C.; Stadlmayr, W.; Weilach, C.; Penner, S.; Lorenz, H.; Havecker, M.; Blume, R.; Rocha, T.; Teschner, D.; Knop-Gericke, A.; et al. Subsurface-controlled CO_2 selectivity of PdZn near-surface alloys in H_2 generation by methanol steam reforming. Angew. Chem. Int. Ed. 2010, 49, 3224–3227.
13. Centi, G.; Perathoner, S. Opportunities and prospects in the chemical recycling of carbon dioxide to fuels. Catal. Today2009, 148, 191–205.
14. Schaefer, M.; Behrendt, F.; Hammer, T. Evaluation of strategies for the subsequent use of CO_2. Front. Chem. Eng. China2010, 4, 172–183.
15. Edwards, J.H.; Maitra, A.M. The chemistry of methane reforming with carbon dioxide and its current and potential applications. Fuel Process. Technol. 1995, 42, 269–289.
16. Rostrup-Nielsen, J.R.; Hansen, J.H.B. CO_2-reforming of methane over transition metals. J. Catal. 1993, 144, 38–49.
17. Besenbacher, F.; Chorkendorff, I.; Clausen, B.S.; Hammer, B.; Molenbroek, A.M.; Nørskov, J.K.; Stensgaard, I. Design of a surface alloy catalyst for steam reforming. Science 1998, 279, 1913–1915.
18. Molenbroek, A.M.; Nørskov, J.K.; Clausen, B.S. Structure and reactivity of Ni-Au nanoparticles catalysts. J. Phys. Chem. B 2001, 105, 5450–5458.
19. Holmblad, P.M.; Larsen, J.H.; Chorkendorff, I.; Nielsen, L.P.; Besenbacher, F.; Stensgaard, I.; Lægsgaard, E.; Kratzer, P.; Hammer, B.; Nørskov, J.K. Designing surface alloys with specific active sites. Catal. Lett. 1996, 40, 131–135.
20. Chin, Y.H.; King, D.L.; Roh, H.S.; Wang, Y.; Heald, S.M. Structure and reactivity investigations on supported bimetallic Au-Ni catalysts used for hydrocarbon steam reforming. J. Catal. 2006, 244, 153–162.
21. Guczi, L.; Stefler, G.; Geszti, O.; Sajó, I.; Pászti, Z.; Tompos, A.; Schay, Z. Methane dry reforming with CO_2: A study on surface carbon species. Appl. Catal. A 2010, 375, 236–246.
22. Lazar, M.D.; Dan, M.; Mihet, M.; Almasan, V.; Rednic, V.; Borodi, G. Hydrogen production by low temperature methane steam reforming using Ag and Au modified alumina supported nickel catalysts. Rev. Roum. Chim. 2011, 56, 637–642.
23. Lazar, M.D.; Mihet, M.; Dan, M.; Almasan, V.; Marginean, P. Preparation and characterization of nickel based multicomponent catalysts. J. Phys. Conf. Ser. 2009, 182.
24. Dan, M.; Mihet, M.; Biris, A.R.; Marginean, P.; Almasan, V.; Borodi, G.; Watanabe, F.; Biris, A.S.; Lazar, M.D. Supported nickel catalysts for low temperature methane steam reforming: comparison between metal additives and support modification. React. Kinet. Mech. Catal. 2012, 105, 173–193.
25. Maniecki, T.P.; Stadnicheno, A.I.; Maniukiewicz, W.; Bawolak, K.; Mierczynski, P.; Boronin, A.I.; Jozwiak, W.K. An active phase transformation on surface of Ni-Au/Al_2O_3 catalyst during partial oxidation of methane to synthesis gas.Kinet. Catal. 2010, 51, 573–578.
26. Maniecki, T.P.; Bawolak, K.; Gebauer, D.; Mierczynski, P.; Jozwiak, W.K. Catalytic activity and physicochemical properties of Ni-Au/Al_3CrO_6 system for partial oxidation of methane to synthesis gas. Kinet. Catal. 2009, 50, 138–144.
27. Maniecki, T.P.; Bawolak, K.; Mierczynski, P.; Jozwiak, W.K. Development of stable and highly active bimetallic Ni-Au catalysts supported on binary oxides $CrAl_3O_6$ for POM reaction. Catal. Lett. 2009, 128, 401–404.
28. Maniecki, T.P.; Bawolak-Olczak, K.; Mierczynski, P.; Maniukiewicz, W.; Jozwiak, W.K. Effect of the chemical composition of $(MgO)_x(Al_2O_3)_y$ support on the catalytic performance of Ni and Ni-Au catalysts for the partial oxidation of methane. Chem. Eng. J. 2009, 154, 142–148.
29. Qi, C.X.; Amphlett, J.C.; Peppley, B.A. Methanol steam reforming over NiAl and Ni (M) Al layered double hydroxides (M = Au, Rh, Ir) derived catalysts. Catal. Lett. 2005, 104, 57–62.
30. Niakolas, D.K.; Ouweltjes, J.P.; Rietveld, G.; Dracopoulos, V.; Neophytides, S.G. Au-doped Ni/GDC as a new anode for SOFCs operating under rich CH_4 internal steam reforming. Int. J. Hydrogen Energy 2010, 35, 7898–7904.
31. Gavrielatos, I.; Drakopoulos, V.; Neophytides, S.G. Carbon tolerant Ni-Au SOFC electrodes operating under internal steam reforming conditions. J. Catal. 2008, 259, 75–84.
32. Parizotto, N.V.; Rocha, K.O.; Damyanova, S.; Passos, F.B.; Zanchet, D.; Marques, C.M.P.; Bueno, J.M.C. Alumina-supported Ni catalysts modified with silver for the steam reforming of methane: Effect of Ag on

the control of coke formation. Appl. Catal. A 2007, 330, 12–22.

33. Parizotto, N.V.; Fernandez, R.F.; Marques, C.M.P.; Bueno, J.M.C. Promoter effect of Ag and La on stability of Ni/Al2O3 catalysts in reforming of methane processes. Stud. Surf. Sci. Catal. 2007, 167, 421–426.

34. Gavrielatos, I.; Montinaro, D.; Orfanidi, A.; Neophytides, S.G. Study of carbon deposition of Ag-doped Ni/YSZ electrodes under internal CH_4 steam reforming conditions. Fuel Cell 2009, 9, 883–890.

35. Jeong, H.; Kang, M. Hydrogen production from butane steam reforming over Ni/Ag loaded $MgAl_2O_4$ catalyst. Appl. Catal. B 2010, 95, 446–455.

36. Xu, Y.; Fan, C.; Zhu, Y.A.; Li, P.; Zhou, X.G.; Chen, D.; Yuan, W.K. Effect of Ag on the control of Ni-catalyzed carbon formation: A density functional theory study. Catal. Today 2012, 186, 54–62.

37. Padeste, C.; Trimm, D.L.; Lamb, R.N. Characterization of Sn doped Ni/Al_2O_3 steam reforming catalysts by XPS. Catal. Lett. 1993, 17, 333–339.

38. Nikolla, E.; Holewinski, A.; Schwank, J.; Linic, S. Controlling carbon surface chemistry by alloying: Carbon tolerant reforming catalyst. J. Am. Chem. Soc. 2006, 128, 11354–11355.

39. Nikolla, E.; Schwank, J.; Linic, S. Promotion of the long-term stability of reforming Ni catalysts by surface alloying. J. Catal. 2007, 250, 85–93.

40. Nikolla, E.; Schwank, J.; Linic, S. Comparative study of the kinetics of methane steam reforming on supported Ni and Sn/Ni alloy catalysts: The impact of the formation of Ni alloy on chemistry. J. Catal. 2009, 263, 220–227.

41. Nikolla, E.; Schwank, J.; Linic, S. Hydrocarbon steam reforming on Ni alloys at solid oxide fuel cell operating conditions. Catal. Today 2008, 136, 243–248.

42. Penkova, A.; Bobadilla, L.; Ivanova, S.; Dominguez, M.I.; Romero-Sarria, F.; Roger, A.C.; Centeno, M.A.; Odriozola, J.A. Hydrogen production by methanol steam reforming on $NiSn/MgO-Al_2O_3$ catalysts: The role of MgO addition.Appl. Catal. A 2011, 392, 184–191.

43. Saadi, S.; Hinnemann, B.; Helveg, S.; Appel, C.C.; Abild-Pedersen, F.; Nørskov, J.K. First-principles investigations of the Ni_3Sn alloy at steam reforming conditions. Surf. Sci. 2009, 603, 762–770.

44. Barcicki, J.; Denis, A.; Grzegorczyk, W.; Nazimek, D.; Borowiecki, T. Promotion of nickel catalysts for the steam reforming of methane. React. Kinet. Catal. Lett. 1976, 5, 471–478.

45. Rogatis, L.D.; Montini, T.; Lorenzut, B.; Fornasiero, P. Ni_xCu_y/Al_2O_3 based catalysts for hydrogen production. Energy Environ. Sci. 2008, 1, 501–509.

46. Khzouz, M.; Wood, J.; Pollet, B.; Bujalski, W. Characterization and activity test of commercial Ni/Al_2O_3, $Cu/ZnO/Al_2O_3$ and prepared Ni-Cu/Al_2O_3 catalysts for hydrogen production from methane and methanol fuels. Int. J. Hydrogen Energy 2013, 38, 1664–1675.

47. Chen, L.C.; Lin, S.D. The ethanol steam reforming over Cu-Ni/SiO_2 catalysts: Effect of Cu/Ni ratio. Appl. Catal. B 2011,106, 639–649.

48. Djaidja, A.; Kiennemann, A.; Barama, A. Effect of Fe or Cu addition on Ni/Mg-Al and Ni/MgO catalysts in the steam-reforming of methane. Stud. Surf. Sci. Catal. 2006, 162, 945–952.

49. Zhang, J.G.; Wang, H.; Dalai, A.K. Development of stable bimetallic catalysts for carbon dioxide reforming of methane. J. Catal. 2007, 249, 300–310.

50. An, W.; Zeng, X.C.; Turner, C.H. First-principles study of methane dehydrogenation on a bimetallic Cu/Ni(111) surface. J. Chem. Phys. 2009, 131, 174702:1–174702:11.

51. Resini, C.; Concepción, M.; Delgado, H.; Presto, S.; Alemany, L.J.; Riani, P.; Marazza, R.; Ramis, G.; Busca, G. Yttria-stabilized zirconia (YSZ) supported Ni-Co alloys (precursor of SOFC anodes) as catalysts for the steam reforming of ethanol. Int. J. Hydrogen Energy 2008, 33, 3728–3735.

52. Gonzalez-delaCruz, V.M.; Pereñiguez, R.; Ternero, F.; Holgado, J.P.; Caballero, A. In situ XAS study of synergic effects on Ni-Co/ZrO_2 methane reforming catalysts. J. Phys. Chem. C 2012, 116, 2919–2926.

53. Borowiecki, T.; Golebiowski, A. Influence of molybdenum and tungsten additives on the properties of nickel steam reforming catalysts. Catal. Lett. 1994, 25, 309–313.

54. Borowiecki, T.; Golebiowski, A.; Stasińska, B. Effects of small MoO_3 additions on the properties of nickel catalysts for the steam reforming of hydrocarbons. Appl. Catal. A 1997, 153, 141–156.

55. Maluf, S.S.; Assaf, E.M. Ni catalysts with Mo promoter for methane steam reforming. Fuel 2009, 88, 1547–1553.

56. Abelló, S.; Bolshak, E.; Montané, D. Ni-Fe catalysts derived from hydrotalcite-like precursors for hydrogen production by ethanol steam reforming. Appl. Catal. A 2013, 450, 261–274.

57. Wang, W.; Su, C.; Ran, R.; Shao, Z.P. A new Gd-promoted nickel catalyst for methane conversion to syngas and as an anode functional layer in a solid oxide fuel cell. J. Power Sour. 2011, 196, 3855–3862.

58. Xu, J.; Chen, L.W.; Tan, K.F.; Borgna, A.; Saeys, M. Effect of boron on the stability of Ni catalysts during steam methane reforming. J. Catal. 2009, 261, 158–165.

CHAPTER 8

A NOVEL APPROACH FOR PREDICTION OF INDUSTRIAL CATALYST DEACTIVATION USING SOFT SENSOR MODELING

Hamed Gharehbaghi and Jafar Sadeghi

Keywords: Soft sensor; catalyst deactivation; data-based modeling; state-dependent-parameter

INTRODUCTION

Due to large-scale application of catalysts in the petroleum and petrochemical processes, the main problem concerning these processes is the deactivation of catalyst. As the catalyst activity is related to the availability of catalyst active area for the reactants, deactivation refers to reducing the activated area of the catalyst or blocking the path of the moving reactants and the products. Mathematical modeling of catalyst deactivation is the main concern of these processes during operation. One of these processes with high potential of deactivation is the disproportonation (DP) of toluene and/or transalkilation (TA) of C_9 aromatics and toluene to C_8 aromatics over zeolites in fixed-bed reactors. DP and TA involve solid-acid catalytic processes which occur over medium-pore zeolites, such as ZSM-5, and large-pore zeolites, such as mordenite, Beta, and Y [1,2,3,4]. For H-mordenite, the temperature is a dominant factor that determines the activity and coking tendency, as the undesired reaction, due to its one-dimensional channel system [5].

The major problem for the commercialization of solid acid catalyzed processes is the rapid deactivation of the catalyst [4,6,7]. Therefore, modeling of the deactivation of the catalyst is a major problem that must be taken into account for the success of such processes. Some studies have already reported the deactivation behavior of the solid acid catalysts quantitatively [8,9,10,11]. These studies have focused on the solution of the set of differential and algebraic equations that describe catalyst deactivation in various reactor systems by experimental data. In these models, however, many simplified conditions have been assumed. Most of these models have been conducted at the laboratory scale under certain conditions with applying no dynamic data [1,2,3,4,5,6,7,12]. Additionally, online investigation of the industrial reactors' parameters (e.g., catalyst deactivation, conversion, product quality, etc.) can be of considerable significance in the identification of these parameters. In the last two decades, researchers began to make use of the vast amounts of data being measured and collected in the industrial processes by constructing predictive models based on such data. In the context of the industrial processes, these predictive models are called soft sensors [13,14]. In chemical plants, soft sensors are widely used to estimate a process variable which is difficult to be measured online. In general, one can broadly classify soft sensors into two types, namely, model-based (white-box) and data-based (black-box).White-box models are dependent on a prior theoretical knowledge and thus often unavailable since industrial processes are commonly too complicated to be analyzed theoretically. These models are developed primarily for the planning and the design of the processing plants and, therefore, usually focus on the description of the ideal steady states of the processes. As a solution, the data-based soft sensors gained increasing popularity in the industrial processes. A wide variety of statistical inference techniques and machine learning techniques have been employed in data-based soft sensors, among which representative examples are principal component regression (PCR), partial least squares (PLS) regression, support vector machines (SVM), and artificial neural networks (ANN) [13,14,15,16,17,18,19,20,21,22].

Reactors, by their nature, are non-stationary because their complexity cannot be fully described using constant-parameter models. The modeling of such systems can be represented by a

Citation: Hamed Gharehbaghi and Jafar Sadeghi, A Novel Approach for Prediction of Industrial Catalyst Deactivation Using Soft Sensor Modeling, doi:10.3390/catal6070093

two-step procedure. First, the dynamic behavior of the system is described by a deterministic mathematical model. Second, the model is simplified to utilize in conventional control theories. The first assumption causes errors in modeling, while the second suffers from the complexity and dimensionality problem. Moreover, digitization in order to use computer facilities for application of the continuous control theories provides another source of errors in modeling and control strategies [23,24]. Hence, soft sensor modeling of these systems requires the use of powerful statistical inference techniques.

Data-based mechanistic (DBM) modeling philosophy, as introduced by Young [25], is an alternative to complicated modeling strategies. The major advantage of this method is balancing between order of the model (complexity) and the model efficiency while, at the same time, good explanation of data and statistical analysis is required for identification and estimation of such systems [25,26,27,28,29,30]. Since the resulting model is simple and directly in digital form, it avoids both the model complication and errors produced through either simplification or digitization of a complex continuous-time model. It is worth mentioning that the resulting model and its parameters are, according to DBM philosophy, meaningful and interpretable while in data-based modeling, which is not mechanistic, this is generally not the case. In this philosophy, the non-stationary aspects of the system are reflected by a simple structure-varying parameter model, whether its parameters are varying with time or as a function of some states of the system. This model is called a state-dependent parameter (SDP) model [26,30]. In this case, all nonlinearities of the model can be estimated in model parameters, while the structure of the model remains linear and simple. In other words, when a system is described by a nonlinear model, such as $y_t = \alpha y_{t-1} + \beta y_{t-1}^2$, the SDP model is written as $y_t = \alpha_t y_{t-1}$ which has a linear structure but its parameter is $\alpha_t = \alpha + \beta y_{t-1}$. This model has the complexity of a nonlinear model while using a linear structure. The parameters are also identified in a non-parametric manner; i.e., when this approach is used, the functionality of the identified parameter can be seen in a graph rather than a mathematical relation.

In this study, a new SDP method for identification and modeling of the kinetics and catalyst activity has been proposed. The reactor modeling and identification have been considered for an industrial TA reactor using the input and output dynamic data. The obtained soft sensor model has a simple structure and appropriate for control purposes. This model also presents a good identification based on effective parameters of the catalyst activity.

KINETIC MODEL

In the present study, the process data of the fixed bed TA reactor of Bou-Ali-Sina Petrochemical Complex, Mahshahr, Iran, have been collected and analyzed daily for two years of reactor's operation. In the TA reactor, benzene and xylenes are produced from toluene and C_9 aromatics through the two main reactions, DP (Equation (1)) and TA (Equation (2)). The DP produces benzene and xylenes from toluene, while the TA forms xylenes from toluene and C_9 aromatics:

Disproportionation:

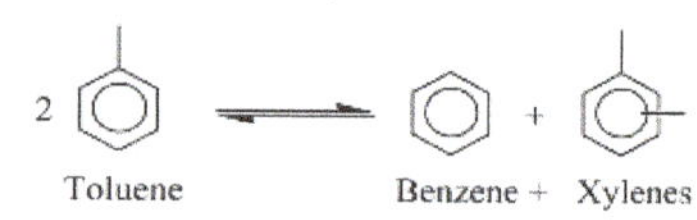

(1)

Transalkylation:

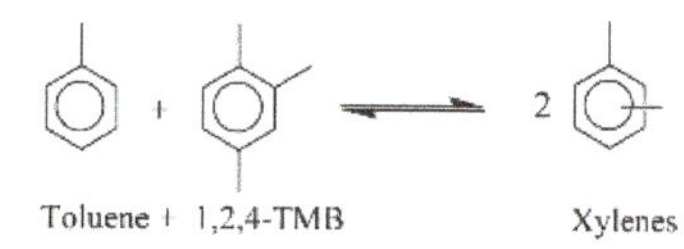

(2)

Different kinetic models have been proposed to describe the catalytic toluene disproportionation and C_9 aromatics transalkylation. Table 1 summarizes the kinetic and deactivation models proposed in the literature, as well as the experimental conditions.

Table 1: Kinetic studies of toluene DP and C_9 aromatics TA over mordenite zeolite catalyst.

Reaction	P(bar)	T(°C)	E (kJ/mol)	Kinetic Models [a]	Deactivation Function	Ref.
DP	1	350–450	60.7	SO	$a = a_0 e^{-0.003t}$	[6]
DP	1	350–450	51.9	RSO	$-\frac{da}{dt} = \frac{k_d P_T^2}{(1+K_B P_B + K_T P_T)^2} a^2$	[31]
DP	28.59	372–352	104.6	RSO	NA	[5]
DP	31.2	347–352	102	RSO	$a = exp(-0.0006.t)\, WHSV^{-1.3781}$	[32]
TA	31.2	347–352	96.77	RSO	$a = exp(-0.0006.t)\, WHSV^{-1.3781}$	[32]
DP	NA	300–400	54.22	RSO	$a = e^{-a_0 t}$	[1]
TA	NA	300–400	7.21	RSO	$a = e^{-a_0 t}$	[1]
DP	5–32	360–431	27.17	RSO	NA	[12]
TA	5–32	360–431	151	RSO	NA	[12]
TA	NA	300–400	17.84	RSO	$a = e^{-a_0 C_{Ar}}$	[33]
DP	NA	300–400	50.4	RSO	$a = e^{-a_0 t}$	[34]

[a] SO: Second Order; RSO: Reversible Second Order.

The reactor contains H-mordenite zeolite catalyst, which is manufactured by Sinopec Tech Company. Its characteristics are as follows (Table 2).

Table 2: Characteristics of mordenite zeolite catalyst.

Appearance	White, Cylindrical
Diameter/Length	1.6–2.0 mm/3–20 mm
Bulk density	0.72 ± 0.05 kg/L
Crush strength	>100 N/cm
Main ingredients	mordenite and alumina
Main components	SiO_2, Al_2O_3

IDENTIFICATION METHODOLOGY

DBM modeling which was initially acquired from the analysis of observational time-series data has only been considered reliable if it can be interpreted in physically meaningful terms. This model is identified from a general and widely-used class of linear/nonlinear models. In this method, the non-stationary aspects (data with variable mean and/or variance) of the system are shown by a simple-structure varying-parameter model, either its parameters are varying with time or as a function of some states of the system. One example of these models is the SDP model [25,26,35].

State Dependent Parameter (SDP) Model

The SDP models are representative of nonlinear systems which have similar structure to the linear time-invariant (LTI) systems, but their parameters are changing as a function of the state variables to accommodate the system's non-linearity [36].

Young [27] introduced the estimation of 'slowly changing' time-variable parameters (TVPs) in the various kinds of linear regression models. One of these TVP models is the Dynamic Auto-Regressive with eXogenous (DARX) variables model (the observed output is dependent on its past values and the past values of exogenous inputs). The DARX model can describe a real dynamic system and the output is a complex response. The model parameters are changing with time, when they are functions of the system variables. The results of the model can behave in a heavily nonlinear or even chaotic manner. Young [25,36] proposed an SDP model of this type as a state dependent parameter DARX (SDARX) model. The SDARX model relates a single input variable u_t to an output variable y_t that can be written conveniently as:

$$y_t = \mathbf{z}_t^T \mathbf{p}_t + e_t \tag{3}$$

More generally, it can be expressed as follows:

$$y_t = \sum_{i=1}^{n+m+1} p_i\left(\mathbf{x}_{1,t}^{(i)}, t\right) z_{i,t} + e_t \text{ with } \begin{cases} e_t = N\left(0, \sigma^2\right) \\ \mathbf{x}_{1,t}^{(i)} = \left[x_{1,t}^{(i)}\ x_{2,t}^{(i)} \ldots x_{ns_i,t}^{(i)}\right] \end{cases} \tag{4}$$

pi(.) is the ith state dependent parameter $\left(p_i = \begin{cases} -a_i & i \leq n \\ b_{i-(m+1)} & i > n \end{cases}\right)$, y_t and $x_{j,t}^{(i)}$ are the observed output and the jth state corresponds to pi(.) at t, respectively. In addition, n+m+1 is the number of model parameters, nsi is the number of states in which pi(.) depends on, $z_{i,t}$ is the ith regressor $\left(z_{i,t} = \begin{cases} y_{t-i} & i \leq n \\ u_{t-\delta+n+1-i} & i > n \end{cases}\right)$, and et is a zero-mean white-noise input with Gaussian normal amplitude distribution and variance σ^2.

The TA reactor is described by mass balances for benzene and C_9 aromatics [37]:

$$\frac{\partial C_B}{\partial t} - D_A \frac{\partial^2 C_B}{\partial l^2} - D_R \left(\frac{\partial^2 C_B}{\partial r^2} + \frac{1}{r}\frac{\partial C_B}{\partial r}\right) + \frac{Q}{S}\frac{\partial C_B}{\partial l} - \Re'_B \rho_x = 0 \tag{5}$$

$$\frac{\partial C_{Ar}}{\partial t} - D_A \frac{\partial^2 C_{Ar}}{\partial l^2} - D_R \left(\frac{\partial^2 C_{Ar}}{\partial r^2} + \frac{1}{r}\frac{\partial C_{Ar}}{\partial r}\right) + \frac{Q}{S}\frac{\partial C_{Ar}}{\partial l} - \Re'_{Ar} \rho_x = 0 \tag{6}$$

Since the reactor has a large volume, axial and radial diffusion terms can be neglected and assumed to be a part of the internal system error. Additionally, since the data are collected daily, the system is supposed to be in steady-state condition; therefore, the accumulation term is ignored and Equations (5) and (6) can be rewritten by following simple forms:

$$\frac{Q}{S}\frac{\partial C_B}{\partial l} - \Re'_B \rho_x = \xi \tag{7}$$

$$\frac{Q}{S}\frac{\partial C_{Ar}}{\partial l} - \Re'_{Ar} \rho_x = \xi \tag{8}$$

where ξ is the error. Discretization of Equations (7) and (8) in backward difference leads to:

$$\frac{Q}{S}\frac{\left(C_{B,L} - C_{B,0}\right)_{t-1}}{\Delta l} = \Re'_B \rho_x + \xi \tag{9}$$

$$\frac{Q}{S}\frac{\left(C_{Ar,L} - C_{Ar,0}\right)_{t-1}}{\Delta l} = \Re'_{Ar} \rho_x + \xi \tag{10}$$

Since the terms related to diffusion (adsorption term) are considered as a part of catalyst deactivation, the reaction rate is considered to be power function of the concentration of the reaction components, temperature-dependent specific reaction rate, and time-dependent deactivation rate as follows:

$$\Re'_B = C_{Tol.}^{\lambda_1} \cdot C_B^{\lambda_2} \cdot C_X^{\lambda_3} \cdot k_0 \cdot \exp\left(\frac{-E_1}{RT}\right) \cdot \exp(-a_1 \cdot t) \tag{11}$$

$$\Re'_{Ar} = C_{Tol.}^{\mu_1} \cdot C_{Ar}^{\mu_2} \cdot C_X^{\mu_3} \cdot k_0 \cdot \exp(\frac{-E_2}{RT}) \cdot \exp(-a_2 \cdot t) \tag{12}$$

Substitution of Equations (11) and (12) into Equations (9) and (10) and rearrangement gives:

$$(C_{B,L} - C_{B,0}) = \frac{S\,L}{Q} k_{0_1} C_{Tol.}^{\lambda_1} C_B^{\lambda_2} C_X^{\lambda_3} \exp(\frac{-E_1}{RT}) \exp(-a_1 t) \tag{13}$$

$$(C_{Ar,L} - C_{Ar,0}) = \frac{S\,L}{Q} k_{0_2} C_{Tol.}^{\mu_1} C_{Ar}^{\mu_2} C_X^{\mu_3} \exp(\frac{-E_2}{RT}) \exp(-a_2 t) \tag{14}$$

Taking the logarithm of both sides of Equation (10) yields:

$$\log(C_{B,L} - C_{B,0}) = \lambda_1 \log(C_{Tol}) + \lambda_2 \log(C_B) + \lambda_3 \log(C_X) + \log(\frac{S \cdot L}{Q} k_{0_1} \rho_x) + \frac{-E_1}{R}\frac{1}{T} + (-a_1 t) \tag{15}$$

$$\log(C_{Ar,L} - C_{Ar,0}) = \mu_1 \log(C_{Tol}) + \mu_2 \log(C_{Ar}) + \mu_3 \log(C_X) + \log(\frac{S \cdot L}{Q} k_{0_2} \rho_x) + \frac{-E_2}{R}\frac{1}{T} + (-a_2 t) \tag{16}$$

Having time-series concentration data of the process, the identification method of Equations (15) and (16) will be discussed in the following section.

It should be noted that in Equations (15) and (16) the terms of the reaction kinetics have been considered based on the following assumptions:

Reaction specific rate is a function of temperature according to the Arrhenius equation;

Due to reversible reactions, the concentration of the reaction components is considered by a power law expression (since all the adsorption terms are included in the deactivation term); and

Catalyst deactivation is a function of time and the concentrations of some other components. This function is assumed to be exponential (see Table 1).

Time-Varying Parameter (TVP) Estimation

SDP estimation is a form of TVP estimation. Normally, when little is known about the nature of the time variability, the model needs to be both simple and flexible. One of the simplest and most-useful class of stochastic, state-space models involves the assumption that the ith parameter (p_t) is defined by a two-dimensional stochastic state vector $x_{i,t} = [l_{i,t} d_{i,t}]^T$, where $l_{i,t}$ and $d_{i,t}$ are the changing level and slope of the associated TVP, respectively. The recursive SDP estimation algorithms are an extension of the stochastic approach to TVP estimation, where the time variations in the parameters are assumed to evolve as one of the generalized random walk (GRW) class of the Gauss-Markov process [25,27]. Early research in the 1960s [38,39] tended to use a simple scalar random walk model for the parameter variations when the parameters are supposed to change the TVP models. The stochastic evaluation of each parameter is described by a GRW process [40]:

$$\mathrm{x}_{i,t} = \mathrm{F}_i \mathrm{x}_{i,t-1} + \mathrm{G}_i \eta_{i,t} \qquad i = 1, 2, ..., m + n + 1 \tag{17}$$

where:

$$\mathrm{F}_i = \begin{bmatrix} \alpha & \beta \\ \gamma & 0 \end{bmatrix} \quad \text{and} \quad \mathrm{G}_i = \begin{bmatrix} \delta & 0 \\ 0 & \varepsilon \end{bmatrix} \tag{18}$$

$\eta_{i,t} = [\eta_{1i,t}\ \eta_{2i,t}]^T$ is a 2 × 1, zero mean, white-noise vector that allows for stochastic variability in the parameters and is assumed to be characterized by a covariance diagonal matrix $Q_{\eta,t}$ that each diagonal element Q_i corresponds to $\eta_{i,t}$. The various, normally constant, parameters in the GRW model (α,β,γ,δ,ε, and the elements of $Q_{\eta,t}$) are often referred to as hyper-parameters. This is to differentiate them from the TVPs that are the main object of the estimation analysis. Having introduced the GRW models for the parameter variations, an overall state space model can be constructed through the aggregation of the subsystem matrices defined by Equation (17), with the 'observation' equation defined by Equation (3):

$$\left.\begin{array}{lll} \text{Observation Equation} & y_t & = \mathrm{H}_t \mathrm{x}_t + e_t \\ \text{State Equation} & \mathrm{x}_t & = \mathrm{F}\mathrm{x}_{t-1} + \mathrm{G}\eta_t \end{array}\right\} \tag{19}$$

Where

$x_t = \left[x_{1,t}^T\ x_{2,t}^T \ldots x_{n+m+1,t}^T\right]^T$ if $p = 2(n+m+1)$, F is a $p \times p$ diagonal matrix with blocks defined by the Fimatrices in Equation (17), G is a p×p block diagonal matrix with blocks defined by the corresponding subsystem matrices Gi in Equation (17), and ηt is a (n+m+1)×1 vector containing the white-noise inputs which is assumed to be independent of the observation noise et and has a covariance diagonal (n+m+1)×(n+m+1) matrix Q that each diagonal element is Q_i corresponding to $\eta_{i,t}$. Additionally, H_t is a row vector of the following form:

$$H_t = \lfloor y_{t-1}\ 0_1\ y_{t-2}\ 0_2 \ldots y_{t-n}\ 0_n\ u_{t-\delta}\ 0_{n+1} \ldots u_{t-\delta-m}\ 0_{n+m+1}\rfloor \tag{20}$$

where 0i, i=1,2,…,n+m+1 are defined as $1\times q_i$ vectors of zero and the term δ is a pure time delay, measured in sampling intervals. Now the relation between scalar observation yt and the state variables defined by Equation (17) can be defined by Equation (19), which represents the model Equation (17), where each parameter is defined as a GRW process. The state space formulation of Equation (17) is particularly well-suited for optimal recursive estimation in which the time variable parameters are estimated sequentially while working through the data in temporal order through two separate passes, "forward pass filtering" and "backward pass smoothing" [25,41].

Forward Pass Recursive LS equations:

$$\text{Prediction}: \begin{array}{l} \hat{x}_{t|t-1} = F\hat{x}_{t-1} \\ P_{t|t-1} = FP_{t-1}F^T + GQ_rG^T \end{array} \tag{21}$$

The estimation is based on all the times before t. To include time t the correction is needed:

$$\text{Correction}: \begin{array}{l} \hat{x}_t = \hat{x}_{t|t-1} + P_{t|t-1}H_t^T\left[1 + H_tP_{t|t-1}H_t^T\right]^{-1}\left(y_t - H_t\hat{x}_{t|t-1}\right) \\ P_t = P_{t|t-1} - P_{t|t-1}H_t^T\left[1 + H_tP_{t|t-1}H_t^T\right]^{-1}H_t^TP_{t|t-1} \end{array} \tag{22}$$

where t=1,2, … ,N and $\hat{x}_0$ and P_0 are the initial value and its covariance matrix, respectively.

Backward Pass Smoothing Equations

Estimated values that are obtained from the forward-pass filtering algorithm are based on the initial conditions, x^0 and P0, and the values are prior to and include sample t. Therefore, they need to be updated sequentially, while working through the data, in reverse temporal order (backward pass smoothing) using a backward-recursive procedure termed fixed interval smoothing (FIS) algorithm, Equation (21), to obtain a smooth estimate based on all the elements of the time series and to remove any lag effect.

$$\begin{array}{l} \hat{x}_{t|N} = F^{-1}\left[\hat{x}_{t+1|N} + GQ_rG^TL_t\right] \\ L_t = \begin{cases} \left[I - P_{t+1}H_{t+1}^TH_{t+1}\right]^T \times \left\{F^TL_{t+1} - H_{t+1}^T\left(y_{t+1} - H_{t+1}\hat{x}_{t+1}\right)\right\} & t < N \\ 0 & t = N \end{cases} \\ P_{t|N} = \begin{cases} \hat{P}_t + \hat{P}_tF^TP_{t+1|t}^{-1}\left[P_{t+1|N} - P_{t+1|t}\right]P_{t+1|t}^{-1}F^T\hat{P} & t < N \\ \hat{P}_t & t = N \end{cases} \end{array} \tag{23}$$

In these algorithms the p×p noise variance ratio (NVR) matrix, Qr, and the p×p matrix $\hat{P}_t$ are defined as:

$$Q_r = \frac{Q}{\sigma^2} \quad \text{and} \quad \hat{P}_t = \frac{P_t^*}{\sigma^2} \tag{24}$$

P_t^* is the error covariance matrix associated with the state estimates which defines the estimated uncertainty in the parameters. For convenience, it is usually assumed that the NVR matrix Q_r is diagonal, although it is not necessary. The NVR parameters that characterize Q_r are unknown prior to the analysis and clearly need to be estimated based on of the time-series data y_t and u_t before the filtering and smoothing algorithms can be utilized. The usual way of dealing with this estimation problem is to formulate it in maximum likelihood (ML) methods of estimation based on Gaussian assumptions for the stochastic disturbances and the application of prediction error decomposition [25,26,39,41]:

$$e_t = y_t - H_t\hat{x}_{t|t-1} \qquad t = 1, 2, \ldots, N \tag{25}$$

It is here assumed that the stochastic white noise inputs to the system have a normal distribution. Now, if p is the number of TVPs being estimated and the first p observations are regarded as fixed, then the log-likelihood function of the series y_{p+1},…,yN can be defined as follows in terms of the prediction error decomposition:

$$\log L = \frac{-(N-p)}{2}\log(2\pi) - \frac{1}{2}\log(\sigma^2) - \frac{1}{2}\sum_{t=p+1}^{N}\log\left(1 + H_tP_{t|t-1}H_t^T\right) - \frac{1}{2\sigma^2}\sum_{t=p+1}^{N}\frac{e_t^2}{1 + H_tP_{t|t-1}H_t^T} \tag{26}$$

It can be shown that $\sigma^2\left(1+\mathbf{H}_t\mathbf{P}_{t|t-1}\mathbf{H}_t^T\right)$ is the variance of e_t, so the last term in Equation (26) is based on the sum of the squares of the normalized one-step-ahead prediction errors. Now, the ML estimate of σ^2, conditional on the hyper-parameters, is given by:

$$\hat{\sigma}^2=\frac{1}{N-p}\sum_{t=p+1}^{N}\frac{e_t^2}{1+\mathbf{H}_t\mathbf{P}_{t|t-1}\mathbf{H}_t^T} \tag{27}$$

By substituting Equation (27) into Equation(26) and removing the constant term, the 'concentrated likelihood' can be expressed as:

$$\log\left(L_c\right)=-\frac{1}{2}\sum_{t=p+1}^{N}\log\left(1+\mathbf{H}_t\mathbf{P}_{t|t-1}\mathbf{H}_t^T\right)+\frac{N-p}{2}\log\left(\hat{\sigma}^2\right) \tag{28}$$

which can then be minimized if it is multiplied by −1. This minimization is accomplished by initiating the optimization with the hyper-parameter estimates either selected by the user or set to some default values. Since p_t is potentially state-dependent, it cannot be assumed that the simple GRW model Equation (17) is appropriate to describe the parametric variation over time, which could be very fast. One obvious requirement with this new approach is that the sorting of data, prior to FIS estimation, must be common to all of the variables in Equation (3). The simplest strategy is to sort according to the ascending order of the 'dependent' variable y_t. In back-fitting algorithm, each parameter is estimated, in turn, based on the 'modified dependent variable' (MDV) series obtained by subtracting all the other terms on the right-hand side of Equation (3) from yt, using the values of the other parameter estimates from the previous iteration. At each such back-fitting iteration, the sorting can then be based on the single variable associated with the current SDP being estimated.

Back-Fitting Algorithm for SDP Models

1. Assume FIS estimation has yielded prior TVP estimates $\hat{p}^0_{i,t|N},\ i=1,2,\ldots,m+n+1,$ of the SDPs.
2. Iterate: i=1,2,...,m+n+1; k=1,2,...,kc
3. Form the MDV: $y_t^i=y_t-\sum_{j\neq i}z_{i,t}\cdot\hat{p}^k_{j,t|N}.$
4. Sort both y_t^i and $z_{i,t}$ according to the ascending order of zi,t.
5. Obtain an FIS estimate $\hat{p}^k_{i,t|N}$ of $p_{i,t}$ in the MDV relationship $y_t^i=p_{i,t}z_{i,t}.$
6 Repeat steps 2–5 until convergence criterion occurs.

The smoothing hyper parameters required for FIS estimation at each stage are optimized by ML.

Sadeghi et al. [42,43] extended this approach to accommodate the multidimensionality of parameters vs. its states. In this modeling approach, the model is optimized and is regarded as non-parametric by ML.

RESULTS AND DISCUSSION

To verify the effectiveness of the SDP modeling, data series of fixed bed industrial TA reactor of the Bou-Ali-Sina Petrochemical Complex are analyzed. A record of 623 data as daily means for two years of reactor operation are used. Overall process data for the input and output concentrations of toluene, C_9 aromatics, olefin, hydrogen, benzene, xylene, and temperature T versus time are shown in Figure 1.

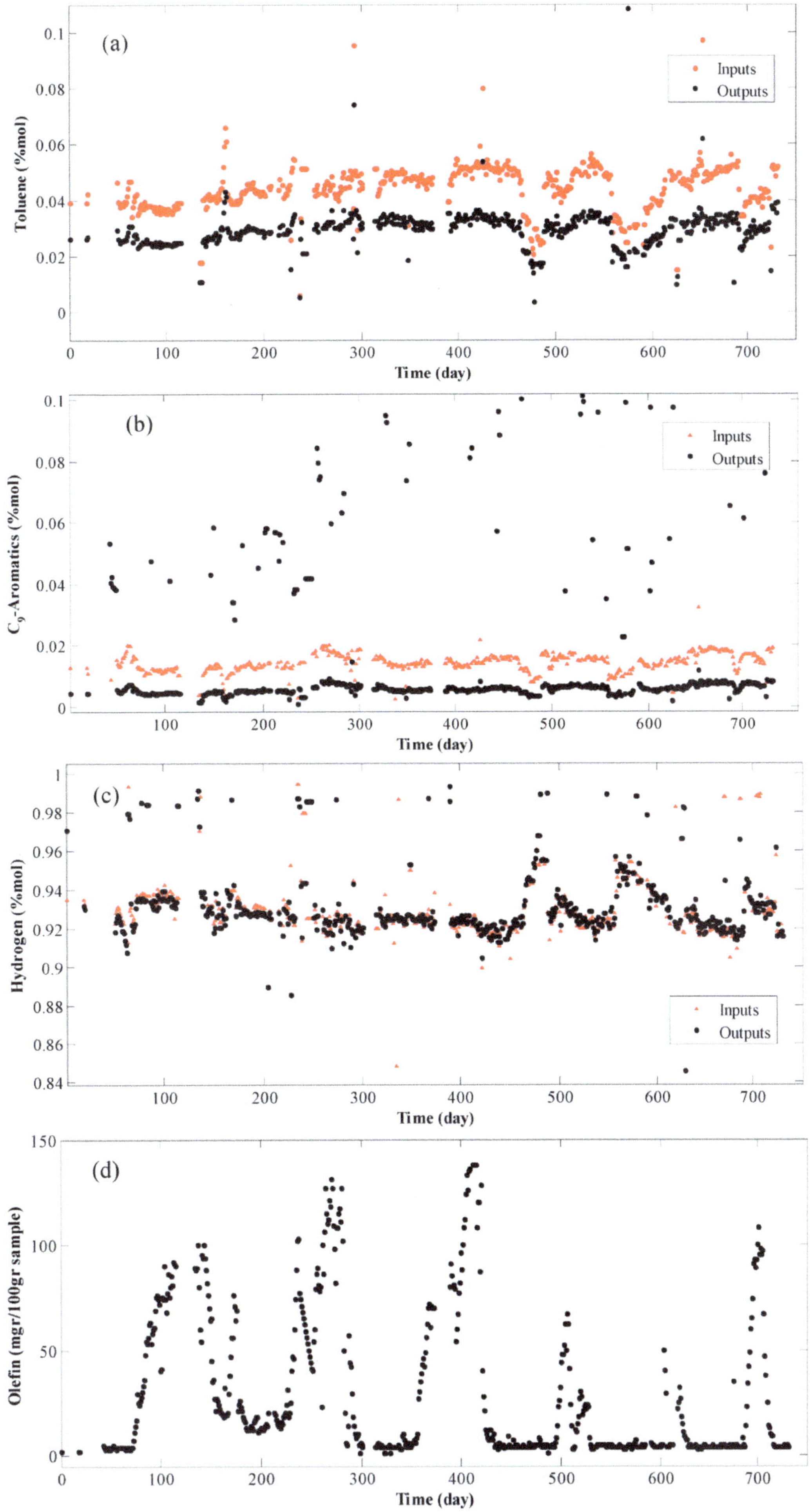
(a)
Inputs
Outputs
Toluene (%mol)
Time (day)
(b)
Inputs
Outputs
C9-Aromatics (%mol)
Time (day)
(c)
Inputs
Outputs
Hydrogen (%mol)
Time (day)
(d)
Olefin (mgr/100gr sample)
Time (day)

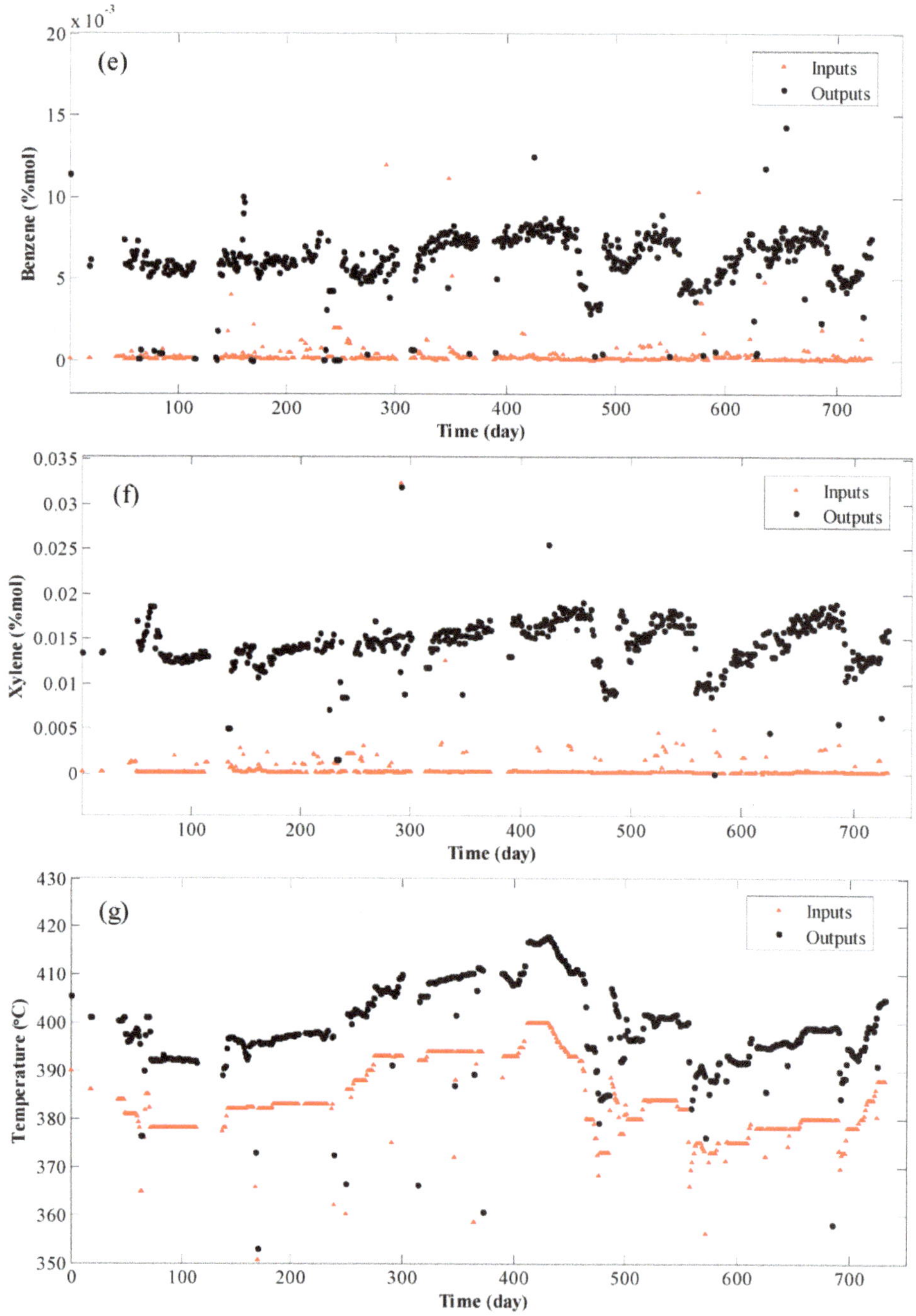

Figure 1: Input and output data of process. (**a**) Toluene; (**b**) C_9 aromatics; (**c**) hydrogen; (**d**) olefin; (**e**) benzene; (**f**) xylene; and (**g**) temperature.

As the figure shows, the data are recorded with high input and output oscillation, without interfering with the working conditions and the dynamics form.

By designing the soft sensor, it is expected that while all of the states or variables are changing simultaneously, the system identification is performed. Moreover, model parameters are expected to be functions of the system state variables. When a complex nonlinear model with constant parameters is replaced by a linear structure with varying parameters, it is clear that the model with linear variations of the state variables gives a better control. The following assumptions are considered:

- Modeling is based on the continuity equation;
- Grey-box model is considered;
- Data are registered as daily average, that is, for each 24 h one data is recorded;
- Since the residence time of the reactor is half an hour and the data are recorded per day, the system is considered stable and the term for the accumulation in the continuity equation is vanishes;
- Due to the high volumetric flow rate of the reactor, the terms for the axial and radial diffusions are considered as part of the system error and hence removed; and
- The hydrogen molar percentage in the feed is about 90%. Therefore, the internal porous mass transfer resistance is disregarded.

Using the SDP modeling and through the investigation of the considered reactor model parameters Equations (15) and (16), it has been shown that most parameters have slight changes. Therefore, the parameters are supposed to be constant and are represented in Table 3.

Table 3: Initial estimates for fixed parameters of models.

Parameter	Mean (95% CL)
λ_1	−0.19 (±0.30)
λ_2	−0.04 (±0.20)
λ_3	1.85 (±0.28)
μ_1	0.10 (±0.27)
μ_2	0.60 (±0.24)
μ_3	1.17 (±0.15)
E_1 (J/kmol· K)	43388 (±921.15)
k_{01} (kg· h/kmol)	2.90 (±1.01)
E_2 (J/kmol· K)	18201 (±490.40)
k_{02} (kg· h/kmol)	0.32 (±0.43)

The only exception is due to the variations of the term for the deactivation of catalyst a, which is considered to be a function of different variables of the system states. The changes have been found to be significant in terms of the time and hydrogen and olefin concentrations. An important issue is the high sensitivity of the catalyst activity to the olefin concentration (lower than 200ppm). Olefin causes coke formation (as the undesired reaction) and reduces the catalyst efficiency; hence, the conversion decreases. In order to compensate for this effect, the system operation needs to increase the temperature; however, increasing the reactor temperature will also increase the deactivation rate of the catalyst. Furthermore, high molar fraction of hydrogen is required to avoid undesired coking reaction and dilution of other components in order to reduce the diffusion resistance and increased temperature result in the recovery of the catalysts. However, the process is then able to perform the required conversion in a lower temperature. Thus, the effect of olefin on the catalyst poisoning, along with the high molar fraction of hydrogen and the increased temperature for the catalyst recovery, cause the catalyst activity to have a roughly monotonic trend over the two-year period of operation. One main advantage of the SDP method to design the soft sensor is that the modeling is exclusively performed by process data and, hence, exact determination of the process information (e.g., reaction mechanism, kinetics, etc.) is not needed. Therefore, in the present study, the parameters related to such problems as the order of concentration in the reaction are found to be different from those recorded in the literature (Table 3). As can be seen, some parameters have values close to zero with high uncertainty. Thus, the related parameters are omitted and the equation is again recovered and identified as follows:

$$\log\left(C_{B,L} - C_{B,0}\right) - \log\left(\frac{S \cdot L}{Q}\rho_x\right) = \lambda_3 \log(C_X) + \frac{-E_1}{R}\frac{1}{T} + \phi_1$$

$$\log\left(C_{Ar,L} - C_{Ar,0}\right) - \log\left(\frac{S \cdot L}{Q}\rho_x\right) = \mu_2 \log(C_{Ar}) + \mu_3 \log(C_X) + \frac{-E_2}{R}\frac{1}{T} + \phi_2 \quad (29)$$

In these equations, the term for the catalyst deactivation multiplied by the pre-exponential factor is replaced by φ. The results of the new identification show a reduction in the uncertainty of the estimated parameters as represented in Table 4.

Table 4: Final estimates for fixed parameters of models.

Parameter	Mean (95% CL)
λ_3	1.7 (±0.15)
μ_2	0.67 (±0.12)
μ_3	1.19 (±0.13)
E_1 (J/kmol· K)	43842 (±918)
E_2 (J/kmol· K)	18084 (±489)

In both models, the xylene concentration plays a decisive role, compared to the other components. Therefore, it shows that with the increase in the xylene concentration in the feed, a higher conversion is obtained. As shown by Table 4, the activation energies of the DP and TA reactions are in an acceptable range and in compliance with the previous studies (seeTable 1). Comparison of the activation energies of the DP and TA reactions shows that the TA reaction is faster. Concerning the catalyst deactivation term in the first model (φ_1), from the investigation of different states in the two dimensions of time and concentration of olefin, it can be observed that olefin causes catalyst toxicity and significantly decreases catalytic activity (see Figure 2). In this context, as shown in Figure 2a–c, it is suggested that for avoiding the catalyst toxicity, olefin concentration in the feed should be lower than 20ppm. However, during the time the catalyst activity is in an oscillatory form, and as mentioned before, it shows an approximately constant activity over the period of study. As shown in Figure 2, the catalyst deactivation term in the first model is changing by a function of olefin concentration and time. The catalyst deactivation term in the second model (φ_2) has been identified as a function of the two states of hydrogen concentration and time. Hydrogen concentration and the increased temperature have recovered the catalyst (Figure 3a–c).

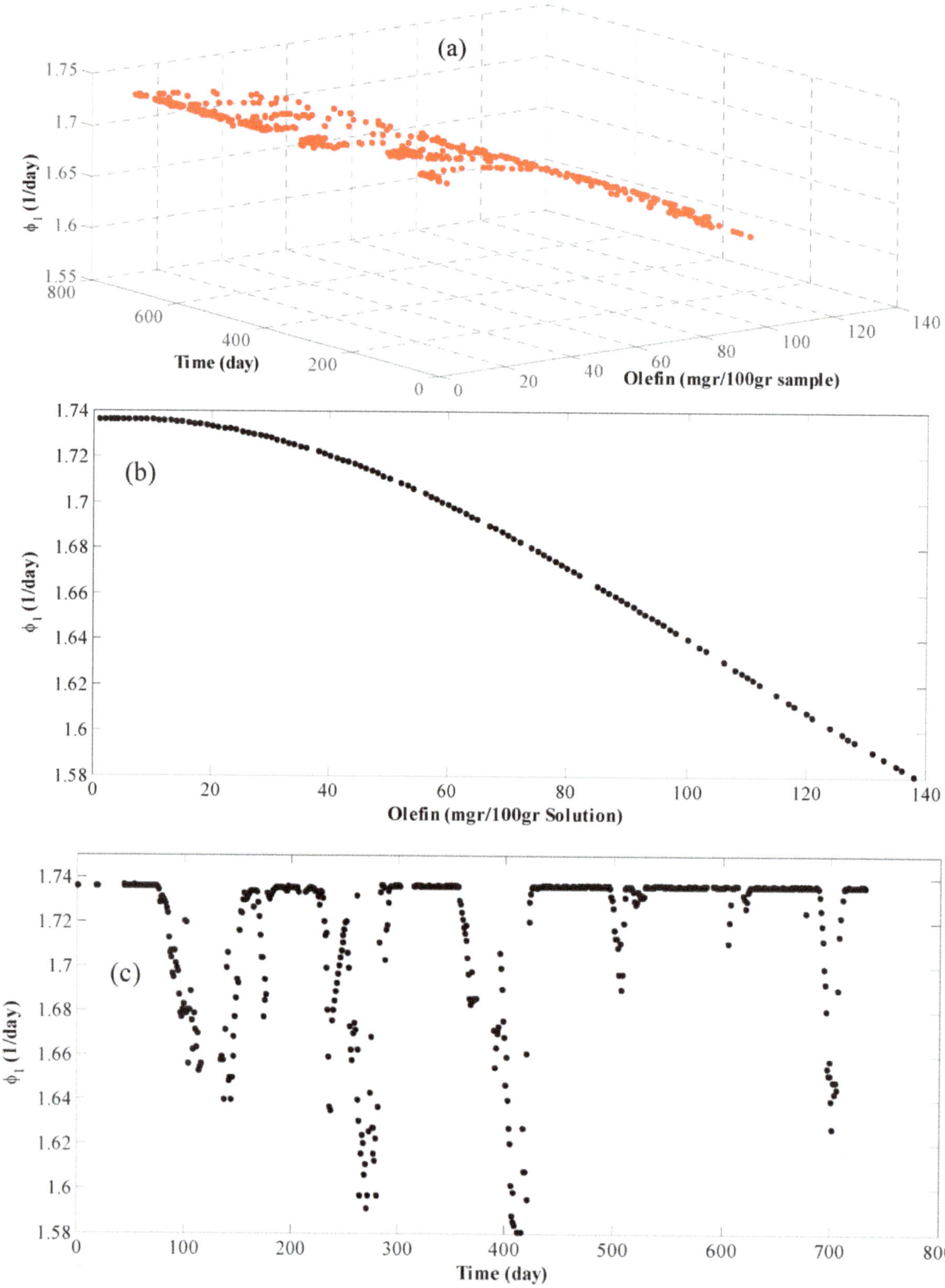

Figure 2: Catalyst deactivation term in the first model; (**a**) based on time and olefin states, simultaneously; (**b**) olefin state; and (**c**) time state.

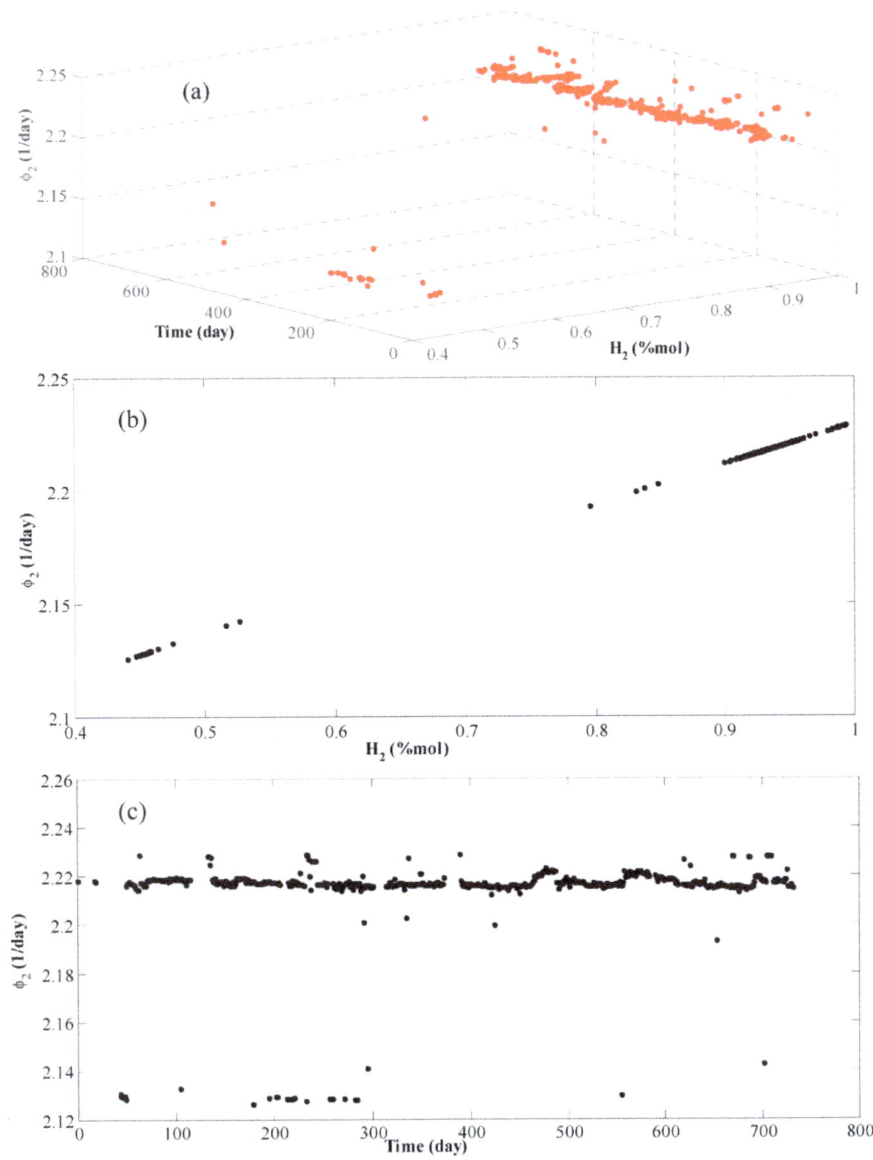

Figure 3: Catalyst deactivation term in the second model, (**a**) based on time and hydrogen states; simultaneously; (**b**) hydrogen state; and (**c**) time state.

Based on the changes in temperature and olefin concentration (Figure 4), it can be concluded that with an increase in the amount of olefin in the inlet flow, the reactor temperature is increased by the operator with a considerable time delay. The time delay causes a reduction in the conversion of the reactor. This can be improved by employing an online prediction method. In other words, by applying the soft sensor, the temperature can be adjusted based on the catalyst activity.

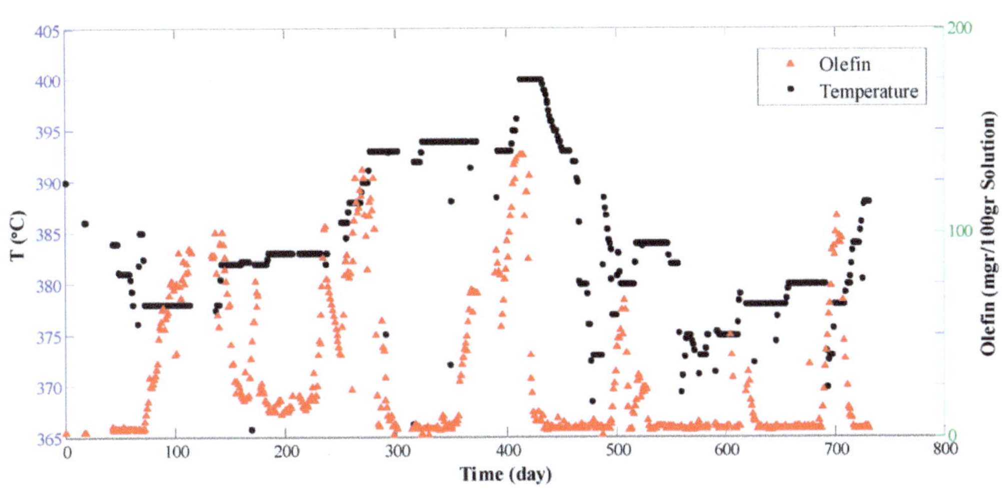

Figure 4: Temperature and olefin concentration vs. time, simultaneously.

It would be a good progress for the industrial processes to be identified and fault-detected using the process data with no extra cost imposed. As shown in Figure 5, the predicted values are in good agreement with the process data which can be considered as an important result of the SDP method. It should be noted that the SDP method with reference to the reduction in the laboratory costs is a great success for petrochemical industries. As a result, it assists in the increase of prediction and fault detection ability.

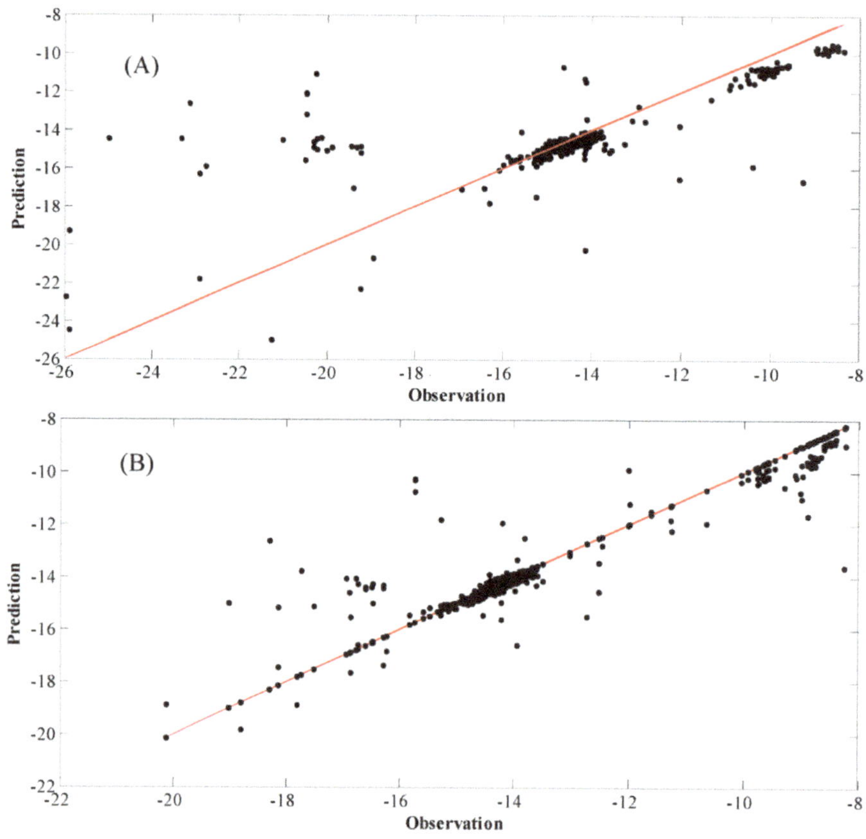

Figure 5: The relationship between observed and predicted y in the SDP method. (**A**) First model; and (**B**) second model.

CONCLUSIONS

A soft sensor is designed for identification of the variations in the catalyst activity using process data of an industrial reactor. Based on the DBM philosophy, the present study is developed in an innovative method for soft sensor design, which demonstrates appropriate capability for analyzing the data with drastic changes. Using the SDP method, the catalyst activity has been estimated non-parametrically as a function of time, olefin concentration, and hydrogen concentration. Other parameters are considered to be constant. Based on the obtained results, it is suggested that the amount of olefin in the inlet flow should be lower than 20 ppm. In the corresponding models, for both reactions, the order of xylene concentration is greater than that of the other components and, thus, for the increase in the conversion it is suggested that the xylene concentration in the inlet flow should be increased through the recycle flow. However, the estimated order of concentration of the components does not show a good agreement with the previous results. The activation energies for the two DP and TA reactions have been found to be 43.8 and 18 kJ/mol, respectively. The model prediction is in good agreement with the process data. The present work can be considered as an advance towards the identification, modeling, and fault detection of a complex industrial process.

ACKNOWLEDGMENT

The authors acknowledge the technical support of Bou-Ali-Sina Petrochemical Company.

CONFLICTS OF INTEREST

The authors declare no conflict of interest.

REFERENCES

1. Waziri, S.M.; Aitani, A.M.; Al-Khattaf, S. Transformation of Toluene and 1,2,4-Trimethylbenzene over ZSM-5 and Mordenite Catalysts: A Comprehensive Kinetic Model with Reversibility. *Ind. Eng. Chem. Res.* 2010, *49*, 6376–6387.
2. Al-Khattaf, S. Catalytic transformation of toluene over a high-acidity Y-zeolite based catalyst. *Energy Fuels* 2006, *20*, 946–954.
3. Wang, T.; Tsai, C.; Huang, S.T. Disproportionation of toluene and of trimethylbenzene and their transalkylation over zeolite beta. *Ind. Eng. Chem. Res.* 1990, *29*, 2005–2012.
4. Tsai, T.-C.; Liu, S.-B.; Wang, I. Disproportionation and transalkylation of alkylbenzenes over zeolite catalysts. *Appl. Catal. A* 1999, *181*, 355–398.

5. Tsai, T.C.; Chenb, W.H.; Lai, C.S.; Liub, S.B.; Wangc, I.; Ku, C.S. Kinetics of Toluene Disproportionation over Fresh and Coked H-mordenite. *Catal. Today* 2004, *97*, 297–302.
6. Bhavlkattl, S.S.; Patwardhan, S.R. Toluene Disproportionation over Nickel-Loaded Aluminum-Deficient Mordenite 2.*Kinetics. Ind. Eng. Chem. Prod. Res. Dev.* 1981, *20*, 106–109.
7. Al-Khattaf, S.; Rabiu, S.; Tukur, N.; Alnaizy, R. Kinetics of toluene methylation over USY-zeolite catalyst in a riser simulator. *Chem. Eng. J.* 2008, *139*, 622–630.
8. De Jong, K.; Mesters, C.; Peferoen, D.; Van Brugge, P.; De Groot, C. Paraffin alkylation using zeolite catalysts in a slurry reactor: Chemical engineering principles to extend catalyst lifetime. *Chem. Eng. Sci.* 1996, *51*, 2053–2060.
9. Taylor, R.J.; Sherwood, D.E. Effects of process parameters on isobutane/2-butene alkylation using a solid acid catalyst.*Appl. Catal. A* 1997, *155*, 195–215.
10. Sahebdelfar, S.; Kazemeini, M.; Khorasheh, F.; Badakhshan, A. Deactivation behavior of the catalyst in solid acid catalyzed alkylation: Effect of pore mouth plugging. *Chem. Eng. Sci.* 2002, *57*, 3611–3620.
11. Simpson, M.F.; Wei, J.; Sundaresan, S. Kinetic analysis of isobutane/butene alkylation over ultrastable HY zeolite. *Ind. Eng. Chem. Res.* 1996, *35*, 3861–3873.
12. Lobao, M.W.N.; Alberton, A.L.; Melo, S.A.B.V.; Embirucu, M.; Monteiro, J.L.F.; Pinto, J.C. Kinetics of Toluene Disproportionation: Modeling and Experiments. *Ind. Eng. Chem. Res.* 2012, *51*, 171–183.
13. Fortuna, L.; Graziani, S.A.; Rizzo, S.; Xibilia, M.G. *Soft Sensors for Monitoring and Control of Industrial Processes*; Springer: London, UK, 2007.
14. Kadleca, P.; Gabrysa, B.; Strandt, S. Data-driven Soft Sensors in the process industry. *Comput. Chem. Eng.* 2009, *33*, 795–814.
15. Kaneko, H.; Arakawa, M.; Funatsu, K. Development of a New Soft Sensor Method Using Independent Component Analysis and Partial Least Squares. *AIChE J.* 2009, *55*, 87–98.
16. Lin, B.; Recke, B.; Schmidt, T.M.; Knudsen, J.K.H.; Jørgensen, S.B. Data-Driven Soft Sensor Design with Multiple-Rate Sampled Data: A Comparative Study. *Ind. Eng. Chem. Res.* 2009, *48*, 5379–5387.
17. Liu, J.; Chen, D.-S.; Shen, J.-F. Development of Self-Validating Soft Sensors Using Fast Moving Window Partial Least Squares. *Ind. Eng. Chem. Res.* 2010, *49*, 11530–11546.
18. Liu, Y.; Gao, Z.; Chen, J. Development of soft-sensors for online quality prediction of sequential-reactor-multi-grade industrial processes. *Chem. Eng. Sci.* 2013, *102*, 602–612.
19. Shang, C.; Yang, F.; Huang, D.; Lyu, W. Data-driven soft sensor development based on deep learning technique. *J. Process Control* 2014, *24*, 223–233.
20. Kaneko, H.; Funatsu, K. Database Monitoring Index for Adaptive Soft Sensors and the Application to Industrial Process. *AIChE J.* 2014, *60*, 160–169.
21. Kaneko, H.; Funatsu, K. Adaptive soft sensor model using online support vector regression with time variable and discussion of appropriate hyperparameter settings and window size. *Comput. Chem. Eng.* 2013, *58*, 288–297.
22. Kaneko, H.; Funatsu, K. Adaptive soft sensor based on online support vector regression and Bayesian ensemble learning for various states in chemical plants. *Chemom. Intell. Lab. Syst.* 2014, *137*, 57–66.
23. Elnashaie, S.S.E.H.; Elshishini, S.S. *Modelling, Simulation and Optimization of Industrial Fixed Bed Catalystic Reactors*; Gordon & Breach Science: Philadelphia, PA, USA, 1993.
24. Manenti, F.; Cieri, S.; Restelli, M.; Bozzano, G. Dynamic modeling of the methanol synthesis fixed-bed reactor. *Comput. Chem. Eng.* 2013, *48*, 325–334.
25. Young, P. *Recursive Estimation and Time-Series Analysis an Introduction for the Student and Practitioner*; Springer: Berlin, Germany, 2011.
26. Young, P. Data-based mechanistic modelling of environmental, ecological, economic and engineering systems.*Environ. Model. Softw.* 1998, *13*, 105–122.
27. Young, P. Data-based mechanistic modelling, generalised sensitivity and dominant mode analysis. *Comput. Phys. Commun.* 1999, *117*, 113–129.
28. Price, L.; Young, P.; Berckmans, D.; Janssens, K.; Taylor, J. Data-based mechanistic modelling (DBM) and control of mass and energy transfer in agricultural buildings. *Ann. Rev. Control* 1999, *23*, 71–82.
29. Young, P. The data-based mechanistic approach to the modelling, forecasting and control of environmental systems.*Ann. Rev. Control* 2006, *30*, 169–182.
30. Taylor, C.J.; Young, P.C.; Chotai, A. *True Digital Control: Statistical Modelling and Non-Minimal State Space Design*; John Wiley & Sons: Hoboken, NJ, USA, 2013.
31. Bharati, S.P.; Bhatia, S. Deactivation Kinetics of Toluene Disproportionation over Hydrogen Mordenite Catalyst. *Ind. Eng. Chem. Res.* 1987, *26*, 1854–1860.
32. Xu, O.; Su, H.; Ji, J.; Jin, X.; Chu, J. Kinetic Model and Simulation Analysis for Toluene Disproportionation and C9-Aromatics Transalkylation. *Chin. J. Chem. Eng.* 2007, *15*, 326–332.
33. Al-Mubaiyedha, U.A.; Ali, S.A.; Al-Khattaf, S.S. Kinetic Modeling of Heavy Reformate Conversion into Xylenes over Mordenite-ZSM5 Based Catalysts. *Chem. Eng. Res. Des.* 2012, *90*, 1943–1955.
34. Odedairo, T.; Balasamy, R.J.S.; Al-Khattaf, R.J. Toluene Disproportionation and Methylation over Zeolites TNU-9,SSZ-33, ZSM-5, and Mordenite Using Different Reactor Systems. *Ind. Eng. Chem. Res.* 2011, *50*, 3169–3183.
35. Beven, K.J.; Leedal, D.T.; Smith, P.J.; Young, P.C. *Identification and Representation of State Dependent Non-Linearitiesin Flood Forecasting Using the DBM Methodology*; Springer: Berlin, Germany, 2012; pp. 341–366.

36. Toivonen, H.; Tötterman, S.; Åkesson, B. Identification of state-dependent parameter models with support vector regression. *Int. J. Control* 2007, *80*, 1454–1470.

37. Trambouze, P.; Landeghem, H.V.; Wauqier, J.P. *Chemical Reactors Design/Engineering/Operation*; Éditions Technip: Paris, France, 1988.

38. Lee, R.C. *Optimal Estimation, Identification, and Control*; MIT Press: Cambridge, MA, USA, 1964.

39. Young, P. An instrumental variable method for real-time identification of a noisy process. *Automatica* 1970, *6*, 271–287.

40. Young, P. *Recursive Estimation and Time-Series Analysis: An Introduction*; Springer-Verlag: New York, NY, USA, 1984.

41. Norton, J. *An Introduction to Identification*; Academic Press: London, UK, 1986.

42. Sadeghi, J.; Tych, W.; Chotai, A.A.; Young, P. Multi-state dependent parameter model identification and estimation for nonlinear dynamic systems. *Electron. Lett.* 2010, *46*, 1265–1266.

43. Tych, W.; Sadeghi, J.; Smith, P.J.; Chotai, A.; Taylor, C.J. *Multi-state Dependent Parameter Model Identification and Estimation*; Springer: London, UK, 2012; pp. 191–210. [

CHAPTER 9 HIGH-THROUGHPUT SCREENING AS A SUPPLEMENTAL TOOL FOR THE DEVELOPMENT OF ADVANCED EMISSION CONTROL CATALYSTS: METHODOLOGICAL APPROACHES AND DATA PROCESSING

Andreas Sundermann and Olga Gerlach

Keywords: Emission control catalysts; high-throughput screening; data processing; automotive catalysis

INTRODUCTION

The development and successful implementation of advanced exhaust aftertreatment systems requires continuous catalyst improvement targeting better low temperature activity, sulfur tolerance and thermal durability associated with the reduction of precious metals and overall cost. Advanced test technologies are required for the improvement and optimization of commercial emission control catalysts [1], as well as for long-term R&D projects aimed at the development of future technologies. In the future, efficient material screening will play an essential role in the achievement of step change improvements in exhaust gas aftertreatment technologies. This is enabled by knowledge-based robot-controlled preparation and dynamic models coupled with information from real operation [2].

During the last decade, high-throughput (HT) methods have become an established tool for the evaluation of heterogeneous catalysts for different chemical processes [3,4,5,6,7,8,9,10,11,12,13,14,15,16]. hte's approach to HT testing is based on a patent filed on 3 March 1998, which relates to parallel reactor systems for testing the activity of solid catalysts simultaneously exposed to gaseous feed streams [17]. Further developments of hte cover the parallel processing of liquid feed and handling of liquid feed products. For example, an improved system for controlling the reaction pressure in parallel reactor systems had been described [18].

Since 2000, hte has applied an HT technology platform in the field of automotive catalysis. Several successful R&D programs demonstrated the power of the rapid testing paradigm when exploring large formulation matrices. Results of fast and precise screening of chemical compositions have been described [19,20]. However, compared to other applications, the number of studies reporting on the use of HT screening for automotive applications is still relatively small [21,22,23,24,25].

In a previous publication, the main features of the hte screening platform for automotive applications were described [26]. Two case studies on (i) steady-state testing of hydrocarbon oxidation and (ii) testing of dynamic oxygen storage capacity (OSC) were given to exemplify the state of maturity of HT technology and its application for the development of exhaust gas catalysts. The intention of the current paper is to provide insight into the actual measurement in an HT reactor system with time resolved, quantitative online analytics, which is fundamentally different from other conventional catalyst screening methodologies. Since the analytical sensors are usually the cost-critical part in an experimental setup, the experimental sequence has been optimized in the HT units to

Citation: Sundermann and Olga Gerlach, High-Throughput Screening as a Supplemental Tool for the Development of Advanced Emission Control Catalysts: Methodological Approaches and Data Processing, doi:10.3390/catal6020023

make the most efficient use of these analyzers. This involves fully-automated operation of 48-fold parallel reactors and a sequence of measurements that minimizes equilibration time during and after changing experimental conditions. As a consequence, experimental results are less straight forward to evaluate than in conventional tests, and adequate software support for data processing is required.

In addition to the data handling aspect of HT testing, this manuscript will describe dynamic test protocols involving feed gas switches that have been successfully implemented on a 48-fold testing unit.

RESULTS AND DISCUSSION

hte has acquired 15 years of experience in operating parallel reactors in the field of environmental catalyst screening. The most important lesson learned in this period is that relevant data can only be generated if oversimplified test conditions and sample preparation methodologies are avoided. This means that activity testing cannot neglect the well-known inhibiting effects of steam and sulfur in the exhaust. HT units are used for the fast primary screening of new materials and for accelerated catalyst evaluation by variation of test conditions (GHSV, T, feed composition, *etc.*) close to the values found in the actual application. An important use related to product development involves screening of large sample libraries for optimization of washcoat formulations, both in terms of composition and processing conditions. These libraries are typically defined by systematic variation of different parameters (e.g., carrier, PGM type, PGM loading, promoter type, promoter quantity, preparation route, calcination temperature, binder, slurry pH, slurry aging, *etc.*) based on the statistical design of experiment methods [27,28]. Even at the initial screening stage, such studies usually take hydrothermal and S-aging into account to avoid costly false positives.

In order to support actual catalyst development, close interaction with engine testing and scale-up is required. Experiments can be simplified to save time and/or costs, but any allowed simplification requires frequent re-evaluation during the progress of a research project. Since environmental catalysis is a rather mature field, every new development needs to be benchmarked against a rather high state of the art. As a consequence, HT technology is often used for incremental improvements on existing catalyst technologies, rather than for application-detached combinatorial screening using very simple conditions and giving more qualitative results. hte's HT workflows can significantly increase the screening capacity of an environmental catalyst development lab, although the sample throughput is still orders of magnitude smaller than what is considered high throughput in pharmaceutical or biochemical fields. In combination with proper statistical tools like DoE, much larger parameter spaces can be screened in a more reliable way than with conventional experimentation.

EXPERIMENTAL SECTION

Description of the HT Test Unit

The key features of hte's 3rd generation HT test reactor shown in Figure 1 can be summarized as follows.

Forty eight reaction and 1 by-pass positions with liners made from stainless steel with an inner diameter of 7 mm for each individual reactor (*cf.* Figure 2).

Liner filling up to 1 mL.

T = 110–575 °C; atmospheric pressure; GHSV (gas hourly space velocity): 30,000–100,000 h^{-1}.

Dosing of gases (NO, NO_2, NH_3, N_2O, HC, CO, CO_2, O_2, N_2) and liquids (H_2O, HC).

Online gas analyzers for NO, NO_2, N_2O, NH_3, HC, CO, CO_2, O_2 and H_2, as well as a mass spectrometer for specific compounds (*m*/*z* 1–512).

Changes in feed gas composition in the s range (e.g., lean/rich cycles).

Flexible process control ("hteControl4") for automated experiments with complex test protocols, as well as unattended and safe 24/7 operation.

Easy change between preconfigured operation modes (diesel oxidation catalyst (DOC), TWC, selective catalytic reduction (SCR), lean NO_x traps (LNT)).

Automated data processing and data management system ("myhte") allowing easy data export to sophisticated data analysis solutions.

Figure 1: Environmental catalysis high-throughput (HT) testing unit.

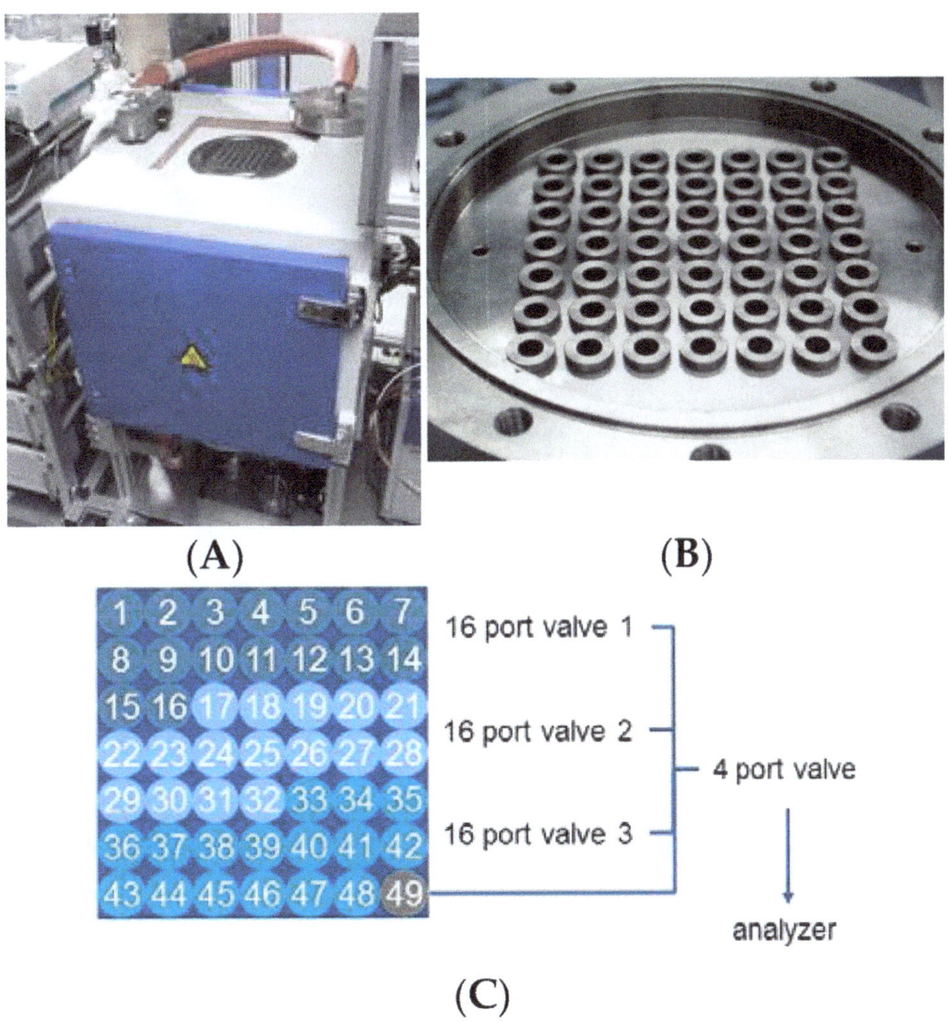

Figure 2: Reactor layout for HT testing unit (**A**). The reactor (**B**) (closeup view) has 48 equivalent positions for catalyst screening and one position (labelled as 49), which is used as a by-pass line to perform periodic measurements of the inlet concentrations. Numbering and relation of positions are shown in part (**C**).

S-Aging

In addition to the actual testing units, the lab is equipped with a sulfur-aging unit having an identical 48-fold reactor block. For modelling the catalyst lifecycle, this is a valuable supplemental tool to perform catalyst evaluation regarding poisoning/regeneration behavior without liner refilling. Thus, it is possible to run sulfur aging studies during primary material screening and, therefore, reduce the number of false positive findings that would occur if more expensive experiments were conducted at a larger scale.

For evaluation of sulfur tolerance (or sulfur resistance) from exposure to sulfur oxides (SO_x), different accelerated S-aging protocols are used. The main characteristics of the S-aging protocols applied at hte can be summarized as follows:

Temperatures: 300–450 °C.

Feed gas: 10–100 ppm SO_2, 10% H_2O in air; optional: 100–200 ppm NO.

Duration: 5–70 h.

After exposure to SO_2, H_2O and NO (if used), the catalysts are cooled in air.

The S-aging conditions are adjusted to the target S loading relevant for the application, and in most cases, this corresponds to 0.5–5 g·S/L for monolith catalysts with 120-g/L washcoat loading. Thermal regeneration is performed at temperatures between 600 and 800 °C using feed gas with 10% H_2O in air.

Test Procedure

Screening protocols are continuously validated and optimization by integrating feedback from scale-up experiments on cores and full-sized parts in associated laboratories.

Catalyst Amount and Shape

The catalysts used for HT testing are shaped to a particle size of 250–500 μm. Usually, this is achieved by formulating the active components into a slurry, milling to a D_{50} < 15 μm, drying under agitation and crushing/sieving after calcination. A particle size fraction of 250–500 μm is typically used as a compromise between pressure drop over the catalyst bed and having a realistic inter-particle diffusion length comparable to that found in a typical washcoat thickness on a coated automotive catalyst monolith. The majority of the catalysts are tested as powders with 100–300 mg diluted to a 1-mL bed volume with corundum of the same particle size fraction. Another option often used for testing of crushed monolith catalysts uses a sieve fraction between 500 and 1000 μm. For proper comparison, catalyst amounts in each reactor are adjusted for the identical content of active components (e.g., PGM (precious group metal)). The quantity of catalyst is selected to represent that found in 1 mL of monolith catalyst, such that space velocities can be calculated with reference to 1 mL of catalyst volume facilitating the transfer and comparison of results with monolith core and full size tests.

Hydrothermal Aging

The catalysts are tested both fresh and after hydrothermal oven aging performed in feeds containing 5%–15% water. For diesel applications, catalysts are kept for 5–24 h at a temperature between 700 and 850 °C in water/air mixture, cooled to 300 °C in the presence of water and further to room temperature in air. Hydrothermal aging for TWC applications is performed at temperatures between 850 and 1150 °C with a duration of 5–24 h. "Rich/lean" aging is also performed with two gas feeds containing 10% water (4% H_2 in N_2 and air) that are switched every 10 min during aging. For quality assurance, furnaces are equipped with time-resolved monitoring of temperature and steam dosing.

Screening Protocols

The screening protocols related to both diesel and gasoline programs can be divided into two groups in terms of HT test rig operation and data processing: (i) steady-state tests; and (ii) dynamic tests with feed switches. Testing protocols and on-line analytics are defined to be as close as possible to the conventional lab testing of automotive catalysts. Simulated exhaust feeds and test conditions are used to mimic selected operating points within regulated driving cycles (e.g., different for light-duty and heavy-duty applications). Due to the 48-fold reactor block of a high thermal mass, only isothermal operation with discrete set points is possible. Thus, no experiments with dynamic temperature ramps can be performed. During testing, all positions are exposed to the same feed all of the time, whereby the measured channel is selected by active flow control of down-stream VICI valves (*cf.* Figure 2). While the catalyst in a selected position is being evaluated, the remaining channels are exposed to a lower flow rate. This methodology has the advantage that all catalysts remain equilibrated close to their steady state for any given operating condition, thus allowing for short equilibration time after a new position is selected, and space velocity is increased to the target value (*cf.* Figure 3 for a typical response of the catalysts to the change in feed conditions). In general an experiment in the parallel reactor has the following test sequence, which is automated in the control software:

Set first experimental condition (temperature, feed gas levels).

Wait until the whole reactor is equilibrated.

Switch to Position 1.

Equilibrate in stationary feed or run dynamic feed switching program.

Repeat Steps 3 and 4 for all 48 reactor positions.

Set next experimental condition (e.g., higher temperature).

Continue with Steps 2–6 until all conditions are evaluated for all 48 reactors.

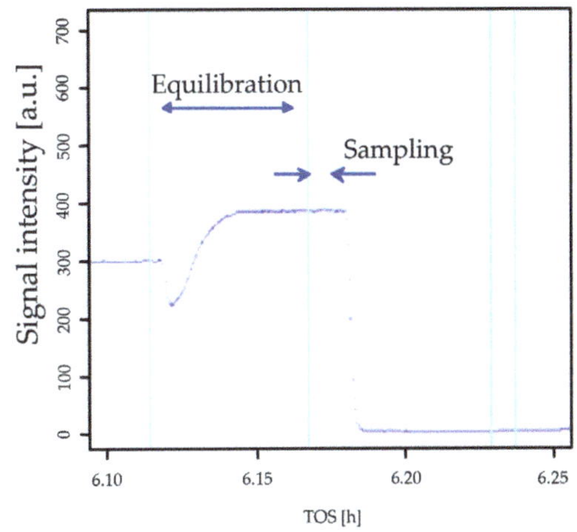

Figure 3: Time dependency of the analytical signal for selected positions in steady-state tests.

Data Processing

Throughout each experiment, the process values from all sensors (temperature, flow, pressure, gas analyzers) are recorded with a frequency of 1 Hz and automatically linked by the control software to the corresponding set points for that condition, most importantly to the reactor position that is being tested. An example of the typical raw data output is shown in Figure 4for a three-way catalyst screening protocol involving a stationary temperature ramp followed by a λ-sweep experiment. Obviously, for direct catalyst comparison, data reduction is required. This data reduction process is automated in the control software, which averages the concentration readings for each position over a predefined time interval (usually 30 s). An example of this sampling time interval after an equilibration period and before selecting the next reactor position is shown in Figure 3. Selecting this sampling interval, as well as a proper equilibration time after switching reactor positions is an important part of test protocol validation and depends on the application. In this example, for each reactor position, a 3-min equilibration time with 30-sec sampling time was defined. The average values (Figure 5) are then transferred into a relational database system ("myhte" data warehouse) and can be retrieved for further processing (e.g., using R, a language and environment for statistical computing [29]). Obviously, more complex test protocols involving the response of a catalyst to transient conditions (e.g., monitoring NO_x breakthrough curves after rich spikes in LNT (lean NO_x trap) testing) involve more complex data reduction steps.

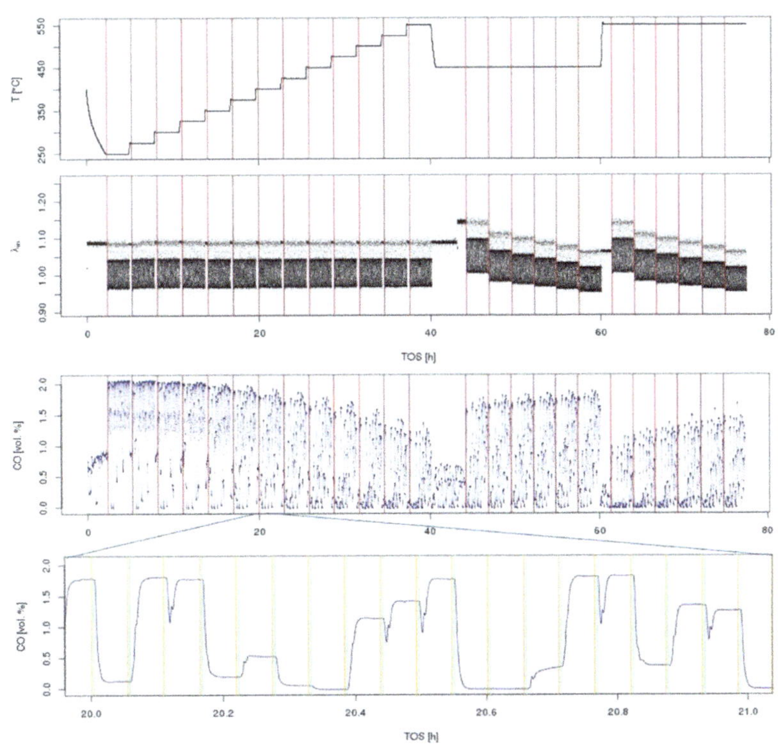

Figure 4: Overview of time-resolved sensor readings from a typical three-way catalyst screening in a simulated gasoline engine exhaust with λ perturbation. The test involves a light-off at λ = 1 at T = 250–550 °C in 25 °C steps followed by a λ sweep at T = 450, 550 °C at 6 different λ values, 1.05, 1.02, 1.01, 1.0, 0.99, 0.98, in the sequence from lean to rich (top: temperature and λ traces; bottom: CO concentration for the whole experiment + zoom into a shorter time interval). Vertical lines in the top three graphs indicate a change in experimental conditions (T or λ); in the bottom graph, they indicate the averaging interval for each reactor position. Throughout the whole experiment, >270,000 data points are collected for each individual sensor.

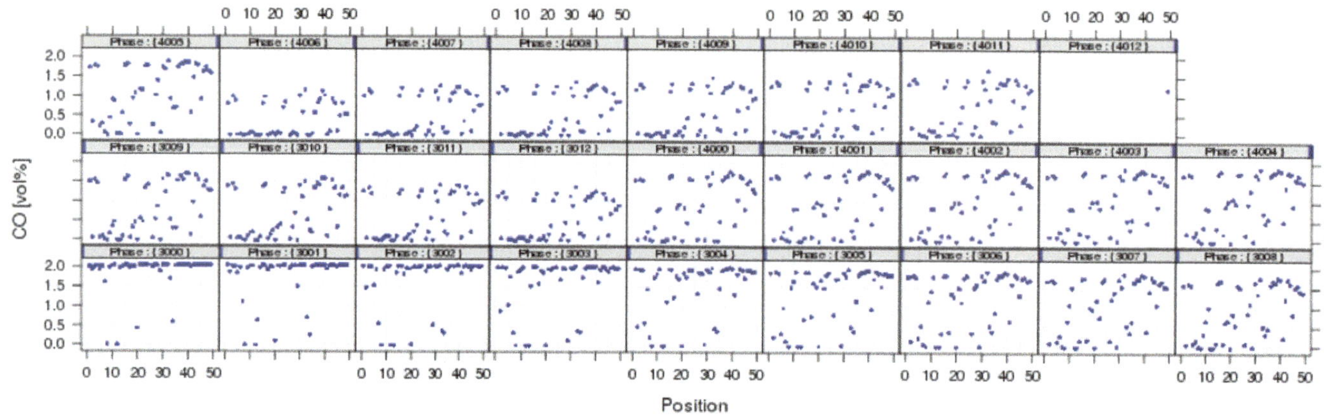

Figure 5: Average CO concentration for each individual reactor position after the initial data reduction step. The individual subplots correspond to the different experimental conditions shown in Figure 4.

Once the results are stored in the database, it is possible to relate individual measurements, such as using the inlet concentration measured for the by-pass line, to calculate conversion. Another important step in many test protocols is the calculation of light-off temperature. From the measurement of individual gas concentrations and the calculation of conversions resulting from the stationary operation, accurate light-off temperatures are obtained from interpolation of test data obtained at the discrete temperature intervals. This interpolation methodology needs to be flexible, robust and must be applicable to all catalysts tested in the same experiment. An automated estimation of light-off temperatures was implemented for hydrocarbon conversions (X-HC) in diesel oxidation catalysts [30]. The algorithm uses logistic functions ($f(T) = a/(1 + \exp(-b(c - T)))$ as a basis and a nonlinear least squares fitting procedure to create the light-off curves shown inFigure 6. In this case, T_{70} (HC) at X-HC = 70% was calculated in a numerical root-finding procedure. The overall process of calculating light-off temperatures has been implemented into hte's automated data processing workflow, so that the data reduction from a light-off experiment on 48 different samples (see Figure 7) only takes minutes and requires little human interaction. Usually, the catalysts tested together in one experiment (in the same plate or several plates tested in a campaign) are not totally unrelated, but rather, are part of a designed experiment. Utilizing the principles of the statistical design of experiments (DoE) [27,28], it is possible to quickly optimize material composition, slurry formulations, thermal activation steps or other treatments of interest. Depending on the problem, factor screening in an early stage or optimization based on previous knowledge, either fractional factorial or response surface design methodologies can be applied. As the capacity of the 48-fold parallel reactor naturally limits the number of samples, split plot designs are utilized when larger libraries of samples have to be spread over several experiments. Sometimes, computer-generated optimal designs, such as D-optimal designs, are required to ensure that these design constraints do not introduce uncontrolled statistical bias.

It must be understood that each high-throughput experiment with its complex test protocols is just a small part of a larger knowledge-finding process. To make this process robust and reliable, the following requirements for using HT testing must be taken into account: (i) continuous back-to-back comparison of stationary HTE and dynamic data to verify that the observed effects are real for the application; (ii) the HT lab must be flexible for re-optimization of test protocols (sometimes requiring modification of the unit); and (iii) there must be close integration of powder testing results with scale-up experiments in research projects.

Quality Assurance

HT testing in a parallel reactor heavily relies on the assumption that all 48 channels are equivalent with respect to temperature and flow distribution. Obviously, this assumption has to be verified for any new test protocol. In Figure 8, the results for such an experiment are shown using the SCR reaction over a Cu-chabazite-based catalyst. In this case, 18 samples from the same catalyst batch were mounted on one plate. The positions were selected to cover the whole range of the reactor block. An experiment at 4 temperature levels was performed, and a satisfactory 2.5% absolute standard deviation in the measured NO_x conversion was found in the light-off region, which is most sensitive to small changes in reaction conditions. Furthermore, there is no systematic trend with respect to position on the 48-fold reactor plate that would indicate problems with dosing stability or temperature distribution over the reactor block.

It is advisable to include a standard sample into every experimental plate as a check that sensor aging or contamination of lines is not causing problems. Such validation should be performed in a periodic fashion, preferably with the same standard sample to protect against creeping loss of precision. An additional measure to ensure robust data output from a parallel reactor is to ensure that factors of the experimental design are not biased with respect to either reactor position or time on stream.

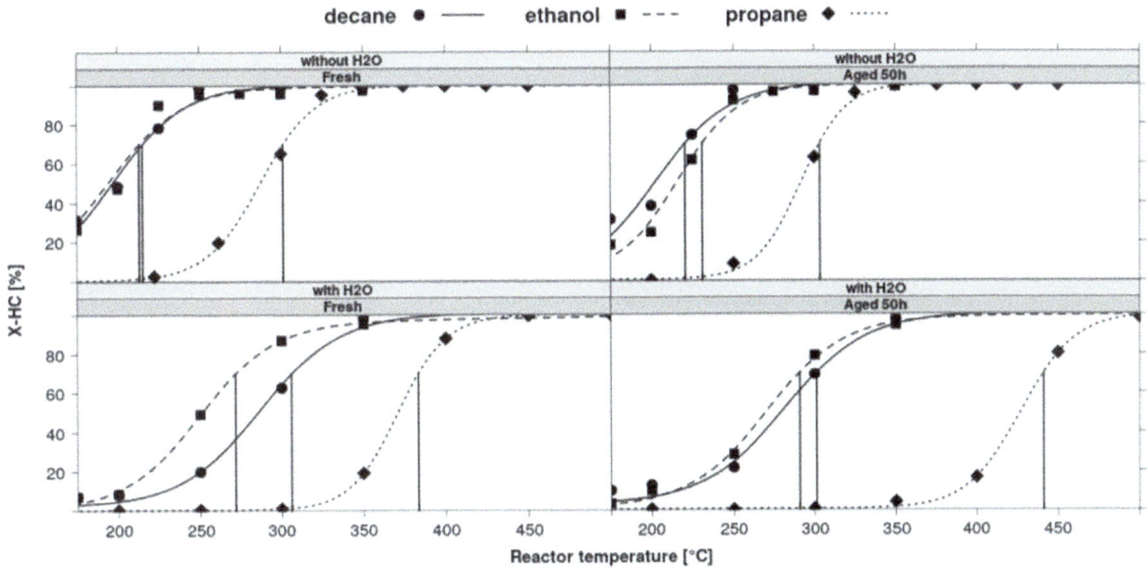

Figure 6: Light-off curves for different hydrocarbons obtained by fitting measured data points. Catalyst: 1% Pd/Al_2O_3 prepared from Pd nitrate as the precursor.

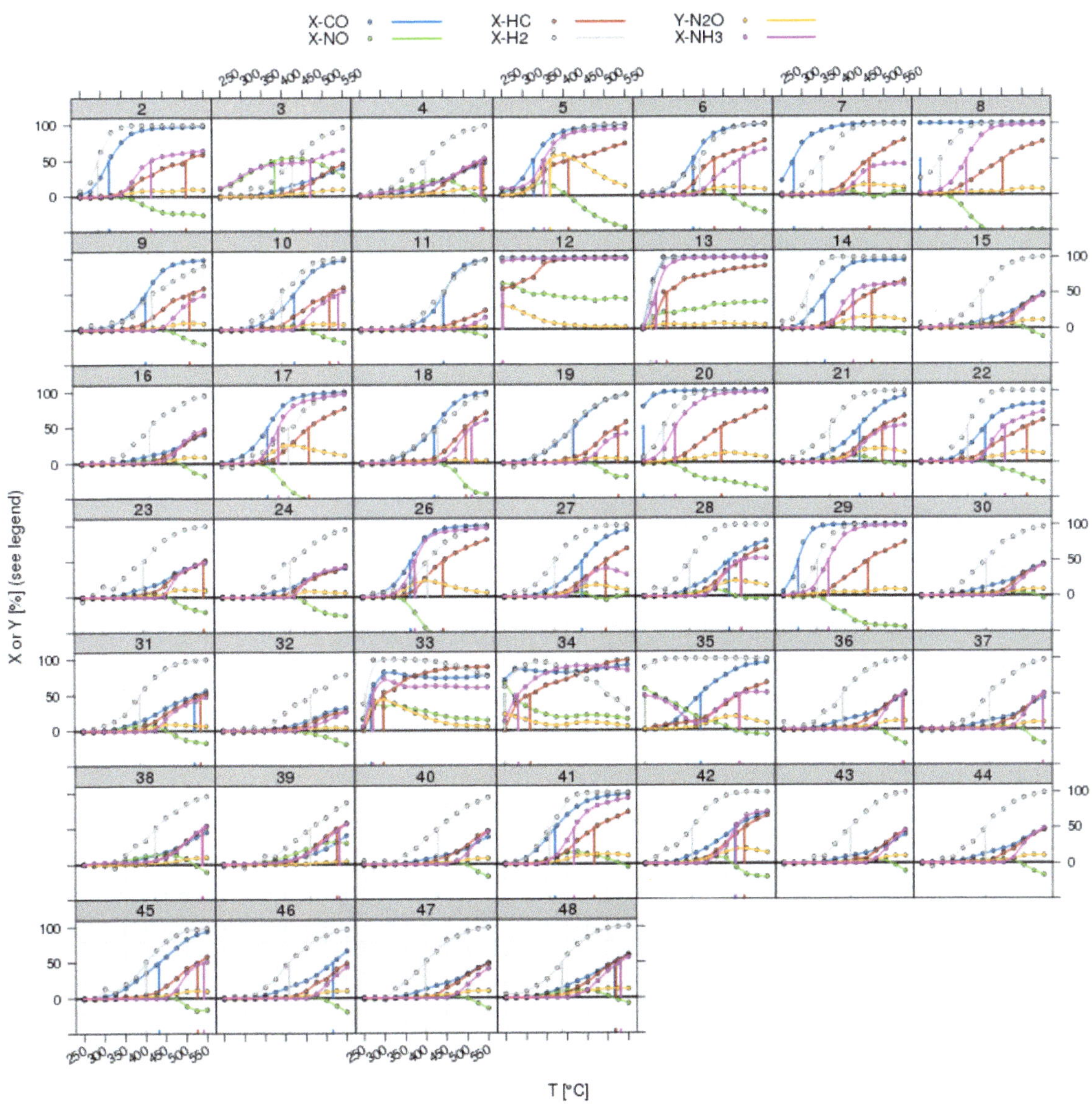

Figure 7: Light-off curves for all pollutants in a three-way catalyst (TWC) experiment with λ perturbation; overview for all catalyst positions in a screening plate (commercial benchmarks have been omitted). The vertical lines indicate the position of T_{50} from the automated fitting procedure.

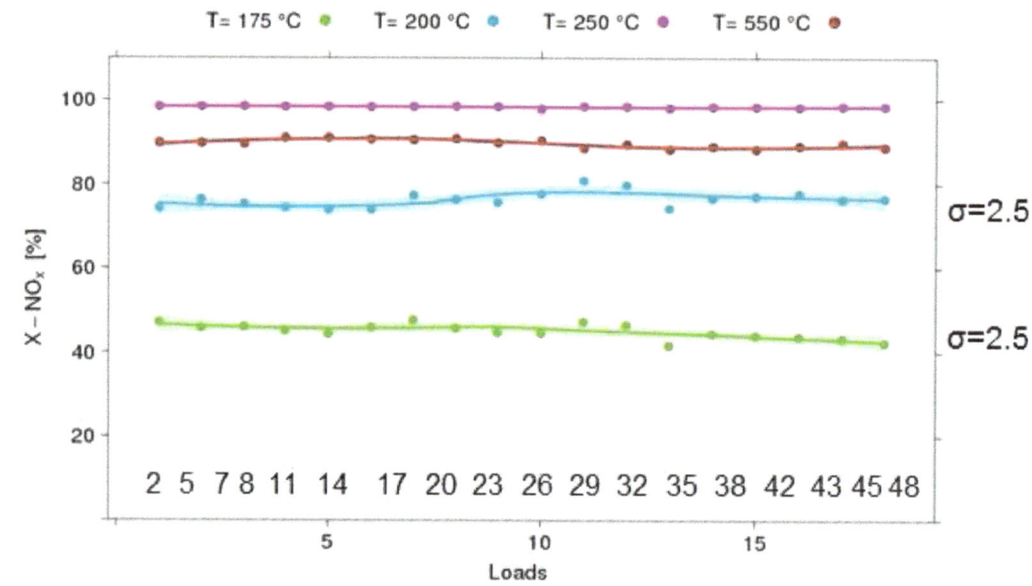

Figure 8: Results of validation experiments on the equivalence of channels (example: 18 loads of the selective catalytic reduction (SCR) Cu-Chabazite catalyst, 120 mg/reactor, 500 ppm NO, 500 ppm NH_3, GHSV 70 kh^{-1}).

Test Protocols

At the, a variety of test protocols are applied on a routine basis. These can be split into steady-state tests and tests with dynamic feed switches. Table 1 summarizes the test conditions for some of the steady-state test protocols used at hte related to catalyst technologies for diesel emission control.

Steady-State Tests

Diesel oxidation catalysts (DOC): Although new diesel cars consist of rather complex system arrangements, including a particulate filter and components for the reduction of NO_x emission, the catalytic functionality to oxidize CO and HC remains an important component of diesel aftertreatment systems. Typical tests involve several stationary light-off experiments on fresh and oven-aged samples, including tests for performance after S-aging and thermal regeneration.

Selective catalytic reduction (SCR): Experiments are performed with a simulated exhaust using NH_3 as the reductant. Samples are tested under standard SCR conditions (*i.e.*, without NO_2 in the feed) and with varying levels of the NO_2/NO_x ratio. This test also involves a dynamic component to measure the NH_3 storage capacity in a cycling feed with and without NH_3.

NH_3 oxidation: An important component of an SCR aftertreatment system, the NH_3 oxidation catalyst ensures removal of unconverted NH_3 downstream of the SCR catalyst.

Table 1: Overview of steady-state test protocols.

Catalyst technology	Feed compositions	GHSV (kh -1)	Temperatures
Diesel oxidation catalyst (DOC)	500–3000 ppm CO, 80–500 ppm-C_1 HC (HC = methane, propene, octane, decane, toluene or mixtures thereof) 50–200 ppm NO 5%–10% O_2, 5%–10% CO_2, 5%–10% H_2O	45–80	up to 12 temperature levels between 120 and 350 °C
Selective catalytic reduction (SRC)	50–1000 ppm NO 50–300 ppm NO_2 50–1000 ppm NH_3 5%–10% O_2, 5%–10% H_2O Optional: 5%–10% CO_2 50–500 ppm-C_1 HC (see DOC)	30–90	5–10 temperature levels between 150 and 575 °C
Ammonia oxidation	50–1000 ppm NH_3 5%–10% O_2, 5%–10% H_2O	30–90	2 temperature levels between 200 and 575 °C

Dynamic tests with feed switches (see Figure 9)

Three protocols are used for the evaluation of TWC (three-way catalyst) functionalities for exhaust aftertreatment in gasoline cars. Two protocols (LNT and NH_3 storage) are related to the evaluation of catalyst technologies used for NO_x reduction in diesel exhaust. Table 2 summarizes the corresponding test conditions.

Table 2: Overview of dynamic test protocols with fast feed switches.

Catalyst technology	Feed compositions	GHSV (kh -1)	Temperatures
Three-way catalysis	1%–3% CO, 1000–4000 ppm-C_1 HC (HC = methane, propene, propane, *iso*-butane, *iso*-octane, toluene or mixtures thereof) 1%–3% H_2 500–2000 ppm NO 0.5%–3 % O_2, 10%–15 % CO_2, 10%–15 % H_2O Oscillating lean/rich perturbations with defined average λ and amplitude (frequency 0.5–0.2 Hz)	45–80	up to 12 temperature levels between 200 and 550 °C
LNT (lean NO_x trap)	50–1000 ppm NO 50–300 ppm NO_2 500–3000 ppm CO, 5%–10% O_2, 5%–10% H_2O Optional: 5%–10% CO_2 50–500 ppm-C_1 HC (see DOC) Rich spikes (3–10 s) with defined λ (by reducing O_2 and increasing CO/H_2)	30–90	5–10 temperature levels between 150 and 575 °C
OSC (oxygen storage capacity)	Alternating gas atmosphere with 2% CO 1% O_2 (variable duration 10/10 s, 30/30 s)	60	2 temperature levels between 350 and 575 °C

TWC (three-way catalysis): Three-way catalysts operate best in an exhaust of an engine operated close to stoichiometry, an air:fuel ratio of 1. As the catalyst has to tolerate excursions from this optimal point, tests with λ perturbations are crucial. Several protocols are used to evaluate the different functionalities of fresh and aged catalysts.

Light-off with average λ = 1 and high frequency (up to 0.5 Hz) λ perturbations.

λ-sweep test at different temperatures. Performed by running light-off tests, but with different average lambda set points, and then plotting conversion as a function of lambda at constant temperature.

OSC test (oxygen storage capacity): specific test for the oxygen-buffer functionality of the catalyst. Involves a feed cycling between CO and O_2 in nitrogen and monitoring CO_2 generation from stored oxygen. Results of this test have been previously reported [26].

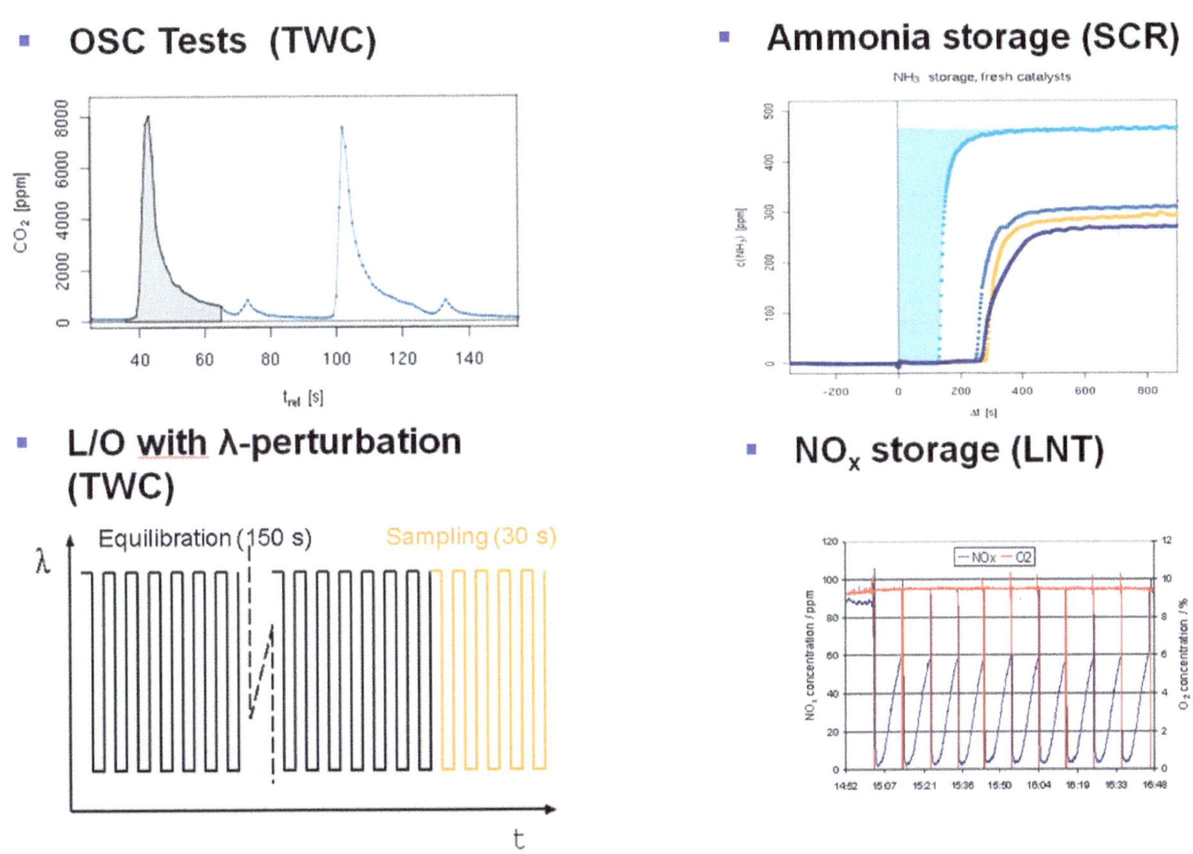

Figure 9: Examples for dynamic test protocols implemented on a 48-fold reactor system. Top left: OSC test, cycling O_2 and CO containing feed and monitoring CO_2release in the absence of O_2 in the gas phase, which is a measure for the oxygen storage capacity at a given temperature. Top right: NH_3 storage test by cycling NH_3 and monitoring the NH_3 breakthrough under SCR reaction conditions (*i.e.*, in the presence of NO, O_2, H_2O, optionally CO_2 and/or HC). Bottom right: lean NO_xtraps (LNT) test with repeated rich regeneration cycles with defined λ. The NO_xbreak-through is monitored and analyzed. Bottom left: TWC testing in a feed with rapid λ oscillations with defined amplitude. After an equilibration time of 150–180 s, the traces from relevant gas analyzers are averaged for 30 s.

LNT (lean NO_x trap): Important for lab testing of NO_x storage catalysts is the ability to generate reproducible rich pulses with defined rich λ and pulse width. Usually, 3–7 pulses are applied to each catalyst. After each rich pulse, the NO_x breakthrough curve is recorded and integrated in several ways to calculate average NO_x conversion or average NO_x storage. Here, it is also important to compare storage in each individual rich/lean cycle and to monitor time on stream effects. The LNT test is usually run at several temperatures. A crucial component of LNT evaluation is activity after S-aging and lean/rich regeneration.

Ammonia storage for SCR to measure the NH_3 storage capacity in a cycling feed with and *w/o* NH_3.

The typical throughput in terms of samples and the quantity of generated data points is summarize for each class of protocols in Table 3.

Table 3: Average throughput for different test protocols.

Topic	Performance characteristics/Protocols	Typical throughput (samples/week)	Data output (data points/week)
DOC	Light-off performance (CO/HC/NO) and sulfur resistance 2–5 light-offruns per sample	45–135	4500–11,250
SCR	Low and high temperature SCR performance 4 protocols (standard. and fast SCR, ammonia oxidation and storage)	30–45 (all protocols) 45–135 (*w/o* NH_3 storage)	2800–4200
TWC	Oxygen storage capacity (OSC) Catalytic performance: light-off, λ-sweep for CO/HC/NO	225 (OSC) 45–90 (L/O and λ-sweep)	900 (OSC) 6000–12,000- (L/O and λ-sweep)
LNT	NO_x efficiencies (lean/rich and lean) and NO_x storage (lean) at 3 temperatures, 5 cycles per position, two sample loads	45	1800–2700

SUMMARY

hte GmbH applies three state-of-the-art test units with 48 parallel reactors for several R&D programs in the field of environmental catalysis. High-throughput experiments have proven to be a powerful tool for rapid testing of a large number of formulations and enable fast and precise exploration of both chemical and experimental parameter spaces. To make best use of HT screening capacity-designed experiments, data processing and statistical methods for catalyst optimization are of high importance. Software and algorithms for data handling have been established and proven to be a powerful tool for handling large datasets. High throughput material testing for several automotive applications has been successfully cross-validated with the results on cores and full-sized monoliths

The results from different R&D projects demonstrate that HT screening yields high data, which can drive a rational approach to catalyst development and optimization beyond the level of pure primary material screening.

ACKNOWLEDGMENTS

Financial support by BASF is greatly acknowledged.

AUTHOR CONTRIBUTIONS

Olga Gerlach and Andreas Sundermann wrote the paper. Both authors conceived and designed the experiments, collected and analyzed the data and wrote the paper. Andreas Sundermann did all work related to the numerical data processing.

CONFLICTS OF INTEREST

The authors declare no conflict of interest.

REFERENCES

1. Heck, R.M.; Farrauto, R.J.; Gulati, S.T. *Catalytic Air Pollution Control, Commercial Technology*, 3rd ed.; John Wiley & Sons: Hoboken, New Jersey, 2009.
2. Deutschmann, O.; Grunwaldt, J.-D. Abgasnachbehandlung in mobilen Systemen: Stand der Technik, Herausforderungen und Perspektiven. *Chem. Ing. Tech.* 2013, *85*, 595–617.
3. Rodemerck, U.; Wolf, D.; Buyevskaya, O.V.; Claus, P.; Senkan, S.; Baerns, M. High-throughput synthesis and screening of catalytic materials: Case study on the search for a low-temperature catalyst for the oxidation of low-concentration propane. *Chem. Eng. J.* 2001, *82*, 3–11.
4. Hendershot, R.J.; Lasko, S.S.; Fellmann, M.F.; Oskarsdottir, G.; Delgass, W.N.; Snively, C.M.; Lauterbach, J. A novel reactor system for high throughput catalyst testing under realistic conditions. *Appl. Catal. A* 2003, *254*, 107–120.
5. Lucas, M.; Claus, P. High throughput screening in monolith reactors for total oxidation reactions. *Appl. Catal. A* 2003,*254*, 35–43.
6. Mills, P.L.; Nicole, J.F. A novel reactor for high-throughput screening of gas-solid catalyzed reactions. *Chem. Eng. Sci.*2004, *59*, 5345–5354.
7. Paul, J.S.; Jacobs, P.A.; Weiss, P.-A.W.; Maier, W.F. Combinatorial discovery of new catalysts for the selective oxidation of isobutene. *Appl. Catal. A* 2004, *265*, 185–193.
8. Zech, T.; Bohner, G.; Klein, J. High-throughput screening of supported catalysts in massively parallel single-bead microreactors: Workflow aspects related to reactor bonding and catalyst preparation. *Catal. Today* 2005, *110*, 58–67.
9. Brooks, C.; Cypes, S.; Grasselli, R.K.; Hagemeyer, A.; Hogan, Z.; Lesik, A.; Streukens, G.; Volpe, A.F., Jr.; Turner, H.W.; Weinberg, W.H.; *et al.* High throughput discovery of CO oxidation/VOC combustion and water-gas shift catalysts for industrial multi-component streams. *Top. Catal.* 2006, *38*, 195–209.
10. Farrusseng, D. High-throughput heterogeneous catalysis. *Surf. Sci. Rep.* 2008, *63*, 487–513.
11. Moehmel, S.; Steinfeldt, N.; Engelschalt, S.; Holena, M.; Kolf, S.; Baerns, M.; Dingerdissen, U.; Wolf, D.; Weber, R.; Bewersdorf, M. New catalytic materials for the high-

temperature synthesis of hydrocyanic acid from methane and ammonia by high-throughput approach. *Appl. Catal. A* 2008, *334*, 73–83.

12. Turner, H.W.; Volpe, A.F., Jr.; Weinberg, W.H. High-throughput heterogeneous catalyst research. *Surf. Sci.* 2009, *603*, 10–12.
13. Gaudillere, C.; Vernoux, P.; Mirodatos, C.; Caboche, G.; Farrusseng, D. Screening of ceria-based catalysts for internal methane reforming in low temperature SOFC. *Catal. Today* 2010, *157*, 263–269.
14. Valtchev, M.; Hammes, M.; Richter, R.; Holtzen, H.; Stowe, K.; Maier, W.F. Corrosion-Resistant Parallel Fixed-Bed Reactors for High-Throughput Screening of New Deacon Reaction Catalysts. *Chem. Eng. Technol.* 2014, *37*, 1251–1260.
15. Zhu, H.B.; Laveille, P.; Rosenfeld, D.C.; Hedhili, M.N.; Basset, J.M. A high-throughput reactor system for optimization of Mo-V-Nb mixed oxide catalyst composition in ethane ODH. *Catal. Sci. Technol.* 2015, *5*, 4164–4173.
16. Serna, P.; Baumes, L.A.; Moliner, M.; Corma, A. Combining high-throughput experimentation, advanced data modeling and fundamental knowledge to develop catalysts for the epoxidation of large olefins and fatty esters. *J. Catal.* 2008, *258*, 25–34.
17. Brenner, A.; Schüth, F.; Schunk, S.A.; Stichert, W. Anordnung zum Testen der katalytischen Aktivität von einem Reaktionsgas ausgesetzten Feststoffen. DE 198 61 316 B4, 6 October 2005.
18. Haas, A.; Strehlau, W.; Brenner, A.; Koechel, O.; Friess, M.; Zech, T. Device and method for pressure and flow control in parallel reactors. WO 2005063372 A2, 17 July 2005.
19. Strehlau, W.; Gerlach, O.; Maier, J.; Gabriel, T. Catalysts for the simultaneous removal of carbon monoxide and hydrocarbons from oxygen-rich exhaust gases and processes for the manufacture thereof. WO 2005102513 A1, 3 November 2005.
20. Strehlau, W.; Gerlach, O.; Maier, J.; Gabriel, T. Catalyst for the treatment of exhaust gases and processes for producing the same. WO 2006120013 A1, 16 November 2006.
21. Hendershot, R.J.; Rogers, W.B.; Snively, C.A.; Ogunnaike, B.A.; Lauterbach, J. Development and optimization of NO_x storage and reduction catalysts using statistically guided high-throughput experimentation. *Catal. Today* 2004, *98*, 375–385.
22. Hendershot, R.J.; Vijay, R.; Snively, C.M.; Lauterbach, J. High-throughput study of the performance of NO_x storage and reduction catalysts as a function of cycling conditions and catalyst composition. *Chem. Eng. Sci.* 2006, *61*, 3907–3916.
23. Iojoiu, E.E.; Bassou, B.; Guilhaume, N.; Farrusseng, D.; Desmartin-Chomel, A.; Lombaert, K.; Bianchi, D.; Mirodatos, C. High-throughput approach to the catalytic combustion of diesel soot. *Catal. Today* 2008, *137*, 103–109.
24. Kern, P.; Klimczak, M.; Heinzelmann, T.; Lucas, M.; Claus, P. High-throughput study of the effects of inorganic additives and poisons on NH_3-SCR catalysts. Part II: Fe-zeolite catalysts. *Appl. Catal. B* 2010, *95*, 48–56.
25. Gärtner, A.; Lenk, T.; Kiemel, R.; Casu, S.; Breuer, C.; Stöwe, K. High-throughput screening approach to identify new catalysts for total oxidation of methane from gas fueled lean burn engines. In Proceedings of the Preprints of the tenth international congress on Catalysis and automotive pollution control, Brussels, Belgium, 28–30 October 2015.
26. Sundermann, A.; Gerlach, O. High-Throughput Screening Technology for Automotive Applications. *Chem. Ing. Tech.* 2014, *86*, 1941–1947.
27. Montgomery, D.C. *Design and Analysis of Experiments*, 8th ed.; John Wiley & Sons: Hoboken, New Jersey, 2012.
28. Box, G.E.P.; Hunter, J.S.; Hunter, W.G. *Statistics for Experimenters: Design, Innovation, and Discovery*, 2nd ed.; Wiley: New York, NY, USA, 2005.
29. Team, R.C. R Foundation for Statistical Computing, Vienna, Austria. Available online: http://www.R-project.org/(accessed on 25 January 2016).
30. Ricketts, J.H.; Head, G.A. A five-parameter logistic equation for investigating asymmetry of curvature in baroreflex studies. *Am. J. Phys.-Regul. Integr. Comp. Phys.* 1999, *277*, R441–R45

CHAPTER 10 EFFECTS OF CATALYST PREPARATION ON HYDROCARBON PRODUCT DISTRIBUTION IN HYDROCRACKING OF THE FISCHER-TROPSCH PRODUCT WITH LOW PT-LOADED CATALYSTS

Toshiaki Hanaoka, Tomohisa Miyazawa, Katsuya Shimura and Satoshi Hirata

Keywords. Biomass; hydrocracking; jet fuel; Pt; microemulsion

INTRODUCTION

In Japan, the annual demand for transportable liquid fuels is approximately 90 billion liters. Taking the reserve amounts of fossil fuel into account, the utilization of liquid fuel derived from biomass would contribute towards decreasing the consumption of fossil fuels because biomass is the only source of renewable energy that can be converted into liquid fuels (e.g., gasoline, light oil, and jet fuel), and the utilization of biomass-based fuel does not substantially increase the level of CO_2 emission.

Among conversion technologies from biomass to liquid fuels, the biomass-to-liquid (BTL) process generally consists of biomass gasification for producing syngas ($CO + H_2$), gas cleaning, Fischer-Tropsch (FT) synthesis, hydrocracking, and distillation. The process has the advantage of utilizing a wider range of feedstocks compared to other conversion technologies since biomass feedstock can be converted into syngas in the gasification step [1]. However, there are no standardized BTL plants in commercial operation anywhere in the world owing to a lack of economic sustainability.

The simulation study of the BTL process by Fujimoto *et al.* showed that the production of steam as a gasifying agent in the gasification step, compression of cleaned gas, and distillation required the dominant amount of input energy [2]. Therefore, biomass gasification without steam, as well as the enhancement of catalyst performance for FT synthesis and hydrocracking, has been reported previously.

Hydrocracking is one of the most important unit operations responsible for increasing the yield of hydrocarbon liquid fuel. In a petrochemical process, Ni-Mo and Co-Mo catalysts are generally employed because the feedstock has to be hydrocracked in the presence of sulfur compounds. In contrast, for the hydrocracking of clean feedstock derived from woody biomass, noble metals provide suitable active sites for the dehydrogenation/hydrogenation reactions [3]. In the hydrocracking of *n*-alkanes over Pt catalysts using SiO_2-Al_2O_3 [4,5], sulfated zirconia [6,7,8], tungsta-zirconia [8,9,10,11,12,13], and zeolite [14,15], higher product yields were obtained under mild conditions. Pt is a promising catalytic metal but is extremely costly. Therefore, the amount of Pt employed as a hydrocracking catalyst should be as low as possible in order to enhance the economics of the BTL process.

Leu *et al.* reported that in the isomerization of *n*-hexane over Pt/H-β catalysts, the catalytic activity and product selectivity were maximized at a Pt content of 0.3–0.5 wt. % [14]. However, in almost all cases, the catalytic activity and product selectivity decreased drastically when the Pt content

Citation: Toshiaki Hanaoka , Tomohisa Miyazawa, Katsuya Shimura and Satoshi Hirata, Effects of Catalyst Preparation on Hydrocarbon Product Distribution in Hydrocracking of the Fischer-Tropsch Product with Low Pt-Loaded Catalysts, doi:10.3390/catal5041983

was less than 0.5 wt. %. On the contrary, high activity and selectivity were obtained above 0.5 wt. % [6,11]. Thus, it is unclear whether the previously reported synergetic effect between Pt and its support materials is applicable for Pt-loaded catalysts having a lower Pt content.

In the isomerization of naphtha with carbon numbers of 5–10, some studies have reported the use of Pt-loaded catalysts with a Pt content of approximately 0.1 wt. % [16,17,18,19,20,21]. However, the performance of catalysts with low Pt content (*i.e.*, less than 0.1 wt. %) in the hydrocracking of the FT product has not been fully discussed. The authors previously reported the preparation of Pt-loaded catalysts using water-in-oil (w/o) microemulsions and the effect of Pt particle size on hydrocracking reaction [22,23]. This preparation method is advantageous as the size of the supported Pt particles is well controlled, regardless of the Pt content.

In this study, a screening study of support materials was performed using Pt-loaded catalysts with low Pt content. The effects of Pt content, particle size, and ratio between Pt and acid sites were investigated. As products, hydrocarbons with carbon numbers less than 8, 9–15 (*i.e.*, corresponding jet fuel), and greater than 16 were focused on, and the effect of structure of Pt-loaded catalysts with low content on product yield was discussed.

RESULTS AND DISCUSSION

Characterization

In the present study, 20 catalysts were used, including 19 catalysts prepared by the impregnation method and the deposition method using w/o microemulsions. The Pt-loaded catalysts prepared by the impregnation method are denoted as I-Pt(*X*)/support. *X* (0.01–0.1) represents the Pt content. Those prepared by the deposition method are denoted as D-Pt(A)/support. A represents the Pt complex forming agent. In the latter method, the Pt content was 0.02 wt. %.

Table 1 shows the surface area, pore volume, and average pore diameter of Pt-loaded catalysts. When zeolite supports were used (Entries 6–20), the Brunauer-Emmett-Teller (BET) surface areas were similar to the micropore areas. In contrast, the BET surface areas were similar to the mesopore areas when other supports were used (Entries 1–5). For these catalysts, no reasonable values of micropore areas were obtained by the *t* method [24]. These results indicated that mesopores were dominant in Entries 1–5, while micropores were dominant in Entries 6–20.

Table 1: Surface area, pore volume, and pore diameter of Pt-loaded catalysts.

Catalyst		BET Surface Area (m^2/g)	Total Pore Volume (cm^3/g)	Average Pore Diameter (nm)	Micropore Area (m^2/g)	Mesopore Area (m^2/g)
Entry	Remarks					
1	I-Pt(0.1)/SiAl8.7	333	0.83	10.0	-	368
2	I-Pt(0.1)/SiAl(28.6)	399	0.84	8.4	-	453
3	I-Pt(0.1)/SiAl(13)	515	0.67	5.2	-	523
4	I-Pt(0.1)/SiAl5.5	426	0.44	4.1	-	339
5	I-Pt(0.1)/S-ZrO_2	108	0.19	7.1	-	123
6	I-Pt(0.1)/FAU110	758	0.55	2.9	918	112
7	I-Pt(0.1)/MOR18	495	0.27	2.2	645	33
8	BEA28	701	0.67	3.8	849	209
9	I-Pt(0.01)/BEA28	719	0.58	3.2	876	211
10	I-Pt(0.02)/BEA28	709	0.57	3.2	864	211
11	I-Pt(0.04)/BEA28	709	0.66	3.7	862	211
12	I-Pt(0.06)/BEA28	719	0.63	3.5	871	215
13	I-Pt(0.1)/BEA28	719	0.62	3.4	874	213
14	I-Pt(0.1)/BEA40	640	0.30	1.9	836	59
15	I-Pt(0.1)/BEA41.6	603	0.31	2.0	784	76
16	I-Pt(0.1)/BEA510	608	0.39	2.6	781	96
17	D-Pt(CTAC)/BEA28	723	0.65	3.6	880	211
18	D-Pt(TPAB)/BEA28	727	0.62	3.4	884	209
19	D-Pt(TEAC)/BEA28	731	0.73	4.0	887	213
20	D-Pt(HTAB)/BEA28	740	0.67	3.6	894	214

The surface area, pore volume, and pore diameter of catalysts prepared using w/o microemulsions (Entries 17–20) were similar to those of impregnated catalysts using BEA28 (Entries 8–13). These properties were independent of the catalyst preparation methods when the same support was employed. BEA28 (Entry 8) was obtained from a calcination process performed at 500 °C for 2 h. The preparation procedure of the Pt-loaded BEA28 catalysts, by either the impregnation or deposition methods, did not significantly affect the structure of the BEA28 support. The Pt-loaded zeolite catalysts in Entries 8–13, and 17–20 had larger average pore diameters and mesopore areas compared to other zeolite catalysts (Entries 6, 7, and 14–16).

Table 2 shows the results of temperature-programmed desorption of ammonia (NH_3-TPD) of Pt-loaded catalysts. The NH_3-TPD analysis is generally employed to estimate the strength and amount of acid sites. NH_3 molecules are likely to be desorbed from acid sites at higher temperatures when the adsorbed NH_3 molecules interact with acid sites more strongly. Therefore, the acid amount and acidity can be estimated from the amount of desorbed NH_3 molecules and desorption temperature. In the present study, the acid sites from which the NH_3 molecules were released, in the temperature range from 100 to 250 °C, were regarded as weak acid sites. Meanwhile, those from which the NH_3 molecules were released, in the range of 250–450 °C and 450–800 °C, were regarded as medium and strong acid sites, respectively. The total amount of NH_3molecules desorbed from 100 to 800 °C was regarded as the acid amount.

Table 2: Acid amounts of Pt-loaded catalysts.

Catalyst		Acid Amount [a] (mmol/g)	Weak [b] (mmol/g)	Medium [c] (mmol/g)	Strong [d] (mmol/g)
Entry	Remarks				
1	I-Pt(0.1)/SiAl8.7	0.43	0.19	0.15	0.09
2	I-Pt(0.1)/SiAl(28.6)	0.80	0.30	0.22	0.28
3	I-Pt(0.1)/SiAl(13)	0.70	0.27	0.23	0.20
4	I-Pt(0.1)/SiAl5.5	0.56	0.25	0.20	0.11
5	I-Pt(0.1)/S-ZrO_2	0.63	0.16	0.18	0.30
6	I-Pt(0.1)/FAU110	0.13	0.05	0.05	0.03
7	I-Pt(0.1)/MOR18	1.18	0.68	0.22	0.28
8	BEA28	1.37	0.65	0.67	0.04
9	I-Pt(0.01)/BEA28	1.38	0.68	0.63	0.07
10	I-Pt(0.02)/BEA28	1.35	0.70	0.61	0.05
11	I-Pt(0.04)/BEA28	1.41	0.70	0.64	0.07
12	I-Pt(0.06)/BEA28	1.36	0.66	0.64	0.06
13	I-Pt(0.1)/BEA28	1.32	0.62	0.61	0.08
14	I-Pt(0.1)/BEA40	1.08	0.52	0.50	0.06
15	I-Pt(0.1)/BEA41.6	0.77	0.34	0.34	0.09
16	I-Pt(0.1)/BEA510	0.14	0.03	0.03	0.08
17	D-Pt(CTAC)/BEA28	1.46	0.70	0.70	0.06
18	D-Pt(TPAB)/BEA28	1.38	0.67	0.65	0.06
19	D-Pt(TEAC)/BEA28	1.38	0.69	0.64	0.05
20	D-Pt(HTAB)/BEA28	1.44	0.71	0.67	0.06

[a] Calculated by assuming 1:1 stoichiometry for the amounts of NH_3 and acid; [b] The acid sites where NH_3 desorbed from 100 to 250 °C; [c] The acid sites where NH_3 desorbed from 250 to 450 °C; [d] The acid sites where NH_3 desorbed from 450 to 800 °C.

The acid amount was largely dependent on the type of catalysts. When MOR18, BEA28, and BEA40 were employed as support materials (Entries 7–14 and 17–20), the acid amounts were more than 1.00 mmol/g, exceeding those of other support materials. In particular, the Pt-loaded BEA28 (Entries 8–13 and 17–20) had the highest acid amount. Similar to the pore structure for these catalysts in Table 1, the acid amount and acidity were independent of the Pt content and catalyst preparation methods.

Screening of the Hydrocracking Reaction

Firstly, a screening study using Pt-loaded catalysts with the same content but different support materials was performed to investigate how different factors associated with Pt-loaded catalysts with low Pt content affect the product yield. Table 3shows the analyses of the FT product as a feedstock of hydrocracking. The feedstock contained hydrocarbons with carbon numbers 5–56. Meanwhile, Table 4 shows the hydrocracking behavior of the FT product over Pt-loaded catalysts employed in this study. The product yields for hydrocarbons with carbon numbers 1–8, 9–15, and greater than 16 were denoted as $Y_{C1–C8}$, $Y_{C9–C15}$, and Y_{C16+}, respectively. The reaction temperature and initial H_2 pressure were set at 250 °C and 0.5 MPa, respectively, with a reaction time of 1 h. The reactor temperature was maintained at 250 ± 6 °C in all experimental runs.

Table 3: Analyses of the FT product as a feedstock of hydrocracking.

Content/% on a Carbon Basis				Elemental Analysis/wt %				
C5–C8	C9–C15 [a]	C16 [b]	C17+	C	H	N	S	O [c]
0.1	9.7	61.7	28.5	84.4	14.8	0.0	0.0	0.8

[a] Corresponding jet fuel; [b] Mixture of product obtained in the FT synthesis and solvent charged at the initial stage of the FT synthesis; [c] By difference.

Table 4: Hydrocracking reaction over Pt-loaded catalysts and Pt particle size by CO chemisorption.

Catalyst		Product Yield/% on a Carbon Basis				d_{pulse} (nm)
Entry	Remarks	Y_{C1-C8}	Y_{C9-C15}	Y_{C16+}	Loss	
1	I-Pt(0.1)/SiAl8.7	0.8	10.2	86.6	2.5	42.5
2	I-Pt(0.1)/SiAl(28.6)	0.3	11.5	89.2	−1.0	56.5
3	I-Pt(0.1)/SiAl(13)	0.2	9.2	90.6	0.0	66.6
4	I-Pt(0.1)/SiAl5.5	0.2	10.3	89.2	0.2	28.8
5	I-Pt(0.1)/S-ZrO_2	0.8	12.9	87.3	−1.0	13.4
6	I-Pt(0.1)/FAU110	1.0	9.8	89.0	0.2	26.3
7	I-Pt(0.1)/MOR18	2.3	13.3	85.8	−1.4	13.0
8	BEA28	19.4	20.7	58.5	1.4	-
9	I-Pt(0.01)/BEA28	24.1	22.4	54.2	−0.7	1.8
10	I-Pt(0.02)/BEA28	26.5	23.9	47.9	1.7	1.9
11	I-Pt(0.04)/BEA28	28.0	24.0	44.3	3.7	2.3
12	I-Pt(0.06)/BEA28	30.9	24.7	42.2	2.2	3.5
13	I-Pt(0.1)/BEA28	37.4	23.1	37.5	2.0	4.6
14	I-Pt(0.1)/BEA40	23.1	22.1	53.2	1.6	3.1
15	I-Pt(0.1)/BEA41.6	15.7	20.6	65.1	−1.4	4.8
16	I-Pt(0.1)/BEA510	4.2	10.6	83.8	1.4	35.4
17	D-Pt(CTAC)/BEA28	29.2	21.5	46.5	2.8	2.6
18	D-Pt(TPAB)/BEA28	27.5	23.9	48.6	0.0	2.0
19	D-Pt(TEAC)/BEA28	25.1	22.6	47.9	4.4	2.3
20	D-Pt(HTAB)/BEA28	27.4	22.7	47.5	2.4	1.0
21	None	0.8	9.4	90.9	−1.0	-

In all experimental runs, the mass balance on a carbon basis was almost 100%; therefore, the loss was only −1.4%–4.4% (Table 4). The individual product yield was largely dependent on the type of Pt catalysts.

Firstly, the effect of the support material on the hydrocracking reaction was investigated using impregnated catalysts, which had a Pt content of 0.1 wt. % (Entries 1–7 and 13–16). I-Pt(0.1)/BEA28, I-Pt(0.1)/BEA40, and I-Pt(0.1)/BEA41.6 (Entries 13–15) exhibited higher Y_{C9-C15} values (20.6%–23.1%) compared to those of other catalysts. In contrast, the product yields obtained in Entries 1–6 were almost the same as the run without a catalyst (Entry 21) and the feedstock composition, as shown in Table 3. Accordingly, the zeolite supports were promising for jet fuel production while the amorphous oxide supports hardly contributed to the hydrocracking of the FT product.

Figure 1 shows the effect of the amount of medium acid site on the individual product yield in the hydrocracking reaction (Entries 1–7 and 13–16 in Table 2 and Table 4, respectively). Increasing the amount of medium acid site led to an increase in Y_{C1-C8} and Y_{C9-C15} but a decrease of Y_{C16+}. The correlation factors associated with Y_{C1-C8}, Y_{C9-C15}, and Y_{C16+} (denoted as R) were 0.898, 0.895, and −0.903, respectively. The R values linked to the acid, weak acid, and strong acid amounts are not shown; but the R values associated with Y_{C1-C8} and Y_{C9-C15} in Figure 1 were closer to 1, while the value with Y_{C16+} was closer to −1, when compared to other values. These results indicated that there was a strong positive correlation between the amount of medium acid site and product yields for Y_{C1-C8} and Y_{C9-C15}, but a strong negative correlation for Y_{C16+}. Moreover, the observations also suggested that other factors, in addition to the medium acid amount, influence the product yield.

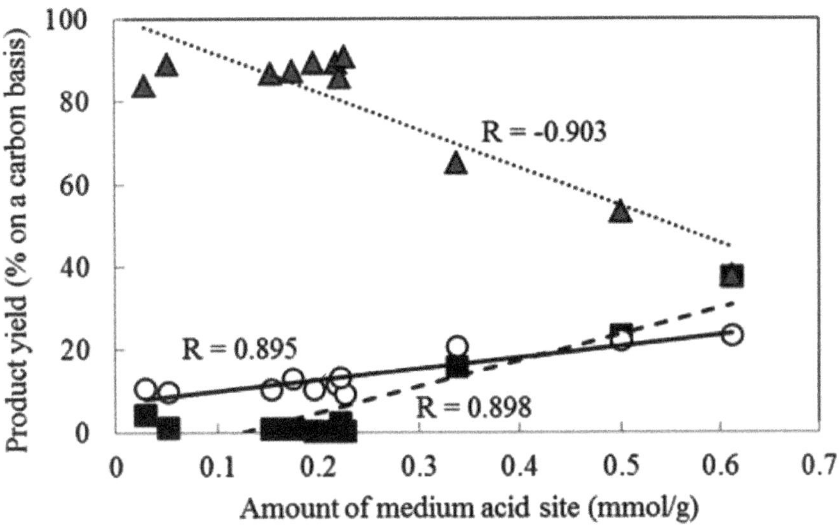

Figure 1: Relationship between acid amount and product yield in the hydrocracking reaction over Pt-loaded catalysts. Entries 1–7 and 13–16 in Table 4, Pt content: 0.1 wt. %, ■: Y_{C1-C8}; ○: Y_{C9-C15}; ▲: Y_{C16+}.

Table 4 also shows the Pt particle size by CO chemisorption (denoted as d_{pulse}) and Figure 2 shows the effect of the d_{pulse} value on the product yield. The Pt catalysts shown in the figure were prepared by the impregnation method. Therefore, the size distribution of the supported Pt particles would not be narrow. However, the d_{pulse} value could be regarded as the average particle size. When the d_{pulse} value increased to approximately 13 nm, Y_{C1-C8} and Y_{C9-C15} decreased while Y_{C16+} increased. These product yields remained almost constant at d_{pulse} values of more than 13 nm. The d_{pulse} values of I-Pt(0.1)/BEA28, I-Pt(0.1)/BEA40, and I-Pt(0.1)/BEA41.6 were 4.6, 3.1, and 4.8 nm, respectively, and they were smaller compared to other catalysts (Table 4). I-Pt(0.1)/MOR18, I-Pt(0.1)/BEA28, and I-Pt(0.1)/BEA40 had higher acid amounts (>1.08 mmol/g, Table 2) and the d_{pulse} value of I-Pt(0.1)/MOR18 (13.0 nm, Table 4) was relatively large compared to the other two catalysts. Although the support materials were largely different, these results are consistent with the trends shown in Figure 1 and Figure 2. Based on the hydrocracking results afforded by I-Pt(0.1)/BEA28, a small Pt particle size and high medium acid amount would be conducive toward improving the catalyst performance.

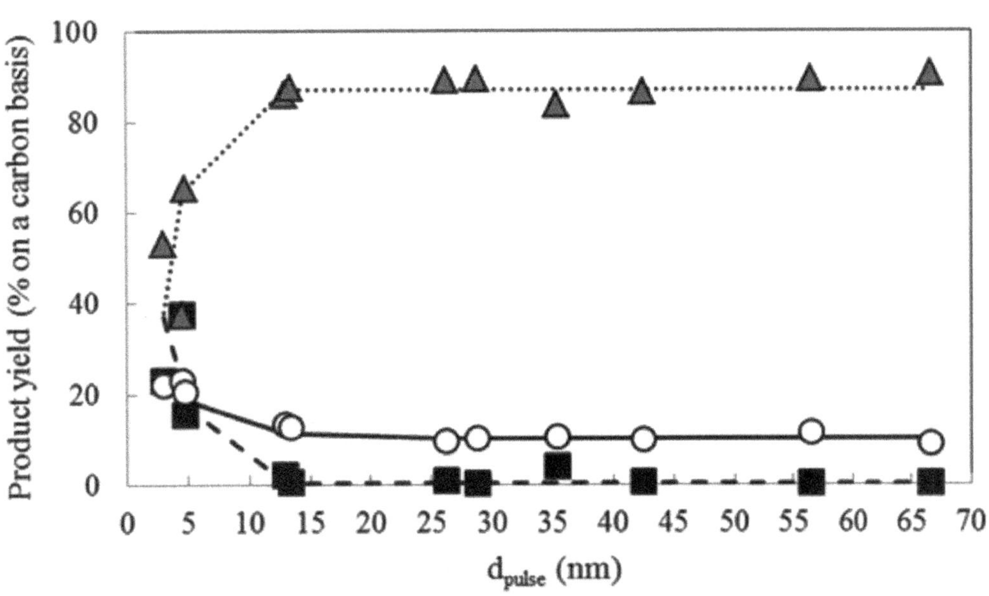

Figure 2: Effect of d_{pulse} on product yield in the hydrocracking reaction over Pt-loaded catalysts. Entries 1–7 and 13–16 in Table 4, Pt content: 0.1 wt. %, ■: Y_{C1-C8}; ○: Y_{C9-C15}; ▲: Y_{C16+}.

Effect of Pt Content on the Hydrocracking Reaction over Pt-Loaded BEA28 Catalysts

In the previous section, BEA28 was established as a promising support. Therefore, it was used to study the catalytic performance of Pt-loaded catalysts with low Pt content. Figure 3 shows the effect of Pt content (0–0.1 wt. %) on the hydrocracking reaction over impregnated Pt-loaded BEA28 catalysts (Entries 8–13 in Table 4). The d_{pulse} values are also shown in Figure 3.

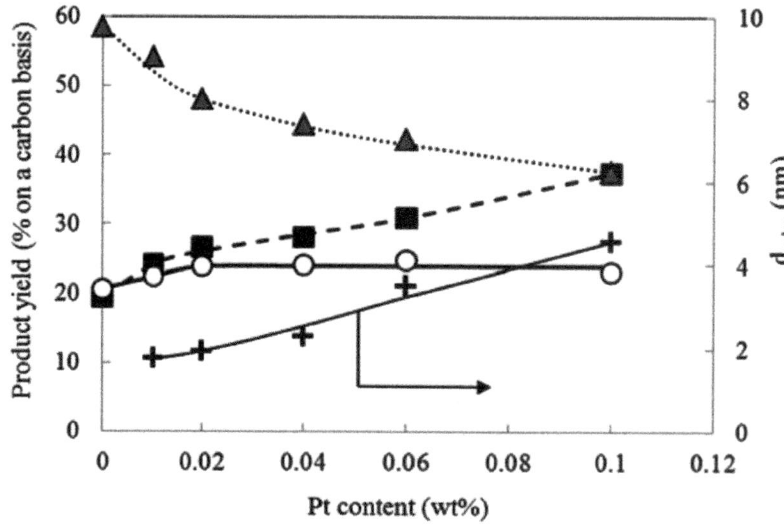

Figure 3: Effect of Pt content on the hydrocracking reaction over Pt-loaded BEA28 catalysts prepared by the impregnation method. Entries 8–13 in Table 4, Pt content: 0–0.1 wt. %, ■: Y_{C1-C8}; ○: Y_{C9-C15}; ▲: Y_{C16+}; +: d_{pulse}.

The d_{pulse} value steadily increased to 4.6 nm as the Pt content reached 0.1 wt. %. This is a general trend for impregnated catalysts. With an increase of the Pt content (*i.e.*, with increasing Pt particle size), Y_{C1-C8} increased from 19.4% to 37.4%, whereas Y_{C16+} decreased from 58.5% to 37.5%. In contrast, Y_{C9-C15} increased to 23.9% against a Pt content of 0.02 wt. % and remained almost constant thereafter. At a Pt content of 0.02 wt. %, the yield of the corresponding jet fuel was almost the same as that for 0.1 wt. % Pt content.

Batalha *et al.* have previously reported a quantified understanding of metal-acid balance involving hydrocracking catalysts. The turn over frequency increased with the ratio of accessible Pt sites (C_{Pt}) to acid sites (C_A), *i.e.*, up to 0.03 and the value remained constant in the range of 0.03 to 0.1 [25]. In the present study, the amount of medium acid site affected the product yield (Figure 1). Figure 5 shows the effect of the ratio of accessible Pt sites to the amount of medium acid site (denoted by C_{Pt}/C_{MA}) on the hydrocracking reaction. The relationship between d_{pulse} and C_{Pt} is as follows [23]:

$$d_{\text{pulse}} = 5.808 \times 10^{-5} \times \frac{W}{C_{\text{Pt}}} \tag{1}$$

where W (wt. %) is Pt content and C_{Pt} (mol/g) is the amount of adsorbed CO per gram of catalyst. The C_{MA} (mol/g) was used in Table 2.

The trend in Figure 4 is similar to that in Figure 3. The correlation factors associated with Y_{C1-C8} and Y_{C16+} were 0.891 and 0.976, respectively, while that with Y_{C9-C15} was 0.419. These results indicated that there was a strong positive correlation between the C_{Pt}/C_{MA} and product yields for Y_{C1-C8} and Y_{C16+}, but a weak correlation for Y_{C9-C15}. In a study about acid-metal balance of a hydrocracking catalyst by Thybaut *et al.*, non-ideal cracking would be favored for higher carbon numbers of feedstock in the range of 8 to 16 [26]. In this case, the kinetic effect became larger than the thermodynamic effect. The feedstock in the present study contained hydrocarbons with carbon numbers of 5 to 56. Therefore, the kinetic effect would affect the product yield.

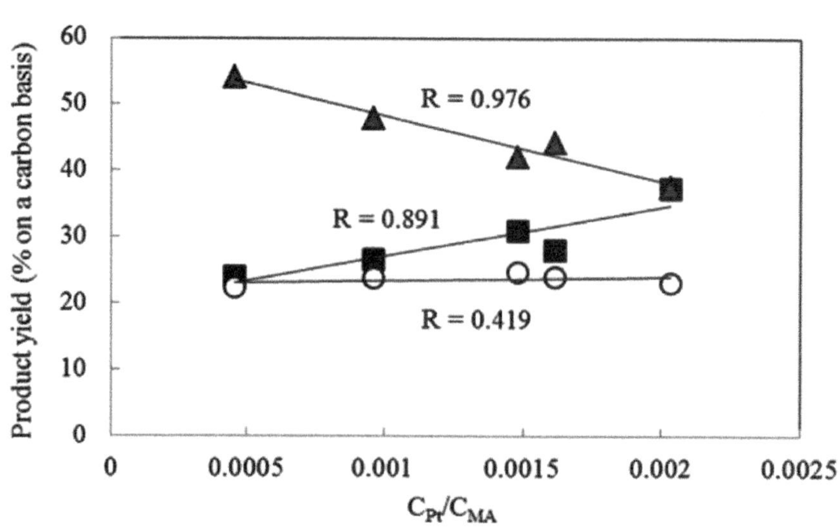

Figure 4: Effect of C_{Pt}/C_{MA} on the hydrocracking reaction over Pt-loaded BEA28 catalysts prepared by the impregnation method. Entries 9–13 in Table 4, Pt content: 0.01–0.1 wt. %, ■: Y_{C1-C8}; ○: Y_{C9-C15}; ▲: Y_{C16+}.

Here, the C_{Pt}/C_{MA} values (0.0005–0.002) were much smaller compared to those (*i.e.*, 0.03–0.1) listed in Ref. [25], indicating that a high jet fuel yield needed a larger C_{Pt}/C_{MA} ratio. To obtain a desirable cracking of longer chains, the enhancement of the cracking rate was required. The higher yield of the corresponding jet fuel was obtained at such a high C_{Pt}/C_{MA} ratio because the turn over frequency of the cracking rate associated with the medium acid sites would increase.

Hydrocarbons with carbon numbers of 9–15 are intermediates during the hydrocracking of longer chain hydrocarbons. With a large C_{Pt}/C_{MA} value, the production of lighter hydrocarbons was favored because of a higher cracking rate. As shown in Figure 3, at more than 0.02 wt. %, an increase in Pt sites would promote hydrogenation/dehydrogenation reactions, followed by a decrease in Y_{C16+}, and an increase in Y_{C1-C8}. In this case, increasing the Pt content had a positive effect on both the Pt particle size as well as the amount of Pt particles. Therefore, these results suggested that there were distinct optimum values of Pt content and particle size for producing the corresponding jet fuel in the hydrocracking reaction over Pt-loaded catalysts. Moreover, an optimum Pt particle density was also anticipated, as reported in a previous study [27]. Therefore, the effect of Pt particle size on the individual product yield was investigated using Pt-loaded BEA28 catalysts with a Pt content of 0.02 wt. %.

Effect of Pt Particle Size on the Hydrocracking Reaction over Pt-Loaded Catalysts with Low Pt Content

Table 5 shows the d_{pulse} values of Pt-loaded BEA28 catalysts prepared using w/o microemulsions. The Pt-complex forming agent and the type of Pt complex synthesized in microemulsion [28] are also shown. The d_{pulse} value was dependent on the type of Pt complex forming agents and varied in the range of 1.0–2.6 nm. The value increased in the order of tetraethylammonium chloride (denoted as TEAC) < tetrapropylammonium bromide (denoted as TPAB) < hexyltrimethylammonium bromide (denoted as HTAB) < cetyltrimethylammonium chloride (denoted as CTAC), as forming agents. Results in Table 1 indicated that Pt-loaded BEA28 catalysts (Entries 8–13 and 17–20) had larger average pore diameters and mesopore areas compared to those of the Pt-loaded BEA41.6 (Entry 15). This suggested that the smaller size of the Pt complex particles synthesized in the liquid phase promotes particle diffusion into the mesopore of the BEA28 support.

Table 5: Pt particles synthesized in w/o microemulsions and their d_{pulse} values.

Catalyst	Pt Complex Forming Agent		Pt Particle Type		d_{pulse} (nm)
	Name	Rational Formula	Name	Rational Formula	
D-Pt(CATC)/BEA28	CTAC	$C_{16}H_{33}(CH_3)_3N^+Cl^-$	$(CTA)_2PtCl_6$	$[C_{16}H_{33}(CH_3)_3N]_2PtCl_6$	2.6
D-Pt(TPAB)/BEA28	HTAB	$C_6H_{13}(CH_3)_3N^+Cl^-$	$(HTA)_2PtCl_6$	$[C_6H_{13}(CH_3)_3N]_2PtCl_6$	2.0
D-Pt(TEAC)/BEA28	TPAB	$(C_3H_7)_4N^+Br^-$	$(TPA)_2PtCl_6$	$[(C_3H_7)_4N]_2PtCl_6$	2.3
D-Pt(HTAB)/BEA28	TEAC	$(C_2H_5)_4N^+Cl^-$	$(TEA)_2PtCl_6$	$[(C_2H_5)_4N]_2PtCl_6$	1.0

The CTAC molecule has a larger hydrophobic volume ($C_{16}H_{33}$– and $(CH_3)_3$–) compared to the TEAC molecule ($(C_2H_5)_4$–). Therefore, the occupied volume of $(TEA)_2PtCl_6$ would be smaller compared to $(CTA)_2PtCl_6$. In addition, the HTAB molecule has a longer hydrophobic group (C_6H_{13}–) than the TPAB molecule. The carbon numbers associated with the hydrophobic group of HTAB (=9) are smaller than that of TPAB (=12). However, the chain of the hydrophobic group (C_6H_{13}–) might fold up to form a smaller occupied volume of $(HTA)_2PtCl_6$ particles compared to the $(TPA)_2PtCl_6$ particles. When the forming agents had a smaller hydrophobic occupied volume, smaller complex particles were more likely to diffuse into the mesopores of the support. As a result, the aggregation of complex particles was inhibited owing to the deposition of the smaller particles on the BEA28 support after calcination and reduction.

Figure 5 shows the effect of d_{pulse} value on product yield in the hydrocracking reaction over Pt-loaded BEA28 catalysts with a Pt content of 0.02 wt. %. The product yields of the impregnated catalyst (Entry 10) are also shown in Figure 5. I-Pt(0.02)/BEA28 and D-Pt(TPAB)/BEA28 had a similar particle size (Table 4), resulting in comparable product yields. With a Pt content of 0.2 wt. %, the size distribution of Pt particles obtained by the impregnation method was broader compared to that afforded by the deposition method using w/o microemulsions [27]. In the present study, the Pt content was 0.02 wt. %. For the impregnation method, the dried precursor was calcined to aggregate the Pt oxide particles. With a very low Pt content, the distance between the Pt oxide particles would be further apart. Therefore, the inhibition of particle aggregation during the calcination process would lead to the formation of small particles, with a relatively narrow size distribution.

The correlation coefficient between the d_{pulse} value and the respective product yield was −0.36–0.08, indicating a lack of correlation between the two variables. In the d_{pulse} range of 1.0–2.6 nm, Y_{C1-C8} was higher than Y_{C9-C15}. These results were similar to those of the Pt-loaded catalysts with a size of less than 4.6 nm (Table 4 and Figure 2). The product yield was independent of the C_{Pt}/C_{MA} value in the range of 0.0005 to 0.0018 (not shown in figure). These results suggested that the cracking reaction was dominant compared to the dehydrogenation/hydrogenation reaction on the Pt sites (within the specified particle range), leading to an excessive cracking reaction.

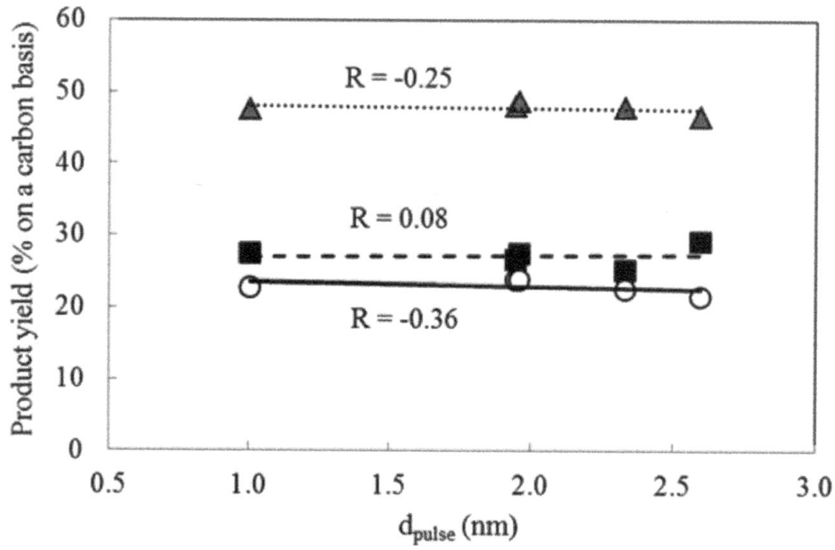

Figure 5: Effect of d_{pulse} on product yield in the hydrocracking reaction over Pt-loaded BEA28 catalysts. Entries 10 and 17–20, Pt content: 0.02 wt. %, ■ Y_{C1-C8}; ○: Y_{C9-C15}; ▲:Y_{C16+}.

We previously reported that a high jet fuel yield was obtained using a Pt particle size of 7.6 nm; in contrast, decreasing the Pt particle size to 2.3 nm increased $Y_{C1-C8,}$ whose values exceeded those of Y_{C9-C15} [22]. Supported Pt particles consist of atoms at different positions, namely the plane, edge, and corner. The ratio of the number of Pt atoms contributing to each position to those contributing only to the surface of a Pt particle was dependent on the particle size [29]. For the hydrocracking of 2-methylpentane, it was suggested that the Pt(111) terrace was the least reactive while the Pt(119) site was the most active cracking surface among the four types of Pt single-crystal surfaces ((111), (557), (119), and (311)) [30]. Increasing the Pt particle size to 6 nm, the ratio of Pt atoms for plane to those for surface increased rapidly, with the effect gradually leveling off as Pt particle size exceeds 6 nm [22]. In this study, when the particle size was less than 2.6 nm, the ratio of Pt atoms consisting of planes to the total Pt atoms was very low. Increasing the Pt particle size to 2.6 nm, the excess cracking reaction was inhibited by a decrease in the Pt sites, which contributes to the edge position such as Pt(119). Therefore, Y_{C9-C15} was lower than Y_{C1-C8}.

However, it was unclear as to why the product yield was independent of the d_{pulse} value. The critical size above which the band structure appears is about 2 nm [31]. In the present study, the Pt-Pt bond responsible for the formation of the supported Pt particles might be affected by the interaction between particle defects and the surface of the zeolite support, followed by an inhibition of the Pt bulk's catalytic activity. Alternatively, the small particles would be likely to be sintered. Therefore, the product yields were independent of the initial Pt particle sizes because the sintering would proceed to form Pt particles with a similar size during the reaction.

In this study, the utilization of support material with a high amount of medium acid sites contributed to an increase in the yield of the corresponding jet fuel by 14% when the Pt content was very low, compared to Pt-loaded amorphous support materials and feedstock. The screening study suggested that a smaller Pt particle size would increase the yield. The d_{pulse} value was controlled in the range of 1.0–2.6 nm, with a Pt content of 0.02 wt. %. Controlling the size within this range led to an increase of Y_{C9-C15} by only 3%, while it favored the increase in the yield of lighter hydrocarbons, as shown in Table 4 andFigure 3. This suggests that it is very important to control the supported Pt particle size and we believe that the precise control of the Pt particle size (*i.e.*, more than 2.6 nm) contributes to the enhancement of corresponding jet fuel yield.

EXPERIMENTAL SECTION

Catalyst Preparation

Pt-loaded catalysts employed in the present study were prepared by the impregnation method and the deposition method using w/o microemulsions. A total of 11 types of supports were employed. Four types of silica-alumina: SiO_2-Al_2O_3 308 (Si/Al = 8.7, Fuji Silysia Chemical, Aichi, Japan, denoted as SiAl8.7), JRC-SAH-1 (28.6% Al_2O_3, JGC Catalysts and Chemicals, Kanagawa, Japan, denoted as SiAl(28.6)), JRC-SAL-3 (13% Al_2O_3, JGC Catalysts and Chemicals, Kanagawa, Japan, denoted as SiAl(13)), and JRC-SAL-4 (Si/Al = 5.5, JGC Catalysts and Chemicals, Kanagawa, Japan, denoted as SiAl5.5); an amorphous metal oxide: sulfated zirconia (Wako Pure Chemical Industries, Osaka, Japan, denoted as S-ZrO_2); four β-type zeolites: HSZ-930NHA (SiO_2/Al_2O_3 = 28, Tosoh, Tokyo, Japan, denoted as BEA28), HSZ-940NHA (SiO_2/Al_2O_3=40, Tosoh, Tokyo, Japan, denoted as BEA40), HSZ-940HOA (SiO_2/Al_2O_3 = 41.6, Tosoh, Tokyo, Japan, denoted as BEA41.6), and HSZ-980HOA (SiO_2/Al_2O_3 = 510, Tosoh, Tokyo, Japan, denoted as BEA510); a mordenite-type zeolite: HSZ-640HOA (SiO_2/Al_2O_3 = 18, Tosoh, Tokyo, Japan, denoted as MOR18); and a Y-type zeolite: HSZ-385HUA (SiO_2/Al_2O_3 = 110, Tosoh, Tokyo, Japan, denoted as FAU110).

The impregnation method is described as follows. The support powder was first immersed in an aqueous solution of $H_2PtCl_6·6H_2O$ (Kishida Chemical Co., Osaka, Japan) and then dried at 105 °C overnight in an oven. The dried precursors were calcined in air at 500 °C for 2 h in a muffle furnace, and subsequently reduced with H_2 (99.999%, 50 mL/min) at 400 °C for 30 min to obtain the impregnated catalysts (I-Pt(X)/

support, where X is the Pt content in wt. %). The Pt content was in the range of 0.01–0.1 wt. %.

The deposition method, carried out according to literature procedures, is described as follows [22,23]. The Pt complex particles were synthesized in the liquid phase and then immobilized onto support materials. The complex particle synthesis and immobilization were performed at 45 °C. Polyoxyethylene (n = 5.5) cetyl ether (Union 67-55R; New Japan Chemical Co., Osaka, Japan, denoted as CE-5.5) was used as a surfactant, whereas the solvent of choice was *n*-hexadecane (Tokyo Chemical Industry Co., Tokyo, Japan). A microemulsion consisting of CE-5.5/*n*-hexadecane/aqueous solution of H_2PtCl_6 (0.019 mol/L) was denoted as ME-A (50 mL), whereas ME-B (50 mL) referred to a microemulsion consisting of CE-5.5/*n*-hexadecane/aqueous solution of a Pt complex-forming agent (0.32 mol/L).

Four types of Pt complex-forming agents were used: tetraethylammonium chloride (Wako Pure Chemical Industries, Osaka, Japan), tetrapropylammonium bromide (Wako Pure Chemical Industries, Osaka, Japan), hexyltrimethylammonium bromide (Tokyo Chemical Industry Co., Tokyo, Japan), and hexadecyltrimethylammonium chloride (Wako Pure Chemical Industries, Osaka, Japan).

The Pt complex particles were synthesized in the liquid phase by mixing ME-A and ME-B. The molar ratio of water to surfactant was six, and the molar ratio of the Pt complex-forming agent to Pt was 10. The particles were synthesized with stirring for 10 min. An ammonia solution (28%, Nacalai Tesque Inc., Kyoto, Japan, 2 mL), ethanol (10 mL), and BEA28 powder (39 g) were added to the solution, and the Pt complex particles were gradually deposited onto the zeolite support with vigorous stirring for 10 min. Next, the liquid-solid mixture was centrifuged at 3000 rpm for 10 min, followed by a separation of the precipitate from the transparent liquid. The precipitate was dried at 105 °C in an oven. The dried precursors were calcined in the muffle furnace at 500 °C for 1 h. After cooling down to room temperature, the calcined precursors were stirred to obtain uniformity, and re-calcined at 500 °C for an additional 1 h. The calcined precursors were reduced with H_2 (99.999%, 50 mL/min) at 400 °C for 30 min to obtain the deposited catalysts (D-Pt(A)/BEA28, where A is the Pt-complex forming agent). The Pt content of the catalysts was 0.02 wt. %.

Characterization

The specific surface area, pore volume, and average pore diameter were estimated by an automatic specific area/pore size distribution measurement system (BELSORP-mini; MicrotracBEL Co., Osaka, Japan). The N_2 adsorption-desorption isotherm at −196 °C was measured under the relative pressures of 0.01 to 0.99. The total specific surface area, pore volume, and average pore diameter were calculated by the BET method [32]. The specific surface area and pore volume of the micropores were obtained by the t method [24]. The parameters for the mesopores were obtained by the Barrett-Joyner-Halenda (BJH) method [33].

The NH_3-TPD analysis was performed using a chemisorption catalyst analyzer (BELCAT-B; MicrotracBEL Co., Osaka, Japan). The catalyst samples (approximately 0.05 g) were prepared at 400 °C for 1 h under a He flow of 50 mL/min. After decreasing the temperature to 100 °C, ammonia was adsorbed onto the sample surface by passing NH_3/He (5/95% (v/v), 50 mL/min) through the sample for 30 min, followed by an evacuation step for 1 h at 100 °C to eliminate the weakly adsorbed ammonia. The NH_3-TPD analysis was then carried out from 100 to 800 °C at a heating rate of 10 °C/min.

The d_{pulse} value was calculated based on the amount of CO molecules adsorbed on the catalyst surface and measured using the chemisorption catalyst analyzer (BELCAT-B; MicrotracBEL Co., Osaka, Japan). The catalyst samples (approximately 0.2 g) were charged into a sample holder and oxidized at 400 °C for 15 min under an O_2 flow (99.999%) of 50 mL/min, and subsequently reduced at 400 °C for 30 min under a H_2 flow of 50 mL/min. The samples were then cooled down to 50 °C under a He flow of 50 mL/min. CO/He (10/90% (v/v)) was intermittently injected into the holder at 50 °C until the amount of CO leaving the holder became constant. The calculation method has been reported in detail previously [23]. The CO pulse analysis was performed three times, and the average value from these replicates was taken as the d_{pulse} value.

Hydrocracking Tests

Hydrocracking tests were performed using an autoclave made of the Inconel alloy (inner volume: 75 mL). The FT product produced by a bench-scale BTL plant (AIST Chugoku, Hiroshima, Japan) was employed as feedstock. The woody biomass gasification was performed with oxygen-enriched air and CO_2 as the gasifying agents to obtain the syngas. Subsequently, the FT synthesis was conducted over a (Co-Mn-Zr)/SiO_2 catalyst prepared by the impregnation method [34,35]. The Co content was 20 wt. %. During the FT synthesis, *n*-hexadecane (Tokyo Chemical Industry Co., Tokyo, Japan, 4 L) was initially charged into the reactor to give a slurry phase reactor. The reaction was performed for 8 h. Table 3 shows the elemental and chemical analyses of the biomass-derived FT product. The FT product comprised of hydrocarbons, with the largest fraction being those with a carbon number of 16. Generally, the product distribution of the FT synthesis reaction follows the Anderson-Schulz-Flory (ASF) distribution. Based on the product distribution, the FT product mixture still contained *n*-hexadecane, which was initially charged into the reactor, because the FT reaction did not go to completion. The target product in the hydrocracking reaction was hydrocarbons with carbon numbers of 9–15. Table 3 lists the content of hydrocarbons with carbon numbers of 9–15, which serve as the feedstock. Therefore, we believe that it is worthwhile to discuss the results of the hydrocracking tests using the FT product.

The FT product exists as a solid at room temperature. A uniform feedstock was obtained by heating the FT product at 50 °C with stirring. The Pt-loaded catalyst (0.2 g) and heated feedstock (5 g) were charged into the autoclave. The gas in the reactor was replaced with hydrogen (99.999%) and the inner pressure was set at 0.5 MPa. The autoclave was heated by an electric furnace, stirring at 300 rpm. The temperatures of the reactor and furnace were recorded with a data acquisition system (NR-250; Keyence Co., Osaka, Japan) The reaction time of the hydrocracking reaction was regarded as 0 min when the reactor temperature reached 245 °C. The reaction temperature and time were set at 250 °C and 1 h, respectively.

After the hydrocracking reaction, the furnace was removed and the reactor was cooled to room temperature with a fan. The gas in the autoclave was purged with N_2 (99.9998%) and collected into a gas collection bag using a wet gas meter (W-NKA-0.5A; Shinagawa Co., Tokyo, Japan). The reaction mixture in the reactor was collected after opening the autoclave.

The inorganic gas and gaseous hydrocarbons with carbon numbers of 1–9 in the collected gas were analyzed using gas chromatographs equipped with a thermal conductivity detector (GC323; GL Science Inc., Tokyo, Japan, Column: Molecular Sieve 5A) and a flame ionization detector (GC353B; GL Sciences Inc., Tokyo, Japan, Column: RT-QPLOT). The products collected from the reactor were analyzed using a chromatograph with a flame ionization detector (GC353B; GL Sciences Inc., Tokyo, Japan, Column: Ultra Alloy-DX30) using 2-methyl naphthalene (Tokyo Chemical Industry Co., Tokyo, Japan) as the internal standard. Carbon tetrachloride (Sigma-Aldrich Co., Tokyo, Japan) and *n*-hexadecane (Tokyo Chemical Industry Co., Tokyo, Japan) were used as solvents.

The yield of the products, *i*, on a carbon basis (Y_i: i = C1–C8, C9–C15, and C16+, respectively) and loss are defined as follows:

$$Y_{C1-C8}\ (\%) = \sum_{n=1}^{8} \frac{C_j}{C_{in}} \times 100 \tag{2}$$

$$Y_{C9-C15}\ (\%) = \sum_{n=9}^{15} \frac{C_j}{C_{in}} \times 100 \tag{3}$$

$$Y_{C16+}\ (\%) = \sum_{n=16}^{56} \frac{C_j}{C_{in}} \times 100 \tag{4}$$

$$\text{Loss}\ (\%) = 100 - Y_{C1-C8} - Y_{C9-C15} - Y_{C16+} \tag{5}$$

where C_j (mol) is the total carbon number of the hydrocarbon product, with a carbon number of *j*, and C_{in} (mol) is the total carbon number contained in the feedstock that was charged into the reactor.

CONCLUSIONS

A screening study using Pt-loaded catalysts with a Pt content of 0.1 wt. % was performed to investigate the effective production of hydrocarbon liquid fuel in the hydrocracking of FT product. Zeolite supports were more favorable for jet fuel production compared to amorphous oxide supports. The small Pt particle size of the supported Pt particles and the high amount of medium acid sites for zeolite supports led to a higher performance of Pt-loaded zeolite catalyst with low Pt content. For the Pt-loaded BEA28 catalysts, the yields of the corresponding jet fuel at 0.02 and 0.1 wt. % were almost the same. The Pt particle size was controlled in the range of 1.0–2.6 nm by preparing Pt-loaded BEA28 catalysts with a Pt content of 0.02 wt. % using w/o microemulsions. There was no correlation between the Pt particle size and product yield. Small Pt particles were beneficial for the production of lighter hydrocarbons such as gasoline.

ACKNOWLEDGMENTS

The authors thank Hideyuki Yokoyama, Chitose Tokifuji, and Maiko Nishida for their assistance with the experiments.

CONFLICTS OF INTEREST

The authors declare no conflict of interest.

REFERENCES

1. Sunde, K.; Brekke, A.; Solberg, B. Environment impacts and costs of woody Biomass-To-Liquid (BTL) production and use—A review. *For. Policy Econ.* 2011, *13*, 591–602.
2. Fujimoto, S.; Yanagita, T.; Ogata, M.; Minowa, T. Evaluation of CO_2 mitigation by BTL biofuels from woody biomass through simulated case studies. *Int. Energy J.* 2008, *9*, 73–80.
3. Calemma, V.; Peratello, S.; Perego, C. Hydroisomerization and hydrocracking of long chain *n*-alkanes on Pt/ amorphous SiO_2-Al_2O_3 catalyst. *Appl. Catal. A* 2000, *190*, 207–218.
4. Kang, J.; Ma, W.; Keogh, R.A.; Shafer, W.D.; Jacobs, G.; Davis, B.H. Hydrocracking and hydroisomerization of *n*-hexadecane, *n*-octacosane and Fischer-Tropsch wax over a Pt/SiO_2-Al_2O_3 catalyst. *Catal. Lett.* 2012, *142*, 1295–1305.
5. Kim, M.Y.; Kim, Y.A.; Jeong, K.E.; Chae, H.J.; Kim, C.U.; Jeong, S.Y.; Han, J.; Park, E.D. Effect of Al content on hydrocracking of *n*-paraffin over Pt/SiO_2-Al_2O_3. *Catal. Commun.* 2012, *26*, 78–82.
6. Keogh, R.A.; Srinivasan, R.; Davis, B.H. The effect of Pt concentration on the activity and selectivity of SO_4^{2-}-ZrO_2catalysts for the hydrocracking and hydroisomerization of *n*-hexadecane. *Appl. Catal. A* 1996, *140*, 47–57.
7. Busto, M.; Vera, C.R.; Grau, J.M. Optimal process conditions for the isomerization-cracking of long-chain *n*-paraffins to high octane isomerizate gasoline over Pt/ SO_4^{2-}-ZrO_2 catalysts. *Fuel Process. Technol.* 2011, *92*, 1675–1684.
8. Grau, J.M.; Yori, J.C.; Parera, J.M. Hydroisomerization-cracking of *n*-octane on Pt/WO_4^{2-}-ZrO_2 and Pt/SO_4^{2-}-ZrO_2. Effect of Pt load on catalyst performance. *Appl. Catal. A* 2001, *213*, 247–257.
9. Busto, M.; Lovato, M.E.; Vera, C.R.; Shimizu, K.; Grau, J.M. Silica supported tungsta-zirconia catalysts for hydroisomerization—Cracking of long alkenes. *Appl. Catal. A* 2009, *335*, 123–131.
10. Busto, M.; Benitez, V.M.; Vera, C.R.; Grau, J.M.; Yori, J.C. Pt-Pd/WO_3-ZrO_2 catalysts for isomerization-cracking of long paraffins. *Appl. Catal. A* 2008, *347*, 117–125.
11. Zhang, S.; Zhang, Y.; Tierney, J.W.; Wender, I. Anion-modified zirconia: Effect of metal promotion and hydrogen reduction on hydroisomerization of *n*-hexadecane and Fischer-Tropsch waxes. *Fuel Process. Technol.* 2001, *69*, 59–71.

12. Zhang, S.; Zhang, Y.; Tierney, J.W.; Wender, I. Hydroisomerization of normal hexadecane with platinum-promoted tungstate-modified zirconia catalysts. *Appl. Catal. A* 2000, *193*, 155–171.

13. Zhou, Z.; Zhang, Y.; Tienery, J.W.; Wender, I. Hybrid zirconia catalysts for conversion of Fischer-Tropsch waxy products to transportation fuels. *Fuel Process. Technol.* 2003, *83*, 67–80.

14. Leu, L.J.; Hou, L.Y.; Kang, B.C.; Li, C.; Wu, S.T.; Wu, J.C. Synthesis of zeolite β and catalytic isomerization of *n*-hexane over Pt/H-β catalysts. *Appl. Catal.* 1991, *69*, 49–63.

15. Hanaoka, T.; Miyazawa, T.; Shimura, K.; Hirata, S. Jet fuel synthesis from Fischer-Tropsch product under mild hydrocracking conditions using Pt-loaded catalysts. *Chem. Eng. J.* 2015, *263*, 178–185.

16. Iliopoulou, E.F.; Heracleous, E.; Delimitis, E.; Lappas, A.A. Producing high quality biofuel: Pt-based hydroisomerization catalysts evaluated using BtL-naphtha surrogates. *Appl. Catal. B* 2014, *145*, 177–186.

17. Heracleous, E.; Iliopoulou, E.F.; Lappas, A.A. Microporous/mesoporous Pt/ZSM-5 catalysts for hydroisomerization of BTL-naphtha. *Ind. Eng. Chem. Res.* 2013, *52*, 14567–14573.

18. Alvarez, F.; Ribeiro, F.R.; Giannetto, G.; Chevalier, F.; Perot, G.; Guisnet, M. Hydroisomerization and hydrocracking of alkanes. 5. Hydroisomerization and hydrocracking of *n*-hexane and *n*-heptane on PtHY catalysts. Effect of the distribution of metallic and acid sites. *Stud. Surf. Sci. Catal.* 1989, *49*, 1339–1348.

19. Alvarez, F.; Giannetto, G.; Guisnet, M.; Perot, G. Hydroisomerization and hydrocracking of *n*-alkanes. 2. *n*-Heptane transformation on a Pt-dealuminated Y zeolite—Comparison with a Pt-Y zeolite. *Appl. Catal.* 1987, *34*, 353–365.

20. Guisnet, M.; Alvarez, F.; Giannetto, G.; Perot, G. Hydroisomerization and hydrocracking of *n*-heptane on PtH zeolites. Effect of the porosity and of the distribution of metallic and acid sites. *Catal. Today* 1987, *1*, 415–433.

21. Alvarez, F.; Ribeiro, F.R.; Perot, G.; Thomazeau, C.; Guisnet, M. Hydroisomerization and hydrocracking of alkanes. 7. Influence of the balance between acid and hydrogenating functions on the transformation of *n*-decane on PtHY catalysts. *J. Catal.* 1996, *162*, 179–189.

22. Hanaoka, T.; Miyazawa, T.; Shimura, K.; Hirata, S. Jet fuel synthesis in hydrocracking of Fischer-Tropsch product over Pt-loaded zeolite catalysts prepared using microemulsions. *Fuel Process. Technol.* 2015, *129*, 139–146.

23. Hanaoka, T.; Miyazawa, T.; Shimura, K.; Hirata, S. Preparation for Pt-loaded zeolite catalysts using w/o microemulsion and their hydrocracking behaviors on Fischer-Tropsch product. *Catalysts* 2015, *5*, 88–105.

24. Lippens, B.C.; Boer, J.H. Studies on pore systems in catalysts. V. The *t* method. *J. Catal.* 1965, *4*, 319–323.

25. Batalha, N.; Pinard, L.; Pouilloux, Y.; Guisnet, M. Bifunctional hydrogenating/acid catalysis: Quantification of the intimacy criterion. *Catal. Lett.* 2013, *143*, 587–591.

26. Thybaut, J.W.; Narasimhan, C.S.L.; Denayer, J.F.; Baron, G.V.; Jacobs, P.A.; Martens, J.A.; Marin, G.B. Acid-metal balance of a hydrocracking catalyst: Ideal *versus* nonideal behavior. *Ind. Eng. Chem. Res.* 2005, *44*, 5159–5169.

27. Hanaoka, T.; Miyazawa, T.; Shimura, K.; Hirata, S. Effect of Pt particle density on the hydrocracking of Fischer-Tropsch products over Pt-loaded zeolite catalysts prepared using water-in-oil microemulsions. *Chem. Eng. J.* 2015, *274*, 256–264.

28. Ikeda, M.; Takeshima, S.; Tago, T.; Kishida, M.; Wakabayashi, K. Preparation of size-controlled Pt catalysts supported alumina. *Catal. Lett.* 1999, *58*, 195–197.

29. Hardeveld, R.; Hartog, F. The statistics of surface atoms and surface sites on metal crystals. *Surface Sci.* 1969, *15*, 189–230.

30. Dauscher, A.; Garin, F.; Maire, G. Correlations between the surface structure of platinum single crystals and hydrocarbon skeletal rearrangement mechanisms: Approach to the nature of the active sites. *J. Catal.* 1987, *105*, 233–244.

31. Stace, T. How small is a solid? *Nature* 1988, *331*, 116–117.

32. Brunauer, S.; Emmett, P.H.; Teller, E. Adsorption of gases in multimolecular layers. *J. Am. Chem. Soc.* 1938, *60*, 309–319.

33. Barrett, E.P.; Joyner, L.G.; Halenda, P.P. The determination of pore volume and area distributions in porous substances. Computations from nitrogen isotherms. *J. Am. Chem. Soc.* 1951, *73*, 373–380.

34. Hanaoka, T.; Miyazawa, T.; Nurunnabi, M.; Hirata, S.; Sakanishi, K. Liquid fuel production from woody biomass via oxygen-enriched air/CO_2 gasification on a bench scale. *J. Jpn. Inst. Energy* 2011, *90*, 1072–1080.

Miyazawa, T.; Hanaoka, T.; Shimura, K.; Hirata, S. Mn and Zr modified Co/SiO_2 catalysts development in slurry-phase Fischer-Tropsch synthesis. *Appl. Catal. A* 2013, *467*, 47–54.

CHAPTER 11 ADVANCES IN MAGNETICALLY SEPARABLE PHOTOCATALYSTS

Gcina Mamba and Ajay Mishra

Keywords: Magnetically separable; ferrites; recyclable photocatalysts; photocatalysis; nanocomposites

INTRODUCTION

Rapid development in terms of industrialisation and population growth over the years has put pressure on sustainable clean water supply. This is partly due to the ever increasing demand and the significant contribution towards compromised fresh water quality that comes with population growth and industrialisation. Similarly, climate change (drought conditions and elevated temperatures) has also impacted negatively on the amount of fresh water available. Consequently, the development of effective water treatment technologies remains an important area of research. This would ensure adequate utilisation of the available fresh water sources, which are constantly polluted through various domestic and industrial activities. Moreover, it would allow efficient treatment of industrial effluent water and enable it to be recycled which would lessen the burden on constant fresh water supply. This is particularly key to the smooth operations of the industries in water scarce localities where the amount of fresh water allocated per industry per day is highly restricted [1].

Over the years, a number of water treatment technologies such as membrane processes [1,2], biological methods [3,4], advanced oxidation processes (AOPs) [5,6], adsorption [7,8], electrochemical methods [9,10], *etc.* have been developed and explored for the destruction/removal of organic, inorganic, and microbial pollutants. These methods have also been explored in various combinations in order to improve their efficiency [11,12,13,14]. Despite the impressive level of success observed with these technologies in water treatment, they often suffer a number of setbacks, which relate to inefficiency, high operational costs, or both. For example, in membrane processes, fouling (organic, inorganic and biofouling) is a major problem that results in the alteration of the membrane flux and rejection properties [15,16,17]. Toxicity of the effluent matrix and the non-biodegradable nature of some pollutants often render biodegradation ineffective. Moreover, some of the pollutants may be transformed into more toxic compounds during biological treatment, resulting in secondary pollution [18,19,20]. The generation of highly concentrated sludge which poses further treatment and disposal challenges is a major setback associated with adsorption and coagulation–flocculation [21].

Advanced oxidation processes such as ozonation, hydrogen peroxide oxidation, and their combination with UV light have been extensively exploited in water treatment; however, problems such as the short half-life of ozone, activation by UV light, incomplete mineralisation of organic compounds, and the costs associated with these tools are usually cause for concern [12,22]. Semiconductor photocatalysis has rapidly emerged as the most attractive AOP

Citation: Gcina Mamba and Ajay Mishra, Advances in Magnetically Separable Photocatalysts, doi:10.3390/catal6060079

in recent years for the destruction of organic, inorganic, and microbial pollutants due to a number of advantages it bears over other water treatment protocols. Such merits include its versatile nature in terms of the type of pollutants that can be removed (organic, inorganic, and microbial) and the medium in which the process is applicable (liquid and aqueous medium) [23,24,25,26,27]. Moreover, the semiconductor photocatalysts can also be exploited in other applications such as sensors [28], solar cells [29], pollutant adsorption [30], catalysts for organic synthesis [31], water splitting [32], *etc.* Apart from its versatile nature, semiconductor photocatalysis has the potential to completely mineralise organic pollutants into carbon dioxide, water, and inorganic ions, thereby eliminating the problem of sludge formation which otherwise cause secondary pollution. Semiconductor photocatalysis provides a stable reaction site/medium since the photocatalyst is not used up during the process and can be re-used several times, and the toxicity of some of the pollutants or their matrices does not affect their destruction by photocatalysis as would be the case with biodegradation [33,34]. In addition, photocatalysis has the advantage of being applicable at ambient conditions (room temperature and standard pressure).

Despite all the highlighted merits associated with semiconductor photocatalysis for water treatment, practical exploitation of this technology has remained a daunting challenge. The first challenge relates to the size of the band gap of the semiconductor since this determines the photoresponse of the photocatalyst. Some of the widely exploited semiconductors such as TiO_2, ZnO, SnO_2, *etc.* have wide band gaps of 3.0–3.2, 3.2, and 3.5 eV, respectively, which means these semiconductors can only be activated by UV light irradiation. However, artificial UV sources are costly, and UV light itself is toxic and requires protective gear when working under it. The sun does supply UV light but it only makes up about 4% of the solar spectrum, making practical exploitation unsustainable [35,36]. The second challenge in semiconductor photocatalysis is the fast recombination rate of the photogenerated charge carriers, which ultimately lowers the photocatalytic activity of the semiconductor [37,38]. Ideally, subsequent to light absorption and electron excitation in the semiconductor, the electrons, which are promoted to the conduction band, and the positive holes, which remain in the valence band of the semiconductor, must remain far apart so that they can react with oxygen and water or hydroxide ions to form the superoxide and hydroxyl radicals, respectively. These are the active species that together with the holes may be responsible for the degradation of the pollutants [39,40]. Thirdly, the slow degradation kinetics also hamper the full exploitation of photocatalysis in water treatment. The residence times are usually long and sometimes accompanied by low degradation efficiencies [41]. Incomplete mineralisation of organic pollutants is another problem usually observed in photocatalytic experiments and may lead to the formation of toxic by-products which are more toxic than the parent compound [42]. Lastly, the photocatalysts are usually applied in powdered form, which may undergo aggregation in the aqueous medium thereby lowering its activity. Moreover, after the treatment has been completed, the photocatalyst needs to be separated and recovered from the water, and this presents an expensive and time-consuming exercise, especially when it comes to nanometre-sized particles. This often leads to inefficient recovery of the photocatalyst, which not only affects its recyclability but also causes secondary pollution and escalates the treatment costs.

Over the past years, tremendous work has been done focusing on improving the visible light response of various photocatalysts, enhancing charge carrier separation and the overall photocatalytic activity. Efficient utilisation of visible light would enable the exploitation of sunlight as a source of energy, which consists of about 40% visible light. Generally, strategies such as doping the semiconductor with metal ions (alkali, alkaline earth, noble, rare earth, and transition metals) [43,44,45,46,47], non-metal ions (oxygen, sulphur, nitrogen, phosphorus, halogens, carbon, and boron) [48,49,50,51], codoping/multidoping (metal + metal, non-metal + metal, non-metal + non-metal) [52,53,54,55], coupling semiconductors with carbon nanomaterials (reduce graphene oxide (RGO), graphene oxide (GO), multiwalled carbon nanotubes (MWCNTs), single-walled carbon nanotubes (SWCNTs), carbon nanospheres (CNS), fullerenes, *etc.*) [56,57,58,59], sensitisation (dye, polymer and surface complex sensitisation) [60,61,62], and coupling two or more semiconductors [63,64,65] have been extensively investigated for improving the photocatalytic activity of various semiconductors. Encouraging results have been reported, especially for the nanocomposite photocatalysts, which showed improved visible light utilisation and efficient charge separation, which could be ascribed to the combined contribution of the individual components of the nanocomposite.

Engineering of materials with high visible light photocatalytic activity that are easily separable and recoverable from the treated water remains the ultimate goal of ongoing research. With the increased usage of nanophotocatalysts, efficient separation and recovery of these materials from water is becoming increasingly difficult. Immobilising the photocatalyst particles on solid supports such as glass, ceramics, polymers, zeolites, sand, *etc.* has been widely investigated and has successfully enabled easy separation of the photocatalyst [66,67,68,69,70,71]. However, in the process of attaining easy separation, the photocatalytic activity of the immobilised semiconductors decreased remarkably. This was due to a decrease in the surface area, poor interaction with the reaction medium, and limited exposure of the photocatalyst to light [72,73]. Since neither the photocatalytic activity nor the easy separation can be sacrificed, researchers turned their interests to magnetic nanoparticles as integral ingredients for developing photocatalysts, which were easily separated from water with an external magnetic field [74,75,76]. Magnetic nanoparticles such as haematite (α-Fe_2O_3), maghemite (γ-Fe_2O_3), magnetite (Fe_3O_4), ferrites (MFe_2O_4, M = Mg, Ni, Zn, Cu, Co, Cd, *etc.*), *etc.* have been successfully coupled with various photocatalysts to induce magnetic behaviour in the composite material and allow easy magnetic separation [39,77,78,79]. Since the photocatalyst remains powdery upon incorporation of the magnetic nanoparticles, this does not compromise its surface area and the distribution of the photocatalyst in the aquatic medium. In addition, some of these magnetic nanoparticles have good visible light absorption and photocatalytic properties, which further enhance the overall pollutant degradation kinetics over the magnetic nanocomposite [39,40,64,80,81,82].

Considering the future prospects of magnetic photocatalysts, the design and preparation of magnetic visible-light-active nanocomposites has emerged as a hot topic in photocatalytic environmental pollution control. Therefore, it is necessary to bring into perspective the current developments in the tailoring and application of such materials in water treatment. A few reviews are available mainly on the synthesis, structure, and various applications of magnetic materials [83,84,85,86,87,88]. This review exclusively discusses the most recent developments in terms of the synthesis and exploitation of various magnetic nanocomposite photocatalysts in water pollution mitigation. Details on the synthesis routes and structure of the various magnetic nanoparticles have been well-reviewed and have been excluded in this review. Furthermore, this review only focusses on iron oxide magnetic nanoparticles resulting in four categories of magnetic nanocomposite photocatalysts: haematite, maghemite, magnetite, and ferrite-based magnetic photocatalysts. Under each category, we discuss the influence of the magnetic nanoparticle on optical properties, charge transfer mechanism, and the overall photocatalytic properties of the nanocomposites. We conclude the review by providing some summary remarks and highlight the future challenges towards tailoring and utilising magnetic photocatalysts in water treatment.

RECENT DEVELOPMENTS IN MAGNETIC PHOTOCATALYSTS

Haematite (α-Fe_2O_3)-Based Magnetic Photocatalysts

As a pristine material, haematite is a narrow band gap semiconductor (2–2.2 eV) with good visible light response, good chemical stability, low-cost, ferromagnetic behaviour, abundantly available, high resistance to corrosion, and environmentally friendly [89,90,91]. On its own, haematite shows poor photocatalytic activity due to the high recombination rate of the charge carriers and poor conductivity [90,91]. Despite its poor photocatalytic activity, the narrow band gap of haematite makes it an ideal sensitiser for improving the visible light response of various semiconductors. Zhao *et al.* developed a nanostructured α-Fe_2O_3-AgBr nonwoven cloth (Figure 1a,b) via a simple electrospinning-calcination method and studied its visible light photocatalytic behaviour towards rhodamine B (RhB) and para-chlorophenol (4-CP) decomposition. For both pollutants, the α-Fe_2O_3-AgBr nonwoven cloth showed superior activity over α-Fe_2O_3 and AgBr clothes individually, attaining 91.8% 60 min and 74.2% in 120 min for RhB and 4-CP decomposition, respectively [81]. Coupling AgBr with α-Fe_2O_3 resulted in a significant red-shift and resulted in the formation of heterojunctions, which facilitated charge separation, thereby improving the overall photocatalytic activity of the α-Fe_2O_3-AgBr nonwoven cloth. In addition, the cloth was easy to handle and recycle during photocatalysis by simple dipping and removal or by using an external magnet. It was found that the composite cloth was stable over four cycles with only a slight decrease in photocatalytic performance [81].

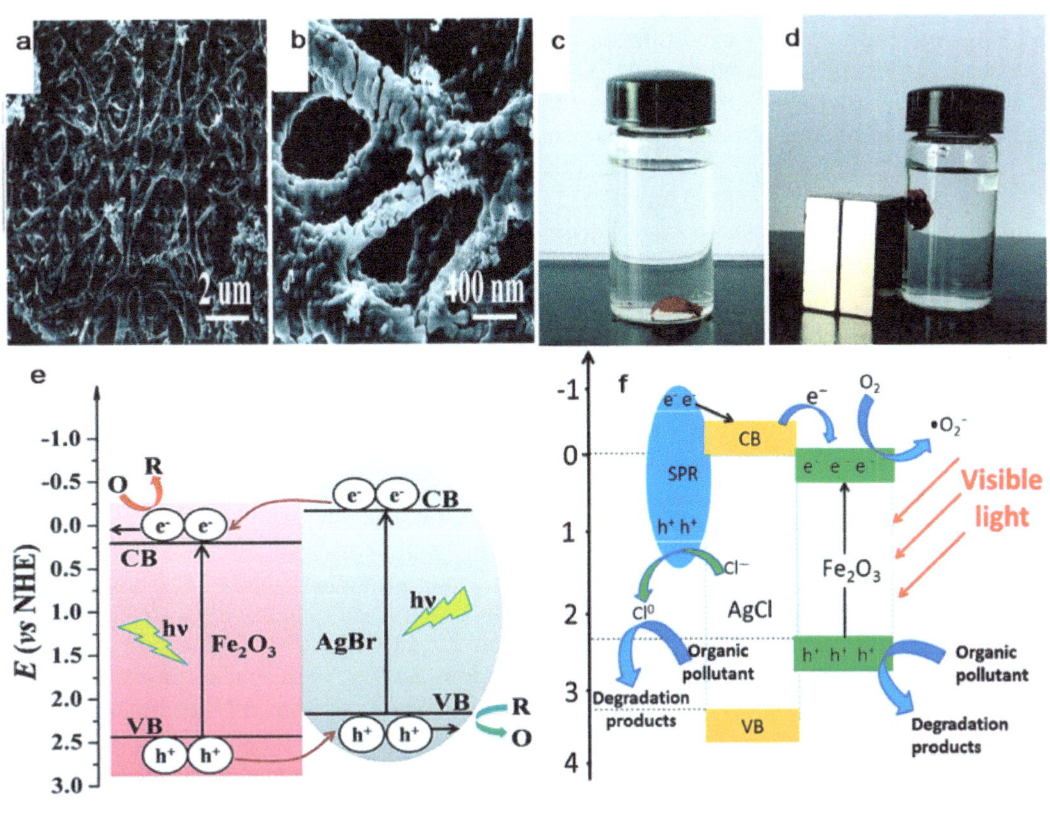

Figure 1: (**a,b**) SEM images of α-Fe_2O_3-AgBr nonwoven cloth; (**c,d**) illustration of magnetic separation of α-Fe_2O_3-AgBr cloth and charge transfer mechanism in (**e**) α-Fe_2O_3-AgBr cloth (Reproduced with permission from [81], Copyright 2015, RSC); and (**f**) Ag/AgCl/Fe_2O_3. Reproduced with permission from [39]. Copyright 2016, RSC.

Xue and coworkers fabricated magnetically separable α-Fe_2O_3/graphitic carbon nitride (g-C_3N_4) nanocomposite photocatalysts with varying α-Fe_2O_3 loadings (5%, 10%, 30%, 50%) and investigated their performance for degradation of RhB. The highest degradation efficiency (96.7% in 4 h) was recorded using the sample containing 10% α-Fe_2O_3, which suggests optimum composition [92]. The introduction of α-Fe_2O_3 nanoparticles into g-C_3N_4 matrix enhanced its visible light utilisation and charge separation efficiency. Moreover, the composite photocatalyst displayed good stability and could be easily separable and recycled using an external magnet [92]. Similarly, core-shell magnetically separable Ag/AgCl/Fe_2O_3 showed remarkable photocatalytic activity towards bisphenol A (BPA) and *E. coli* under visible light exposure. At a 5% α-Fe_2O_3loading, the composite photocatalyst reached the highest activity, which was 13 times that of Ag/AgCl for BPA degradation, and complete destruction of *E. coli* was observed after just 30 min of irradiation [39]. The nanocomposite displayed good stability and sufficient magnetic response to an external magnet.

In terms of charge transfer mechanism, the path followed by the electrons and holes depends on the semiconductor coupled with haematite and how their band structures match. In any case, under visible light illumination, haematite is excited and electrons accumulate in the conduction band while leaving positive holes in the valence band. In the case of α-Fe_2O_3-AgBr, both semiconductors are excited by visible light, and the electrons from AgBr transfer to the conduction band of α-Fe_2O_3, while the holes transfer in the opposite direction (Figure 1e). The holes will directly attack the organic pollutants to form the degradation products, while the electrons may react with adsorbed oxygen leading to the eventual formation of hydroxyl radicals, which are strong oxidising agents [81]. In the Ag/AgCl/Fe_2O_3 ternary nanostructure, the charge transfer mechanism (Figure 1f) is a bit complex. Only Ag and α-Fe_2O_3 are excited upon visible light absorption, and the plasmon-generated electrons in Ag will be injected into the conduction band of AgCl and further transfer to the conduction band of α-Fe_2O_3. Meanwhile, the holes in α-Fe_2O_3 can directly decompose the pollutants, while the holes left in Ag react with Cl^- ions to form Cl^0, which then attacks the pollutants [39].

Other haematite-based magnetically separable nanostructured photocatalysts such as graphene (GR)-α-Fe_2O_3-ZnO [93], α-Fe_2O_3/TiO_2 nanofibres [94], α-Fe_2O_3/RGO [95], and α-Fe_2O_3/ZnO [96] have been fabricated, have shown remarkable visible light utilisation, and have been easily separable using an external magnetic field. Coupling the various semiconductors with α-Fe_2O_3resulted in synergy between the materials, leading to enhanced visible light response, improved charge separation, and an overall increase in photocatalytic activity. Moreover, the incorporation of haematite allowed for the easy separation of the photocatalysts after the water treatment. However, α-Fe_2O_3 generally shows weak ferromagnetic behaviour, and, most importantly, its magnetic properties largely depend on the synthesis method, which influences its particle size and shape [92]. Several α-Fe_2O_3-incorporating heterostructures such as α-Fe_2O_3/Ag_3VO_4 [91], g-C_3N_4/Ag/α-Fe_2O_3 [97], g-C_3N_4/α-Fe_2O_3 [98,99,100], α-Fe_2O_3/TiO_2 [101] Au/g-C_3N_4/α-Fe_2O_3 [102], and α-Fe_2O_3/CdS [90] have been fabricated, showing enhanced visible light photocatalytic properties, and no magnetic behaviour was reported. Therefore, when the main aim of incorporating the iron oxide nanoparticles is to induce magnetic response, haematite may not be the ideal choice.

Maghemite (γ-Fe_2O_3)-based Magnetic Photocatalysts

Maghemite is another important form of iron oxide but, unlike haematite, it shows good magnetic properties with saturation magnetisation values of up to 76 emu/g [103]. Consequently, the incorporation of maghemite into various photocatalyst matrices has been explored to fabricate magnetically separable composite photocatalysts for easy recovery and recycling [104,105]. Interestingly, despite the sensitivity to an external magnetic field, when the magnetic field is removed, the γ-Fe_2O_3-incorporating nanocomposites do not retain any significant residual magnetism that would otherwise cause problems by creating large clusters of magnetised particles during water treatment [106]. Maghemite shows good visible light response owing to its narrow band gap (2.2 eV) and possesses good adsorption properties. However, its photocatalytic activity is poor as a result of the high recombination rate of the photogenerated charge carriers and photodissolution [74,107]. Two approaches are usually adopted when coupling maghemite with other semiconductors. In the first approach, maghemite is directly coupled with other semiconductors to form heterojunctions, and it contributes towards visible light response of the nanocomposite photocatalyst as well as the overall charge transfer process [74,108,109]. According to the second approach, maghemite and the other semiconductor do not form heterojunctions (not in direct contact), but they are separated by an inert layer or shell usually silica (SiO_2). This is aimed at preventing photodissolution of the iron oxide and charge transfer between the two materials, which could otherwise result in the recombination on the maghemite surface. In this case, maghemite nanoparticles do not play any part in the photocatalysis process but function solely in the magnetic separation of the photocatalyst from the reaction medium [104,106,110].

A magnetically separable quaternary γ-Fe_2O_3/N and Fe-codoped TiO_2 nanocomposite was prepared using a sol-gel method (as illustrated in Figure 2a) and employed for the decomposition of reactive blue 4 (RB 4) under visible light illumination. Significantly higher photocatalytic activity was observed over the quaternary nanocomposite photocatalyst compared to N,Fe-TiO_2, N-TiO_2, and TiO_2, reaching 100% degradation in 180 min (Figure 2b) [111]. The remarkable improvement in activity of the nanocomposite could be explained in terms of the combined contribution of codoping TiO_2 and coupling with γ-Fe_2O_3, which resulted in significant band gap narrowing, efficient visible light response, and efficient charge separation and transfer. Furthermore, the multicomponent photocatalyst demonstrated good stability and recyclability over four cycles without any significant loss in activity or magnetic properties, which allowed for easy magnetic separation [111]. Yu and coworkers probed the adsorption and UV photocatalytic oxidation of As(III) over TiO_2 decorated on magnetic (γ-Fe_2O_3) mesoporous SBA-15 nanocomposite (γ-Fe_2O_3/SBA-15/TiO_2) derived from simple inner-pore hydrolysis combined with solvent evaporation route [112]. The nanocomposite photocatalyst displayed a dual function of adsorption and photocatalytic degradation. Firstly,

As(III) ions were adsorbed onto the photocatalyst and then upon UV irradiation, undergo photocatalytic oxidation to As(V), which is less toxic and also removable by the photocatalyst. In addition, the photocatalyst showed good stability and could be regenerated for five cycles using NaOH (0.01 M) and still maintained over 90% As(III) removal and good magnetic properties for easy separation [112]. The incorporation of the SBA-15 ensured good dispersibility of the nanoparticles and improved the surface area of the nanocomposite. All three components of the nanocomposite were equally important in the overall removal of As(III).

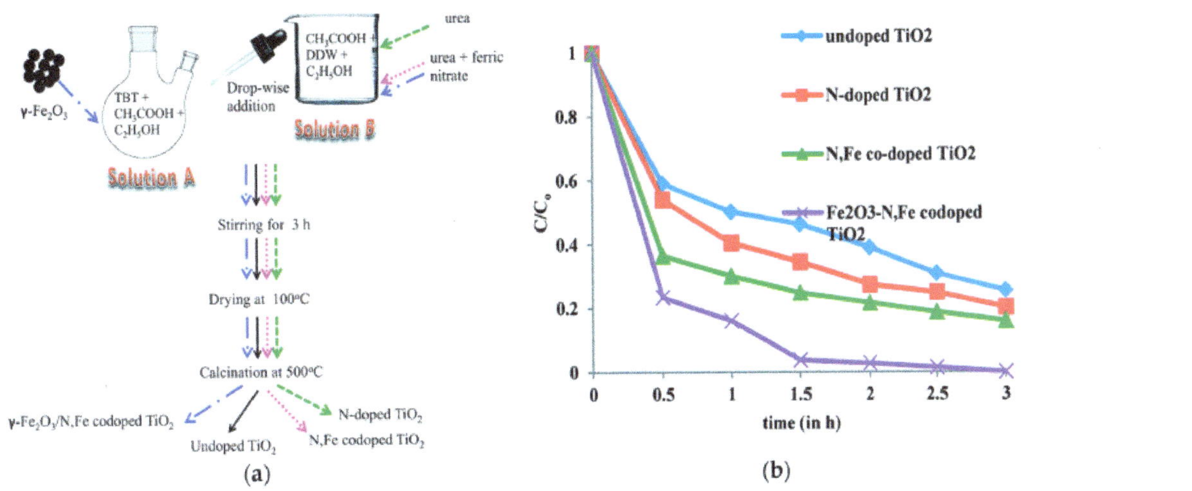

Figure 2: (**a**) Schematic illustration of the sol-gel preparation of γ-Fe_2O_3/N.Fe-TiO_2; and (**b**) photocatalytic degradation of RB 4 over γ-Fe_2O_3/N.Fe-TiO_2 and the controls. Reproduced with permission from [111]. Copyright 2015, RSC.

Improved optical and photocatalytic properties have also been reported for other magnetic photocatalysts such as ZnO/γ-Fe_2O_3 [113], g-C_3N_4/γ-Fe_2O_3 [109], and γ-Fe_2O_3/TiO_2 [74] towards the degradation of various pollutants including RhB, methylene blue (MB), and methyl orange (MO). Moreover, these nanostructures showed good stability and could be magnetically separated and recycled without a significant loss in activity. However, the maghemite nanoparticle loading needs to be carefully controlled in order to strike a good balance between sufficient magnetism to facilitate separation and the detrimental effect of recombination of the charge carriers on the maghemite nanoparticles. This means that, while enjoying the luxury of easy separation, the detrimental effect of recombination must not be ignored in order to positively benefit from the incorporation of the maghemite nanoparticles.

According to the proposed charge transfer route in γ-Fe_2O_3/N,Fe-TiO_2 (Figure 3a), both maghemite and codoped titania absorb visible light, and electrons are excited from the valence band to the conduction band in γ-Fe_2O_3, leaving holes in the valence band. Similarly, in the codoped titania, electrons are excited from the valence band of TiO_2 to the impurity states formed by Fe doping and also from N sub-band gap states to the conduction band of TiO_2. Subsequently, electrons transfer from the conduction band of maghemite to the Fe impurity states in TiO_2, while the holes migrate to the N states in TiO_2. The electrons and holes will be captured by molecular oxygen and water to form the superoxide and hydroxyl radicals, respectively [111]. These radical species are important oxidising agents for pollutant decomposition.

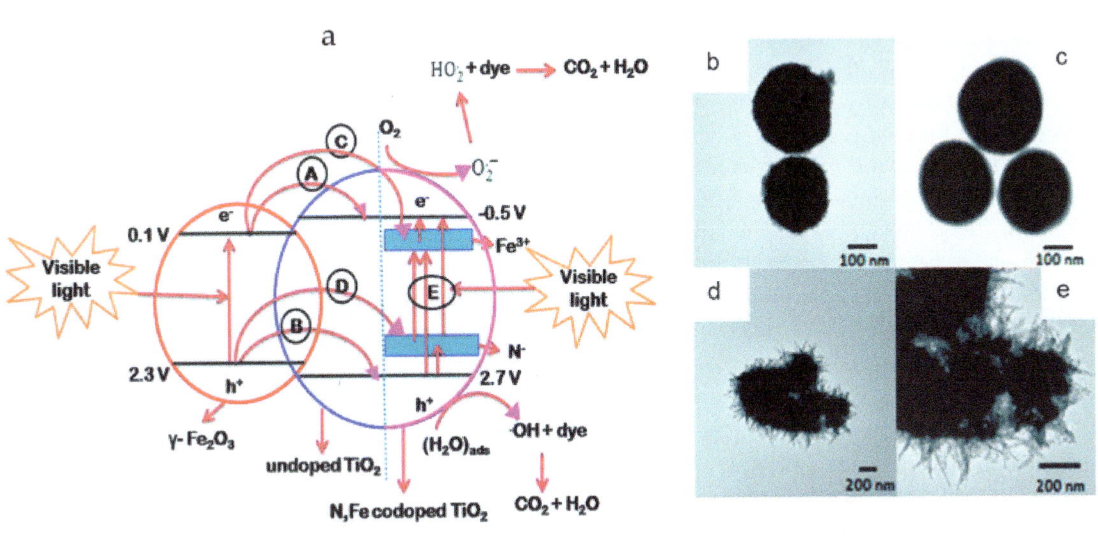

Figure 3: (**a**) Charge transfer mechanism in γ-Fe_2O_3/Fe,N-TiO_2(Reproduced with permission from [111], Copyright 2015, RSC) and (**b–e**) SEM images of urchin-like γ-Fe_2O_3@SiO_2@TiO_2composite microparticles. Reproduced with permission from [114]. Copyright 2015, RSC.

The strategy of preventing the formation of heterojunctions between maghemite and the semiconductor coupled with it has been widely explored using silica as the barrier/shell surrounding the magnetic core. For example, Szeto *et al.* used a layer-by-layer assembly route to fabricate urchin-like γ-Fe_2O_3@SiO_2@TiO_2 composite microparticles (Figure 3b–e) and observed enhanced phenol decomposition under UV irradiation [114]. The tailored composite photocatalyst was shown to possess a large surface area, a prolonged life of the charge carriers, remarkable permeability, and significantly higher photocatalytic activity compared to commercial P25. In addition, incorporation of the maghemite nanoparticles ensured easy and efficient separation of the photocatalyst using an external magnetic field [114]. Similar results have been reported for other magnetic heterostructures such as γ-Fe_2O_3@SiO_2@Ce-TiO_2 [110], γ-Fe_2O_3@SiO_2@AgBr:Ag [104], and γ-Fe_2O_3@SiO_2@TiO_2 [106], whereby a silica coating separates the maghemite nanoparticles from direct contact with the other components of the heterostructure. In these nanocomposites, the maghemite nanoparticles were only incorporated to facilitate the magnetic separation and play no part in the charge transfer process and overall photocatalytic decomposition of the pollutants. Although this approach is logical and has yielded some positive results, a critical comparison between the heterojunction type of photocatalyst and the core-shell structure of the same nanocomposite is rarely documented. This could map a way forward in terms of the most efficient approach towards engineering the maghemite-based magnetic photocatalyst and benefitting from the properties of this material.

Magnetite (Fe_3O_4)-based Magnetic Photocatalysts

Magnetite is by far one of the most widely exploited iron oxide magnetic nanoparticle in the fabrication of magnetic photocatalysts due to its low toxicity, good magnetic properties, biocompatibility, and remarkable adsorption properties [115,116]. The magnetisation saturation for magnetite can reach values of up to 92 emu/g, making it the ideal candidate for synthesis of magnetically separable photocatalysts [103]. However, unlike maghemite, magnetite shows no photocatalytic properties, and, like maghemite, the formation of heterojunctions with other semiconductors in nanocomposite photocatalysts is undesirable because magnetite may suppress the overall photocatalytic efficiency by acting as a recombination centre for the charge carriers [117,118]. Silica is commonly utilised to enclose the magnetite core, which is wrapped with the other semiconductor nanoparticles to form the magnetic composite photocatalysts. The silica barrier prevents electron transfer between magnetite and the other semiconductor, additionally preventing photodissolution of the iron oxide [119,120].

Despite the possible negative effect of magnetite in heterojunction nanocomposites, several reports on successful tailoring and improvement in photocatalytic activity over heterojunction nanocomposites have emerged, and careful control of the Fe_3O_4 loading in the nanocomposite is an important point of consideration. For example, Zhao *et al.* observed enhanced photocatalytic activity towards ampicillin (AMP) over Fe_3O_4/TiO_2/Ag with a sea urchin-like morphology, both under visible and UV light irradiation. The composite photocatalyst reached 98.7% and 91.5% AMP degradation after 360 min of UV and visible light illumination, respectively [116]. Furthermore, the nanocomposite possessed good magnetic behaviour with a magnetisation saturation of 26.5 emu·g^{-1} (Figure 4), which was high enough to facilitate magnetic separation (Figure 4 insert). In addition to the degradation of AMP, the nanocomposite photocatalyst showed remarkable photocatalytic properties towards *S. aureus* (G^+), *E. coli* (G^-), and *A. niger* [116]. In another example, improved MO decomposition was realised over Fe_3O_4/N-TiO_2/Ag hollow nanospheres, reaching 99.5% in 80 min of visible light exposure when the Ag loading was 1.0% [121]. Ma and coworkers employed a low-temperature crystallisation method to wrap Fe_3O_4 spheres with TiO_2 nanoparticles and subsequently immobilised Au nanoparticles on the core-shell Fe_3O_4/TiO_2 to yield a Fe_3O_4/TiO_2/Au plasmonic magnetic photocatalyst. RhB was used as a model pollutant to evaluate the photocatalytic properties of the nanocomposite, and it showed superior performance over Fe_3O_4 and Fe_3O_4/TiO_2, with good magnetic properties (magnetisation saturation (Ms), 44.6 emu·g^{-1}), and recyclability [122]. The heterojunctions formed between Fe_3O_4 and titania allow electrons to transfer between the two materials, which improves charge separation. However, the Fe_3O_4 loading needs to be kept in check as excess Fe_3O_4 will cause detrimental effects. Therefore, Fe_3O_4 nanoparticles have a dual function in the heterojunction nanostructures—electron trapping and enabling magnetic separation.

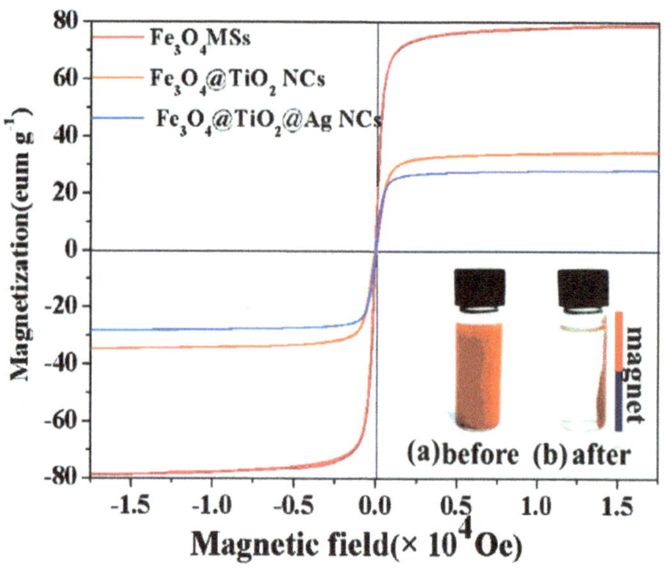

Figure 4: Magnetization curves and insert showing magnetic separation of Fe_3O_4/TiO_2/Ag. Reproduced with permission from [116]. Copyright 2016, RSC.

Li *et al.* fabricated Fe_3O_4 magnetic core coated with SiO_2 prior to the deposition of TiO_2 and surface modification with lysine, and the quaternary nanocomposite photocatalyst (Fe_3O_4@SiO_2@TiO_2-lysine) showed multifunctional properties as it was employed as an adsorbent, photocatalyst, and sensor for dissolved organic and inorganic phosphorus in seawater [123]. The photocatalyst displayed remarkable stability and could be magnetically separated and recycled for 10 cycles without any significant loss in its adsorption and photocatalytic properties [123]. Other core-shell Fe_3O_4@SiO_2 nanostructures decorated with TiO_2 have been fabricated and evaluated for the decomposition of various pollutants such as RhB, 2-chlorophenol (2-CP), phenol, MB, and MO. These materials showed improved photocatalytic properties, good stability, and sufficiently high magnetic response that enabled separation using an external magnetic field [124,125,126,127]. Table 1 provides a summary of the synthesis and photocatalytic properties of various other magnetite-incorporating titania-based photocatalysts towards water treatment applications. In the magnetic core-shell nanostructures, the Fe_3O_4 nanoparticles encapsulated with the silica shell do not partake in the charge transfer process but only in the magnetic separation of the photocatalyst.

Table 1: Summary of the photocatalytic properties of Fe_3O_4-incorporating TiO_2-based photocatalysts.

Sample	Preparation Method	Application	Catalyst Dose	Pollutant Concentration	Degradation Efficiency	Saturation Magnetisation	Refere
Fe_3O_4/GO/Ce-TiO_2	Low-temperature sol-gel/ ultrasonication	Degradation of tetracycline (TC)/Vis/300 W Xe lamp/400 nm UV filter	0.050 g	25 ppm	82.9% in 60 min	No data	[33]
Fe_3O_4/RGO/TiO_2	Reduction-deposition	Degradation of MB and tetrabromobisphenol A (TBBPA)/UV/230 W Hg lamp	0.025 g	10 ppm	99.5% in 60 min/TBBPA, 95.9% in 60 min/MB	1.14 emu/g	[128
$_3O_4$/P(MAA-DVB)/TiO_2	Magnetic field induced assembly/precipitation-polymerisation	Degradation of RhB/UV/250 W Hg lamp	0.040 g	4.43 ppm	71.5% in 120 min	35.2 emu/g	[129
Fe_3O_4@SiO_2@N-TiO_2	Sol-gel	Degradation of phenol/Vis/15 W florescent lamp	0.800 g	100 ppm	46% in 480 min	~2 emu/g	[127
Fe_3O_4/chitosan/TiO_2	Hydrothermal/crosslinking	Degradation of MB/UV/8 W UV lamp	1.00 g	1.28 ppm	93% in 40 min	4.2 emu/g	[130
Fe_3O_4/TiO_2	Reverse microemulsion/sol-gel	Degradation of RhB/UV/50 W Xe lamp	0.004 g	22.12 ppm	100% in 100 min	21.68 emu/g	[131
Ag_3PO_4/TiO_2/Fe_3O_4	*In situ* hydrolysis/deposition	Degradation of acid orange 7 (AO 7)/50 mW diode blue laser and *E. coli*/Vis/300 W Xe lamp/420 nm UV filter	0.100 g	5.25 ppm/AO7 and 10^7 CFU/mL/*E. coli*	~100% in 2.5 min/AO 7, 99.8% in 5 min/*E. coli*	No data	[38]
Fe_3O_4/TiO_2/Bi_2O_3	Sol-gel	Degradation of MO/Simulated solar light/350 W Xe lamp	0.200 g	No data	69% in 150 min	No data	[132
Fe_3O_4/TiO_2/Au	Sol-gel/hydrothermal	Degradation of MB and *E. coli*/UV/4 × 9 W black lights	0.015 g/MB and 0.010 g/*E. coli*	8.00 ppm/MB and 10^8 CFU/mL/*E. coli*	78% in 4 h/MB, 89.3% in 60 min/*E. coli*	No data	[133
\g@Fe_3O_4@SiO_2@TiO_2	Solvothermal	Degradation of MB and Cr(VI) reduction/Vis/500 W Xe lamp/425 UV filter	0.020 g	50 ppm/MB and 22.24 ppm/$K_2Cr_2O_7$	99.9% in 4 h/Cr(VI), ~90 mg/g/MB	13.92 emu/g	[134
e_3O_4@SiO_2@TiO_2@GO	Reverse microemulsion/sol-gel/amide conjugation	Degradation of RhB/UV/400 W column high pressure Hg lamp	0.050 g	8.85 ppm	92% in 120 min	16.90 emu/g	[120
Fe_3O_4@C@TiO_2	Solvothermal/calcination	Degradation of RhB/UV/125 W high pressure Hg lamp	0.020 g	10 ppm	~100% in 80 min	6.04 emu/g	[135
Sample	Preparation Method	Application	Catalyst Dose	Pollutant Concentration	Degradation Efficiency	Saturation Magnetisation	Reference
Fe_3O_4@SiO_2@TiO_2	Stobber process/sol-gel/chemical precipitation	Degradation of RhB/UV/50 W high pressure Hg lamp	0.010 g	8.85 ppm	100% in 10 min	30.60 emu/g	[136]
Fe_3O_4/TiO_2	Hydrothermal	Degradation of MB/UV/9 W UV lamp	0.002 g	1.00 ppm	100% in 10 min	No data	[75]
Fe_3O_4@TiO_2@GR	Sol-gel/assembly route	Degradation of 2,4-dichlorophenoxy-acetic acid (2,4-D)/simulated solar light/500 W Xe arc lamp	0.020 g	20 ppm	100% in 40 min	13.00 emu/g	[137]
WO_3/TiO_2/Fe_3O_4	Sol-gel	Degradation of direct blue 71 (DB 71)/Vis/200 W Xe lamp	0.030 g	50 ppm	98% in 35 min	18.20 emu/g	[138]

Chidambaram and coworkers prepared a novel plasmonic, magnetic Fe_3O_4/ZnO/Ag nanocomposite and employed it for the photocatalytic degradation of MB under visible light illumination. Complete MB removal was observed after 6 h of visible light exposure on Fe_3O_4/ZnO/Ag, while only 20% MB removal was realised over Fe_3O_4/ZnO during the same irradiation time. The improved activity could be explained in terms of improved charge separation and the surface plasmon resonance (SPR) effect of Ag nanoparticles, which is responsible for the visible light activity of the nanocomposite photocatalyst [139]. Fe_3O_4 nanoparticles played a dual role of facilitating charge separation and enabling magnetic separation. In addition, the photocatalyst showed great stability over 10 cycles without much loss in activity or magnetic properties. Upon visible light irradiation, the SPR-generated electrons are injected into the conduction band of ZnO and further transfer to the Fe_3O_4 nanoparticles. Subsequently, the electrons reduce Fe^{3+} to Fe^{2+}, which in turn is oxidised back to Fe^{3+} upon reaction with adsorbed oxygen forming the superoxide radicals. Meanwhile, the SPR-generated holes in Ag could react with water to form the hydroxyl radicals and together with the superoxide radicals are responsible for pollutant degradation [139]. Yan *et al.* observed higher visible light photocatalytic performance over urchin-like Fe_3O_4@SiO_2@ZnO/CdS core-shell microspheres (Figure 5a,b) towards RhB compared to Fe_3O_4@SiO_2@ZnO microspheres. Coating Fe_3O_4 with SiO_2 prevented the formation of heterojunctions with ZnO and CdS, and the nanocomposite showed good magnetic properties with magnetisation saturation value of 18.97 emu/g [140]. On visible light exposure, only CdS nanoparticles were excited and played the role of a sensitiser for ZnO, with electrons transferred to the conduction band of ZnO while the holes remained in the valence band of CdS (Figure 5c), resulting in efficient charge separation and the formation of the oxidising species [140].

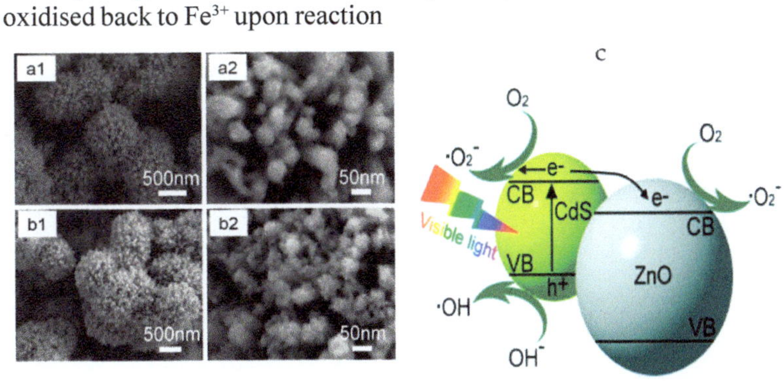

Figure 5: (**a**,**b**) SEM images of urchin-like Fe_3O_4@SiO_2@ZnO/CdS microspheres and (**c**) charge transfer route in Fe_3O_4@SiO_2@ZnO/CdS. Reproduced with permission from [140]. Copyright 2016, RSC.

Habibi-Yangjeh's group has recently prepared and investigated the visible light photocatalytic properties of various magnetic nanocomposites-based on Fe_3O_4 and ZnO such as ZnO/AgI/Fe_3O_4[141], ZnO/AgBr/Fe_3O_4/Ag_3VO_4 [142], ZnO/Ag_3VO_4/Fe_3O_4 [143], Fe_3O_4/ZnO/AgCl [144], and Fe_3O_4/AgBr-ZnO [145] towards model pollutants such as RhB, phenol, MB, and MO. All of the nanocomposite photocatalysts showed good photocatalytic properties, which can be credited to efficient charge separation and improved visible light absorption as a result of the heterojunctions formed between ZnO and the co-catalysts. Moreover, it was shown that despite the formation of the composite photocatalysts, the presence of Fe_3O_4 nanoparticles ensured that the photocatalysts still retained sufficient magnetic response to allow for efficient separation using an external magnetic field. Obviously, the magnetisation saturation of Fe_3O_4 decreased sharply upon the formation of the nanocomposites due to their interaction and coverage by the non-magnetic components of the photocatalysts [141,142,143,144,145]. Under visible light irradiation, ZnO remains inactive and only the co-catalyst with a narrow band gap can be excited. For example, in ZnO/AgI/Fe_3O_4 and ZnO/AgBr/Fe_3O_4/Ag_3VO_4, only AgI, AgBr, and Ag_3VO_4, respectively, are excited, and electrons are promoted to the conduction band, while the holes remain in the valence band. Subsequently, the electrons transfer to the conduction band of ZnO where they react with adsorbed oxygen to form the radical species (superoxide and hydroxyl radicals). Accordingly, the holes are directly involved in the degradation of the pollutants since they are not positive enough to oxidise water for the hydroxyl radicals [141,142]. The electrons and holes occupied different sites in the nanocomposite photocatalysts, which ensured efficient charge separation and the formation of the reactive species for pollutant degradation.

Owing to its meta-free nature, polymeric structure, non-toxicity, good visible light response, good photocatalytic properties, and high thermal and chemical stability, tailoring graphitic carbon nitride (g-C_3N_4)-based magnetic photocatalysts is an important exploit towards potential practical application of these materials in water treatment. Using a combination of calcination and co-precipitation, Yang and coworkers prepared Fe_3O_4/g-C_3N_4 nanocomposite photocatalyst and probed its photocatalytic behaviour towards 2,4,6-trichlorophenol (2,4,6-TCP) under visible light illumination [77]. In a similar work, Jia *et al.* fabricated Fe_3O_4/g-C_3N_4 via a hydrothermal route and studied its visible light photocatalytic performance towards RhB decomposition [146]. In both studies, higher photocatalytic performance was observed over the nanocomposites compared to Fe_3O_4 and g-C_3N_4 individually, reaching 95.5% in 60 min [146] and 96.5% in 100 min [77] for RhB and 2,4,6-TCP, respectively. Coupling g-C_3N_4 with Fe_3O_4 improved charge separation since only g-C_3N_4 is excited by visible light, and the electrons in the conduction band quickly transfer to Fe_3O_4, where they are captured by oxygen to form the radical species. Meanwhile, the holes remain in the valence band of g-C_3N_4 and directly oxidise the pollutants to form the degradation products. Therefore, the holes and electrons occupy different locations in the nanocomposite. Moreover, magnetic separation was made possible by the presence of the Fe_3O_4 nanoparticles [77,146].

Habibi-Yangjeh's group investigated the visible light photocatalytic performance of novel magnetic nanocomposites: g-C_3N_4/Fe_3O_4/BiOI [147], g-C_3N_4/Fe_3O_4/AgCl [148], and g-C_3N_4/AgBr/Fe_3O_4 [149] towards the decomposition of RhB, MO, and MB. Significantly higher photocatalytic activity was observed over the ternary heterostructures compared to single and binary nanocomposites, and this could be ascribed to the synergistic effect of the various components of the ternary nanocomposites, resulting in improved visible light response and charge separation efficiency. Moreover, the presence of Fe_3O_4 in the heterostructures endowed them with magnetic response, which allowed separation using an external magnetic field [147,148,149]. In terms of charge transfer, only g-C_3N_4 is excited by visible light in the case of g-C_3N_4/Fe_3O_4/AgCl, while both g-C_3N_4 and BiOI are excited in g-C_3N_4/Fe_3O_4/BiOI, and the electrons are promoted to their conduction bands, leaving positive holes in the valence bands. This is followed by the transfer of electrons from the conduction band of g-C_3N_4, which is more negative, to the conduction bands of AgCl and BiOI, which are less negative, whereby photoreduction reactions involving adsorbed oxygen take place and result in the formation of the radical species [147,148]. Accordingly, in both materials, the holes accumulate in the valence band of g-C_3N_4 where they directly oxidise the pollutants to form the degradation products.

Generally, the incorporation of Fe_3O_4 nanoparticles in the photocatalysts can serve as a means to prolong the life of the photogenerated charge carriers by acting as electron traps and induce magnetic response in the nanocomposites, which enables easy separation using an external magnetic field. In both magnetic core-shell and heterojunction nanostructures, the magnetic influence of the Fe_3O_4 nanoparticles was significant despite showing lower magnetisation saturation compared to neat Fe_3O_4 due to the presence of the non-magnetic components of the photocatalysts. In the heterojunction nanostructures, Fe_3O_4 nanoparticles played a dual role of charge separation and magnetic separation, but the detrimental effect of recombination becomes a problem if the optimum Fe_3O_4 loading is not carefully controlled. Meanwhile, the silica shell in the core-shell nanostructures serve as a barrier to prevent the formation of heterojunctions and charge transfer, limiting the role of the Fe_3O_4 nanoparticles only to magnetic separation of the photocatalysts. A host of other Fe_3O_4-based magnetic photocatalyst have been designed, prepared, and exploited in water treatment applications and are summarised in Table 2.

Table 2: Summary of the photocatalytic properties of various Fe_3O_4-based photocatalyst.

Sample	Preparation Method	Application	Catalyst Dose	Pollutant Concentration	Degradation Efficiency	Magnetisation Saturation	Reference
Fe_3O_4@SiO_2@CdS	Stobber method/chelating assisted growth	Degradation of MB and TC/Vis	0.010 g	10 ppm/MB and 100 ppm/TC	80% in 21 min/TC, 94% in 140 min/MB	22.00 emu/g	[63]
Fe_3O_4(TAMH)/ZnO	Solvothermal	Degradation of phenol/Vis/ 575 W MSR metal halide lamp	0.325 g	20 ppm	71.3% in 150 min	No data	[150]
Fe_3O_4@mSiO_2@BiOBr	Solvothermal	Degradation of MB/Vis/500 W Xe lamp/420 nm UV filter	0.100 g	20 ppm	96% in 120 min	40.00 emu/g	[151]
Fe_3O_4/$BaTiO_3$	Solvothermal/sol-gel	Degradation of MO and orange II/UV/150 W UV lamp	0.002 g	10 ppm	71.2% in 20 h/MO, 43.7% in 20 h/ orange II	60.50 emu/g	[152]
Fe_3O_4/Cr_2O_3	Wet chemical/ultrasonication	Degradation of 4-CP/UV/12 W low-pressure Hg lamp	0.100 g	1.29 ppm	100% in 150 min	~20 emu/g	[153]
Cu_2O/chitosan/Fe_3O_4	Precipitation-reduction	Degradation of Reactive Brilliant red X-3B (X-3B)/Vis/500 W W-halogen lamp	0.100 g	50 ppm	99.7% in 50 min	15.1 emu/g	[154]
Fe_3O_4/AgBr	Precipitation route	Degradation of MO/Vis/300 W Xe arc lamp/420 nm UV filter	0.100 g	20 ppm	85% in 12 min	No data	[155]
RGO/Ag/AgCl/Fe_3O_4	Solvothermal/deposition-precipitation	Degradation of MB and RhB/Vis/500 W Xe arc/10 cm water filter	0.100 g	10 ppm	97.4% in 100 min/MB, 97.9% in 120 min/RhB	18.8 emu/g	[156]
Metalloporphyrin/Fe_3O_4	Covalent conjugation	Degradation of AO 7/Vis/125 W W-halogen lamp	0.444 g	17.5 ppm	69.0% in 5 h	61.45 emu/g	[157]
Au(Ag)/AgCl/Fe_3O_4@PDA@Au	Solvothermal/galvanic replacement	Degradation of MB/Vis	0.004 g	5 ppm	100% in 20 min	5.40 emu/g	[158]
Fe_3O_4@resorcinol-formaldehyde-Ag	*In situ* polymerisation/reduction	Inactivation of *S. aureus*, *E. coli* and degradation of 4-nitrophenol (4-NP), MB and RhB/Vis/No data on light source	0.010 g	16 ppm/ RhB/MB/4-NP and 10^7 CFU/mL/ bacteria	1.5 < OD in 6 h/ *E. coli*, ~1.0 OD in 6 h, *S. aureus*, ~100% in 3 min/ 4-NP, ~100% in 4 min/MB and ~100% in 3 min/RhB	~30 emu/g	[26]
Poly(p-phenylenediamine)-Fe_3O_4	Chemical oxidation polymerisation	Degradation of bromocresol green (BG), blue (BB), purple (BP), RhB, neutral red (NR), MB, Sudan III (SIII), MO and Congo red (CR)/UV (500 W Hg)/Vis (500 W Xe lamp/420 nm UV filter)	0.025 g	50 ppm	~95% in 1100 min/ BB/UV, ~80% in 1100 min/BB/Vis, ~100% in 1100 min/ BG/UV, ~90% in 1100 min/BG/Vis	No data	[159]
Polypyrrole (PPY)/ Fe_3O_4/ZnO	*In situ* polymerisation	Degradation of MB/UV/8 W germicidal lamp	0.200 g	10 ppm	85.2% in 4 h	No data	[160]
Fe_3O_4/GO/Ag_3PO_4	Co-precipitation/ ultrasonication	Degradation of MB/Vis/250 W halogen lamp/400 nm UV filter	0.025 g	20 ppm	100% in 10 min	12.56 emu/g	[161]
RGO/Fe_3O_4	*In situ* chemical synthesis	Degradation of MB/natural sunlight/bright sunny days/ 9.00 am to 2.00 pm	0.002 g	10 ppm	100% in 60 min	30.30 emu/g	[162]
ZrO_2/Fe_3O_4/chitosan	Co-precipitation/refluxing	Reduction of Cr(VI) and degradation of 4-CP/ natural sunlight	0.050 g/ Cr(III), 0.010 g/ 4-CP	70 ppm/$K_2Cr_2O_7$, 20 ppm/4-CP	88.6% in 180 min/4-CP, 90.2% in 180 min/Cr(VI)	42.00 emu/g	[115]
Cu_2O/Fe_3O_4	Solvothermal/precipitation	Degradation of MO/Vis/500 W Xe lamp/420 nm UV filter	0.100 g	30 ppm	90% in 90 min	41.70 emu/g	[163]
$FeWO_4$/Fe_3O_4	Hydrothermal	Degradation of MB/UV-Vis/ 500 W Xe lamp	0.020 g	20 ppm	97.1% in 60 min	9.00 emu/g	[164]
Fe_3O_4@carbon quantum dots (CQDs)	Hydrothermal	Degradation of MB/Vis75 W Xe lamp/420 UV filter	0.001 g	1.00 ppm	94.4% in 30 min	33.80 emu/g	[165]
BiOBr@SiO_2@Fe_3O_4	Hydrothermal/ Stobber method	2,2-bis(4-hydroxyphenyl)propane (BPA)/500 W Xe lamp/ 420 UV filter	0.100 g	20 ppm	87.0% in 50 min	No data	[119]

Ferrites (MFe_2O_4)-based Magnetic Photocatalysts

The spinel ferrites with the chemical formula MFe_2O_4, where M is a cation such as Mg^{2+}, Cu^{2+}, Zn^{2+}, Ni^{2+}, Co^{2+}, *etc.*, are popular materials due to their narrow band gaps, which endow them with remarkable visible light response. Moreover, ferrites show good chemical, thermal, and ferromagnetic properties, numerous oxygen vacancies, and a high density of surface hydroxyl groups [166,167]. In terms of the crystal structure of ferrites, the oxygen atoms form a face-centred cubic (fcc) close packing, while the M^{2+} and Fe^{2+} cations occupy either octahedral or tetrahedral interstitial sites [168]. Generally, ferrites have attracted tremendous interests in various applications such as fabrication of memory devices, biotechnology, catalysts for organic reactions, water splitting, pollution remediation, lithium ion batteries, *etc.* [169,170,171,172]. Among the ferrites, zinc ferrite ($ZnFe_2O_4$) is one of the most attractive ferrite due to its photochemical and thermal stability, low toxicity, narrow band gap (1.9 eV), natural abundance, good magnetic properties, and easy preparation [64,173,174]. However, as a pristine photocatalyst, $ZnFe_2O_4$ show low photocatalytic activity (low quantum efficiency) due to the high recombination rate of the photogenerated charge carriers, unconvincing photoelectric conversion, and low valence band potential [64].

In an attempt to improve the photocatalytic performance of $ZnFe_2O_4$, laser ablation obtained $ZnO_x(OH)_x$ and FeO_x colloids were used as precursors to uniformly deposit $ZnFe_2O_4$ nanoparticles on RGO sheets to form a magnetic RGO/$ZnFe_2O_4$ nanocomposite. The RGO/$ZnFe_2O_4$nanocomposite exhibited improved charge separation and photocatalytic activity towards MB degradation under visible light irradiation [175]. In a similar work by Yang *et al.*, $ZnFe_2O_4$nanocrystals were confined within an interconnected graphene network, which served as dispersing agents, as transport channels for electrons, and as electron scavengers and reservoirs to minimise recombination. Enhanced visible light photocatalytic degradation of MB was observed over the nanocomposite photocatalyst [176]. In the RGO/$ZnFe_2O_4$ nanocomposite, electrons are excited by visible light from the valence band to the conduction band of $ZnFe_2O_4$, and these electrons quickly transfer to the RGO skeleton where they are trapped by adsorbed oxygen to form superoxide radicals (Figure 6a). The holes are confined to the valence band of $ZnFe_2O_4$ where they directly oxidise the pollutants since they are not positive enough to react with water to form the hydroxyl radicals [175,176]. More interestingly, the RGO/$ZnFe_2O_4$ nanocomposites showed good stability and recyclability and were easily separable using an external magnetic field owing to the good magnetic response (Figure 6b). The presence of RGO ensured that the electrons and holes are located at different parts of the nanocomposite, thereby ensuring efficient charge separation.

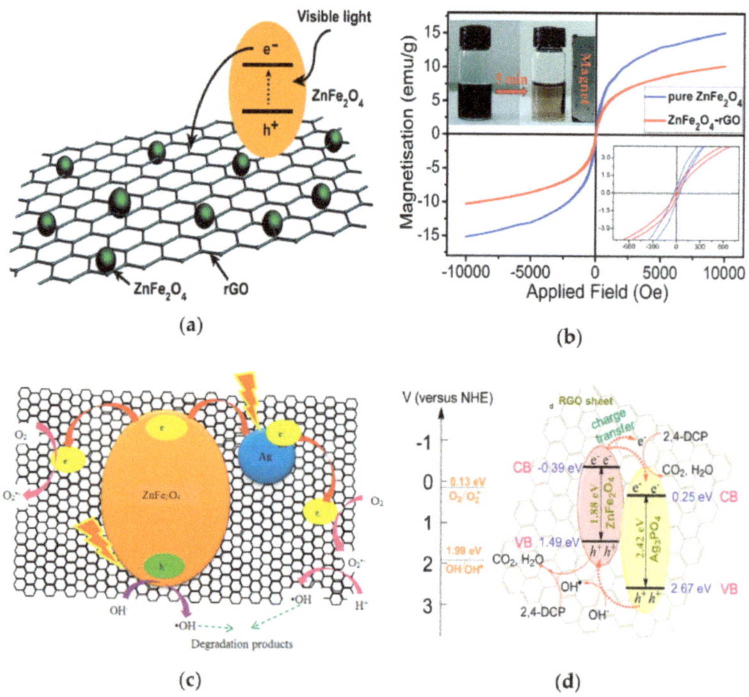

Figure 6: (**a**) Charge transfer route in $RGO/ZnFe_2O_4$; (**b**) magnetisation curves for pure $ZnFe_2O_4$ and $RGO/ZnFe_2O_4$, insert showing magnetic separation of $RGO/ZnFe_2O_4$(Reproduced with permission from [175], Copyright 2014, RSC); charge transfer mechanism in (**c**) $ZnFe_2O_4$–Ag/RGO (Reproduced with permission from [177], Copyright 2016, Wiley) and (**d**) $RGO/ZnFe_2O_4/Ag_3PO_4$. (Reproduced with permission from [178], Copyright 2016, ACS.

Enhanced visible light photocatalytic degradation of 17α-ethinylestradiol (EE 2) was observed over magnetic $ZnFe_2O_4$–Ag/RGO nanocomposite prepared via a hydrothermal method. In terms of degradation kinetics, the nanocomposite photocatalyst showed the highest activity, which was 14.6 and 5.6 times that of $ZnFe_2O_4$ and $ZnFe_2O_4$/RGO, respectively. This could be credited to the combined contribution of RGO, Ag, and $ZnFe_2O_4$ in the nanocomposite, resulting in an improved surface area, efficient charge separation and transportation, minimal aggregation, and improved visible light utilization [177]. Chen and coworkers used a combination of solvothermal and *in situ*precipitation routes to tailor a magnetic $RGO/ZnFe_2O_4/Ag_3PO_4$ nanocomposite and examined its photocatalytic behaviour towards 2,4-DCP under visible light exposure. The ternary nanocomposite exhibited significantly superior activity over pure Ag_3PO_4 [178]. In the $ZnFe_2O_4$–Ag/RGO heterostructure, electrons excited by visible light from the valence band of $ZnFe_2O_4$ to its conduction band transfer to the Ag nanoparticles and, further, to the RGO skeleton where they are captured by adsorbed oxygen to form the radical species (Figure 6c). The holes directly attack the pollutant to form the degradation products [177]. Similarly, in the $RGO/ZnFe_2O_4/Ag_3PO_4$ ternary heterostructure, a heterojunction type of charge transfer mechanism was the most plausible route (Figure 6d). Both $ZnFe_2O_4$ (band gap, 1.92 eV) and Ag_3PO_4 (band gap, 2.42 eV) are excited by visible light, and electrons are promoted to the conduction bands leaving positive holes in the valence bands. The conduction band potential (−0.39 eV) and valence band potential (+1.49 eV) of $ZnFe_2O_4$ are more negative than those of Ag_3PO_4; therefore, electrons transfer to the conduction band of Ag_3PO_4 and RGO, while the holes transfer in the opposite direction to the valence band of $ZnFe_2O_4$ [178]. This results in efficient charge separation and the formation of the reactive species.

Several other interesting $ZnFe_2O_4$-based magnetic nanocomposite photocatalysts such as $ZnFe_2O_4/ZnO/Ag_3PO_4$ [64], $ZnFe_2O_4$/ZnO [179], N-$TiO_2/ZnFe_2O_4$ [180], $ZnFe_2O_4/Ag_3PO_4$ [38], and Ag/ZnO/$ZnFe_2O_4$ [181] have recently been fabricated, and their photocatalytic properties were examined for the degradation of various pollutants in water. Notably, coupling the various semiconductors with $ZnFe_2O_4$ not only improved the overall visible response and photocatalytic properties of the photocatalysts but also introduced magnetic behaviour, which then allowed for easy separation using an external magnetic field.

Nickel ferrite ($NiFe_2O_4$) is another commonly exploited ferrite with a narrow band gap (1.7 eV), good magnetic properties, and high chemical and thermal stability. Structurally, $NiFe_2O_4$ has a typical inverse spinel structure and is ferromagnetic in nature. Its magnetism stems from the magnetic moment of antiparallel spins between Fe^{3+} cations occupying tetrahedral sites and Ni^{2+}cations sitting at octahedral sites [172,182,183]. Pure $NiFe_2O_4$ exhibits very low photocatalytic activity due to a high recombination rate of the charge carriers and the aggregation of the nanoparticles. However, $NiFe_2O_4$ is an attractive semiconductor to couple with other semiconductors or carbon nanomaterials and form nanocomposite photocatalysts with improved activity and, most importantly, could easily be separated using a magnet [167,184,185]. For example, magnetically separable AgBr/$NiFe_2O_4$ nanocomposite photocatalyst prepared via a combination of hydrothermal and an ultrasound-assisted precipitation method showed improved photocatalytic performance towards RhB under visible light illumination [186]. Ji and coworkers observed a significant improvement in the photocatalytic activity of magnetic g-$C_3N_4/NiFe_2O_4$towards MB degradation, reaching 87% in 4 h of visible light irradiation [172]. In another example, magnetic $Ag_3PO_4/NiFe_2O_4$ nanocomposite tailored using an *in situ* precipitation method exhibited better visible light-driven photocatalytic activity towards MB compared to Ag_3PO_4 and $NiFe_2O_4$, individually [184]. $NiFe_2O_4$-based magnetic nanocomposites incorporating carbon nanomaterials such as Pd-$NiFe_2O_4$/RGO [27] and $NiFe_2O_4$/MWCNTs [187] have also shown remarkable enhancement in photocatalytic activity owing to the synergy between the carbon nanomaterials and $NiFe_2O_4$, which resulted in improved optical properties, charge separation, and transportation and specific surface area [27,187].

In all the $NiFe_2O_4$-incorporating photocatalysts, $NiFe_2O_4$ is excited upon the absorption of visible light. The charge transfer process depends on the other semiconductors involved in the nanocomposites. For example in AgBr/$NiFe_2O_4$ and Ag_3PO_4/$NiFe_2O_4$, in addition to $NiFe_2O_4$, AgBr and Ag_3PO_4 are also excited by visible light, and electrons are promoted to their conduction bands, leaving holes in the valence bands. The electrons transfer from the conduction band of $NiFe_2O_4$, since it is more negative, to the conduction bands of AgBr and Ag_3PO_4. Accordingly, the holes transfer from the valence bands of AgBr and Ag_3PO_4 to the valence band of $NiFe_2O_4$ where they are involved in the degradation of the pollutants [184,186]. Meanwhile, in g-C_3N_4/$NiFe_2O_4$(both semiconductors are visible active), the electrons transfer from the conduction band of g-$C_3N_4$to the conduction band of $NiFe_2O_4$, and the holes transfer in the opposite direction to the valence band of g-C_3N_4 (Figure 7a) [172]. Charge separation efficiency is greatly improved upon coupling $NiFe_2O_4$ with the other semiconductors since the holes and electrons are located at different sites in the nanocomposites. Moreover, the presence of $NiFe_2O_4$ nanoparticles in the nanocomposites endowed them with good magnetic response, which allowed for easy magnetic separation.

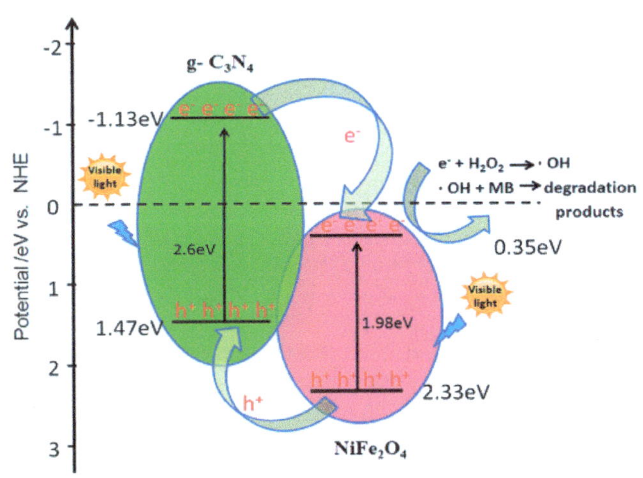

Figure 7: Charge transfer route in g-C_3N_4/$NiFe_2O_4$. Reproduced with permission from [172]. Copyright 2015, RSC.

Cobalt ferrite ($CoFe_2O_4$) is another important magnetic material with several attractive properties such as a crystallisation temperature similar to that of TiO_2, remarkable electromagnetic behaviour, high cubic magnetocrystalline anisotropy, good chemical stability, and high mechanical hardness [188,189,190,191]. Coupling $CoFe_2O_4$ with other semiconductors, carbon nanomaterials and conducting polymers are some of the strategies explored in order to benefit from its optical and magnetic properties. Pristine $CoFe_2O_4$ shows poor photocatalytic efficiency due to the high recombination rate of the charge carriers [61,192,193]. Kim *et al.* prepared hollow cobalt ferrite-polyaniline ($CoFe_2O_4$-PANI) nanofibres using the electrospinning method, calcination, and *in situ*chemical oxidative polymerisation, in succession. It was observed that the nanocomposite was magnetically separable and also showed significantly higher visible light photocatalytic degradation of MO, which was 80 times the activity of neat $CoFe_2O_4$ [61]. Hydrothermally prepared magnetic $CoFe_2O_4$/graphene (GR) nanocomposites with varying GR loadings exhibited improved visible light photocatalytic properties towards MB degradation compared to pristine $CoFe_2O_4$. At an optimum GR loading of 10%, the nanocomposite reached 100% MB degradation in 60 min, while the neat $CoFe_2O_4$ could only manage a low 34% degradation in twice the irradiation time [192]. The enhanced activity could be ascribed to the formation of heterojunctions between $CoFe_2O_4$ and GR, which allowed for the efficient separation of the charge carriers. Upon excitation by visible light, the electrons in the conduction band of $CoFe_2O_4$ transfer to GR, where they are trapped by molecular oxygen adsorbed on the surface [192]. A similar scenario was highlighted as responsible for the improved activity of $CoFe_2O_4$-PANI, whereby the electrons are transferred to the conduction band of $CoFe_2O_4$ from PANI while the holes transfer in the opposite direction to PANI [61]. The electrons and holes are kept apart, which ensures their availability to form the active species responsible for pollutant degradation.

Enhanced visible light photocatalytic decomposition of MB was reported over $CoFe_2O_4$decorated CdS nanorods ($CoFe_2O_4$/CdS, prepared via a soft chemical method. The composite nanorods exhibited better activity than CdS and $CoFe_2O_4$, individually, and were magnetically separable [193]. Three-dimensional (3D) urchin-like TiO_2, prepared hydrothermally, was decorated with $CoFe_2O_4$ using a co-precipitation strategy to yield the nanocomposite photocatalyst $CoFe_2O_4$/TiO_2 with improved photocatalytic properties compared to the individual materials. Upon UV light illumination, the nanocomposite reached 98.9% MB degradation, while 79.9% MB decomposition was realised when the urchin-like TiO_2 was used as the photocatalyst [194]. Moreover, the nanocomposite photocatalyst showed good stability and recyclability, reaching 93.8% MB degradation in the fifth cycles while still maintaining good magnetic properties [194]. Impressive photocatalytic activity and versatility was observed for $CoFe_2O_4$/BiOX (X = Cl, Br and I) nanocomposites towards the UV and visible light photocatalytic decomposition of MO, RhB, MB, and their mixture (MO + MB + RhB) [195]. Similarly, Xu and coworkers examined the photocatalytic properties and versatile nature of the Ag/AgCl/$CoFe_2O_4$ nanocomposite by employing the photocatalyst for the decomposition of MO, bisphenol A (BPA), ciprofloxacin (CIP), and the deactivation of *E. coli*. The nanocomposite photocatalyst displayed outstanding photocatalytic performance towards both coloured (MO) and colourless (BPA and CIP) pollutants as well as microbial pollutants, reaching 93.4% in 150 min for MO and 100% in 90 min for CIP, and almost all of the *E. coli* were deactivated in just 30 min of visible light illumination [196]. In all the $CoFe_2O_4$-incorporating nanocomposites, the enhancement in activity could be ascribed to the formation of a heterojunction between $CoFe_2O_4$ and the other semiconductor, which allowed efficient charge separation and formation of the reactive species responsible for pollutant degradation. In addition, the presence of $CoFe_2O_4$ endowed the nanocomposite photocatalysts with sufficient magnetic sensitivity for easy separation using an external magnetic field [193,194,195,196]. Similar results have been reported for other $CoFe_2O_4$-based magnetic photocatalysts such as $CoFe_2O_4$/MCM-41/TiO_2 [197], Fe,N-TiO_2/$CoFe_2O_4$ [198], and BiOBr/$CoFe_2O_4$ [65].

A host of other ferrite-based magnetic nanocomposite photocatalysts (Table 3) have been designed and tailored for the removal of various pollutants in water, for which their photocatalytic properties were also examined. Enhanced photocatalytic performance has been observed upon the incorporation of the ferrites. Moreover, the presence of these ferrites induced magnetic response in the photocatalyst, which aids the separation process for easy recovery and recycling of the photocatalysts. Despite the decrease in magnetisation saturation of the nanocomposites compared to the neat ferrites, the photocatalyst retained enough magnetic response to enable separation using an external magnet.

Table 3: Summary of the photocatalytic properties of ferrite-based photocatalysts.

Sample	Preparation Method	Application	Catalyst Dose	Pollutant Concentration	Degradation Efficiency	Magnetisation Saturation	Reference
P25/$CoFe_2O_4$/RGO	Hydrothermal	Degradation of MB, MO, neutral dark yellow (NDY)/Vis/500 W Xe lamp/420 nmUV filter	0.010 g	40 ppm	~90 mg/g in 60 min/MB, ~20 mg/g in 60 min and 10 mg/g in 60 min	~35 emu/g	[169]
$Mn_xMg_{1-x}Fe_2O_4$ ($0.0 \leq x \leq 0.5$)	Microwave-assisted combustion	Degradation of 4-CP/Vis/ 8 × 8 W medium pressure Hg lamps	0.030 g	200 ppm	~98% in 300 min	64.67 emu/g	[199]
Ag/TiO_2/$NiFe_2O_4$	Photodeposition/solvothermal	Degradation of MO/Vis/100 W LED	0.100 g	10 ppm	100% in 120 min	No data	[200]
TiO_2@SiO_2@Ni-Cu-ZnFe_2O_4	Chemical co-precipitation	Degradation of MB/simulated solar light/35 W Xe arc lamp	0.400 g	10 ppm	83.9% in 6 h	37.45 emu/g	[201]
$CoFe_2O_4$/TiO_2	Co-precipitation	Degradation of Reactive Red 120 (RR 120)/Vis/ 150 W W-halogen lamp	0.400 g	14.70 ppm	$4.98 \times 10^{-9} S^{-1}$	~0.2 emu/g	[202]
Ag/$NiFe_2O_4$	Combustion	Degradation of MB/Vis/300 W Xe lamp/450 nm UV filter	0.025 g	20 ppm	~80% in 120 min	~20 emu/g	[80]
$NiFe_2O_4$/TiO_2-SiO_2	Modified sol-gel/solvothermal	Degradation of cyanide/Vis/150 W blue fluorescent lamp/420 nm UV filter	No data	100 ppm	100% in 60 min	42.7 emu/g	[203]
$MgFe_2O_4$-ZnO	Solution method/chemical co-precipitation	Degradation of RhB/Vis/500 W Xe lamp/420 nm UV filter	0.050 g	4.40 ppm	100% in 120 min	21.37 emu/g	[204]
PANI-$CoFe_2O_4$-TiO_2	Hydrothermal/*in situ* chemical deposition	Degradation of MB/UV (500 W Xe lamp)/Vis/500 W Xe lamp/400 nm UV filter	0.100 g	50 ppm	0.0962/min/UV and 0.0110/min/Vis	11.4 emu/g	[205]
$Bi_{25}FeO_{40}$-RGO	Hydrothermal	Degradation of MB/Vis/500 W Xe lamp/400 nm UV filter	0.080 g	28.80 ppm	92.8% in 180 min	10.50 emu/g	[206]
BiOCl-$SrFe_{12}O_{19}$	Hydrothermal	Degradation of MB/UV/Vis/No data on light source	0.400 g	10 ppm	99% in 50 min/UV, 67.8% in 8 h/Vis	15.13 emu/g	[207]
$CuFe_2O_4$@C_3N_4	Self-assembly route	Degradation of orange II/Vis/500 W Xe lamp/ 420 nm UV filter	0.010 g	9.81 ppm	100% in 90 min	No data	[40]
$Co_{0.6}Zn_{0.4}Mn_xFe_{2-x}O_4$ ($x = 0.2$–1.0)	Sol-gel auto-combustion	Degradation of MO/Vis/400 W Hg lamp	0.100 g	30 ppm	94% in 60 min	39.51 emu/g	[208]
$Ni_{0.5}Zn_{0.5}Fe_2O_4$/$Zn_{0.95}Ni_{0.05}$	Egg albumen assisted sol-gel	Degradation of RhB/natural sunlight/sunny days/ 11 am to 3 pm.	0.010 g	20 ppm	~90% in 4 h	8.00 emu/g	[209]
$CdFe_2O_4$/GR	Hydrothermal	Degradation of MB/Vis/500 W Xe lamp/420 nm UV filter	0.100 g	10 ppm	89.2% in 4 h	14.26 emu/g	[82]
$Mn_xZn_{1-x}Fe_2O_4$-TiO_2 ($0.0 \leq x \leq 0.5$)	Auto-combustion	Degradation of 4-CP/UV/ 8 × 8 W low pressure Hg lamps	0.030 g	200 ppm	~98% in 270 min	45.17 emu/g	[210]
$Co_{0.6}Zn_{0.4}Cu_xFe_{2-x}O_4$	Sol-gel auto-combustion	Degradation of MO/Vis/400 W Hg lamp	0.100 g	30 ppm	86% in 60 min	38.02 emu/g	[211]
$Co_xZn_{1-x}Fe_2O_4$-GR	Chemical co-precipitation/ultrasonication	Degradation of MB/Vis/No data on light source	0.100 g	5 ppm	>95% in 60 min	No data	[212]
Sr-TiO_2/$Ni_{0.6}Zn_{0.4}Fe_2O_4$	Combustion/sol-gel	Degradation of BPA/UV (10 W low pressure Hg lamp)/Vis/500 W Xe arc lamp	0.150 g	10 ppm	100% in 4 h/UV, 90% in 4 h/Vis	19.04 emu/g	[213]
$CoFe_2O_4$-GR	Combustion	Degradation of MB/Vis/300 W Xe lamp/450 nm UV filter	0.025 g	20 ppm	100% in 120 min	5.3 emu/g	[166]
$MnFe_2O_4$/g-C_3N_4/TiO_2	Chemical impregnation	Degradation of MO/Simulated solar light/150 W Xe arc lamp	0.050g	10 ppm	99.3% in 180 min	0.065 emu/g	[214]
$Ni_{1-x}Co_xFe_2O_4$	Hydrothermal	Degradation of malachite green (MG)/natural sunlight (200–250 Wm^{-2})	0.025 g	0.365 ppm	~100% in 60 min	~50.7–64.2 emu/g.	[215]
$Mn_xZn_{1-x}Fe_2O_4$/β-Bi_2O_3	Dip-calcination	Degradation of RhB/Simulated solar light/300 W Xe lamp	0.200g	10 ppm	99.1% in 150 min	7.01 emu/g	[216]
$Zn_{1-x}Co_xFe_2O_4$	Hydrothermal	Degradation of MB/Natural sunlight	0.050 g	3.20 ppm	~88% in 270 min	~50 emu/g	[217]
$SrFeO_{3-x}$/g-C_3N_4	Sintering method	Degradation of chloramphenicol (CAP) and crystal violet (CV)/Vis/ 150 W Xe arc lamp	0.010 g	10 ppm	91.3% in 96 h/CAP, 99.9% in 12 h/CV	~0.17 emu/g	[41]
$CoFe_2O_4$/GR/CdS	Solvothermal	Degradation of MB/daylight/40 W daylight lamp	0.025 g	20 ppm	80% in 180 min	18.00 emu/g	[218]
ZnO/$CoFe_2O_4$	Co-precipitation	Degradation of direct blue 71 (BD 71)/150 W W-halogen lamp	0.160 g	38.64 ppm	~100% in 30 min	~0.040 emu/g	[219]

SUMMARY REMARKS AND FUTURE OUTLOOK

The development of efficient visible light photocatalytic degradation of pollutants and easy recovery and recyclability of the photocatalyst is an exciting prospect in water treatment. The incorporation of magnetic nanoparticles such as haematite, maghemite, magnetite, and the ferrites into various photocatalyst matrices provides an attractive strategy to enhance the optical properties, charge separation efficiency, and the overall photocatalytic activity. Most interestingly, the presence of the magnetic nanoparticles endow the nanocomposite photocatalyst with sufficient magnetic response to enable separation of the photocatalyst using an external magnetic field. However, careful control of the magnetic nanoparticle, especially loading maghemite and magnetite, is important in ensuring a balance between good magnetic response and the detrimental effect of recombination that occurs on the surface of the nanoparticles. Alternatively, silica is used as a barrier between the magnetic core and the other semiconductor coupled with maghemite or magnetite. This prevents charge transfer between the coupled materials that could otherwise result in photodissolution and recombination. The incorporation of the ferrites in various photocatalysts provides more beneficial properties than just good magnetic response; they contribute towards visible light absorption, charge separation, and the photocatalytic performance of the nanocomposite photocatalyst. Therefore, the ferrites are the most attractive nanoparticles compared to haematite, maghemite, and magnetite, in terms of their photocatalytic performance and multifaceted contribution in the magnetic nanocomposites. However, a systematic, comparative study of the photocatalytic performance of the ferrites, haematite, magnetite, and maghemite is seldom reported, and this makes it difficult to map a way forward in terms of exploiting the best materials for photocatalytic applications.

Despite the promising results regarding visible-light-active magnetic photocatalysts, there are still some challenges that possibly hinder the practical exploitation of these materials. The most obvious problem relates to the reaction kinetics; most of the degradation experiments are slow (may take several hours) despite the improvements in visible light absorption of the photocatalyst, and this presents a massive challenge for practical applications. Secondly, work still needs to be done to develop synthesis routes that will ensure uniform shape and size of the magnetic particles as well as their uniform distribution within the nanocomposite matrix in order to ensure good magnetic response and efficient recovery. Strong contact between the magnetic nanoparticles and the other semiconductor particles is crucial in ensuring minimal leaching out and loss of the photocatalytic activity and magnetic response. In the magnetic core-shell structures, careful control of the thickness of the silica shell is important, as this will affect the magnetic response of the composite material. Extensive theoretical studies are needed in order to understand the interactions between the magnetic nanoparticles and the other semiconductor nanoparticles in the nanocomposites; this can help design and tailor materials with good heterojunctions that promote efficient charge separation. Moreover, theoretical studies can provide a good idea of the optimum loadings of the magnetic nanoparticles in the photocatalyst to ensure good magnetic response and minimal recombination. In most cases, the magnetisation saturation of the magnetic nanoparticles decrease significantly upon coupling with the non-magnetic components of the nanocomposite photocatalyst, which has a negative bearing on the separation efficiency and recovery of the photocatalyst. Work still needs to be done to ensure that nanocomposites retain sufficiently high magnetic response in order to enable efficient magnetic separation.

ACKNOWLEDGMENTS

Financial support from the College of Science, Engineering and Technology (CSET) of the University of South Africa is highly appreciated.

CONFLICTS OF INTEREST

The authors declare no conflict of interest.

REFERENCES

1. Fersi, C.; Gzara, L.; Dhahbi, M. Treatment of textile effluents by membrane technologies.*Desalination* 2005, *185*, 399–409.
2. Pendergast, M.M.; Hoek, E.M.V. A review of water treatment membrane nanotechnologies.*Energy Environ. Sci.* 2011, *4*, 1946–1971.
3. Wang, H.; Zheng, X.W.; Su, J.Q.; Tian, Y.; Xiong, X.J.; Zheng, T.L. Biological decolorization of the reactive dyes Reactive Black 5 by a novel isolated bacterial strain *Enterobacter* sp. EC3. *J. Hazard. Mater.* 2009, *171*, 654–659.
4. Kumari, K.; Abraham, T.E. Biosorption of anionic textile dyes by nonviable biomass of fungi and yeast. *Bioresour. Technol.* 2007, *98*, 1704–1710.
5. Soutsas, K.; Karayannis, V.; Poulios, I.; Riga, A.; Ntampegliotis, K.; Spiliotis, X.; Papapolymerou, G. Decolorization and degradation of reactive azo dyes via heterogeneous photocatalytic processes. *Desalination* 2010, *250*, 345–350.
6. Wu, J.; Doan, H.; Upreti, S. Decolorization of aqueous textile reactive dye by ozone. *Chem. Eng. J.* 2008, *142*, 156–160.
7. Harsha, N.; Krishna, K.V.S.; Renuka, N.K.; Shukla, S. Facile synthesis of g–Fe_2O_3nanoparticles integrated $H_2Ti_3O_7$ nanotubes structure as a magnetically recyclable dye-removal catalyst. *RSC Adv.* 2015, *5*, 30354–30362.
8. Li, K.; Chai, F.; Zhao, Y.; Guo, X. Facile synthesis of magnetic Fe_3O_4/$CeCO_3OH$ composites with excellent adsorption capability for small cationic dyes. *RSC Adv.* 2015, *5*, 94397–94404.
9. Song, S.; Fan, J.; He, Z.; Zhan, L.; Liu, Z.; Chen, J.; Xu, X. Electrochemical degradation of azo dye C.I. Reactive Red 195 by anodic oxidation on Ti/SnO_2-Sb/PbO_2 electrodes. *Electrochim. Acta* 2010, *55*, 3606–3613.
10. Faouzi, A.M.; Nasr, B.; Abdellatif, G. Electrochemical degradation of anthraquinone dye Alizarin Red S by anodic oxidation on boron-doped diamond. *Dyes Pigments* 2007, *73*, 86–89.
11. Rodrigues, C.S.D.; Madeira, L.M.; Boaventura, R.A.R. Synthetic textile dyeing wastewater treatment

by integration of advanced oxidation and biological processes—Performance analysis with costs reduction. *J. Environ. Chem. Eng.* 2014, *2*, 1027–1039.

12. Basha, C.A.; Selvakumar, K.V.; Prabhu, H.J.; Sivashanmugam, P.; Lee, C.W. Degradation studies for textile reactive dye by combined electrochemical, microbial and photocatalytic methods. *Sep. Purif. Technol.* 2011, *79*, 303–309.
13. Berberidou, C.; Avlonitis, S.; Poulios, I. Dyestuff effluent treatment by integrated sequential photocatalytic oxidation and membrane filtration. *Desalination* 2009, *249*, 1099–1106.
14. Khouni, I.; Marrot, B.; Moulin, P.; Ben, R. Decolourization of the reconstituted textile effluent by different process treatments: Enzymatic catalysis, coagulation/flocculation and nanofiltration processes. *Desalination* 2011, *268*, 27–37.
15. Trzcinski, A.P.; Stuckey, D.C. Inorganic fouling of an anaerobic membrane bioreactor treating leachate from the organic fraction of municipal solid waste (OFMSW) and a polishing aerobic membrane bioreactor. *Bioresour. Technol.* 2016, *204*, 17–25.
16. Schulz, M.; Soltani, A.; Zheng, X.; Ernst, M. Effect of inorganic colloidal water constituents on combined low-pressure membrane fouling with natural organic matter (NOM). *J. Membr. Sci.* 2016, *507*, 154–164.
17. Hegab, H.M.; Elmekawy, A.; Barclay, T.G.; Michelmore, A.; Zou, L.; Saint, C.P.; Ginic-Markovic, M. Effective in-situ chemical surface modification of forward osmosis membranes with polydopamine-induced graphene oxide for biofouling mitigation. *Desalination* 2016,*385*, 126–137.
18. Gottlieb, A.; Shaw, C.; Smith, A.; Wheatley, A.; Forsythe, S. The toxicity of textile reactive azo dyes after hydrolysis and decolourisation. *J. Biotechnol.* 2003, *101*, 49–56.
19. Kariyajjanavar, P.; Jogttappa, N.; Nayaka, Y.A. Studies on degradation of reactive textile dyes solution by electrochemical method. *J. Hazard. Mater.* 2011, *190*, 952–961.
20. Faouzi Elahmadi, M.; Bensalah, N.; Gadri, A. Treatment of aqueous wastes contaminated with Congo Red dye by electrochemical oxidation and ozonation processes. *J. Hazard. Mater.*2009, *168*, 1163–1169.
21. Mohammad, N.; Arami, M. Degradation and toxicity reduction of textile wastewater using immobilized titania nanophotocatalysis. *J. Photochem. Photobiol. B* 2009, *94*, 20–24.
22. Aksu, Z.; Dönmez, G. Combined effects of molasses sucrose and reactive dye on the growth and dye bioaccumulation properties of *Candida tropicalis*. *Process Biochem.* 2005, *40*, 2443–2454.
23. Yu, W.; Xu, D.; Peng, T. Enhanced photocatalytic activity of g-C_3N_4 for selective CO_2reduction to CH_3OH via facile coupling of ZnO: A direct Z-scheme mechanism. *J. Mater. Chem. A* 2015, *3*, 19936–19947.
24. Lu, D.; Zhang, G.; Wan, Z. Visible-light-driven g-C_3N_4/Ti^{3+}-TiO_2 photocatalyst co-exposed {001} and {101} facets and its enhanced photocatalytic activities for organic pollutant degradation and Cr(VI) reduction. *Appl. Surf. Sci.* 2015, *358*, 223–230.
25. Li, G.; Nie, X.; Chen, J.; Jiang, Q.; An, T.; Wong, P.K.; Zhang, H.; Zhao, H.; Yamashita, H. Enhanced visible-light-driven photocatalytic inactivation of *Escherichia coli* using g-C_3N_4/TiO_2 hybrid photocatalyst synthesized using a hydrothermal-calcination approach.*Water Res.* 2015, *86*, 17–24.
26. Zhong, Y.; Ni, Y.; Li, S.; Wang, M. Chain-like Fe_3O_4@ resorcinol-formaldehyde resins-Ag composite microstructures: Facile construction and applications in antibacterial and catalytic fields. *RSC Adv.* 2016, *6*, 15831–15837.
27. Li, Y.; Zhang, Z.; Pei, L.; Li, X.; Fan, T.; Ji, J.; Shen, J.; Ye, M. Multifunctional photocatalytic performances of recyclable Pd-$NiFe_2O_4$/reduced graphene oxide nanocomposites via different co-catalyst strategy. *Appl. Catal. B Environ.* 2016, *190*, 1–11.
28. Zeng, B.; Zhang, L.; Wan, X.; Song, H.; Lv, Y. Fabrication of α-Fe_2O_3/g-C_3N_4 composites for cataluminescence sensing of H_2S. *Sens. Actuators Chem.* 2015, *211*, 370–376.
29. Wang, G.; Kuang, S.; Zhang, J.; Hou, S.; Nian, S. Graphitic carbon nitride/multiwalled carbon nanotubes composite as Pt-free counter electrode for high-efficiency dye-sensitized solar cells. *Electrochim. Acta* 2016, *187*, 243–248.
30. Wang, M.; Yang, X.; Bi, W. Application of magnetic graphitic carbon nitride nanocomposites for the solid-phase extraction of phthalate esters in water samples. *J. Sep. Sci.* 2015, *38*, 445–452.
31. Baig, R.B.N.; Verma, S.; Varma, R.S.; Nadagouda, M.N. Magnetic Fe@g-C_3N_4: A photoactive catalyst for the hydrogenation of alkenes and alkynes. *ACS Sustain. Chem. Eng.* 2016, *4*, 1661–1664.
32. Preethi, V.; Kanmani, S. Photocatalytic hydrogen production using Fe_2O_3-based core shell nano particles with ZnS and CdS. *Int. J. Hydrogen Energy* 2014, *39*, 1613–1622.
33. Cao, M.; Wang, P.; Ao, Y.; Wang, C.; Hou, J.; Qian, J. Visible light activated photocatalytic degradation of tetracycline by a magnetically separable composite photocatalyst: Graphene oxide/magnetite/cerium-doped titania. *J. Colloid Interface Sci.* 2016, *467*, 129–139.
34. Bao, X.; Qiang, Z.; Chang, J.; Ben, W.; Qu, J. Synthesis of carbon-coated magnetic nanocomposite (Fe_3O_4@C) and its application for sulfonamide antibiotics removal from water. *J. Environ. Sci.* 2014, *26*, 962–969.
35. Zhou, L.; Deng, J.; Zhao, Y.; Liu, W.; An, L.; Chen, F. Preparation and characterization of N–I co-doped nanocrystal anatase TiO_2 with enhanced photocatalytic activity under visible-light irradiation. *Mater. Chem. Phys.* 2009, *117*, 522–527.
36. Nolan, N.T.; Synnott, D.W.; Seery, M.K.; Hinder, S.J.; van Wassenhoven, A.; Pillai, S.C. Effect of N-doping on the photocatalytic activity of sol-gel TiO_2. *J. Hazard. Mater.* 2012, *211–212*, 88–94.

37. Pany, S.; Parida, K.M. A facile *in situ* approach to fabricate N,S-TiO_2/g-C_3N_4 nanocomposite with excellent activity for visible light induced water splitting for hydrogen evolution. *Phys. Chem. Chem. Phys.* 2015, *17*, 8070–8077.
38. Ge, M.; Chen, Y.; Liu, M.; Li, M. Synthesis of magnetically separable Ag_3PO_4/$ZnFe_2O_4$composite photocatalysts for dye degradation under visible LED light irradiation. *J. Environ. Chem. Eng.* 2015, *3*, 2809–2815.
39. Xu, Y.; Huang, S.; Xie, M.; Li, Y.; Jing, L.; Xu, H. Core-shell magnetic Ag/AgCl@Fe_2O_3photocatalysts with enhanced photoactivity for eliminating bisphenol A and microbial contamination. *New J. Chem.* 2016, *40*, 3413–3422.
40. Yao, Y.; Lu, F.; Zhu, Y.; Wei, F.; Liu, X.; Lian, C.; Wang, S. Magnetic core-shell $CuFe_2O_4$@C_3N_4hybrids for visible light photocatalysis of Orange II. *J. Hazard. Mater.* 2015, *297*, 224–233.
41. Lin, H.-P.; Chen, C.-C.; Lee, W.W.; Lai, Y.-Y.; Chen, J.; Chen, Y.; Fu, J.-Y. Synthesis of a $SrFeO_{3-x}$/g-C_3N_4 heterojunction with improved visible-light photocatalytic activities in chloramphenicol and crystal violet degradation. *RSC Adv.* 2015, *6*, 2323–2336.
42. Li, G.; Nie, X.; Gao, Y.; An, T. Can environmental pharmaceuticals be photocatalytically degraded and completely mineralized in water using g-C_3N_4/TiO_2 under visible light irradiation?—Implications of persistent toxic intermediates. *Appl. Catal. B Environ.* 2016, *180*, 726–732.
43. Zhang, J.; Hu, S.; Wang, Y. A convenient method to prepare a novel alkali metal sodium doped carbon nitride photocatalyst with a tunable band structure. *RSC Adv.* 2014, *4*, 62912–62919.
44. Jin, R.; Hu, S.; Gui, J.; Liu, D. A convenient method to prepare novel rare earth metal Ce-doped carbon nitride with enhanced photocatalytic activity under visible light. *Bull. Korean Chem. Soc.* 2015, *36*, 17–23.
45. Akpan, U.G.; Hameed, B.H. Enhancement of the photocatalytic activity of TiO_2 by doping it with calcium ions. *J. Colloid Interface Sci.* 2011, *357*, 168–178.
46. Inturi, S.N.R.; Boningari, T.; Suidan, M.; Smirniotis, P.G. Visible-light-induced photodegradation of gas phase acetonitrile using aerosol-made transition metal (V, Cr, Fe, Co, Mn, Mo, Ni, Cu, Y, Ce, and Zr) doped TiO_2. *Appl. Catal. B Environ.* 2014, *144*, 333–342.
47. Li, Z.; Wang, J.; Zhu, K.; Ma, F.; Meng, A. Ag/g-C_3N_4 composite nanosheets: Synthesis and enhanced visible photocatalytic activities. *Mater. Lett.* 2015, *145*, 167–170.
48. Hu, S.; Ma, L.; You, J.; Li, F.; Fan, Z.; Wang, F.; Liu, D.; Gui, J. A simple and efficient method to prepare a phosphorus modified g-C_3N_4 visible light photocatalyst. *RSC Adv.* 2014, *4*, 21657–21663.
49. Sun, H.; Wang, S.; Ang, H.M.; Tadé, M.O.; Li, Q. Halogen element modified titanium dioxide for visible light photocatalysis. *Chem. Eng. J.* 2010, *162*, 437–447.
50. Dong, G.; Ai, Z.; Zhang, L. Efficient anoxic pollutant removal with oxygen functionalized graphitic carbon nitride under visible light. *RSC Adv.* 2014, *4*, 5553–5560.
51. Sharotri, N.; Sud, D. Ultrasound-assisted synthesis and characterization of visible light responsive nitrogen-doped TiO_2 nanomaterials for removal of 2-Chlorophenol. *Desalin. Water Treat.* 2016, *57*, 8776–8788.
52. Reszczyńska, J.; Grzyb, T.; Sobczak, J.W.; Lisowski, W.; Gazda, M.; Ohtani, B.; Zaleska, A. Visible light activity of rare earth metal doped (Er^{3+}, Yb^{3+} or Er^{3+}/Yb^{3+}) titania photocatalysts. *Appl. Catal. B: Environ.* 2015, *163*, 40–49.
53. Kaur, N.; Kaur, S.; Singh, V. Preparation, characterization and photocatalytic degradation kinetics of Reactive Red dye 198 using N, Fe codoped TiO_2 nanoparticles under visible light.*Desalination Water Treat.* 2016, *57*, 9237–9246.
54. Xu, X.; Zhou, X.; Zhang, L.; Xu, L.; Ma, L.; Luo, J.; Li, M.; Zeng, L. Photoredox degradation of different water pollutants (MO, RhB, MB, and Cr(VI)) using Fe-N-S-tri-doped TiO_2nanophotocatalyst prepared by novel chemical method. *Mater. Res. Bull.* 2015, *70*, 106–113.
55. Antonopoulou, M.; Vlastos, D.; Konstantinou, I. Photocatalytic degradation of pentachlorophenol by N-F-TiO_2: Identification of intermediates, mechanism involved, genotoxicity and ecotoxicity evaluation. *Photochem. Photobiol. Sci.* 2015, *14*, 520–527.
56. Mamba, G.; Mbianda, X.Y.; Mishra, A.K. Photocatalytic degradation of the diazo dye naphthol blue black in water using MWCNT/Gd,N,S-TiO_2 nanocomposites under simulated solar light. *J. Environ. Sci.* 2015, *33*, 219–228.
57. Mamba, G.; Mamo, M.A.; Mbianda, X.Y.; Mishra, A.K. Nd,N,S-TiO_2 decorated on reduced graphene oxide for a visible light active photocatalyst for dye degradation: Comparison to its MWCNT/Nd,N,S-TiO_2 analogue. *Ind. Eng. Chem. Res.* 2014, *53*, 14329–14338.
58. Mamba, G.; Mbianda, X.Y.; Mishra, A.K. Enhanced visible light photocatalytic degradation of eriochrome black T and eosin blue shade in water using tridoped titania decorated on SWCNTs and MWCNTs: Effect of the type of carbon nanotube incorporated. *Mater. Chem. Phys.* 2015, *149–150*, 734–742.
59. Apostolopoulou, V.; Vakros, J.; Kordulis, C.; Lycourghiotis, A. Preparation and characterization of [60] fullerene nanoparticles supported on titania used as a photocatalyst.*Colloids Surf. A Physicochem. Eng. Asp.* 2009, *349*, 189–194.
60. Shang, J.; Zhao, F.; Zhu, T.; Li, J. Photocatalytic degradation of rhodamine B by dye-sensitized TiO_2 under visible-light irradiation. *Sci. China Chem.* 2011, *54*, 167–172.
61. Nam, K.; Jung, H.; Lee, W. Hollow cobalt ferrite-polyaniline nanofibers as magnetically separable visible-light photocatalyst for photodegradation of methyl orange. *J. Photochem. Photobiol. A Chem.* 2016, *321*, 257–265.
62. Kim, G.; Choi, W. Charge-transfer surface complex of EDTA-TiO_2 and its effect on photocatalysis under visible light. *Appl. Catal. B Environ.* 2010, *100*, 77–83.

63. Shi, W.; Lu, D.; Wang, L.; Teng, F.; Zhang, J. Core-shell structured Fe_3O_4@SiO_2@CdS nanoparticles with enhanced visible-light photocatalytic photocatalytic activities. *RSC Adv.*2015, *5*, 106038–106043.

64. Li, J.; Liu, Z.; Zhu, Z. Enhanced photocatalytic activity in $ZnFe_2O_4$-ZnO-Ag_3PO_4 hollow nanospheres through the cascadal electron transfer with magnetical separation. *J. Alloys Compd.* 2015, *636*, 229–233.

65. Jiang, R.; Zhu, H.; Li, J.; Fu, F.; Yao, J.; Jiang, S.; Zeng, G.-M. Fabrication of novel magnetically separable BiOBr/$CoFe_2O_4$ microspheres and its application in the efficient removal of dye from aqueous phase by an environment-friendly and economical approach. *Appl. Surf. Sci.*2016, *364*, 604–612.

66. Mi, J.; Hwan, Y.; Keun, B.; Ho, S.; Soo, B.; Bin, J. Novel immobilization of titanium dioxide (TiO_2) on the fluidizing carrier and its application to the degradation of azo-dye. *J. Hazard. Mater.* 2006, *134*, 230–236.

67. Fernhndez, A.; Lassaletta, G.; Jimknez, V.M.; Justo, A. Preparation and characterization of TiO_2 photocatalysts supported on various rigid supports (glass, quartz and stainless steel). Comparative studies of photocatalytic activity in water purification. *Appl. Catal. B Environ.*1995, *7*, 49–63.

68. Carlson, P.J.; Pretzer, L.A.; Boyd, J.E. Solvent deposition of titanium dioxide on acrylic for photocatalytic application. *Ind. Eng. Chem. Res.* 2007, *46*, 7970 7976.

69. Arabatzis, I.M.; Antonaraki, S.; Stergiopoulos, T.; Hiskia, A. Preparation, characterization and photocatalytic activity of nanocrystalline thin film TiO_2 catalysts towards 3,5-dichlorophenol degradation. *J. Photochem. Photobiol. A Chem.* 2002, *149*, 237–245.

70. Anpo, M.; Zhang, S.G.; Mishima, H.; Matsuoka, M.; Yamashita, H. Design of photocatalysts encapsulated within the zeolite framework and cavities for the decomposition of NO into $N_2$and O_2 at normal temperature. *Catal. Today* 1997, *39*, 159–168.

71. Rao, N.N.; Chaturvedi, V. Photoactivity of TiO_2-coated pebbles. *Ind. Eng. Chem. Res.* 2007, *46*, 4406–4414.

72. Gad-allah, T.A.; Kato, S.; Satokawa, S.; Kojima, T. Treatment of synthetic dyes wastewater utilizing a magnetically separable photocatalyst ($TiO_2/SiO_2/Fe_3O_4$): Parametric and kinetic studies. *Desalination* 2009, *244*, 1–11.

73. Wang, W.; Yao, J. Synthesis of magnetically separable Sn doped magnetite/silica core-shell structure and photocatalytic property. *Mater. Res. Bull.* 2010, *45*, 710–716.

74. Belessi, V.; Lambropoulou, D.; Konstantinou, I.; Zboril, R.; Tucek, J.; Jancik, D. Structure and photocatalytic performance of magnetically separable titania photocatalysts for the degradation of propachlor. *Appl. Catal. B Environ.* 2009, *87*, 181–189.

75. Linley, S.; Leshuk, T.; Gu, F.X. Synthesis of magnetic rattle-type nanostructures for use in water treatment. *ACS Appl. Mater. Interfaces* 2013, *5*, 2540–2548.

76. Kumar, S.; Kumar, B.; Baruah, A.; Shanker, V. Synthesis of magnetically separable and recyclable g-C_3N_4–Fe_3O_4 hybrid nanocomposites with enhanced photocatalytic performance under visible-light irradiation. *J. Phys. Chem. C* 2013, *117*, 26135–26143.

77. Yang, J.; Chen, H.; Gao, J.; Yan, T.; Zhou, F.; Cui, S.; Bi, W. Synthesis of Fe_3O_4/g-C_3N_4nanocomposites and their application in the photodegradation of 2,4,6-trichlorophenol under visible light. *Mater. Lett.* 2016, *164*, 183–189.

78. Shih, Y.; Su, C.; Chen, C.; Dong, C. Synthesis of magnetically recoverable ferrite (MFe_2O_4, M = Co, Ni and Fe)-supported TiO_2 photocatalysts for decolorization of methylene blue. *Catal. Commun.* 2015, *72*, 127–132.

79. Pang, Y.L.; Lim, S.; Ong, H.C.; Chong, W.T. Synthesis, characteristics and sonocatalytic activities of calcined α-Fe_2O_3 and TiO_2 nanotubes/α-Fe_2O_3 magnetic catalysts in the degradation of Orange G. *Ultrason. Sonochem.* 2016, *29*, 317–327.

80. Zhang, D.; Pu, X.; Du, K.; Moon, Y.; Jeong, J.; Cai, P.; Kim, S., II; Seo, H.J. Combustion synthesis of magnetic Ag/$NiFe_2O_4$ composites with enhanced visible-light photocatalytic properties. *Sep. Purif. Technol.* 2014, *137*, 82–85.

81. Zhao, H.; Zhang, L.; Gu, X.; Li, S.; Li, B.; Wang, H. Fe_2O_3-AgBr nonwoven cloth with hierarchical nanostructures as efficient and easily recyclable macroscale photocatalysts. *RSC Adv.* 2015, *5*, 10951–10959.

82. Zhang, D.; Wang, Q.; Wang, L.; Zhang, L. Magnetically separable $CdFe_2O_4$/graphene catalyst and its enhanced photocatalytic properties. *J. Mater. Chem. A* 2015, *3*, 3576–3585.

83. Liu, J.; Qiao, S.Z.; Hu, Q.H.; Qing, G.; Lu, M. Magnetic nanocomposites with mesoporous structures: Synthesis and applications. *Small* 2011, *7*, 425–443.

84. Polshettiwar, V.; Luque, R.; Fihri, A.; Zhu, H.; Bouhrara, M.; Basset, J. Magnetically recoverable nanocatalysts. *Chem. Rev.* 2011, *111*, 3036–3075.

85. Zhang, D.; Zhou, C.; Sun, Z.; Wu, L.-Z.; Tung, C.-H.; Zhang, T. Magnetically recyclable nanocatalysts (MRNCs): A versatile integration of high catalytic activity and facile recovery.*Nanoscale* 2012, *4*, 6244–6255.

86. Wang, D.; Astruc, D. Fast-growing field of magnetically recyclable nanocatalysts. *Chem. Rev.*2014, *114*, 6949–6985.

87. Linley, S.; Leshuk, T.; Gu, F.X. Magnetically separable water treatment technologies and their role in future advanced water treatment: A patent review. *CLEAN Soil Air Water* 2013, *41*, 1152–1156.

88. Yao, H.; Fan, M.; Wang, Y. Magnetic titanium dioxide based nanomaterials: Synthesis, characteristics, and photocatalytic application in pollutant degradation. *J. Mater. Chem. A*2015, *3*, 17511–17524.

89. Liu, D.; Li, Z.; Wang, W.; Wang, G.; Liu, D. Hematite doped magnetic TiO_2 nanocomposites with improved photocatalytic activity. *J. Alloy Compd.* 2016, *654*, 491–497.

90. Shi, Y.; Li, H.; Wang, L.; Shen, W.; Chen, H. Novel α-Fe_2O_3/CdS cornlike nanorods with enhanced photocatalytic performance. *ACS Appl. Mater. Interfaces* 2012, *4*, 4800–4806.

91. Xu, Y.; Jing, L.; Chen, X.; Ji, H.; Xu, H.; Li, H. Novel visible-light-driven Fe_2O_3/Ag_3VO_4composite with enhanced photocatalytic activity toward organic pollutants degradation.*RSC Adv.* 2016, *6*, 3600–3607.

92. Xu, Y.; Huang, S.; Xie, M.; Li, Y.; Xu, H.; Huang, L.; Zhang, Q.; Li, H. Magnetically separable Fe_2O_3/g-C_3N_4 catalyst with enhanced photocatalytic activity. *RSC Adv.* 2015, *5*, 95727–95735.

93. Kumar, S.V.; Huang, N.M.; Yusoff, N.; Lim, H.N. High performance magnetically separable graphene/zinc oxide nanocomposite. *Mater. Lett.* 2013, *93*, 411–414.

94. Wang, H.; Fei, X.; Wang, L.; Li, Y.; Xu, S. Magnetically separable iron oxide nanostructures-TiO_2 nanofibers hierarchical heterostructures: Controlled fabrication and photocatalytic activity. *New J. Chem.* 2011, *35*, 1795–1802.

95. Muthukrishnaraj, A.; Vadivel, S.; Kamalakannan, V.P.; Balasubramanian, N. α-Fe_2O_3/reduced graphene oxide nanorod as efficient photocatalyst for methylene blue degradation. *Mater. Res. Innov.* 2016, *19*, 258–264.

96. Wu, W.; Zhang, S.; Xiao, X.; Zhou, J.; Ren, F.; Sun, L.; Jiang, C. Controllable synthesis, magnetic properties, and enhanced photocatalytic activity of spindle-like mesoporous α-Fe_2O_3/ZnO core-shell heterostructures. *ACS Appl. Mater. Interfaces* 2012, *4*, 3602–3609.

97. Xue, J.; Ma, S.; Zhou, Y.; Zhang, Z.; Liu, X. Fabrication of porous g-C_3N_4/Ag/Fe_2O_3composites with enhanced visible light photocatalysis performance. *RSC Adv.* 2015, *5*, 58738–58745.

98. Liu, X.; Jin, A.; Jia, Y.; Jiang, J.; Hu, N.; Chen, X. Facile synthesis and enhanced visible-light photocatalytic activity of graphitic carbon nitride decorated with ultrafine Fe_2O_3nanoparticles. *RSC Adv.* 2015, *5*, 92033–92041.

99. Theerthagiri, J.; Senthil, R.A.; Priya, A.; Madhavan, J.; Michael, R.J.V.; Ashokkumar, M. Photocatalytic and photoelectrochemical studies of visible-light-active α-Fe_2O_3-g-C_3N_4nanocomposites. *RSC Adv.* 2014, *4*, 38222–38229.

100. Liu, Y.; Yu, Y.X.; Zhang, W. Photoelectrochemical study on charge transfer properties of nanostructured Fe_2O_3 modified by g-C_3N_4. *Int. J. Hydrogen Energy* 2014, *39*, 9105–9113.

101. Peng, L.; Xie, T.; Lu, Y.; Fan, H.; Wang, D. Synthesis, photoelectric properties and photocatalytic activity of the Fe_2O_3/TiO_2 heterogeneous photocatalysts. *Phys. Chem. Chem. Phys.* 2010, *12*, 8033–8041.

102. Pawar, R.C.; Pyo, Y.; Ahn, S.H.; Lee, C.S. Photoelectrochemical properties and photodegradation of organic pollutants using hematite hybrids modified by gold nanoparticles and graphitic carbon nitride. *Appl. Catal. B Environ.* 2015, *176–177*, 654–666.

103. Yamauraa, M.; Camilo, R.L.; Sampaio, L.C.; Macedo, M.A.; Nakamurad, M.; Toma, H.E. Preparation and characterization of (3-aminopropyl) triethoxysilane-coated magnetite nanoparticles. *J. Magn. Magn. Mater.* 2004, *279*, 210–217.

104. Tian, B.; Wang, T.; Dong, R.; Bao, S.; Yang, F.; Zhang, J. Core-shell structured–Fe_2O_3@SiO_2@AgBr: Ag composite with high magnetic separation efficiency and excellent visible light activity for acid orange 7 degradation. *Appl. Catal. B Environ.* 2014, *147*, 22–28.

105. Idris, A.; Hassan, N.; Rashid, R.; Ngomsik, A. Kinetic and regeneration studies of photocatalytic magnetic separable beads for chromium (VI) reduction under sunlight. *J. Hazard. Mater.* 2011, *186*, 629–635.

106. Yu, X.; Liu, S.; Yu, J. Superparamagnetic γ-Fe_2O_3@SiO_2@TiO_2 composite microspheres with superior photocatalytic properties. *Appl. Catal. B Environ.* 2011, *104*, 12–20.

107. Lin, Y.; Chang, C. Magnetic mesoporous iron oxide/carbon aerogel photocatalysts with adsorption ability for organic dye removal. *RSC Adv.* 2014, *4*, 28628–28631.

108. Zhang, Y.; Zhang, Y.; Tan, J. Novel magnetically separable AgCl/iron oxide composites with enhanced photocatalytic activity driven by visible light. *J. Alloy. Compd.* 2013, *574*, 383–390.

109. Ye, S.; Qiu, L.-G.; Yuan, Y.-P.; Zhu, Y.-J.; Xia, J.; Zhu, J.-F. Facile fabrication of magnetically separable graphitic carbon nitride photocatalysts with enhanced photocatalytic activity under visible light. *J. Mater. Chem. A* 2013, *1*, 2929–3188.

110. He, M.; Li, D.; Jiang, D.; Chen, M. Magnetically separable γ-Fe_2O_3@SiO_2@Ce-doped TiO_2core-shell nanocomposites: Fabrication and visible-light-driven photocatalytic activity. *J. Solid State Chem.* 2012, *192*, 139–143.

111. Kaur, N.; Shahi, S.K.; Singh, V. Synthesis, characterization and photocatalytic activity of magnetically separable γ-Fe_2O_3/N,Fe codoped TiO_2 heterojunction for degradation of Reactive Blue 4 dye. *RSC Adv.* 2015, *5*, 61623–61630.

112. Yu, L.; Yang, X.; Wang, D. TiO_2 incorporated in magnetic mesoporous SBA-15 by a facile inner-pore hydrolysis process toward enhanced adsorption-photocatalysis performances for As(III). *J. Colloid Interface Sci.* 2015, *448*, 525–532.

113. Liu, Y.; Yu, L.; Hu, Y.; Guo, C.; Wen, X.; Lou, D. A magnetically separable photocatalyst based on nest-like γ-Fe_2O_3/ZnO double-shelled hollow structures with enhanced photocatalytic activity. *Nanoscale* 2012, *4*, 183–187.

114. Szeto, W.; Li, J.; Huang, H.; Xuan, J.; Leung, D.Y.C. Novel urchin-like Fe_2O_3@SiO_2@TiO_2microparticles with magnetically separable and photocatalytic properties. *RSC Adv.* 2015, *5*, 55363–55371.

115. Kumar, A.; Guo, C.; Sharma, G.; Pathania, D.; Naushad, M.; Kalia, S.; Dhiman, P. Magnetically recoverable ZrO_2/Fe_3O_4/chitosan nanomaterials for enhanced sunlight driven photoreduction of carcinogenic Cr(VI) and dechlorination and mineralization of 4-chlorophenol from simulated waste water. *RSC Adv.* 2016, *6*, 13251–13263.

116. Zhao, Y.; Tao, C.; Xiao, G.; Wei, G.; Li, L.; Liu, C.; Su, H. Controlled synthesis and photocatalysis of sea urchin-like Fe_3O_4@TiO_2@Ag nanocomposites. *Nanoscale* 2016, *8*, 5313–5326.

117. Li, G.; Wong, K.H.; Zhang, X.; Hu, C.; Yu, J.C.; Chan, R.C.Y.; Wong, P.K. Degradation of Acid Orange 7 using magnetic AgBr under visible light: The roles of oxidizing species.*Chemosphere* 2009, *76*, 1185–1191.

118. Álvarez, P.M.; Jaramillo, J.; López-pi, F.; Plucinski, P.K. Preparation and characterization of magnetic TiO_2 nanoparticles and their utilization for the degradation of emerging pollutants in water. *Appl. Catal. B Environ.* 2010, *100*, 338–345.

119. Zhang, L.; Wang, W.; Sun, S.; Sun, Y.; Gao, E.; Zhang, Z. Elimination of BPA endocrine disruptor by magnetic BiOBr@SiO_2@Fe_3O_4 photocatalyst. *Appl. Catal. B Environ.* 2014, *148–149*, 164–169.

120. Chen, F.; Yan, F.; Chen, Q.; Wang, Y.; Han, L.; Chen, Z.; Fang, S. Fabrication of Fe_3O_4@SiO_2@TiO_2 nanoparticles supported by graphene oxide sheets for the repeated adsorption and photocatalytic degradation of rhodamine B under UV irradiation. *Dalton Trans.* 2014, *43*, 13537–13544.

121. Wang, H.; Wang, W. Synthesis of Fe_3O_4/N-TiO_2/Ag hollow nanospheres and their application as recyclable photocatalysts. *J. Cluster Sci.* 2016, *27*, 403–415.

122. Ma, J.; Guo, S.; Guo, X.; Ge, H. A mild synthetic route to Fe_3O_4@TiO_2-Au composites: Preparation, characterization and photocatalytic activity. *Appl. Surf. Sci.* 2015, *353*, 1117–1125.

123. Li, S.; Liang, W.; Zheng, F.; Zhou, H.; Lin, X. Lysine surface modified Fe_3O_4@SiO_2@TiO_2microspheres-based preconcentration and photocatalysis for *in situ* selective determination of nanomolar dissolved organic and inorganic phosphorus in seawater. *Sens Actuators B Chem.* 2016, *224*, 48–54.

124. Hongfei, L.I.U.; Shengfu, J.I.; Yuanyuan, Z.; Ming, L.I. Porous TiO_2-coated magnetic core-shell nanocomposites: Preparation and enhanced photocatalytic activity. *Chin. J. Chem. Eng.*2013, *21*, 569–576.

125. Ma, J.; Guo, S.; Guo, X.; Ge, H.-G. Liquid-phase deposition of TiO_2 nanoparticles on core-shell Fe_3O_4@SiO_2 spheres: Preparation, characterization, and photocatalytic activity. *J. Nanopart. Res.* 2015, *17*, 1–11.

126. Rashid, J.; Barakat, M.A.; Ruzmanova, Y.; Chianese, A. Fe_3O_4/SiO_2/TiO_2 nanoparticles for photocatalytic degradation of 2-chlorophenol in simulated wastewater. *Environ. Sci. Pollut. Res.* 2015, *22*, 3149–3157.

127. Hamzezadeh-Nakhjavani, S.; Tavakoli, O.; Akhlaghi, S.P. Efficient photocatalytic degradation of organic pollutants by magnetically recoverable nitrogen-doped TiO_2 nanocomposite photocatalysts under visible light irradiation. *Environ. Sci. Pollut. Res.* 2015, *22*, 18859–18873.

128. Cao, M.; Wang, P.; Ao, Y.; Wang, C.; Hou, J.; Qian, J. Photocatalytic degradation of tetrabromobisphenol A by a magnetically separable graphene-TiO_2 composite photocatalyst: Mechanism and intermediates analysis. *Chem. Eng. J.* 2015, *264*, 113–124.

129. Li, C.; Tan, J.; Fan, X.; Zhang, B.; Zhang, H.; Zhang, Q. Magnetically separable one dimensional Fe_3O_4/P(MAA-DVB)/TiO_2 nanochains: Preparation, characterization and photocatalytic activity. *Ceram. Int.* 2015, *41*, 3860–3868.

130. Xiang, Y.; Wang, H.; He, Y.; Song, G. Efficient degradation of methylene blue by magnetically separable Fe_3O_4/chitosan/TiO_2 nanocomposites. *Desalin. Water Treat.* 2015, *55*, 1018–1025.

131. Su, W.; Zhang, T.; Li, L.; Xing, J.; He, M.; Zhong, Y.; Li, Z. Synthesis of small yolk-shell Fe_3O_4@TiO_2 nanoparticles with controllable thickness as recyclable photocatalysts. *RSC Adv.*2014, *4*, 8901–8906.

132. Dong, X.; Shao, Y.; Zhang, X.; Ma, H.; Zhang, X.; Shi, F.; Ma, C.; Xue, M. Synthesis and properties of magnetically separable Fe_3O_4/TiO_2/Bi_2O_3 photocatalysts. *Res. Chem. Intermed.*2014, *40*, 2953–2961.

133. Li, C.; Younesi, R.; Cai, Y.; Zhu, Y.; Ma, M.; Zhu, J. Photocatalytic and antibacterial properties of Au-decorated Fe_3O_4@mTiO_2 core-shell microspheres. *Appl. Catal. B Environ.* 2014, *156–157*, 314–322.

134. Su, J.; Zhang, Y.; Xu, S.; Wang, S.; Ding, H.; Pan, S. Highly efficient and recyclable triple-shelled Ag@Fe_3O_4@SiO_2@TiO_2 photocatalysts for degradation of organic pollutants and reduction of hexavalent chromium ions. *Nanoscale* 2014, *6*, 5181–5192.

135. Chen, L.; Li, L.; Wang, T.; Zhang, L.; Xing, S. A novel strategy to fabricate multifunctional Fe_3O_4@C@TiO_2 yolk-shell structures as magnetically recyclable photocatalysts. *Nanoscale*2014, *6*, 6603–6608.

136. Chi, Y.; Yuan, Q.; Li, Y.; Zhao, L.; Li, N.; Li, X.; Yan, W. Magnetically separable Fe_3O_4@SiO_2@TiO_2-Ag microspheres with well-designed nanostructure and enhanced photocatalytic activity. *J. Hazard. Mater.* 2013, *262*, 404–411.

137. Tang, Y.; Zhang, G.; Liu, C.; Luo, S.; Xu, X.; Chen, L.; Wang, B. Magnetic TiO_2-graphene composite as a high-performance and recyclable platform for efficient photocatalytic removal of herbicides from water. *J. Hazard. Mater.* 2013, *252–253*, 115–122.

138. Shojaei, F.A.; Shams-Nateri, A.; Ghomashpasand, M. Comparative study of photocatalytic activities of magnetically separable WO_3/TiO_2/Fe_3O_4 nanocomposites and TiO_2, WO_3/$TiO_2$and TiO_2/Fe_3O_4 under visible light irradiation. *Superlattice Microstruct.* 2015, *88*, 211–224.

139. Chidambaram, S.; Pari, B.; Kasi, N.; Muthusamy, S. ZnO/Ag heterostructures embedded in Fe_3O_4 nanoparticles for magnetically recoverable photocatalysis. *J. Alloy Compd.* 2016, *665*, 404–410.

140. Yang, J.; Wang, J.; Li, X.; Wang, D.; Song, H. Synthesis of urchin-like Fe_3O_4@SiO_2@ZnO/CdS core-shell microspheres for the repeated photocatalytic degradation of rhodamine B under visible light. *Catal. Sci. Technol.* 2016.

141. Shekofteh-Gohari, M.; Habibi-Yangjeh, A. Ultrasonic-assisted preparation of novel ternary ZnO/AgI/Fe_3O_4 nanocomposites as magnetically separable visible-light-driven photocatalysts with excellent activity. *J. Colloid Interface Sci.* 2016, *461*, 144–153.

142. Shekofteh-Gohari, M.; Habibi-Yangjeh, A. Novel magnetically separable $ZnO/AgBr/Fe_3O_4/Ag_3VO_4$ nanocomposites with tandem n–n heterojunctions as highly efficient visible-light-driven photocatalysts. *RSC Adv.* 2015, *6*, 2402–2413.

143. Shekofteh-Gohari, M.; Habibi-Yangjeh, A. Ternary $ZnO/Ag_3VO_4/Fe_3O_4$ nanocomposites: Novel magnetically separable photocatalyst for efficiently degradation of dye pollutants under visible-light irradiation. *Solid State Sci.* 2015, *48*, 177–185.

144. Shekofteh-Gohari, M.; Habibi-Yangjeh, A. Novel magnetically separable Fe_3O_4@ZnO/AgCl nanocomposites with highly enhanced photocatalytic activities under visible-light irradiation. *Sep. Purif. Technol.* 2015, *147*, 194–202.

145. Shekofteh-Gohari, M.; Habibi-Yangjeh, A. Facile preparation of Fe_3O_4@AgBr-ZnO nanocomposites as novel magnetically separable visible-light-driven photocatalysts. *Ceram. Int.* 2015, *41*, 1467–1476.

146. Jia, X.; Dai, R.; Sun, Y.; Song, H.; Wu, X. One-step hydrothermal synthesis of $Fe_3O_4/g-C_3N_4$nanocomposites with improved photocatalytic activities. *J. Mater. Sci.* 2016, *27*, 3791–3798.

147. Mousavi, M.; Habibi-Yangjeh, A. Magnetically separable ternary $g-C_3N_4/Fe_3O_4/BiOI$ nanocomposites: Novel visible-light-driven photocatalysts based on graphitic carbon nitride.*J. Colloid Interface Sci.* 2016, *465*, 83–92.

148. Akhundi, A.; Habibi-Yangjeh, A. Novel magnetic $g-C_3N_4/Fe_3O_4/AgCl$ nanocomposites: Facile and large-scale preparation and highly efficient photocatalytic activities under visible-light irradiation. *Mater. Sci. Semicond. Process.* 2015, *39*, 162–171.

149. Akhundi, A.; Habibi-Yangjeh, A. Novel magnetically separable $g-C_3N_4/AgBr/Fe_3O_4$nanocomposites as visible-light-driven photocatalysts with highly enhanced activities.*Ceram. Int.* 2015, *41*, 5634–5643.

150. Feng, X.; Lou, X. The effect of surfactants-bound magnetite (Fe_3O_4) on the photocatalytic properties of the heterogeneous magnetic zinc oxides nanoparticles. *Sep. Purif. Technol.* 2015,*147*, 266–275.

151. Li, W.; Tian, Y.; Li, P.; Zhang, B.; Zhang, H.; Geng, W.; Zhang, Q. Synthesis of rattle-type magnetic mesoporous and investigation of its photoactivity in the degradation of methylene blue. *RSC Adv.* 2015, *5*, 48050–48059.

152. Li, X.; Wang, G.; Cheng, Y. Preparation and characteristics of $Fe_3O_4-BaTiO_3$ heterostructural nanocomposite as photocatalyst. *Res. Chem. Intermed.* 2015, *41*, 3031–3039.

153. Singh, K.K.; Senapati, K.K.; Borgohain, C.; Sarma, K.C. Newly developed $Fe_3O_4-Cr_2O_3$magnetic nanocomposite for photocatalytic decomposition of 4-chlorophenol in water. *J. Environ. Sci.* 2015.

154. Cao, C.; Xiao, L.; Chen, C.; Cao, Q. Magnetically separable Cu_2O/chitosan–Fe_3O_4nanocomposites: Preparation, characterization and visible-light photocatalytic performance.*Appl. Surf. Sci.* 2015, *333*, 110–118.

155. Cao, Y.; Li, C.; Li, J.; Li, Q.; Yang, J. Magnetically separable Fe_3O_4/AgBr hybrid materials: Highly efficient photocatalytic activity and good stability. *Nanoscale Res. Lett.* 2015, *10*, 952.

156. Zhong, S.; Jiang, W.; Han, M.; Liu, G.; Zhang, N.; Lu, Y. Graphene supported silver@silver chloride and ferroferric oxide hybrid, a magnetically separable photocatalyst with high performance under visible light irradiation. *Appl. Surf. Sci.* 2015, *347*, 242–249.

157. Jing, J.; Zhang, Y.; Feng, J.; Li, W.; Yu, W.W. Facile preparation and high performance of magnetically separable metalloporphyrin. *Chem. Eng. J.* 2015, *263*, 385–391.

158. Wang, B.; Zhang, M.; Li, W.; Wang, L.; Zheng, J.; Gan, W. Fabrication of Au(Ag)/AgCl/Fe_3O_4@PDA@Au nanocomposites with enhanced visible-light-driven photocatalytic activity. *Dalton Trans.* 2015, *44*, 17020–17025.

159. Yang, S.; Ye, C.; Song, X.; He, L.; Liao, F. Theoretical calculation based synthesis of a poly(p-phenylenediamine)–Fe_3O_4 composite: A magnetically recyclable photocatalyst with high selectivity for acid dyes. *RSC Adv.* 2014, *4*, 54810–54818.

160. An, L.; Wang, G.; Shi, X.; Su, M.; Gao, F.; Cheng, Y. Recyclable Fe_3O_4/ZnO/PPy composite photocatalyst: Fabrication and photocatalytic activity. *Russ. J. Phys. Chem. A* 2014, *88*, 2419–2423.

161. Wang, C.; Cao, M.; Wang, P.; Ao, Y.; Hou, J.; Qian, J. Preparation of a magnetic graphene oxide-Ag_3PO_4 composite photocatalyst with enhanced photocatalytic activity under visible light irradiation. *J. Taiwan Inst. Chem. Eng.* 2014, *45*, 1080–1086.

162. Peik-See, T.; Pandikumar, A.; Ngee, H.; Ming, N. Magnetically separable reduced graphene oxide/iron oxide nanocomposite materials for environmental remediation. *Catal. Sci. Technol.*2014, *4*, 4396–4405.

163. Li, Z.; Wen, Y.; Shang, J.; Wu, M.; Wang, L.; Guo, Y. Magnetically recoverable Cu_2O/Fe_3O_4composite photocatalysts: Fabrication and photocatalytic activity. *Chin. Chem. Lett.* 2014, *25*, 287–291.

164. Cao, X.; Chen, Y.; Jiao, S.; Fang, Z.; Xu, M.; Liu, X.; Li, L.; Pang, G.; Feng, S. Magnetic photocatalysts with a p–n junction: Fe_3O_4 nanoparticle and $FeWO_4$ nanowire heterostructures. *Nanoscale* 2014, *6*, 12366–12370.

165. Wang, H.; Wei, Z.; Matsui, H.; Zhou, S. Fe_3O_4/carbon quantum dots hybrid nanoflower for highly active and recyclable visible-light driven. *J. Mater. Chem. A* 2014, *2*, 15740–15745.

166. Zhang, D.; Pu, X.; Gao, Y.; Su, C.; Li, H.; Li, H. One-step combustion synthesis of $CoFe_2O_4$-graphene hybrid materials for photodegradation of methylene blue. *Mater. Lett.* 2013, *113*, 179–181.

167. Zhu, H.; Jiang, R.; Huang, S.; Yao, J.; Fu, F.; Li, J. Novel magnetic $NiFe_2O_4$/multi-walled carbon nanotubes hybrids: Facile synthesis, characterization, and application to the treatment of dyeing wastewater. *Ceram. Int.* 2015, *41*, 11625–11631.

168. Bai, S.; Shen, X. One-pot solvothermal preparation of magnetic reduced graphene oxide-ferrite hybrids for organic dye removal. *Carbon* 2012, *50*, 2337–2346.

169. Sun, J.; Fu, Y.; Xiong, P.; Sun, X.; Xu, B.; Wang, X. A magnetically separable P25/$CoFe_2O_4$/graphene catalyst with enhanced adsorption capacity and visible-light-driven photocatalytic activity. *RSC Adv.* 2013, *3*, 22490–22497.

170. Xiong, P.; Zhu, J.; Wang, X. Cadmium sulfide-ferrite nanocomposite as a magnetically recyclable photocatalyst with enhanced visible-light-driven photocatalytic activity and photostability. *Ind. Eng. Chem. Res.* 2013, *52*, 17126–17133.

171. Fu, Y.; Chen, Q.; He, M.; Wan, Y.; Sun, X.; Xia, H.; Wang, X. Copper ferrite-graphene hybrid: A multifunctional heteroarchitecture for photocatalysis and energy storage. *Ind. Eng. Chem. Res.* 2012, *51*, 11700–11709.

172. Ji, H.; Jing, X.; Xu, Y.; Yan, J.; Li, H.; Li, Y.; Huang, L.; Zhang, Q.; Xu, H.; Li, H. Magnetic g-C_3N_4/$NiFe_2O_4$ hybrids with enhanced photocatalytic activity. *RSC Adv.* 2015, *5*, 57960–57967.

173. Zhang, W.; Wang, M.; Zhao, W.; Wang, B. Magnetic composite photocatalyst $ZnFe_2O_4$/$BiVO_4$: Synthesis, characterization, and visible-light photocatalytic activity. *Dalton Trans.* 2013, *42*, 15464–15474.

174. Yao, Y.; Cai, Y.; Lu, F.; Qin, J.; Wei, F.; Xu, C.; Wang, S. Magnetic $ZnFe_2O_4$-C_3N_4 hybrid for photocatalytic degradation of aqueous organic pollutants by visible light. *Ind. Eng. Chem. Res.* 2014, *53*, 17294–17302.

175. Wu, S.; Wang, P.; Cai, Y.; Liang, D.; Ye, Y.; Tian, Z.; Liu, J.; Linag, C. Reduced graphene oxide anchored magnetic $ZnFe_2O_4$ nanoparticles with enhanced visible-light photocatalytic activity. *RSC Adv.* 2015, *5*, 9069–9074.

176. Yang, D.; Feng, J.; Jiang, L.; Wu, X.; Sheng, L.; Jiang, Y.; Wei, T.; Fan, Z. Photocatalyst interface engineering: Spatially confined growth of $ZnFe_2O_4$ within graphene networks as excellent visible-light-driven photocatalysts. *Adv. Funct. Mater.* 2015, *25*, 7080–7087.

177. Khadgi, N.; Li, Y.; Upreti, A.R.; Zhang, C.; Zhang, W.; Wang, Y.; Wang, D. Enhanced photocatalytic degradation of 17 α-ethinylestradiol exhibited by multifunctional $ZnFe_2O_4$-Ag/rGO nanocomposite under visible light. *Photochem. Photobiol.* 2016, *92*, 238–246.

178. Chen, X.; Dai, Y.; Guo, J.; Liu, T.; Wang, X. Novel magnetically separable reduced graphene oxide (RGO)/$ZnFe_2O_4$/Ag_3PO_4 nanocomposites for enhanced photocatalytic performance toward 2,4-dichlorophenol under visible light. *Ind. Eng. Chem. Res.* 2016, *55*, 568–578.

179. Kulkarni, S.D.; Kumbar, S.; Menon, S.G.; Choudhari, K.S.; Santhosh, C. Magnetically separable core-shell $ZnFe_2O_4$@ZnO nanoparticles for visible light photodegradation of methyl orange. *Mater. Res. Bull.* 2016, *77*, 70–77.

180. Yao, Y.; Qin, J.; Chen, H.; Wei, F.; Liu, X.; Wang, J.; Wang, S. One-pot approach for synthesis of N-doped TiO_2/$ZnFe_2O_4$ hybrid as an efficient photocatalyst for degradation of aqueous organic pollutants. *J. Hazard. Mater.* 2015, *291*, 28–37.

181. Wu, S.; Shen, X.; Zhu, G.; Zhou, H.; Ji, Z.; Chen, K. Synthesis of ternary Ag/ZnO/$ZnFe_2O_4$porous and hollow nanostructures with enhanced photocatalytic activity. *Appl. Catal. B Environ.* 2016, *184*, 328–336.

182. Ren, A.; Liu, C.; Hong, Y.; Shi, W.; Lin, S.; Li, P. Enhanced visible-light-driven photocatalytic activity for antibiotic degradation using magnetic $NiFe_2O_4$/Bi_2O_3 heterostructures. *Chem. Eng. J.* 2014, *258*, 301–308.

183. Fu, Y.; Chen, H.; Sun, X.; Wang, X. Graphene-supported nickel ferrite: A magnetically separable photocatalyst with high activity under visible light. *Am. Inst. Chem. Eng. J.* 2012,*58*, 3298–3305.

184. Patil, S.S.; Tamboli, M.S.; Deonikar, V.G.; Umarji, G.G.; Ambekar, J.D.; Kulkarni, M.V.; Kolekar, S.S.; Kale, B.B.; Patil, D.R. Magnetically separable Ag_3PO_4/ $NiFe_2O_4$ composites with enhanced photocatalytic activity. *Dalton Trans.* 2015, *44*, 20426–20434.

185. Kim, S.H.; Kim, D.; Kwak, S.B.; Han, B.G.; Um, M.; Kang, M. Synthesis of magnetically separable core@ shell structured $NiFe_2O_4$@TiO_2 nanomaterial and its use for photocatalytic hydrogen production by methanol/ water splitting. *Chem. Eng. J.* 2014, *243*, 272–279.

186. Ge, M.; Hu, Z. Novel magnetic AgBr/$NiFe_2O_4$ composite with enhanced visible light photocatalytic performance. *Ceram. Int.* 2016, *42*, 6510–6514.

187. Xiong, P.; Fu, Y.; Wang, L.; Wang, X. Multi-walled carbon nanotubes supported nickel ferrite: A magnetically recyclable photocatalyst with high photocatalytic activity on degradation of phenols. *Chem. Eng. J.* 2012, *195–196*, 149–157.

188. Li, C.; Wang, J.; Wang, B.; Ru, J.; Lin, Z. A novel magnetically separable TiO_2/$CoFe_2O_4$nanofiber with high photocatalytic activity under UV-vis light. *Mater. Res. Bull.* 2012, *47*, 333–337.

189. Wilson, A.; Mishra, S.R.; Gupta, R.; Ghosh, K. Preparation and photocatalytic properties of hybrid core-shell reusable $CoFe_2O_4$-ZnO nanospheres. *J. Magn. Magn. Mater.* 2012, *324*, 2597–2601.

190. Pongwan, P.; Inceesungvorn, B.; Phanichphant, S.; Wetchakun, N. Synthesis and characterization of a magnetically separable $CoFe_2O_4$/TiO_2 nanocomposite for the photomineralization of formic acid. *Ferroelectrics* 2013, *453*, 133–140.

191. Wetchakun, N.; Chaiwichain, S.; Wetchakun, K. Synthesis and characterization of novel magnetically separable $CoFe_2O_4$/CeO_2 nanocomposite photocatalysts. *Mater. Lett.* 2013, *113*, 76–79.

192. Gan, L.; Shang, S.; Wah, C.; Yuen, M.; Jiang, S.; Hu, E. Hydrothermal synthesis of magnetic $CoFe_2O_4$/graphene nanocomposites with improved photocatalytic activity. *Appl. Surf. Sci.*2015, *351*, 140–147.

193. Singh, S.; Khare, N. Magnetically separable, $CoFe_2O_4$ decorated CdS nanorods for enhanced visible light driven photocatalytic activity. *Mater. Lett.* 2015, *161*, 64–67.

194. Haw, C.; Chiu, W.; Rahman, S.A.; Khiew, P.; Radiman, S.; Shukor, R.A.; Hamid, M.A.A.; Ghazali, N. The design of new magnetic-photocatalyst nanocomposites ($CoFe_2O_4$-TiO_2) as smart nanomaterials for recyclable-photocatalysis applications. *New J. Chem.* 2016, *40*, 1124–1136.
195. Choi, Y.I.; Kim, Y.; Cho, D.W.; Kang, J.; Leung, K.T.; Sohn, Y. Recyclable magnetic $CoFe_2O_4$/BiOX (X = Cl, Br and I) microflowers for photocatalytic treatment of water contaminated with methyl orange, rhodamine B, methylene blue, and a mixed dye. *RSC Adv.*2015, *5*, 79624–79634.
196. Xu, Y.; Zhou, T.; Huang, S.; Xie, M.; Li, H.; Xu, H. Preparation of magnetic Ag/AgCl/$CoFe_2O_4$composites with high photocatalytic and antibacterial ability. *RSC Adv.* 2015, *5*, 41475–41483.
197. Shi, B.; Wan, J.; Liu, C.; Yu, X. Synthesis of $CoFe_2O_4$/MCM-41/TiO_2 composite microspheres and its performance in degradation of phenol. *Mater. Sci. Semicond. Process.* 2015, *37*, 241–249.
198. Gaikwad, P.N.; Hankare, P.P.; Wandre, T.M.; Garadkar, K.M.; Sasikala, R. Photocatalytic performance of magnetically separable Fe,N co-doped TiO_2-cobalt ferrite nanocomposite.*Mater. Sci. Eng. B* 2016, *205*, 40–45.
199. Manikandan, A.; Durka, M.; Seevakan, K.; Antony, S.A. A novel one-pot combustion synthesis and opto-magnetic properties of magnetically separable spinel $Mn_xMg_{1-x}Fe_2O_4$($0.0 \leq x \leq 0.5$) nanophotocatalysts. *J. Supercond. Novel Magn.* 2015, *28*, 1405–1416.
200. Tio, A.; Nife, O.; Nicola, D.; Vanags, M. Ag sensitized TiO_2 and $NiFe_2O_4$ three-component nanoheterostructures: Synthesis, electronic structure and strongly enhanced visible light.*RSC Adv.* 2016, *6*, 18834–18842.
201. Chen, C.-C.; Fu, Y.-P.; Hu, S.-H. Characterizations of TiO_2/SiO_2/Ni-Cu-Zn ferrite composite for magnetic photocatalysts. *J. Am. Chem. Soc.* 2015, *98*, 2803–2811.
202. Sathishkumar, P.; Viswanathan, R. $CoFe_2O_4$/TiO_2 nanocatalysts for the photocatalytic degradation of Reactive Red 120 in aqueous solutions in the presence and absence of electron acceptors. *Chem. Eng. J.* 2013, *220*, 302–310.
203. Kadi, M.W.; Mohamed, R.M. Environmental remediation of aqueous cyanide by photocatalytic oxidation using a $NiFe_2O_4$/TiO_2-SiO_2 core-shell nanocomposite. *Desalin. Water Treat.* 2016, *56*, 1940–1948.
204. Su, R.N.; Lv, P.; Li, M.; Zhang, X.; Li, M.; Niu, J. Fabrication of $MgFe_2O_4$-ZnO heterojunction photocatalysts for application of organic pollutants. *Mater. Lett.* 2014, *122*, 201–204.
205. Leng, C.; Wei, J.; Liu, Z. Facile synthesis of PANI-modified $CoFe_2O_4$-TiO_2 hierarchical flower-like nanoarchitectures with high photocatalytic activity. *J. Nanopart. Res.* 2013, *15*, 1643.
206. Sun, A.; Chen, H.; Song, C.; Jiang, F.; Wang, X.; Fu, Y. Magnetic $Bi_{25}FeO_{40}$-graphene catalyst and its high visible-light photocatalytic performance. *RSC Adv.* 2013, *3*, 4332–4340.
207. Xie, T.; Xu, L.; Liu, C.; Yang, J.; Wang, M. Magnetic composite BiOCl-$SrFe_{12}O_{19}$:A novel p–n type heterojunction with enhanced photocatalytic activity. *Dalton Trans.* 2014, *43*, 2211–2220.
208. Bhukal, S.; Bansal, S.; Singhal, S. Magnetic Mn substituted cobalt zinc ferrite systems: Structural, electrical and magnetic properties and their role in photo-catalytic degradation of methyl orange azo dye. *Physica B* 2014, *445*, 48–55.
209. Qasim, M.; Asghar, K.; Raj, B.; Prathapani, S.; Khan, W.; Naqvi, A.H.; Das, D. Magnetically recyclable $Ni_{0.5}Zn_{0.5}Fe_2O_4$/$Zn_{0.95}Ni_{0.05}O$ nano-photocatalyst: Structural, optical, magnetic and photocatalytic properties. *Spectrochim. Acta A* 2015, *137*, 1348–1356.
210. Manikandan, A.; Antony, S.A. Magnetically separable $Mn_xZn_{1-x}Fe_2O_4$; ($0.0 \leq x \leq 0.5$) nanostructures: Structural, morphological, opto-magnetic and photocatalytic properties.*Synth. React. Inorg. Met-Org. Nano-Met. Chem.* 2016.
211. Shakir, I.; Sarfraz, M.; Ali, Z.; Aboud, M.F.A. Magnetically separable and recyclable graphene-$MgFe_2O_4$ nanocomposites for enhanced photocatalytic applications. *J. Alloy. Compd.* 2016, *660*, 450–455.
212. Bhukal, S.; Singhal, S. Magnetically separable copper substituted cobalt-zinc nano-ferrite photocatalyst with enhanced photocatalytic activity. *Mater. Sci. Semicond. Process.* 2014, *26*, 467–476.
213. Liu, F.; Xie, Y.; Yu, C.; Liu, X.; Dai, Y. Novel hybrid Sr-doped TiO_2/magnetic $Ni_{0.6}Zn_{0.4}Fe_2O_4$for enhanced separation and photodegradation of organics under visible light. *RSC Adv.*2015, *5*, 24056–24063.
214. Vignesh, K.; Suganthi, A.; Min, B.-K.; Kang, M. Photocatalytic activity of magnetically recoverable $MnFe_2O_4$/g-C_3N_4/TiO_2 nanocomposite under simulated solar light irradiation. *J. Mol. Catal. A Chem.* 2014, *395*, 373–383.
215. He, H.-Y. Photocatalytic degradations of dyes on magnetically separable $Ni_{1-x}Co_xFe_2O_4$nanoparticles synthesized by a hydrothermal process. *Part. Sci. Technol.* 2016, *34*, 143–151.
216. Zhang, Z.; Xu, L.; Liu, C. Preparation and characterization of composite magnetic photocatalyst $Mn_xZn_{1-x}Fe_2O_4$/β-Bi_2O_3. *RSC Adv.* 2015, *5*, 79997–80004.
217. He, H.; Yan, Y.; Huang, J.; Lu, J. Rapid photodegradation of methyl blue on magnetic $Zn_{1-x}Co_xFe_2O_4$ nanoparticles synthesized by hydrothermal process. *Sep. Purif. Technol.* 2014,*136*, 36–41.
218. Shi, Y.; Zhou, K.; Wang, B.; Jiang, S.; Qian, X.; Gui, Z.; Yuen, R.K.K.; Hu, Y. Ternary graphene-$CoFe_2O_4$/CdS nanohybrids: Preparation and application as recyclable photocatalysts. *J. Mater. Chem. A* 2014, *2*, 535–544.
219. Sathishkumar, P.; Pugazhenthiran, N. ZnO supported $CoFe_2O_4$ nanophotocatalysts for the mineralization of Direct Blue 71 in aqueous environments. *J. Hazard. Mater.* 2013, *252–253*, 171–179

CHAPTER 12 BIOCATALYSIS FOR BIOBASED CHEMICALS

Rubén de Regil and Georgina Sandoval

Keywords: Biocatalysis; hydrolases; transferases; polymers; biodiesel; fine chemicals; prebiotics

INTRODUCTION

It is becoming evident that in order to sustain the standard way of life of the developed and in-development world, it will be necessary to make some adjustments either to our consumption habits or to our sources of supplies of energy and materials. In the latter case, Biotechnology, as a diversified discipline in which chemistry, physics, biology, optics, electro-magnetism and thermodynamics converge, possess the knowledge tools to play a relevant and vital role to address the challenges of growth, ageing, employment, limited sources of raw materials, energy and water supply and living standard. Industrial biobased processes will increasingly become the *de facto*alternative to build a sustainable economy. It is estimated that, by 2030, the products of white biotechnology and bioenergy will account for 30% of industrial production worth €300 bn [1].

Conversion of biomass by Biocatalysis is likely to become standard technology which will contribute significantly to open up access to large feedstock supplies for bioprocesses and the production of transport fuels. On the health and ageing field, personalized nutrition, tailor-made medicine will become a reality in the coming years thanks to novel, specific biotech drugs and regenerative medicine obtained by more efficient and greener processes.

Biocatalysis refers to the transformation of substances of chemical or biological origin through the use of the enzymes produced by diverse living organisms. Enzymes may carry out reactions in a free, or immobilized form, or within the living cell in which they naturally live.

Citation: Rubén de Regil and Georgina Sandoval, Biocatalysis for Biobased Chemicals, doi:10.3390/biom3040812

Isolated enzymes obtained as purified and concentrated extracts, whether used immobilized onto a support or in free form, may produce neater products than when the whole microorganism cell is used for the biotransformation. This is because of the absence of other kinds of enzymes from the internal cell machinery, which may subsequently modify the product, and the absence of internal cellular components during the lysis-extraction step. This advantage of isolated enzymes has a positive impact on yield and purification costs. However, isolated enzymes are known to be less stable in a purified form than within the cell. The lower stability of free enzymes has been improved to some extent by immobilization and crosslinking techniques, the results of which vary depending on the type of enzyme, pH memory, extent of crosslinking, immobilization support and the procedure itself.

Whole microorganism cells, obtained in sufficient amount by fermentation, are used instead of isolated enzymes when their enzymes become highly unstable or non-functional outside the inner environment of the cell, so catalysis is confined within the cell and products are later extracted from it.

Chemical synthesis in the food, feed, industrial and pharma sectors is the field to which Biocatalysis has contributed the most, but it has also contributed to the bioremediation sector, another field in which enzymes are having a significant role.

The relative success of Biocatalysis applied to Synthetic Chemistry has been due mainly to the high enantio- and regioselectivities that enzymes exhibit towards their substrates, because enzymes speed up chemical processes that would otherwise run very slowly or without selectivity, and because enzymes function under mild reaction conditions. These advantages have allowed chemists to avoid the burden of group-protecting procedures, saving time, materials (including the harsh, dangerous or toxic ones) and energy costs. Other advantages of enzymes are that they are easy to control and biodegradable. Thus, Biocatalysis has proved in many cases to be a superior pathway than conventional chemical synthesis pathways, not only in the simplicity of accomplishing the reactions, but also from an economical and environmental point of view.

An important landmark in Biocatalysis was the use of organic solvents as reaction media, as it was thought that enzymes may only work in aqueous medium, naturally. Organic solvents became an important aid in dissolving organic, hydrophobic molecules, which account for much of the library of organic chemicals used in Synthetic Chemistry. The possibility to carry out reactions in organic media favored the reactions rate precisely because of the better solubility of reactants. Purification steps became shorter and easier because the avoidance of surfactants needed in aqueous media and the use of low boiling point solvents easy to evaporate. Moreover, hydrolytic, secondary, water-dependent reactions responsible for lower product yields, and growth of microbes in containers and pipes, is avoided because of the absence of water and the harsh effects of solvents to living cells.

However, Biocatalysis still needs to overcome some limitations such as the enzyme costs of production and the relative narrow stability window of many enzymes under diverse reaction conditions, which restricts their applications onto broader industrial processes. Chemicals used in industrial settings are mostly derived from fossil oil, whereas active ingredients in pharmaceutical processes may be new and/or structurally complex. In general, a vast number of chemicals used in industry are artificial, new man-made chemicals for which enzymes did not evolve to work with and thus are not suitable for the synthetic needs in industry. So there are high discrepancies between the enzyme´s function in nature and the functions needed in the transformation industry. High throughput enzyme screening and protein engineering techniques, particularly Directed Evolution, are some of the tools being employed to advance faster in this aim to find enzymes for artificial substrates, and to create enzymes with new capabilities.

In this review we want to present some of the most representative examples of industrial chemicals produced *in vitro* through enzymatic catalysis, and hence, *in vivo* fermentation products are not included.

ENZYMES IN FOOD INDUSTRY

Applications of enzymes in the food industry are found in almost every sector of this industry: confectionary and sweeteners, fruit and vegetables, dairy, brewery and beverages, meat, dietary and the nutraceuticals niche [2]. However, most of these applications are of hydrolytic nature focused in debranching, improving solubility, clarification, and hydrolysis aimed at diverse goals depending on the food being treated.

Prebiotics and Sweeteners

There are some examples of food chemicals being synthesized by enzymes. One of them is Disaccharide Difructose Anhydride III (DFA III), which is a non-reducing, non-cariogenic sweetener with probiotic properties [3]. DFA III can be synthesized by enzyme inulase II (EC 2.4.1.93), (Figure 1), which was first identified in *Arthrobacter ureafaciens* [4] and later in other bacteria [5,6,7]. However, native enzymes are not thermotolerant enough to resist high temperatures required for a large scale process. Thus, the *ift*-gene encoding for the inulase II from Buo141, a thermotolerant strain, was cloned and expressed into *E. coli* XL1-blue and further modified and immobilized to obtain the biocatalyst which was reported to have 1.7×10^6 U/L [3].

Prebiotics is another field in which enzymes are being used. Oligosaccharides are non-digestible saccharide polymers containing 3–10 monomeric sugar units that are found in low amounts in human milk [8,9], onion, garlic, banana, soybean and chicory [10]. They are classified as prebiotics because they selectively promote the growth of bifidobacteria and lactobacilli in the intestine, which are regarded to have a beneficial effect to human health [10,11,12,13,14].

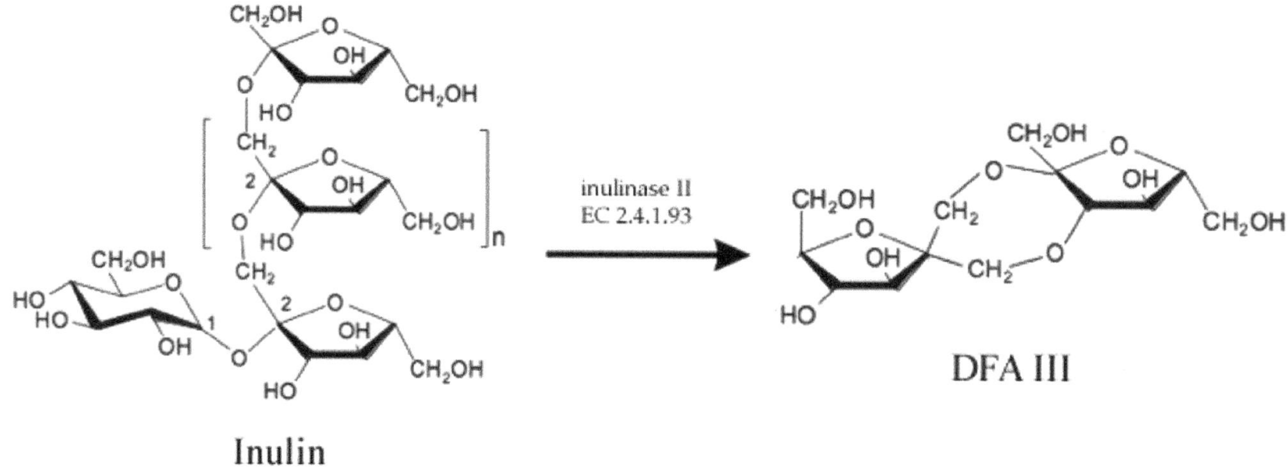

Figure 1: Enzymatic synthesis of Difructose Anhydride III (DFA III) from Inulin using enzyme inulase II, an inulin fructo-transferase.

Oligosaccharides composed either of fructose or galactose units are named fructo-oligosaccharides (FOS) and galacto-oligosaccharides (GOS) respectively.

Galacto-oligosaccharides (GOS) can be synthesized from lactose when this sugar acts as the acceptor and transgalactosylation is catalyzed by β-galactosidase (Figure 2). In addition to the prebiotic activity, GOS have also been reported to contribute to (i) reduction of serum cholesterol and lipid level; (ii) synthesis of B-complex vitamins; (iii) enhance absorption of dietary calcium [14,15,16]; (iv) protection from infection and decrease of pathogenic bacteria; (v) stimulate absorption of some minerals [17]. However, these health promoting properties vary depending on the chemical composition, structure and degree of polymerization and these features depend, in turn, on the origin of the β-galactosidase [18].

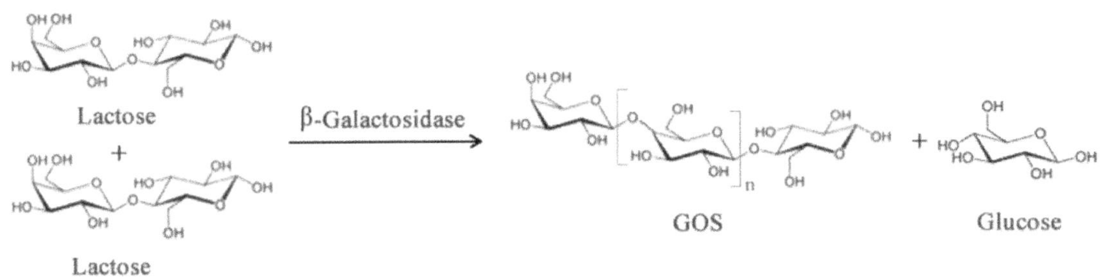

Figure 2: Simplified enzymatic reaction to produce Galacto-oligosaccharides (GOS) from lactose using β-galactosidase. Subscript *n* may range from 0–6 though most native enzymes produce GOS with *n* between 1–2.

The most recurring means of the synthesis of GOS is by enzymatic catalysis from lactose using glycosyltransferases (EC 2.4) or glycoside hydrolases (EC 3.2.1) [19]. Glycosyltransferases are not used in the synthesis of GOS due to their low availability, the need for nucleotide sugar substrates and prohibitive prices. Instead, glycoside hydrolases are used, despite the fact that they are less stereoselective than the former [20]. Glycoside hydrolases from archeas (*Sulfolobus solfactaricum* and*Pyrococcus furiosus*), bacteria (*Saccharopolyspora* sp., *Bifidobacterium* sp., *Thermotoga* sp., *Thermus* sp.,*Bacillus* sp., *Geobacillus* sp., *Caldicellulosiruptor sp.*, *Lactobacillus* sp., *Streptococcus* sp., *Enterobacter* sp.,*Escherichia* sp.) and yeasts (*Aspergillus* sp., *Penicillinum* sp., *Talaromyces* sp., *Trichoderma* sp.,*Kluyveromyces* sp., *Sirobasidium* sp., *Sterigmatomyces* sp., *Rhodotorula* sp., *Sporobolomyces* sp., *Rhizopus*sp.) have been tested for the synthesis of GOS [19,21].

Milk whey is a lactose-rich source which has been regarded in the past as a waste by-product from cheese production highly contaminant. So β-galactosidases have become a valued natural tool to produce a beneficial product out from a contaminant waste in a single step.

As a general finding, reports show that GOS yields average 30%–35% of total sugars [22,23] and are directly proportional to the initial lactose concentration. The major product is the trisaccharide accounting for as much as 80% of total GOS synthesized [12,24]. It was estimated that at least 3,500 tons of GOS are enzymatically synthesized from whey lactose, out of 30,000 tons of total world annual production [25]. They are commercialized to enrich mainly infant formulas.

Just like GOS, Fructo-oligosaccharides (FOS) have low caloric value and have much the same prebiotic properties due to their indigestibility in the upper gastrointestinal tract: they are non-cariogenic and have ability to promote the activity of beneficial colonic lactic acid bacteria and to modulate the intestinal immune response [26,27,28]. However, unlike GOS, FOS are naturally occurring sugars biosynthesized by numerous plants such as asparagus, sugar beet, onion, artichoke, *etc.* [29,30]. Two main types of FOS can be found in nature, the inulin type, which are fructose polymers with linkages β(2–1) biosynthesized by inulosucrases (E.C. 2.4.1.9), and the levan type, which are fructose polymers with linkages β(2–6) biosynthesized by levansucrases (E.C. 2.4.1.10) [31,32,33].

Two different classes of FOS mixtures are produced commercially, one is based on inulin by controlled enzymatic hydrolysis, and the other is based on sucrose by a transfructosylation processes with fructosyltransferases, as shown in Figure 3 (β-fructofuranosidase, E.C. 3.2.1.26 or β-d-fructosyltransferase, EC 2.4.1.9) [34,35]. FOS from inulin hydrolysis contain longer fructose chains (Degree of Polymerization, DP 2–9) and may have either a fructose or glucose unit at the end of the chain. On the other hand, FOS from sucrose transfructosylation contain shorter polymer chains (DP 2–4) and always end with a glucose unit [35].

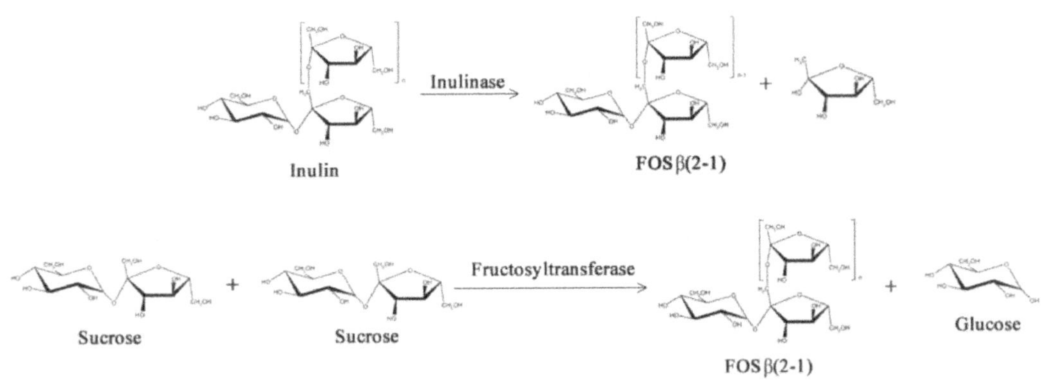

Figure 3: Simplified enzymatic reactions used commercially to obtain Fructo-oligosaccharides (FOS).

A commercial enzyme Pectinex Ultra SP-L was used immobilized onto Eupergit C to synthesize FOS from a 600 g/L of sucrose solution. Reactions were carried out at 65 °C in pH 6.5 during 24 h with a yield of 57% FOS of total sugars [36].

Despite the big production numbers that prebiotics have, the overall European market for prebiotics is still at the beginning stage, with the $87-million fructan (inulin and fructo-oligosaccharide) segment as the most developed part. In US the prebiotics market is estimated to reach US$198.3 million in 2014 [37].

Structured Lipids

Specialty fats and oils are lipids with special functional or nutritional effects in the human body. Among these, structured lipids (SLs) have a predominant importance within this field [38,39]. SLs are tailor-made fats and oils which incorporate specific new fatty acids to triacylglycerols, or have different composition and positional distribution of existing fatty acids, within its glycerol backbone. Among the various types of SLs, the MLM (medium, long, medium fatty acid chain length esterified at the *sn*-1, *sn*-2 and *sn*-3 positions of the glycerol backbone respectively) have been a subject of great interest because of their ability to provide quick energy through the pancreatic lipase hydrolytic release of the medium chain fatty acids which become rapidly oxidized by the liver [40]. The remaining long chain *sn*-2-monoacylglyceride is left to be absorbed through the lymphatic system [41]. This long chain monoacylglyceride can be an omega-6 fatty acid or any other poly unsaturated fatty acid (PUFA) which will provide the already known health benefits to the body [42].

SLs are claimed to have a modulation effect on the immune system, to improve the lipid clearance in the blood [43], and as a fat for special nutritional feeding [44]. Fatty acids at the *sn*-2 position are more readily absorbed *in vivo* than those at the *sn*-1 or 3 positions, thus, SLs of the type MLM that include an essential fatty acid in the *sn*-2 position are desirable nutrition sources for people who suffer from malabsorption or a pancreatic condition [44,45].

SLs can be obtained either chemically or enzymatically using lipases [42]. The high regioselectivity of lipases makes Biocatalysis particularly well suited for the synthesis of SLs, avoiding the synthesis of unwanted by-products, which lower the yields [46]. However, enzymatic synthesis of SLs has yet to be optimized in order to avoid the high costs of large-scale purification of unreacted substrates (Free fatty acids, Triacyglicerides, Diacylgricerides, Esters) which must be removed upon completion of reaction [47].

Two basic approaches have been followed to carry out the enzymatic synthesis of MLM SLs: (1) Transesterification and (2) Acidolysis [48]. Lipase transesterifications for the synthesis of MLM SLs have been performed using different vegetal and fish oils as source of essential long PUFAs, and coconut and palm kernel oil as a source of medium chain fatty acids (C6:0–C12:0) to be incorporated in the *sn*-1 and *sn*-3 positions [42,49]. Thus, to synthesize MLM SLs, *sn*-1 and *sn*-3, regiospecific lipases are favored. In Figure 4 these two approaches are depicted using a *sn*-1,3-specific lipase.

Casas-Godoy *et al.* [50] incorporated caprilyc and capric acid into olive oil using *Yarrowia lipolytica* immobilized on Accurel. Their optimized reaction allowed them to obtain an incorporation rate of 25% and 21% for capylic and capric acid respectively within 2 days at 40 °C and 5% (w/w) enzyme load.

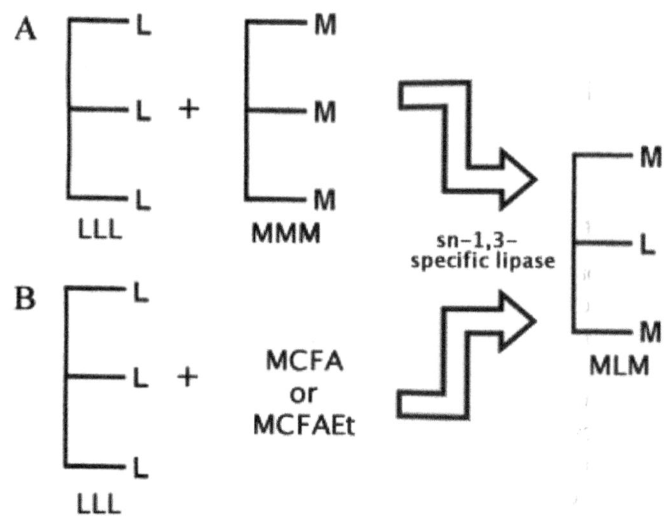

Figure 4: The main two approaches to synthesize MLM Structured Lipids enzymatically: (**A**) Transesterification and (**B**) Acidolysis. Subproducts like LML Structured lipids may also be obtained, but only the product of interest is shown. L = Long Chain; M = Medium Chain; MCFA = Medium Chain Fatty Acid; MCFAEt = Medium Chain Fatty Acid Ester (see also [48]).

In another work, castor oil was dehydrated, isomerized and then transesterified with methyl laurate using *Thermomyces lanuginosa* and *Carica papaya*. The incorporation and transesterification yields were 59% and 88% for *T. lanuginosa* and 44% and 67% for *C. papaya*, respectively. This was achieved in 72 h at 60 °C with a 10% (w/w) enzyme load. Transesterification in the *sn*-2 position was observed and accounted for less than 8% [51].

Palm oil trolein has been subject to enrichment in its *sn*-2 position with DHA and ARA through acidolysis using Novozym 435. The research Group achieved an incorporation of 11% of DHA at*sn*-2 position (17% total) and 5.5% of ARA at *sn*-2 position (7% total) in 24 h at 60 °C and 10% enzyme load [52]. In a similar work, Khodadadi *et al.* [49] transesterified palm oil with tricaprylin in order to modify the former. Their results showed that Linolenic acid chains in palm oil are more easily transesterified than Linoleic or Oleic. Their best results were around 20% incorporation of caprylic chains into *sn*-1, *sn*-2 and *sn*-3 positions of which around a half corresponded to *sn*-2 position. The experiments were carried out with Novozym 435, Lipozyme TL-IM, Lipozyme RM-IM and Amano DF, though the best results were with Novozym 435, 24 h of reaction time, 50 °C and 1% enzyme load. Tecelão *et al.* [53] recently achieved 21% and 8% of incorporation of oleic acid and DHA respectively, to tripalmitin using *Carica papaya* latex at 60 °C, in a solvent-free system during 24 h.

Nagao *et al.* [54] achieved almost 100% of fatty acid incorporation with a DHA-rich by-product waste from a tuna oil industrial, hydrolysis process employed to extract DHA. Their reaction conditions were 30 °C, 120 h, vacuum at 2 kPa and 10% of Novozym 435 load. The degree of esterification with DHA was 51% and 17% at *sn*-1,3 and *sn*-2 position respectively.

ENZYMES IN THE BULK CHEMISTRY INDUSTRY

Although industrial biotransformations are mainly used for the production of fine chemicals, there are a few examples where Biocatalysis is also used to produce commodity chemicals such as acrylamide, and biodiesel. The use of biocatalysts employed to assist in synthetic routes to complex molecules of industrial interest is growing steadily [55,56,57].

Biodiesel

The lipase-catalyzed synthesis of biodiesel offers advantages over conventional methods for its production: lipases are able to work in gentle conditions and with a variety of triglyceride substrates, including waste oils and fats, corrosion problems present in chemical synthesis are avoided in the lipase-catalyzed synthesis, easy recovery of biocatalyst and glycerol, high levels of Free Fatty Acids (FFA) may also be esterified by lipases, and low environmental impact [58,59]. Furthermore, the separation and purification step of biodiesel is easier [60,61], resulting in a more attractive technical and environmental process. Numerous lipases have been studied for biodiesel production, with diverse triglyceride substrates and acyl acceptors. The use of edible oils is today not an option due to the food *vs.* fuel issue, which also makes them a more expensive raw material. Research studies so far have converted a series of non-edible fats and oils replacing the traditional alkaline transesterification. Process flow, reactor design and cost performance have been some of the main elements studied. It has been shown that high productivity, enzyme reuse and low reaction times can be achieved [62,63,64]. Nevertheless, enzymatic biodiesel still faces technical and economic challenges and has room for improvement to make enzymatic biodiesel a more attractive option for industrial production.

The most used method to produce enzymatic biodiesel is by transesterification of oil with an alcohol as acyl acceptor (Figure 5). Acyl acceptors, pose a challenge because the most available and cheap ones for industrial production and those who meet international specifications for biodiesel, are methanol and ethanol, which exert a strong denaturating action towards lipases [65]. Nevertheless, relatively successful efforts to overcome this inconvenience have emerged, like a pioneering work on stepwise alcohol addition [66], acyl acceptor alterations [67,68] and solvent engineering [69,70] which have sorted out this problem with relatively efficacy. Isopropanol [71], Isobutanol [72], 1-butanol [71], 2-butanol [72], 2-ethyl-1-hexanol [73], methyl acetate [74] and ethyl acetate [68] have been reported to be used as acyl acceptors. It was later documented that glycerol, a by-product of biodiesel synthesis, exerted lipase inhibition as well, which is more likely due to mass transfer limitation in the immobilized lipase [75].

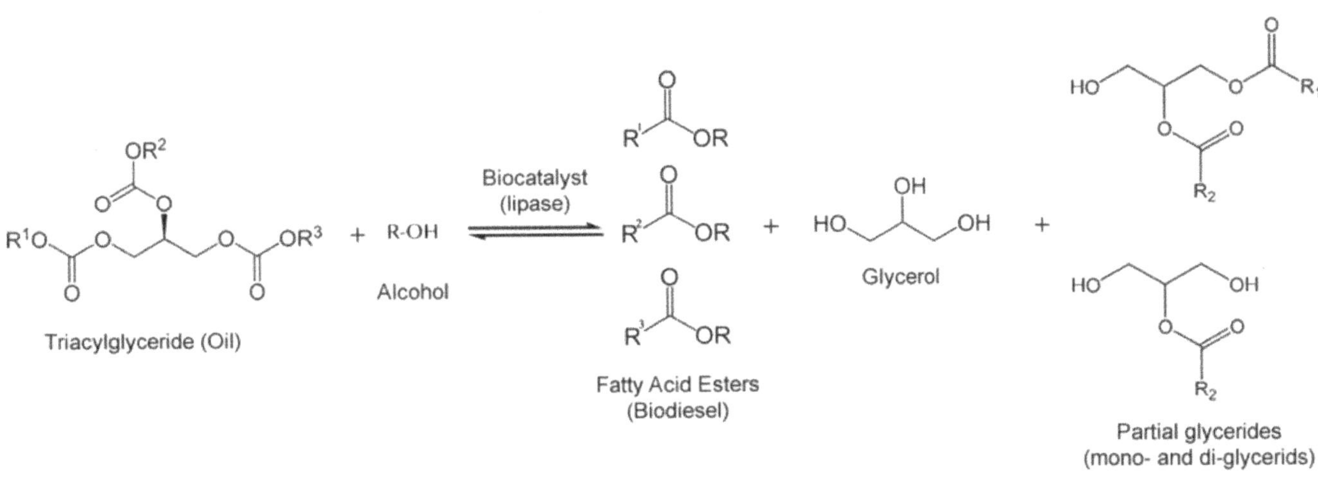

Figure 5: Enzymatic reaction of transesterification of oils with alcohol to produce fatty acid esters (biodiesel).

Free and immobilized lipases of fungal and yeast origin such as *Mucor* sp., *Rhizopus* sp., *Candida*sp., *Aspergillus* sp., *Thermomyces* sp., *Penicillinum* sp. *and Pseudomonas* sp. have shown to be good biocatalysts for biodiesel synthesis as yields for most of them are reported to be 80%–95% [58]. Enzyme immobilization is a cost which greatly influences the economic viability when scaling-up industrial biodiesel projects; thus, alternative methods of preparing the enzyme is a current active topic in the field [76].

Organic solvents have been used to solve problems like substrate solubility and mass transfer [63]. Hydrophobic organic solvents have been used because they improve trygliceride and fatty substrates solubility [77], but leave viscous glycerol insoluble, which poses mass transfer limitations as stated above. Despite the denaturing effects of hydrophilic organic solvents over enzymes [78,79], some solvents like 1,4-dioxane and *tert*-butanol have been proven to generate high transesterification yields [69,70,80]. Incubation of lipases in alcohols with carbon number ≥ 3 has been documented to improve enzymatic activity in the synthesis of methyl esters [81]. Acetone has also been tested; however, the maximum yields of biodiesel have reached 40% [82].

The use of waste animal fat from slaughterhouses as raw material for the production of biodiesel by transesterification with ethanol was recently evaluated by Rivera *et al.* [69]. Enzymes N435 and Lipozyme RM IM at 45 °C at a 2% (w/w) enzyme load, in an organic (*tert*-butanol) and solvent-free systems were used. The biodiesel yields were 80% for the solvent-free system after 2 days of reaction, and 65% for the organic system after 1 day of reaction.

Recently, solvents like supercritical carbon dioxide have been used and biodiesel yields accounted for as much as 81% at 60 °C with a 15% (w/w) enzyme load during 4 h [83]. Ionic liquids have also been used for biodiesel synthesis [84,85] and one of the best results was achieved after 12 h at 50 °C in 1-ethyl-3-methylimidazolium trifluoromethane-sulfonate ([C2mim][OTf]) with an 80 % yield [86].

Intracellular lipases in the form of whole-cell biocatalysts, present a feasible alternative for biodiesel production as well. Their main advantage is the lack of laborious and costly extraction and purification procedures needed when an isolated, free or immobilized, enzyme is used [87]. In order to use whole-cell Biocatalysis, immobilization is advocated. One of the most accepted whole-cell biocatalyst systems for industrial applications is the use of filamentous fungi immobilized onto biomass support particles (BSPs) [88,89,90]. A*spergillus* sp. and *Rhizopus* sp., have been identified as robust for whole-cell immobilization, and thus have become the most widely studied fungi for biodiesel production [58]. Oils from yeasts and fungi are an alternative source of oils for biodiesel, as they can accumulate long-chain triacylglycerols with 16 and 18 carbons. The oil contents obtained from several yeast strains such as *Cryptococcus*, *Lipomyces*, *and Rhodotorula* species have reached 60%–70% of their dry weight [91].

Industrial Polymers

Biocatalysis has opened new synthetic strategies for organic chemists because it has been applied to synthetic chemistry in order perform stereochemical complex reactions in a relatively simple, straightforward way, which otherwise would require laborious protection-deprotection steps, toxic reactants and/or high-temperature/pressure procedures [92,93]. Thus, these advantages of using enzymes have been adopted in the field of polymer synthesis, in which most polymers are difficult to produce or to control by conventional chemical methods. Polymers with specific structures can be prepared enzymatically. In contrast, attempts to attain similar levels of polymer structural control by conventional chemical methods may prove to be a true challenge. From an environmental stand point, polymers derived from enzyme-mediated catalysis, whether polyesters, polyphenols, polysaccharides, proteins, or other polymers, are in most cases biodegradable, a feature highly sought after these days.

The most common type of enzymes which are capable to catalyze polymerization reactions are: Transferases, Oxidoreductases and Hydrolases.

Transferases

Transferases transfer a glycosyl group from a sugar nucleotide donor to specific nucleophilic acceptors. Despite they impose a high degree of regio- and stereochemical control to the glycosidic bond they form, and thus having a great synthetic potential for interesting polymeric materials, their use for synthesis *in vitro* is limited and even prohibitive by both, the availability of enzymes due to expression and solubility, and by the high costs of expensive activated donor sugars such as uridine diphosphate. They are very sensitive biocatalysts so their isolation and use on a larger scale is not practical, and they are mostly reserved for specialty research endeavors [94].

Oxidoreductases

On the other hand, Oxidoreductases and particularly Hydrolases are more robust enzymes, less sensitive and easier to obtain and use *in vitro*. Polymers typically produced by these three types of enzymes are shown in Table 1.

Table 1: Common enzymes and the typical polymers synthesized by them.

Enzymes	Polymers synthesized
Oxidoreductases	
Peroxidases *Laccases* *Tyrosinases* *Glucose oxidases*	Polyphenols, polyanilines, vinyl polymers
Transferases	
Glycosyl transferases *Acyl transferases*	Polysaccharides, cyclic oligosaccharides, polyesters
Hydrolases	
Glycosidases *Lipases* *Peptidases* *Proteases*	Polysaccharides, polyesters, polycarbonates, polyamides, polythioesters, polyphosphates

Polyaromatics

Polyaromatics are widely found in nature such as lignin, and flavonoid compounds. Much of their role in nature is as structural component conferring structural strength to living systems, in the case of lignin, and as a bioactive molecule in the case of flavonoids. Phenol is the single aromatic most important compound for industrial applications. Current polymeric materials produced commercially from phenolic compounds as industrial plastics include Bakelite and poly-2,6-dimethyl-1,4-phenylene oxide) (PPO) which bear good toughness and temperature-resistant properties. However, the oxidative polymerization of phenol with a conventional catalyst usually gives insoluble products with uncontrolled structure [95]. Moreover, as Bakelite and PPO are phenol-formaldehyde resins, the toxicity of formaldehyde has brought limitations on their industrial production.

On the other hand, an enzyme-catalyzed polymerization, offers not only the typical benefits of Biocatalysis, but the possibility to have a better control over polymerization [96].

In a recent work, Salvachúa *et al.* [97] used Versatil peroxidase (VP) to crosslink several monomeric lignans and peptides. Results show lignan oligomers of a maximum DP of 9 with a Number Average Molecular Weight Mn = 3,300. As per the reactions with peptides, peptide oligomers with a maximum DP of 11 and a Mn = 6,500 were obtained. Hetero-oligomers between lignans and peptides were also synthesized by VP. Oligomers of a DP = 6 and a Mn = 2,300 were obtained where the ratio peptide: lignan was 1:5. The reactions were done at room temperature with 1.5 U mL^{-1} of VP, 0.1 mM H_2O_2 and 0.1 mM Mn^{2+}, during 24 h.

A peroxidase-catalyzed polymerization was performed under mild reaction conditions, using an aqueous buffer alcohol resulting in a soluble polyphenol with M_n of 3,000–6,000 [98,99]. The catalysts included horseradish peroxidase (HRP) and soybean peroxidase (SBP). The polymer solubility increased with increasing the oxyphenylene unit content (32%–59%) which was controlled by varying the methanol amount [100], and the solution pH [101]. Monodisperse polymer particles (250 nm diameter) were synthesized from phenol and subsituted phenols like m-cresol, p-cresol and polyphenylphenol, with HRP in a dispersion system 1,4-dioxane/buffer [102,103]. The mechanism of a peroxidase-catalyzed polymerization of phenols comprises three stages: Radical formation, Radical transfer and Radical coupling, shown in Figure 6.

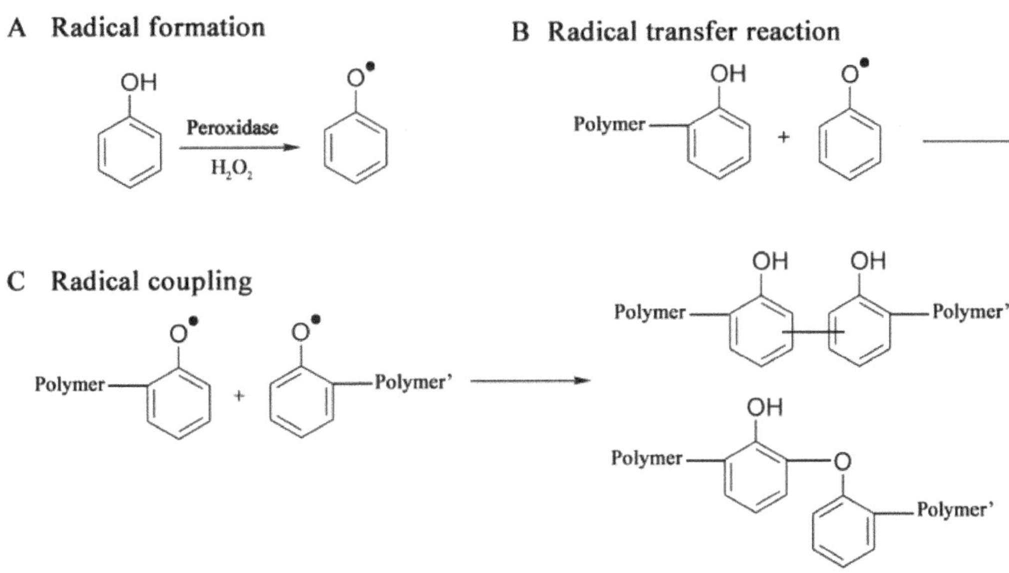

Figure 6: Mechanism of phenol polymer formation (see also [104]).

It was revealed that meta-substituted phenol polymerizations catalyzed by HRP and SBP give rise both to higher yields when small and large-bulky substituents are used, respectively [105]. Cardanol, a natural phenol derived from cashew nut shell liquid (CNSL) with a C15 unsaturated alkyl meta-substituent, has good perspectives for industrial utilization such as resins, friction lining materials, and surface coatings. Therefore, an enzymatic approach to develop derivatives from this compound is attractive. Kobayashi *et al.* [106] obtained an artificial polymer structurally similar to "urushi", a Japanese lacquer, through an SBP-polymerization of Cardanol. The polymer had a molecular weight between 2,000–4,000 Da., soluble in polar organic solvents and possessed a tough and hard property as a film with a glossy surface finish. Kim *et al.* [107] also synthesized polycardanol with SBP and found it had better anti-biofouling activity to *Pseudomonas fluorescens*compared to polypropylene.

Various other m-substituted phenols have been polymerized by HRP and SBP, for example meta-alkylphenols, meta-halogenated phenols, and meta-phenylphenol [108]. These enzymatically synthesized m-substituted polymers were applied to positive-type photoresists for printed wire boards, because of their high solubility toward alkaline solution and high thermal stability [108,109].

Ortho-substituted phenols have been polymerized with SBP affording a variety of oligophenols (dimers to pentamers) and some of their oxidation products, including quinones and demethylated quinones. Some of these are considered to serve as biologically important compounds with therapeutic potentials [110]. In separate reaction systems, it was possible to synthesize poly(phenylene oxide) (PPO) from two different *o*-substituted phenols, 2,6-dimethylphenol and 3,5-dimethoxy-4-hydroxybenzoic acid (syringic acid), using HRP, SBP and a laccase [111,112,113]. The polymers were formed exclusively of 2,6-dimethyl- or 2,6-dimethoxy-1,4-oxyphenylene units respectively, which are similar in structure to a widely used high-performance engineering plastic PPO.

Para-alkylphenols were polymerized by HRP in an aqueous 1,4-dioxane solution to synthesize polymers with Mn of several thousands, whereby the polymer yield increased as the n-alkyl chain length increased from 1 to 5 [102,114]. Oxidative polymerization of natural *para*-substituted phenols like tyrosine ethyl ester or methyl ester or phenol derivative hydroquinone-β-D-glucopyranoside (arbutin) [115] were carried out using HRP catalyst in an aqueous buffer solution. A mixture of phenylene and oxyphenylene units was observed in the tyrosine polymer, whereas the arbutin polymer, after degycosilation, consisted of C-C *o*-position hydroquinone units. Resulting polymers had a molecular weight from 1,500–4,000. Reactions catalyzed by HRP proceeded, involving radical species, yet took place chemoselectively. *Para*-phenylphenol is one well-studied *para*-substituted phenols in peroxidase-catalyzed polymerizations. Polymers up to 26,000 Da have been synthesized with HRP [99]. However, the polymerization of *para*-phenylphenol lacks precise structure control due to the existence of both stabilized *ortho*- and *para*-single electron radical species, causing different kinds of linkages and leading to a complex structural polymer [116].

Other numerous reports manage different aspects of the enzymatic polymerization of phenols: use of mixtures of aqueous-organic solvents with 1,4-dioxane, methanol, DMF [117], or total organic systems (isooctane) [118], ionic liquids [119], using cyclodextrins as a phenol solubilizing agent [120], incorporating PEG as a template aid [121], or carrying out the reaction in micelles and reverse micelles [122] or in capsules [123].

An important group of natural phenols widely distributed throughout the plant kingdom are flavonoids. Flavonoids are benzo-γ-pyrone compounds consisting of phenolic and pyrane rings. Their biological and pharmacological effects as antioxidant, antimutagenic, anticarcinogenic, antiviral and antiinflammatory have been extensively reviewed [124,125,126,127].

Flavanols or commonly known as catechins are the major group of polyphenols present in green tea. The main catechins in green tea are (+)-catechin, (−)-epicatechin (EC), (−)-epigallocatechin (EGC), (−)-epicatechin gallate (ECG), and (−)-epigal-locatechin gallate (EGCG). Poly(cathechin) has been synthesized by HRP in aqueous-organic solvents having yields of DMF-soluble polymers around 30% and molecular weights of 3,000 [128]. It has also been polymerized by a laccase derived from *Myceliophthore* (ML) in a mixture of acetone-buffer pH 5 rendering an 8,000 Da polymer [129]. In this case, as in other phenol polymerizations [100], the organic solvent/buffer ratio greatly affected the solubility of the synthesized polymer. The antioxidant properties of catechins and ECGC, not only remained, but increased in a concentration dependent manner in poly(catechin) and poly(ECGC) synthesized by HRP and ML [129,130]. Superoxide anion scavenging activity of poly(ECGC) was much superior than monomeric ECGC and poly(catechin) [130]. Inhibitory activity of Xanthine oxidase, which was negligible in monomeric catechin or ECGC, was markedly observed in poly(catechin) and in poly(ECGC), inclusive was higher than allopurinol, a commercial inhibitor frequently used for gout treatment [129,131]. Recently, a laccase was used to polymerize catechin in order to attach it to biomedical catheters in order to reduce bacterial biofilm formation [132].

Rutin has also been polymerized in organic-aqueous systems by Laccase ML-mediated catalysis. Yields superior to 70% were mostly observed and molecular weights between 7–9 kDa were registered [133].

Flavonols (quercetin) and isoflavones were also subject to an enzymatic oxidative polymerization by HRP and SBP in aqueous-organic solvent systems. Weight Average Molecular Weight Mw and Number Average Molecular Weight Mn for poly(quercetin) was 10,000 and 2,500 respectively, with a yield around 50%. Poly(rutin), poly(catechin), poly(daidzein), poly(formononetin) and others were also obtained in similar numbers [128].

Functionalization of polymers with flavonoids has also been investigated. Quercetin-functionalized chitosan has been the most studied system to yield antimicrobial products [134,135,136,137]. However, naringin-PVDF, morin-resin, catechin-inulin systems have also been reported [138,139,140].

Hydrolases

Hydrolases are the enzyme class with most widespread use and applications in the world. Around 75% of all commercialized industrial enzymes are hydrolytic in action. Despite proteases and carbohydrases are the most demanded subclass of enzymes accounting for around 40% of world's enzyme sales, lipases remain as the biocatalyst subclass most widely used in organic chemistry [141,142].

Lipases have been used to catalyze reactions in polycondensations of dicarboxylic acids with diols [143], ring-opening polymerization (ROP) of lactones [144], cyclic carbonates [145] and polycondensations of hydroxycarboxylic acids [146]. More examples are described in the following sections.

Polyesters

Polyesters occupy the 4th place among the most important biomacromolecules in living systems. They are macromolecules with good biodegradability, biocompatibility and permeability, and thus are highly suitable for biomedical applications, such as implant biosorbable materials, for tissue engineering and delivery vehicles for drug or genes [147]. Polyesters are very valuable industrial materials that are broadly used like poly(ethylene terephthalate) (PET), poly(butylene succinate) (PBS), poly(ε-caprolactone) (PCL), and poly(lactic acid) (PLA).

Two major approaches can be taken to perform enzymatic polyester synthesis: (1) polycondensation and (2) ring-opening polymerization (ROP). PET and PBS are produced in large-scale by polycondensation whereas PCL and PLA are produced via ROP. Classical cationic, anionic or metallic chemical synthesis of polyesters may render them unsuitable for biomedical applications due to the toxicity of these reagents. Enzyme catalyzed synthesis of esters has circumvented this and other problems since enzymatic reactions do not require high energy input (temperature or pressure), most enzymes can carry out regio-, chemo- and enantiospecific reactions with no need of protecting groups, exhibit high catalytic activity, they may be reused and as a consequence of all this, reactions are environmentally and health safer than the conventional chemical route [148].

Polycondensations

Lipase-Catalyzed Polycondensations

In lipase-catalyzed polycondensations dicarboxylic acids are esterified with diols to produce polyesters. Reactions have been realized mostly in organic solvents, in solvent-free media and in aqueous systems. However, there have been studies in ionic liquids [149] and supercritical fluids as well [150]. In the early years of exploration of these systems, Okumura *et al.* [151] set a series of reactions with dicarboxylic acids and diols in aqueous systems using a lipase from *Aspergillus niger*and he obtained oligomers of 3–7 units. Extracting water molecules from the system, either with molecular sieves or applying vacuum, seemed to improve the polymer size up to a DP of 20 in an adipic acid-1,4-butanediol reaction [152], and up to a Mn = 77 kDa in an adipic acid-1,6-hexanediol reaction [153]. Organic solvents with high boiling points were identified to favor the polyester synthesis. In an aim to study natural hydroxy-acids, Gómez-Patiño *et al.* [154] obtained oligomers from tomato cuticle monomer, 10,16-dihydroxyhexadecanoic acid and its methyl ester. *Candida antarctica* B was the best performer among the 5 lipases tested in organic media reactions obtaining the largest oligomer with a DP = 3 and Mw = 1,206.

Reactions between dicarboxylic acids and diols conducted in water with *Candida antarctica*(N435), but with a continuous dehydrating process, are claimed to give rise to good yields [155].

Polycondensations with structurally different monomers have been realized, for example siloxane-containing diacids with diols (Mw = 20,000) [156], diacids with glycerol (Mw = 2,000–6,000) [157,158], thio-glycerol with diacids (Mw = 170,000) [159], diacids with a variety of sugars, sorbitol, erythritol, xylitol, ribitol, mannitol, glucitol, galactitol yielded polyesters with Mw = 10 kDa for galactitol and 75 kDa for Mannitol. The regioselecitivy of the sugar esterifications was mostly consistent in all the sugars to be on the primary hydroxyl groups [160].

Activation of dicarboxylic acids has been employed by means of esterification the carboxylic groups with methanol or ethanol [161], by esterification [162] or by vinyl activation [163]. Azim *et al.* [164], in an interesting strategy, firstly polymerized succinic acid with 1,4-butanediol to obtain a cyclic oligomer, which later was repolymerized by ring-opening to obtain a high molecular weight polymer (Mw = 130 kDa). Activation of dicarboxylic acids by means of esterification the carboxylic groups with halogenated alcohols like 2-chloroethanol, 2,2,2-trifluroethanol and 2,2,2-trichloroethanol lead to an increase of Mw by a factor of 5 [162]. However, the halogenated alcohols produced as a consequence of the reaction may damage the biocatalyst. Activation of carboxylic groups by vinyl esterification produced in most cases polyesters with Mw around 25 kDa [103,165,166].

Through reaction optimizations, Yao *et al.* [167] were able to achieve a polyester Mw = 16.6 kDa using 1,8-octanediol, adipic acid and L-malic acid as monomers. The incorporation of a sugar to be used as a third monomer is a likely means to modify the properties of conventional polymers, and an interesting strategy to prepare a variety of functional materials.

Recently it was reported the polymerization of β-alanine by a lipase catalyzed polycondensation reaction between β-alanine esters. Lipase B from *Candida antarctica* aggregated in CLEAs was the biocatalyst that converted β-alanine esters into polymers in high yield (up to 90%) and with high DP (up to 53 units). The best results were obtained by using methyl esters and methyl-tert-butyl ether (MTBE) as the solvent medium at 50 °C for 16 h [168].

Protease-Catalyzed Polycondensations

Proteases, like some other hydrolases, catalyze not only hydrolytic reactions (peptide hydrolysis in this case) but also peptide bond formation under appropriate conditions to give polypeptides. There exist several oligopeptides with varied biologically active properties that have been synthesized widely using proteases as catalysts via condensation reactions [169]. Some examples include, aspartame synthesized by thermolysin [170], oxytocin by papain, thermolysin, and chymotrypsin, and somatostatin by thermolysin, and chymotrypsin [171]. Various attempts to synthesize long-chain polyaminoamides were done by a protease-catalyzed polycondensation using chymotrypsin, trypsin, subtilisin and papain. However, only lower molecular weight polymers were obtained [172].

Polyamides

It was recently reported the synthesis of polyamides by lipase Novozym 435 in high yields (93%) and a short lapse of time (30 min). Authors attribute this high productivity to the use of "designed" monomers of the type: ω-amino-α-alkoxy acetic acid ethyl ester, which bears an ω-amino group and an oxygen atom in β position. The reactions were performed in bulk at 80 °C with an enzyme load of 45% (v/v) and the resultant polyamides had a DP = 15 (PDI = 1.6) and a Mw between 3,000–4,000 [173].

Vinyl Polymers

One field that has expanded tremendously in the past two decades is the enzyme-initiated radical polymerization of aromatic and vinyl monomers [149]. Vinyl monomers studied for enzymatic polymerization are basically classified into (meth)acrylic type and styryl type, which both involve the formation of a radical species. HRP and Laccase peroxidases have been reported to perform well in the meth(acrylic)-type polymerization of acrylamide. In a HRP-mediated free radical polymerization of vinyl monomers, the HRP-mediator system has three components: enzyme, oxidant (H_2O_2) and an initiatior, such as β-diketones which are commonly used. On the other hand, when several oxidoreductases, such as laccase, lipoxidase, and sarcosine oxidase, were used as catalysts, the acrylamide polymerization proceeded even without H_2O_2 or β-diketones. This was the case for laccase from *Pycnoporus coccineus* which induced the acrylamide polymerization in water at 50–80 °C under argon to produce poly(acrylamide) with Mn = 1×10^6 , Mw/Mn ≈ 2 for 4–24 h and up to 81% yields [174]. Since the laccase-mediator system (LMS) may induce the radical polymerization of vinyl monomer with no H_2O_2 and is environmentally benign, different conditions for polymerization with commercial laccase *Myceliophthora thermophila* were further examined and optimal conditions were found as slightly acidic reaction media at around 50 °C using β-diketones and O_2 as the initiator and oxidant respectively. The molecular weight of poly(acrylamide) was Mn 6–28 × 10^4; Mw/Mn = 2.5–3.2 and could be controlled through the ratio of monomer to enzyme [175].

Poly(acrylamide) polymers of Mn of 1.5–4.6 × 10^5 with Mw/Mn 2.0–2.4, and acrylamide conversion between 70%–90% were synthesized by HRP in water at room temperature with 2,4-pentanedione as an initiator intermediate [176].

Enzyme-catalyzed vinyl polymerizations have been demonstrated in recent years with significant control of polymer molecular weight and yield depending on reaction conditions [149]. Resins of poly(sodium acrylate) are used as water-absorbent materials for cleaning surfaces, in water and oil conditioning, personal care products, and disposable materials for medical applications [177].

Nylons

Polyamides, also called nylons, display improved physical properties compared with polyolefins and polyesters due to the directionally specific inter-chain hydrogen bonds and significantly enhanced melting points [178]. Nylons can be synthesized either by polycondensation or by ROP. Ragupathy *et al.* [179] presented two and three step polycondensation and ROP methods to enzymatically produce nylon-8,10, nylon-8,13, nylon-6,13, and nylon-12,13. He used 1,8-diaminooctane (DAO) and diethyl sebacate (DES) for the polycondensations and DAO and lactone ethyleneridecanoate (ETD) for ROP. Novozym 435 in diphenyl ether at 150 °C carried out the polymerizations. Monomer conversion reached 97% and 90% and polyamides synthesized had an Mw of 5,000 and 8,000 for polycondensation and ROP reactions respectively.

Acrylate and Styrene Polymerization

Methyl-methacrylate (MMA) was polymerized by en enzyme-catalyzed reaction using the ternary system (enzyme, oxidant, and initiator) in water and water-miscible organic cosolvents such as DMF, acetone, dioxane, and THF. Soybean peroxidase in aqueous solution afforded a 48% yield of poly(methyl methacrylate) (PMMA) with Mn = 9.3 × 10^5 and Mw/Mn = 6.8. When HRP II (type II) was used, a 45% yield of PMMA was obtained with Mn = 6.3 × 10^5 and Mw/Mn = 3.0. It was found that yields of PMMA increased when cosolvents with low dielectric constant like dioxane and THF were used [180].

Respect to styrenic polymerizations, HRP in water was used to polymerize styrene and some derivatives. The polymerization was investigated using several β-diketones as initiators, and proceeded in a mixed solution of H_2O/THF = 3 (v/v). Polystyrene synthesized using 2,5-cyclopentanedione as initiator resulted in 60% yield with Mn and Mw/Mn of 6.7 × 10^4 and 1.9 respectively. Polymer yield, Mn and Mw/Mn of the resulting polystyrene depended on the type of initiator used. Synthesis of polymers from styrene derivatives, 4-methylstyrene and 2-vinylnaphthalene were also investigated and afforded polystyrenes in a >90% yield of polymer with a Mn of 1.15 × 10^5 and Mw/Mn = 2.28 [181].

Ring-Opening Polymerizations (ROP)

Among the two major synthetic polymerization approaches, ring-opening polymerization (ROP) has been most extensively studied to polyester synthesis.

Polymerization of lactones via ring-opening catalysis carried out by lipases has been studied since the early 1990s. This catalytic activity for ROP of lactone monomers has been the research field on diverse lipases such as *Pseudomonas fluorescens*, *Pseudomonas cepacia*, *Candida rugosa*, *Candida cylindracea*, *Candida antarctica*, *Aspergillus niger*, *Mucor miehei*, *Penicillium roqueforti*, *Rhizopus japanicus*. *Yarrowia lipolytica*, *Carica papaya* and porcine pancreas lipase [18 2,183,184,185,186,187,188,189]. Cutinase from *Humicola insolens* (HiC) has also been reported to carry out ROP [190].

Poly(1,4-dioxan-2-one) (polyDO) is a desirable biocompatible polymer with good flexibility and tensile strength which might be considered for biomedical applications. This polyester was synthesized by ROP of 1,4-dioxan-2-one by lipase B from *Candida antarctica* at 60 °C, resulting in a polymer of Mw = 41,000 [191]. In another study, Jiang *et al.* [192] claimed to synthesize by ring-opening copolymerization a polyester from 1,4-dioxan-2-one (DO) with pentadecalactone (PDL) to give a copolyester of poly(DO-co-PDL) with Mw > 30,000. They used *Candida antarctica* B lipase in toluene or diphenyl ether at 70 °C for 26 h.

Lipase 2 from *Yarrowia lipolytica* and *Carica papaya* latex were recently employed to polymerize ε-CL in bulk. Lipase 2 from *Y. lipolytica* was found to be very efficient as catalyst for several reactions [193] and latex from *C. papaya* is a low-cost auto-immobilized biocatalyst. The polymerization yields were 74% and 40% with an Mw of 1,350 and 1,100 respectively. Reactions were realized at 150 °C and 6 h for *Y. lipolytica* and 70 °C and 24 h for *C. papaya* [182].

As to macrolides, lipase *Pseudomonas fluorescens* was able to polymerize 11-undecanolide (12-membered, UDL), 12-dodecanolide (13-membered, DDL), 15-pentadecanolide (16-membered, PDL), and 16-hexadecanolide (17-membered, HDL) [114,186,187,188,189,194,195,196]. Lipases from*Candida cylindracea* and *Pseudomonas fluorescens* catalyzed the ROP of UDL in bulk with quantitative yields affording polyesters of Mn of 23,000 and 25,000 [197,198]. In another study, ROP by lipases CA, lipase CC, lipase PC, lipase PF, or PPL in bulk at 45–75 °C for 5 days was done over HDL, the largest unsubstituted lactone monomer studied so far. The study showed the synthesis of polyHDL with Mn reaching to 5,800 in high yields [199]. It was discovered that a larger ring monomer, like PDL showed a greater reactivity towards polymerization than a smaller monomers like 1,4-polydioxan-2-one (DO) [192].

Ebata *et al.* obtained a poly(ε-CL) from a cyclic dimer of ε-CL (14-memberd) that was polymerized at 70 °C in toluene by lipase B from *Candida antarctica*, affording quantitatively a poly(ε-CL) with Mn of 18,000 [200]. This research Group, under the same strategy used to produce oligomers which are then subject to subsequent polymerization, claimed that a poly(butylene-succinate) (PBS) polyester of Mw 130,000 was obtained from cyclic dimers of butylene-succinate of Mn = 390. The reaction was carried out in toluene at 120 °C using 40% (w/w) of lipase *Candida antarctica* B for 24 h [201].

It was found that among ε-CL substituted monomers, ω-methyl ε-CL showed the least reactivity towards the lipase-catalytic ROP [202]. On the other hand, α-methyl or δ-methyl ε-CL afforded polymers of Mw = 8,400–11,000 with yields of 74% and 93% respectively [203].

An equilibrium phenomenon was observed between the synthesis of cyclic and linear polymers in a lipase-catalyzed ROP reaction of lactones. Cyclic oligomers in addition to major linear polyesters coexist in equilibrium [204,205]. This cyclic-linear equilibrium is solvent-dependent, for example Novozym 435 was used to catalyze the ROP of ε-CL at 60 °C in bulk or in an organic solvent. In bulk, the polymer products coexisted as a cyclic dimer of ε-CL and linear poly(ε-CL) in 2% and in 98% abundance, respectively. When the reaction was carried out in in acetonitrile, THF or 1,4-dioxane the same proportion was observed, as for example the equilibrium yields in acetonitrile were 70% for the cyclic oligomers (17% of dilactone and 53% of cyclic oligomers) and 30% for linear poly(ε-CL). When the reaction was carried out in isooctane, the proportion inverted to 3% and 97%, respectively, as in the bulk reaction [204].

Another hydrolase enzyme, a depolymerase, was found to catalyze ROP reactions *in vitro*. This depolymerase is a poly(hydroxybutyrate) (PHB) (EC 3.1.1.75), and was extracted from *Pseudomonas stutzeri*, *Alcaligenes faecalis* and *Pseudomonas lemoignei*. *A. faecalis* PHB depolymerase was assayed with a series of small, medium and large lactones and the best polimerization activity (93% yield) was obtained with β-butyrolactone, resulting in the formation of polyesters with a Mw = 16,000. On the other hand, medium and large lactones (ε-caprolactone, 11-undecanolide, and 12-dodecanolide), which are readily polymerized by lipases, were scarcely polymerized by PHB depolymerase [206].

Commodity Chemicals

One of the most noteworthy cases of biocatalytic production of a commodity chemical is the bioconversion of acrylonitrile to acrylamide [207]. This reaction, schematically simplified in Figure 7, is carried out by a nitrile hydratase whose catalytic activity transforms a cyanide group into an amide. Mitsubishi Rayon Co., Ltd. (Tokyo, Japan) currently produces over 20,000 tons annually of acrylamide using a third-generation biocatalyst, *Rhodococcus rhodochrous* J1, developed for commercial use by Nitto Chemical Industries. Acrylamide is produced from acrylonitrile in a continuous system of fixed-bed reactors at 10 °C with polyacrylamide-immobilized J1 cells. The process achieves conversion of acrylonitrile to acrylamide in ~99.9% yield, and the catalyst productivity is >7,000 g of acrylamide per g dry cell weight [208].

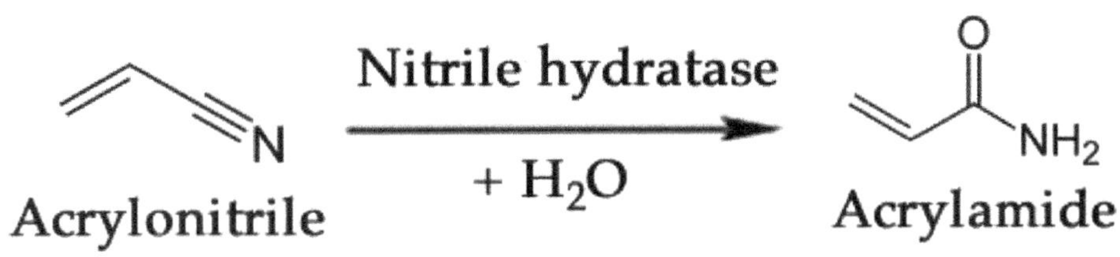

Figure 7: Enzymatic conversion of acrylonitrile to acrylamide by means of a nitrile hydratase.

Lonza Guangzhou Fine Chemicals produces nicotinamide through a chemoenzymatic process shown in Figure 8 [209,210]. They start from 2-methyl-1,5-diaminopentane, a byproduct from production of nylon-6,6, which is converted to 3-methylpyridine, which in turn is ammoxidized. The resulting 3-cyanopyridine is hydrolyzed to nicotinamide using immobilized *Rhodococcus rhodochrous* J1 cells [211,212]. The Plant's capacity is 3,400 tons per year of nicotinamide. The enzymatic process affords the desired amide at >99.3% selectivity at 100% conversion, whereas the chemical process requires caustic hydrolysis of 3-cyanopyridine but also produces 4% nicotinic acid as byproduct.

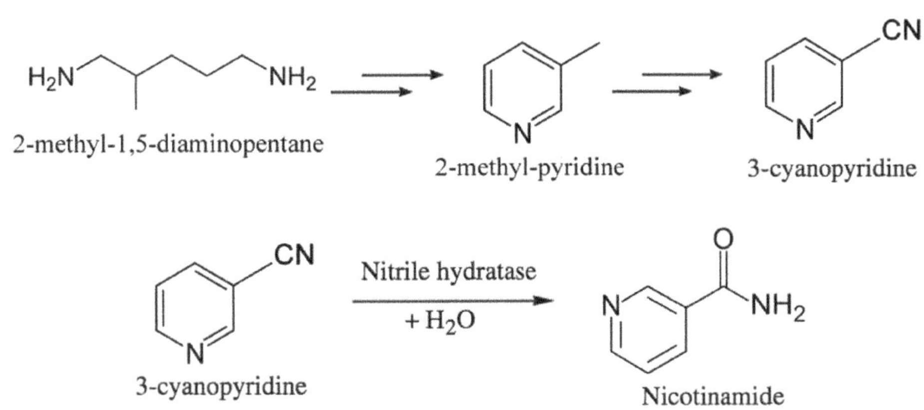

Figure 8: Chemoenzymatic route to obtain nicotinamide developed by Lonza [209,210].

A commodity chemical which is aimed to substitute to some extent the universally oil-derived terephthalic acid is 2,5-furandicarboxylic acid (FDC) [213]. Terephthalic acid is one of the main raw materials to produce polyesters. FDC relies in the supply of 5-hydroxymethylfurfural (HMF), which is the product of the dehydration of hexoses. Oxidation of HMF yields FDC and may be realized by using heterogeneous or electrochemical catalysis [214,215]. However, two biocatalytic processes have been devised to help in having a reliable and sustainable supply for HMF: a whole-cell biocatalytic approach using a recombinant *Pseudomonas putida* hosting an oxidoreductase from*Cupriavidus basilensis*. This process is able to produce 30 g/L of 2,5-furandicarboxylic acid from HMF having a yield of 0.97 mol/mol [216]. The productivity of the whole process is 0.21 g/(L·h) under aerobic fed-batch conditions. The second approach consists of the use of a chloroperoxidase and *C. antarctica* lipase B [217,218].

ENZYMES IN THE FINE CHEMICALS INDUSTRY

The role of Biocatalysis in the pharmaceutical and fine-chemical industries is clearly expanding. There are estimated to be around 150 implemented biocatalytic processes in industry, and the majority of these are in the pharmaceutical sector [219]. As stereoisomerism is quite relevant in pharmaceutics and fine chemistry, and a single swap in hydrogen position may mean a great difference in the bioactive function of a compound, Biocatalysis fits well to meet these challenges. Besides the enantio-, regio- and chemoselectivity that enzymes exhibit, their ability to perform complex catalysis procedures in a simple step, which otherwise, under a chemical approach, might require laborious treatments, toxic reagents or high energy input, has provided a strong basis for their usage in the fine chemicals industry. Some examples of the contribution of Biocatalysis to this field are described below.

A precursor of aspartame, *N*-(benzyloxycarbonyl)-L-aspartyl-L,-phenylalanine methyl ester (Z-APM) has been synthesized from *N*-(benzyloxycarbonyl)-L-aspartic acid (Z-L-ASP) and L-phenylalanine methyl ester (L-PM) with thermolysin [220]. By reverting the peptide-bond hydrolytic nature of thermolysin, this precursor is synthesized on a multi-thousand ton per year scale by a DSM/Tosoh joint venture (Holland Sweetener Company, Geleen, The Netherlands).

Pyrethroids now constitute the majority of commercial household insecticides. A novel industrial application of a lyase enzyme is to produce a pyrethroid intermedieate, (*S*)-*m*-phenoxy-benzaldehyde cyanohydrin (sPBC). Hydroxynitrile lyase, is an enzyme that catalyzes the addition of HCN to aldehydes and ketones [221]. Thus, the production of sPBC from *m*-phenoxybenzaldehyde by hydroxynitril lyase has been carried out on a multi-ton scale by DSM.

L-carnitine is a quaternary ammonium compound biosynthesized from the amino acids lysine and methionine [222]. Within mammal cells, L-carnitine is required for the transport of fatty acids from the cytosol into the mitochondria during the lipids oxidation [223]. Due to this role in lipid oxidation, L-carnitine is often advertised to improve fat metabolism, reduce fat mass, and increase muscle mass, which is widely available as an over-the-counter nutritional. L-carnitine is produced by Lonza (Basel, Switzerland) on an industrial scale using an enzyme from strain *Agrobacterium*HK1349 in a whole cell biotransformation process [224]. The process actually comprises a dehydrogenation of γ-butyrobetaine to 4-(trimethylamino)-butenoic acid followed by selective addition of a water molecule by L-carnitine lyase.

L-*tert*-leucine is an important chiral building block and intermediate of many ligands in chemo-catalysis in the pharmaceutical industry [225]. Menzel *et al.* [226] presented a process to produce L-*tert*-leucine in a whole-cell Biocatalysis system using a recombinant *E. coli* host expressing the two enzymes required by the process: a leucine hydrogenase and a formate dehydrogenase. The former catalyzes the main reaction, reductive amination of an α-keto acid, and the latter serves as a NADH regenerator. A yield of 84% in 24 h and an enantiomeric purity of ee > 99% was obtained. The process has been scaled up to tons level. Degussa-Hüls has used a similar whole-cell Biocatalysis system to produce L-*tert*-leucine, which has taken to a commercial scale [227].

Lonza produces 5-methylpyrazine-2-carboxylic acid on a commercial scale from the *p*-xylene analogue 2,5-dimethylpyrazine. This compound is used as a blood-lowering drug of the sulfenyl urea class commercialized as Glycotrol. The synthesis is carried out in *P. putida* ATCC33015 as a whole-cell biocatalyst, expressing a series of enzymes (a monooxygenase and two dehydrogenases) [224].

Japan Energy (Saitama, Japan) produces alkyloxiranes and phenyloxiranes on an industrial scale, especially the chiral building block 2-(trifluoromethyl)oxirane from 3,3,3-trifluoropropene through an oxidative reaction, using a whole-cell biocatalyst from *Nocardia corallina* [228].

Vanillin, the main component of vanillin extract, a natural flavoring, whose demand has long exceeded the supply, should be synthesized chemically from guaiacol, in order meet demand. Nevertheless, a whole-cell biocatalytic synthesis of vanillin from glucose has now been elaborated. Glucose is transformed into vanillic acid by a recombinant *E. coli* biocatalyst under fed-batch fermentor conditions. Then reduction of vanillic acid to vanillin is carried out by aryl aldehyde dehydrogenase isolated from *Neurospora crassa.* The biocatalytic route avoids the use of carcinogenic reagents and non-renewable petroleum derivatives (guaiacol) [229].

Aresta *et al.* used a carboxylase in the synthesis of 4-hydroxy benzoic acid, which is an intermediate for the synthesis of preservatives. The carboxylase enzyme is extracted from the bacteria *Thauera aromatica*, and it acts over the phenyl-phosphate moeity carboxylating it with 100% selectivity and 90% yield at 300 °K, P_{CO2} = 0.1 MPa [230].

Alfuzosin, a quinazoline derivative, against Benign prostatic hyperplasia (BPH), is synthesized through intermediate compund tetrahydro-*N*-[3-(methylamino)-propyl]-2-furan-carboxamide. This intermediate was traditionally produced in a three-steps chemical route from 2-tetrahydrofuroic acid involving a series of chemicals and distillation. The enzyme-catalyzed reaction with lipase B from *Candida antacrtica*, afforded the intermediate in two steps in which the same enzyme catalyzed the two reactions: esterification and amidation. This biocatalytic procedure simplyfied and made more environmentally friendly the previous chemical synthesis [231].

Another important intermediate in the synthesis of antiviral nucleosides such as 3'-deoxy-3'-azi-dothymidine (AZT), is thymidine. Production of thymidine is currently based in the hydrolysis of DNA from natural sources. A new biocatalytic approach for its production has been developed. It consists of transfering the 2'-deoxy-ribose moiety of 2'-deoxyinosine to thymine. The purine nucleoside phosphorylase of *Bacillus stearothermophilus* was used to carry out this reaction. A xantine oxidase is added in order to convert the concomitant hypoxantine produced to uric acid, and thus draw the thermodynamic equilibrium towards the synthesis of thymidine (Figure 9). The yield of thymidine of this whole-cell biocatalytic route reached 68% under mild conditions [232].

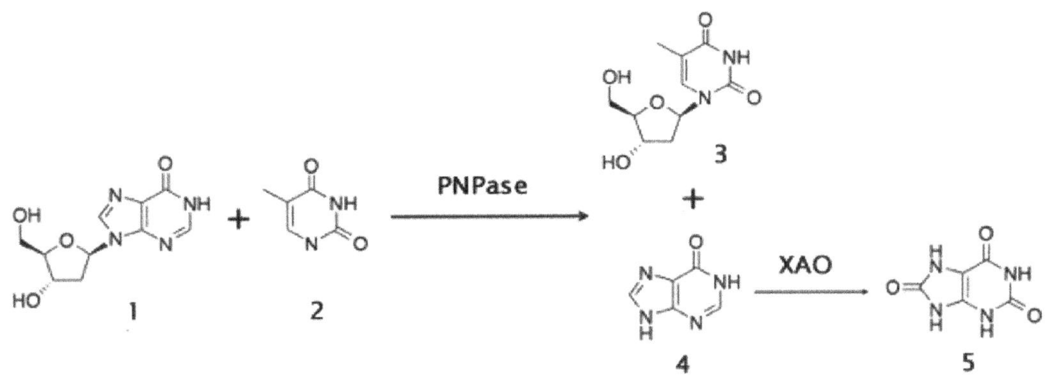

Figure 9: Biocatalytic synthesis of thymidine using a Purine Nucleoside Phosphorilase (PNPase) and the biocatalytical removal of hypoxantine with a Xanthine Oxidase (XAO) to obtain uric acid. Compounds: 1: 2'deoxyinosine; 2: Thymine; 3: Thymidine; 4: Hypoxanthine; 5: Uric acid (see also [233]).

Many catalytic functions that enzymes perform in the pharmaceutical industry is in resolution of racemic mixtures, in which one enantiomer is acylated or hydrolyzed selectively affording an easier separation of both enantiomers. One recent example of this is the resolution of pro-drug (R,S)-2-bromophenylacetic acid octyl ester. Rivera *et al.* [234] accomplished its resolution with high enantioselectivity using lipases embedded in crude latex from *Carica papaya*. An enantioselectivity E-value E > 200 was achieved using decane as solvent, 50 mM of substrate and 50 mg/mL enzyme/reaction medium. Furthermore, the E-value doubled after purifying latex and removing proteases.

CONCLUSIONS

As outlined above, Biocatalysis is currently employed in a number of processes and products in diverse fields, and certainly new areas of application will be added. Advances in computational power have enabled the advent of bioinformatics, genomic sequencing and powerful analytic methods in physics, chemistry and molecular biology, which have allowed us to understand the dynamics, *in vivo* and *in vitro*, of biomacromolecules. The convergence of these tools has led to an improved access to more biocatalysts and to a deeper knowledge of them. In the case of enzymes, it has allowed us to explore their behavior, stability, specificity and even to begin to modify their very own nature through protein engineering. Larger availability of biocatalysts with superior qualities has been the main force behind the development of new industrial biocatalytic processes. The continued progress and interest in enzymes, serves as stimulus to make further efforts and ensure a steady success in meeting new synthetic challenges.

As per the environmental and economic benefits already reaped by the use of Biocatalysis in industry, it is possible to tell that the use of Biocatalysis is only expected to grow. This might be achieved with isolated enzymes but will probably be more successful using genetically engineered organisms where new pathways might be designed. The development of more robust, versatile, efficient enzymes via protein engineering, faster optimization of reaction conditions, microscale processing and better capabilities in design and machining new equipment will open new perspectives for the manufacture of many more products via Biocatalysis.

Future developments in Biocatalysis for industrial applications should be directed into the expansion of capabilities of current biocatalysts, to obtain new biocatalytic activities able to transform the vast amount of non-natural compounds found in industry and to achieve productivities comparable to current chemical processes. There is no doubt that modern biological tools, such as protein engineering and directed evolution will play an important role in meeting these challenges. As biodiversity remains largely unexplored, developing high-throughput screening methods to discover new efficient enzymes will also impact upon the development of biocatalyzed process for production of biobased chemicals.

ACKNOWLEDGMENTS

We would like to thank to CONACYT for the scholarship provided to first author and for project CB 2008-01-104429.

CONFLICTS OF INTEREST

The authors declare no conflict of interest.

REFERENCES

1. Council of European Union. *En Route to the Knowledge-Based Bio-Economy*; German Presidency of the Council of European Union: Cologne, Germany, 2007.
2. Sandoval, G.; Plou, F.G. *Obtención enzimática de compuestos bioactivos a partir de recursos naturales iberoamericanos*; Consejo Superior de Instigaciones Cientificas (CSIC): Madrid, Spain, 2012; pp. 1–336.
3. Jahnz, U.; Schubert, M.; Baars, H.; Klaus, V. Process for producing the potential food ingredient DFA III from inulin-screening, genetic engineering, fermentation and immobilisation of inulase II. *Int. J. Pharm.* 2003, *256*, 199–206.
4. Uchiyama, T.; Niwa, S.; Tanaka, K. Purification and properties of *Arthrobacter ureafaciens*inulase II. *Biochim. Biophys. Acta* 1973, *315*, 412–420.
5. Kim, G.E.; Lee, S. Efficient production of DFA III (di-D-fructofuranose-dianhydride) from chicory root. In *Abstracts of the World Congress on Biotechnology*; DECHEMA: Berlin, Germany, 2000; Volume 272.

6. Yokota, A.; Hirayama, S.; Enomoto, K.; Miura, Y.; Takao, S.; Tomita, F. Production of inulin fructotransferase (depolymerizing) by *Arthrobacter* sp. H65–7 and preparation of DFA III from inulin by the enzyme. *J. Ferm. Bioeng* 1991, *72*, 258–261.
7. Kawamura, M.; Takahashi, S.; Uchiyama, T. Purification and some properties of inulin fructotransferase (depolymerizing) from Arthrobacter ilicis. *Agric. Biol. Chem.* 1988, *52*, 3209–3210.
8. Gosling, A.; Stevens, G.W.; Barber, A.R.; Kentish, S.E.; Gras, S.L. Recent advances refining galactooligosaccharide production from lactose. *Food Chem.* 2010, *121*, 307–318.
9. Rastall, R.A.; Gibson, G.R.; Gill, H.S.; Guarner, F.; Klaenhammer, T.R.; Pot, B.; Reid, G.; Rowland, I.R.; Sanders, M.E. Modulation of the microbial ecology of the human colon by probiotics, prebiotics and synbiotics to enhance human health: An overview of enabling science and potential applications. *FEMS Microbiol. Ecol.* 2005, *52*, 145–152.
10. Sanz, J.I. Production of Glactooligo-Sacchariddes from Lactose by Immobilized β-Galactosidase and Posterioir Chromatographic Spearation. Ph.D. Thesis, Ohio State University, Columbus, OH, USA, 2009.
11. Gullón, B.; Gómez, B.; Martínez-Sabajanesb, M.; Yáñez, R.; Parajó, J.C.; Alonso, J.L. Pectic oligosaccharides: Manufacturc and functional propcrtics. *Trends Food Sci. Technol.* 2013, *30*, 153–161.
12. Neri, D.F.M.; Balcão, V.M.; Costa, R.S.; Rocha, I.C.A.P.; Ferreira, E.M.F.C.; Torres, D.P.M.; Rodrigues, L.R.M.; Carvalho, L.B., Jr.; Teixeira, J.A. Galacto-oligosaccharides production during lactose hydrolysis by free *Aspergillus oryzae* β-galactosidase and immobilized on magnetic polysiloxane-polyvinyl alcohol. *Food Chem.* 2009, *115*, 92–99.
13. Barreteau, H.; Delattre, C.; Michaud, P. Production of oligosaccharides as promising new food additive generation. *Food Technol. Biotechnol.* 2006, *44*, 323–333.
14. Gaur, R.; Pant, H.; Jain, R.; Khare, S.K. Galacto-oligosaccharide synthesis by immobilized*Aspergillus oryzae* β-Galactosidase. *Food Chem.* 2006, *97*, 426–430.
15. Whisner, C.M.; Weaver, C.M. Galacto-oligosaccharides: Prebiotic effects on calcium absorption and bone health. In *Nutritional Influences on Bone Health*; Burckhardt, P., Dawson-Hughes, B., Weaver, C.M., Eds.; Springer: London, UK, 2013; pp. 315–323.
16. Perugino, G.; Trincone, A.; Rossi, M.; Moracci, M. Oligosaccharide synthesis by glycosynthases. *Trends Biotech.* 2004, *22*, 31–37.
17. Dias, L.G.; Veloso, A.C.A.; Correia, D.M.; Rocha, O.; Torres, D.; Rocha, I.; Rodrigues, L.R.; Peres, A.M. UV spectrophotometry method for the monitoring of galacto-oligosaccharides production. *Food Chem.* 2009, *113*, 246–252.
18. Rodríguez, B.; Poveda, A.; Jiménez, J.; Ballesteros, A.O.; Plou, F.J. Galacto-oligosaccharide synthesis from lactose solution or skim milk using the β-galactosidase from *Bacillus circulans*.*J. Agric. Food Chem.* 2012, *60*, 6391–6398.
19. Torres, D.; Gonçalves, M.D.P.F.; Teixeira, J.A.; Rodrigues, L. Galacto-oligosaccharides: Production, properties, applications, and significance as prebiotics. *Compr. Rev. Food Sci. F.*2010, *9*, 438–454.
20. Tzortzis, G.; Vulevic, J. Galacto-oligosaccharide prebiotics. In *Prebiotics and Probiotics Science and Technology*; Charalampopoulos, D., Rastall, R.A., Eds.; Springer: New York, NY, USA, 2009; pp. 207–244.
21. Park, H.Y.; Kim, H.J.; Lee, J.K.; Kim, D.; Oh, D.K. Galactooligosaccharide production by a thermostable β-galactosidase from *Sulfolobus solfataricus*. *World J. Microb. Biot.* 2008, *24*, 1553–1558.
22. Rodriguez, B.; de Abreu, M.A.; Fernandez, L.; de Beer, R.; Poveda, A.; Jimenez, J.; Haltrich, D.; Ballesteros, A.O.; Fernandez, M.; Plou, F.J. Production of galacto-oligosaccharides by the β-galactosidase from *Kluyveromyces lactis*: Comparative analysis of permeabilized cells *versus*soluble enzyme. *J. Agric. Food Chem.* 2011, *59*, 10477–10484.
23. Hsu, C.A.; Lee, S.L.; Chou, C.C. Enzymatic production of galactooligosaccharides by β-galactosidase from *Bifidobacterium longum* BCRC 15708. *J. Agric. Food Chem.* 2007, *55*, 2225–2230.
24. Hansson, T.; Adlercreutz, P. Optimization of galactooligosaccharide production from lactose using β-glycosidases from hyperthermophiles. *Food Biotechnol.* 2001, *15*, 79–97.
25. UBIC Europe Marketing Consulting. In *The World GOS Market*; UBIC Europe Press: Sierre, Switzerland, 2010.
26. Padalino, M.; Perez-Conesa, D.; López-Nicolás, R.; Frontela-Saseta, C.; Berruezo, G. Effect of fructooligosaccharides and galactooligosaccharides on the folate production of some folate-producing bacteria in media cultures or milk. *Int. Dairy J.* 2012, *27*, 27–33.
27. Fanaro, S.; Boehm, G.; Garssen, J.; Knol, J.; Mosca, F.; Stahl, B.; Vigi, V. Galacto-oligosaccharides and long-chain fructo-oligosaccharides as prebiotics in infant formulas: A review. *Acta Paediatr. Suppl.* 2005, *94*, 22–26.
28. Watzl, B.; Girrbach, S.; Roller, M. Inulin, oligofructose and immunomodulation. *Br. J. Nutr.*2005, *93*, S49–S55.
29. Bornet, F.R.; Brouns, F.; Tashiro, Y.; Duvillier, V. Nutritional aspects of short-chain fructooligosaccharides: Natural occurrence, chemistry, physiology and health implications.*Dig. Liver Dis.* 2002, *34*, S111–S120.
30. Yun, J.W. Fructooligosaccharides—Occurrence, preparation, and application. *Enzyme Microb. Tech.* 1996, *19*, 107–117.
31. Arrizon, J.; Urias-Silvas, J.E.; Sandoval, G.; Mancilla-Margalli, N.A.; Gschaedler, A.C.; Morel, S.; Monsan, P. Production and bioactivity of fructan-type oligosaccharides. In *Food Oligosaccharides: Production, Analysis and Bioactivity*; Moreno, F.J., Sanz, M.L., Eds.; Wiley-Blackwell: Hoboken, NJ, USA, 2013; in press.

32. Beine, R.; Morarua, R.; Nimtz, M.; Na'amniehc, S.; Pawlowskic, A.; Buchholz, K.; Seibel, J. Synthesis of novel fructooligosaccharides by substrate and enzyme engineering. *J. Biotechnol.*2008, *138*, 33–41.
33. Castillo, E.; López-Munguía, A. Synthesis of levan in water-miscible organic solvents. *J. Biotechnol.* 2004, *114*, 209–217.
34. Vega, R.J.; Zúniga, M.E. Potential application of commercial enzyme preparations for industrial production of short-chain fructooligosaccharides. *J. Mol. Catal. B Enzym.* 2012, *76*, 44–51.
35. Singh, R.S.; Singh, R.P. Production of fructooligosaccharides from inulin by endoinulinases and their prebiotic potential. *Food Technol. Biotech.* 2010, *48*, 435–450.
36. Tanriseven, A.; Aslan, Y. Immobilization of Pectinex Ultra SP-L to produce fructooligosaccharides. *Enzyme Microb. Technol.* 2005, *36*, 550–554.
37. Guío, F.; Rodríguez, M.A.; Alméciga, C.J.; Sánchez, O.F. Recent trends in fructooligosaccharides production. *Recent Pat. Food Nutr. Agric.* 2009, *1*, 221–230.
38. Jala, R.C.R.; Hu, P.; Yang, T.; Jiang, Y.; Zheng, Y; Xu, X. Lipases as biocatalysts for the synthesis of structured lipids. In *Lipases and Phospholipases. Methods in Molecular Biology*; Sandoval, G., Ed.; Springer-Humana Press: New York, USA, 2012; Volume 861, Chapter 23.
39. Xu, X.; Akoh, C.C. Enzymatic production of Betapol and other specialty fats. In *Lipid Biotechnology*; Marcel Dekker: New York, NY, USA, 2002; pp. 461–478.
40. Odle, J. New insights into the utilization of medium-chain triglycerides by the neonate: Observations from a piglet model. *J. Nutr.* 1997, *127*, 1061–1067.
41. Bugaut, M. Occurrence, absorption and metabolism of short chain fatty acids in the digestive tract of mammals. *Comp. Biochem. Physiol. B Biochem. Mol. Biol.* 1987, *86*, 439–472.
42. Osborn, H.T.; Akoh, C. Structured lipids: Novel fats with medical, nutraceutical, and food applications. *Compr. Rev. Food Sci. F.* 2002, *1*, 110–120.
43. Xu, X.; Hoy, C.E.; Balchen, S.; Adler-Nissen, J. Specific-Structured Lipid: Nutritional Perspectives and Production Potentials. In Proceedings of International Symposium on the Approach to Functional Cereals and Oils, CCOA, Beijing, China, 9–14 November 1997.
44. Trivedi, R.; Singh, R.P. Modification of oils and fats to produce structured lipids. *J. Oleo Sci.*2005, *54*, 423–430.
45. Hellner, G.; Tőke, E.R.; Nagy, V.; Szakács, G.; Poppe, L. Integrated enzymatic production of specific structured lipid and phytosterol ester compositions. *Process Biochem.* 2010, *45*, 1245–1250.
46. Lee, K.T.; Akoh, C.C. Characterization of enzymatically synthesized structured lipids containing eicosapentaenoic, docosahexaenoic, and caprylic acids. *J. Am. Oil Chem. Soc.* 1998,*75*, 495–499.
47. Akoh, C.C. Structured lipids. In *Food Lipids Chemistry, Nutrition, and Biotechnology*; Akoh, C.C., Min, D.B., Eds.; Marcel Dekker: New York, NY, USA, 1998; pp. 699–727.
48. Iwasaki, Y.; Yamane, T. Enzymatic synthesis of structured lipids. *J. Mol. Catal. B Enzym.* 2000,*10*, 129–140.
49. Khodadadi, M.; Aziz, S.; St-Louis, R.; Kermash, S. Lipase-catalyzed synthesis and characterization of flaxseed oil-based structured lipids. *J. Funct. Foods* 2013, *5*, 424–433.
50. Casas-Godoy, L.; Marty, A.; Sandoval, G.; Ferreira-Dias, S. Optimization of medium chain length fatty acid incorporation into olive oil catalyzed by immobilized Lip2 from *Yarrowia lipolytica*. *Biochem. Eng. J.* 2013, *77*, 20–27.
51. Villeneuve, P.; Barouh, N.; Baréa, B.; Piombo, G.; Figueroa-Espinoza, M.C.; Turon, M.C.; Pina, M.; Lago, R. Chemoenzymatic synthesis of structured triacylglycerols with conjugated linoleic acids (CLA) in central position. *Food Chem.* 2007, *100*, 1443–1452.
52. Nagachinta, S.; Akoh, C.C. Enrichment of palm olein with long chain polyunsaturated fatty acids by enzymatic acidolysis. *LWT Food Sci. Technol.* 2012, *46*, 29–35.
53. Tecelão, C.; Rivera, I.; Sandoval, G.; Ferreira-Dias, S. *Carica papaya* latex: A low-cost biocatalyst for human milk fat substitutes production. *Eur. J. Lipid Sci. Technol.* 2012, *114*, 266–276.
54. Nagao, T.; Watanabe, Y.; Maruyama, M.; Momokawa, Y.; Kishimoto, N.; Shimada, Y. One-pot enzymatic synthesis of docosahexaenoic acid-rich triacylglycerols at the sn-1(3) position using by-product from selective hydrolysis of tuna oil. *New Biotechnol.* 2011, *28*, 7–13.
55. Buchholtz, K.; Volker, K.; Borscheuer, U.T. *Biocatalysts and Enzyme Technology*; Wiley-VCH: Berlin, Germany, 2005; pp. 1–465.
56. Bornscheuer, U.T.; Kazlauskas, R.J. *Hydrolases in Organic Synthesis—Regio- and Stereoselective Biotransformations*; Wiley-VCH: Berlin, Germany, 1999; pp. 1–355.
57. Cao, L. *Carrier-Bound Immobilized Enzymes*; Wiley-VCH: Berlin, Germany, 2005; pp. 1–578.
58. Gog, A.; Roman, M.; Toşa, M.; Paizs, C.; Irimie, F.D. Biodiesel production using enzymatic transesterification—Current state and perspectives. *Renew. Energ.* 2012, *39*, 10–16.
59. Véras, I.C.; Silva, F.A.; Ferrão-Gonzales, A.D.; Moreau, V.H. One-step enzymatic production of fatty acid ethyl ester from high-acidity waste feedstocks in solvent-free media. *Bioresour. Technol.* 2011, *102*, 9653–9658.
60. Fan, X.; Niehus, X.; Sandoval, G. Lipases as biocatalyst for biodiesel production. In *Lipases and Phospholipases. Methods in Molecular Biology*; Sandoval, G., Ed.; Springer-Humana Press: New York, NY, USA, 2012; Volume 861, Chapter 27.
61. Fukuda, H.; Kondo, A.; Noda, H. Biodiesel fuel production by transesterification of oils. *J. Biosci. Bioeng.* 2001, *92*, 405–416.
62. Maleki, E.; Aroua, M.K.; Sulaiman, N.M.N. Improved yield of solvent free enzymatic methanolysis of palm and jatropha oils blended with castor oil. *Appl. Energ.* 2013, *104*, 905–909.

63. Hama, S.; Kondo, A. Enzymatic biodiesel production: An overview of potential feedstocks and process development. *Bioresour. Technol.* 2013, *135*, 386–395.

64. Al-Zuhair, S.; Hasan, M.; Ramachandran, K.B. Kinetics of the enzymatic hydrolysis of palm oil by lipase. *Process Biochem.* 2003, *38*, 1155–1163.

65. Salis, A.; Pinna, M.; Monduzzi, M.; Solinas, V. Biodiesel production from triolein and short chain alcohols through biocatalysis. *J. Biotechnol.* 2005, *119*, 291–299.

66. Shimada, Y.; Watanabe, Y.; Samukawa, T.; Sugihara, A.; Noda, H.; Fukuda, H.; Tominaga, Y. Conversion of vegetable oil to biodiesel using immobilized *Candida antarctica* lipase. *J. Am. Oil Chem. Soc.* 1999, *76*, 789–793.

67. Li, Q.; Xu, J.; Du, W.; Li, Y.; Liu, D. Ethanol as the acyl acceptor for biodiesel production.*Renew. Sust. Energ. Rev.* 2013, *25*, 742–748.

68. Modi, M.K.; Reddy, J.R.; Rao, B.V.; Prasad, R.B. Lipase mediated conversion of vegetable oils into biodiesel using ethyl acetate as acyl acceptor. *Bioresour. Technol.* 2007, *98*, 1260–1264.

69. Rivera, I.; Villanueva, G.; Sandoval, G. Biodiesel production from animal grease wastes by enzymatic catalysis. *Grasas Aceites* 2009, *60*, 468–474.

70. Royon, D.; Daz, M.; Ellenrieder, G.; Locatelli, S. Enzymatic production of biodiesel from cotton seed oil using t-butanol as a solvent. *Bioresour. Technol.* 2007, *98*, 648–653.

71. Abigor, R.D.; Uadia, P.O.; Foglia, T.A.; Haas, M.J.; Jones, K.C.; Okpefa, E.; Obibuzor, J.U.; Bafor, M.E. Lipase-catalysed production of biodiesel fuel from some Nigerian lauric oils.*Biochem. Soc. Trans.* 2000, *28*, 979–981.

72. Nelson, L.; Foglia, T.; Marmer, W.N. Lipase-catalyzed production of biodiesel. *J. Am. Oil Chem. Soc.* 1996, *73*, 1191–1195.

73. Linko, Y.Y.; Lamsä, M.; Huhtala, A.; Rantanen, O. Lipase biocatalysis in the production of esters. *J. Am. Oil Chem. Soc.* 1995, *72*, 1293–1299.

74. Du, W.; Xu, Y.; Liu, D.; Zeng, J. Comparative study on lipase-catalyzed transformation of soybean oil for biodiesel production with different acyl acceptors. *J. Mol. Catal. B Enzym.*2004, *30*, 125–129.

75. Watanabe, Y.; Shimada, Y.; Sugihara, A.; Noda, H.; Fukuda, H.; Tominaga, Y. Continuous production of biodiesel fuel from vegetable oil using immobilized *Candida antarctica* lipase.*J.Am. Oil Chem. Soc.* 2000, *77*, 355–360.

76. Zhang, B.; Weng, Y.; Xu, H.; Mao, Z. Enzyme immobilization for biodiesel production. *Appl. Microbiol. Biotechnol.* 2012, *93*, 61–70.

77. Soumanou, M.M.; Bornscheuer, U.T. Lipase-catalyzed alcoholysis of vegetable oils. *Eur. J. Lipid Sci. Technol.* 2003, *105*, 656–660.

78. Zhu, X.; Zhou, T.; Wu, X.; Cai, Y.; Yao, D.; Xie, C.; Liu, D. Covalent immobilization of enzymes within micro-aqueous organic media. *J. Mol. Catal. B Enzym* 2011, *3–4*, 145–149.

79. Zaks, A.; Klibanov, A.M. Enzyme-catalyzed processes in organic solvents. *Proc. Natl. Acad. Sci. USA* 1985, *82*, 3192–3196.

80. Iso, M.; Chen, B.; Eguchi, M.; Kudo, T.; Shrestha, S. Production of biodiesel fuel from triglycerides and alcohol using immobilized lipase. *J. Mol. Catal. B Enzym.* 2001, *16*, 53–58.

81. Chen, J.W.; Wu, W.T. Regeneration of immobilized *Candida antarctica* lipase for transesterification. *J. Biosci. Bioeng.* 2003, *95*, 466–469.

82. Nie, K.; Xie, F.; Wang, F.; Tan, T. Lipase catalyzed methanolysis to produce biodiesel: optimization of the biodiesel production. *J. Mol. Catal. B Enzym.* 2006, *43*, 142–147.

83. Ciftci, O.N.; Temelli, F. Enzymatic conversion of corn oil into biodiesel in a batch supercritical carbon dioxide reactor and kinetic modeling. *J. Supercrit. Fluids.* 2013, *75*, 172–180.

84. Lin, Y.C.; Yang, P.M.; Chen, S.C.; Lin, J.F. Improving biodiesel yields from waste cooking oil using ionic liquids as catalysts with a microwave heating system. *Fuel Process. Technol.* 2013,*115*, 57–62.

85. Earle, M.J.; Plechkova, N.V.; Seddon, K.R. Green synthesis of biodiesel using ionic liquids.*Pure Appl. Chem.* 2009, *81*, 2045–2057.

86. Ha, S.H.; Lan, M.N.; Lee, S.H.; Hwang, S.M.; Koo, Y.M. Lipase-catalyzed biodiesel production from soybean oil in ionic liquids. *Enzyme Microb. Technol.* 2007, *41*, 480–483.

87. Koda, R.; Numata, T.; Hama, S.; Tamalampudi, S.; Nakashima, K.; Tanaka, T; Ogino, C.; Fukuda, H.; Kondo, A. Ethanolysis of rapeseed oil to produce biodiesel fuel catalyzed by*Fusarium heterosporum* lipase-expressing fungus immobilized whole-cell biocatalysts. *J. Mol. Catal. B Enzym.* 2010, *68*, 101–104.

88. Adachi, D.; Koha, F.; Hamab, S.; Ogino, C.; Kondo, A. A robust whole-cell biocatalyst that introduces a thermo- and solvent-tolerant lipase into *Aspergillus oryzae* cells: Characterization and application to enzymatic biodiesel production. *Enzyme Microb. Technol.*2013, *52*, 331–335.

89. Fukuda, H.; Hama, S.; Tamalampudi, S.; Noda, H. Whole-cell biocatalysts for biodiesel fuel production. *Trends Biotechnol.* 2008, *26*, 668–673.

90. Atkinson, B.; Black, G.M.; Lewis, P.J.S.; Pinches, A. Biological particles of given size, shape, and density for use in biological reactors. *Biotechnol. Bioeng.* 1979, *21*, 193–200.

91. Meng, X.; Yang, J.; Xu, X.; Zhang, L.; Nie, Q.; Xian, M. Biodiesel production from oleaginous microorganisms. *Renew. Energ.* 2009, *34*, 1–5.

92. Turner, N.J.; Truppo, M.D. Biocatalysis enters a new era. *Curr. Opin. Chem. Biol.* 2013, *17*, 212–214.

93. Woodley, J. New opportunities for biocatalysis: Making pharmaceutical processes greener.*Trends Biotechnol.* 2008, *26*, 321–327.

94. Kittl, R.; Withers, S.G. New approaches to enzymatic glycoside synthesis through directed evolution. *Carbohyd. Res.* 2010, *345*, 1272–1279.

95. Hay, A.S. Polymerization by oxidative coupling: Discovery and commercialization of PPO®.*J. Polym. Sci. Pol. Chem.* 1998, *36*, 505–517.
96. He, F.; Li, S.; Garreau, H.; Vert, M.; Zhuo, R. Enzyme-catalyzed polymerization and degradation of copolyesters of ε-caprolactone and γ-butyrolactone. *Polymer* 2005, *46*, 12682–12688.
97. Salvachúa, D.; Prieto, A.; Mattinen, M.L.; Tamminen, T.; Liitiä, T.; Lille, M.; Willför, S.; Martínez, A.T.; Martínez, M.J.; Faulds, C.B. Versatile peroxidase as a valuable tool for generating new biomolecules by homogeneous and heterogeneous cross-linking. *Enzyme Microb. Technol.* 2013, *52*, 303–311.
98. Oguchi, T.; Tawaki, S.; Uyama, H.; Kobayashi, S. Enzymatic synthesis of soluble polyphenol.*Bull. Chem. Soc. Jpn.* 2000, *73*, 1389–1396.
99. Dordick, J.S.; Marletta, M.A.; Klibanov, A.M. Polymerization of phenols catalyzed by peroxidase in nonaqueous media. *Biotech. Bioeng.* 1987, *30*, 31–36.
100. Oguchi, T.; Tawaki, S.; Uyama, H.; Kobayashi, S. Soluble polyphenol. *Macromol. Rapid Commun.* 1999, *20*, 401–403.
101. Mita, N.; Oguchi, T.; Tawaki, S.; Uyama, H.; Kobayashi, S. Control of structure and molecular weight of polyphenols in enzymatic oxidative polymerization. *Polymer Prepr.* 2000, *41*, 223–224.
102. Kurioka, H.; Komatsu, I.; Uyama, H.; Kobayashi, S. Enzymatic oxidative polymerization of alkylphenols. *Macromol. Rapid Commun.* 1994, *15*, 507–510.
103. Uyama, H.; Kurioka, H.; Kobayashi, S. Preparation of polyphenol particles by dispersion polymerization using enzyme as catalyst. *Chem. Lett.* 1995, *24*, 795–796.
104. Uyama, H. Enzymatic polymerization. In *Future Directions in Biocatalysis*; Matsuda, T., Ed.; Elsevier Science: Cambridge, MA, USA, 2007; pp. 1–2.
105. Reihmann, M.; Ritter, H. Synthesis of phenol polymers using peroxidases. *Adv. Polym. Sci.*2006, *194*, 1–49.
106. Kobayashi, S.; Uyama, H.; Ikeda, R. Artificial Urushi. *Chem. Eur. J.* 2001, *7*, 4755–4760.
107. Kim, Y.H.; An, E.S.; Song, B.K.; Kim, D.S.; Chelikani, R. Polymerization of cardanol using soybean peroxidase and its potential application as anti-biofilm coating material. *Biotechnol. Lett.* 2003, *25*, 1521–1524.
108. Tonami, H.; Uyama, H.; Kobayashi, S.; Kubota, M. Peroxidase-catalyzed oxidative polymerization of m-substituted phenol derivatives. *Macromol. Chem. Phys.* 1999, *200*, 2365–2371.
109. Kadota, J.; Fukuoka, T.; Uyama, H.; Hasegawa, K.; Kobayashi, S. New Positive-type photoresists based on enzymatically synthesized polyphenols. *Macromol. Rapid Commun.*2004, *25*, 441–444.
110. Antoniotti, S.; Santhanam, L.; Ahuja, D.; Hogg, M.G.; Dordick, J.S. Structural diversity of peroxidase-catalyzed oxidation products of o-methoxyphenols. *Org. Lett.* 2004, *6*, 1975–1978.
111. Mita, N.; Tawaki, S.I.; Uyama, H.; Kobayashi, S. Laccase-catalyzed oxidative polymerization of phenols. *Macromol. Biosci.* 2003, *3*, 253–257.
112. Ikeda, R.; Sugihara, J.; Uyama, H.; Kobayashi, S. Poly(2,6-dihydroxy-l,4-oxyphenylene synthesis of a new poly(phenylene oxide) derivative. *Polym. Bull.* 1997, *38*, 273–277.
113. Ikeda, R.; Sugihara, J.; Uyama, H.; Kobayashi, S. Enzymatic oxidative polymerization of 2,6-dimethylphenol. *Macromolecules* 1996, *29*, 8702–8705.
114. Uyama, H.; Kurioka, H.; Sugihara, J.; Komatsu, I.; Kobayashi, S. Oxidative polymerization of p-alkylphenols catalyzed by horseradish peroxidase. *J. Polym. Sci. Pol. Chem.* 1997, *35*, 1453–1459.
115. Wang, P.; Martin, B.D.; Parida, S.; Rethwisch, D.G.; Dordick, J.S. Multienzymic synthesis of poly(hydroquinone) for use as a redox polymer. *J. Am. Chem. Soc.* 1995, *117*, 12885–12886.
116. Akkara, J.A.; Senecal, K.J.; Kaplan, D.L. Synthesis and characterization of polymers produced by horseradish peroxidase in dioxane. *J. Polym. Sci. Pol. Chem.* 1991, *29*, 1561–1574.
117. Akita, M.; Tsutsumi, D.; Kobayashi, M.; Kise, H. Structural change and catalytic activity of horseradish peroxidase in oxidative polymerization of phenol. *Biosci. Biotech. Bioch.* 2001, *65*, 1581–1588.
118. Angerer, P.S.; Studer, A.; Witholt, B.; Li, Z. Oxidative polymerization of a substituted phenol with ion-paired horseradish peroxidase in an organic solvent. *Macromolecules* 2005, *38*, 6248–6250.
119. Eker, B.; Zagorevski, D.; Zhu, G.; Linhardt, R.J.; Dordick, J.S. Enzymatic polymerization of phenols in room-temperature ionic liquids. *J. Mol. Catal. B Enzym.* 2009, *59*, 177–184.
120. Mita, N.; Tawaki, S.; Uyama, H.; Kobayashi, S. Enzymatic oxidative polymerization of phenol in an aqueous solution in the presence of a catalytic amount of cyclodextrin. *Macromol. Biosci.*2002, *3*, 127–130.
121. Kim, Y.J.; Uyama, H.; Kobayashi, S. Regioselective synthesis of poly(phenylene) as a complex with poly(ethylene glycol) by template polymerization of phenol in water. *Macromolecules*2003, *36*, 5058–5060.
122. Kommareddi, N.S.; Tata, M.; Karayigitoglu, C.; John, V.T.; McPherson, G.L.; Herman, M.F.; Oconnor, C.J.; Lee, Y.S.; Akkara, J.A.; Kaplan, D.L. Enzymatic polymerizations using surfactant microstructures and the preparation of polymer-ferrite composites. *Appl. Biochem. Biotechnol.* 1995, *51–52*, 241–252.
123. Ghan, R.; Shutava, T.; Patel, A.; John, V.T.; Lvov, Y. Enzyme-catalyzed polymerization of phenols within polyelectrolyte microcapsules. *Macromolecules* 2004, *37*, 4519–4524.
124. Marín, F.R.; Frutos, M.J.; Pérez-Alvarez, J.A.; Martinez-Sánchez, F.; del Río, J.A. Flavonoids as nutraceuticals: Structural related antioxidant properties and their role on ascorbic acid preservation. *Stud. Nat. Prod. Chem.* 2002, *26*, 741–778.
125. Di Carlo, G.; Mascolo, N.; Izzo, A.A.; Capasso, F. Flavonoids: Old and new aspects of a class of natural therapeutic drugs. *Life Sci.* 1999, *65*, 337–353.

126. Rice-Evans, C.A.; Miller, N.J.; Paganga, G. Structure-antioxidant activity relationships of flavonoids and phenolic acids. *Free Radic. Biol. Med.* 1996, *20*, 933–956.

127. *Flavonoids in Biology and Medicine III—Current Issues in Flavonoids Research*; Das, N.P., Cheeseman, K.H., Eds.; Informa Healthcare: London, UK, 1991; Volume 14, pp. 77–78.

128. Mejias, L.; Reihmann, M.H.; Sepulveda-Boza, S.; Ritter, H. New polymers from natural phenols using horseradish or soybean peroxidase. *Macromol. Biosci.* 2002, *2*, 24–32.

129. Kurisawa, M.; Chung, J.E.; Uyama, H.; Kobayashi, S. Laccase-catalyzed synthesis and antioxidant property of poly(catechin). *Macromol. Biosci.* 2003, *3*, 758–764.

130. Kurisawa, M.; Chung, J.E.; Uyama, H.; Kobayashi, S. Oxidative coupling of epigallocatechin gallate amplifies antioxidant activity and inhibits xanthine oxidase activity. *Chem. Commun.*2004.

131. Kurisawa, M.; Chung, J.E.; Kim, Y.J.; Uyama, H.; Kobayashi, S. Amplification of antioxidant activity and xanthine oxidase inhibition of catechin by enzymatic polymerization.*Biomacromolecules* 2003, *4*, 469–471.

132. Gonçalves, I.; Matamá, T.; Cavaco-Paulo, A.; Silva, C. Laccase coating of catheters with poly(catechin) for biofilm reduction. *Biocatal. Biotransform* 2013. in press.

133. Kurisawa, M.; Chung, J.E.; Uyama, H.; Kobayashi, S. Enzymatic synthesis and antioxidant properties of poly(rutin). *Biomacromolecules* 2003, *4*, 1394–1399.

134. Božič, M.; Gorgieva, S.; Kokol, V. Laccase-mediated functionalization of chitosan by caffeic and gallic acids for modulating antioxidant and antimicrobial properties. *Carbohyd. Polym.*2012, *87*, 2388–2398.

135. Brzonova, I.; Steiner, W.; Zankel, A.; Nyanhongo, G.S. Enzymatic synthesis of catechol and hydroxyl-carboxic acid functionalized chitosan microspheres for iron overload therapy. *Eur. J. Pharm. Biopharm.* 2011, *79*, 294–303.

136. Fras-Zemljič, L.; Kokol, V.; Čakara, D. Antimicrobial and antioxidant properties of chitosan-based viscose fibres enzymatically functionalized with flavonoids. *Text. Res. J.* 2011, *81*, 1532–1540.

137. Sousa, F.; Guebitz, G.M.; Kokol, V. Antimicrobial and antioxidant properties of chitosan enzymatically functionalized with flavonoids. *Process Biochem.* 2009, *44*, 749–756.

138. Pina-Luis, G.; Rosquete-Pina, G.; Valdés, A.C.; Ochoa, A.; Rivero, I.; Díaz-García, M.E. Morin functionalized Merrifield's resin: A new material for enrichment and sensing heavy metals.*React. Funct. Polym.* 2012, *72*, 61–68.

139. Donato, L.; Chiappetta, G.; Drioli, E. Naringin-imprinted polymer layer using photo polymerization method. *Sep. Sci. Technol.* 2011, *46*, 1555–1562.

140. Spizzirri, U.G.; Parisi, O.I.; Iemma, F.; Cirillo, G.; Puoci, F.; Curcio, M.; Picci, N. Antioxidant-polysaccharide conjugates for food application by eco-friendly grafting procedure. *Carbohyd. Polym.* 2010, *79*, 333–334.

141. Jaeger, K.E.; Eggert, T. Lipases for biotechnology. *Curr. Opin. Biotechnol.* 2002, *13*, 390–397.

142. Sharma, R.; Chisti, Y.; Banerjee, U.C. Production, purification, characterization and applications of lipases. *Biotechnol. Adv.* 2001, *19*, 627–662.

143. Yu, Y.; Wu, D.; Liu, C.; Zhao, Z.; Yang, Y.; Li, Q. Lipase/esterase-catalyzed synthesis of aliphatic polyesters via polycondensation: A review. *Process Biochem.* 2012, *47*, 1027–1036.

144. Yang, Y.; Yu, Y.; Zhang, Y.; Liu, C.; Shi, W.; Li, Q. Lipase/esterase-catalyzed ring-opening polymerization: A green polyester synthesis technique. *Process Biochem.* 2011, *46*, 1900–1908.

145. He, F.; Wang, Y.P.; Liu, G.; Jia, H.L.; Feng, J.; Zhuo, R.X. Synthesis, characterization and ring-opening polymerization of a novel six-membered cyclic carbonate bearing pendent allyl ether group. *Polymer* 2008, *49*, 1185–1190.

146. Runge, M.; O'Hagan, D.; Haufe, G. Lipase-catalyzed polymerization of fluorinated lactones and fluorinated hydroxycarboxylic acids. *J. Polym. Sci. Pol. Chem.* 2000, *38*, 2004–2012.

147. Zinck, P. One-step synthesis of polyesters specialties for biomedical applications. *Rev. Environ. Sci. Biotechnol.* 2009, *8*, 231–234.

148. Kobayashi, S.; Makino, A. Enzymatic polymer synthesis: An opportunity for green polymer chemistry. *Chem. Rev.* 2009, *109*, 5288–5353.

149. Uyama, H.; Takamoto, T.; Kobayashi, S. Enzymatic synthesis of polyesters in ionic liquids.*Polym. J.* 2002, *34*, 94–96.

150. Chaudhary, A.K.; Beckman, E.J.; Russell, A.J. Rational control of polymer molecular weight and dispersity during enzyme-catalyzed polyester synthesis in supercritical fluids. *J. Am. Chem. Soc.* 1995, *117*, 3728–3733.

151. Okumura, S.; Iwai, M.; Tominaga, T. Synthesis of ester oligomer by *Aspergillus niger* lipase.*Agr. Biol. Chem. (Tokyo)* 1984, *48*, 2805–2813.

152. Binns, F.; Roberts, S.M.; Taylor, A.; Williams, C.F. Enzymic polymerization of unactivated diol/diacid system. *J. Chem. Soc. Perkin Trans. 1* 1993, *1*, 899–904.

153. Linko, Y.Y.; Seppala, J. Producing high molecular weight biodegradable polyesters. *Chem. Tech.* 1996, *26*, 25–31.

154. Gómez-Patiño, M.B.; Cassani, J.; Jaramillo-Flores, M.E.; Zepeda-Vallejo, L.G.; Sandoval, G.; Jimenez-Estrada, M.; Arrieta-Baez, D. Oligomerization of 10,16-dihydroxyhexadecanoic acid and methyl 10,16-dihydroxyhexadecanoate catalyzed by lipases. *Molecules* 2013, *18*, 9317–9333.

155. Kobayashi, S.; Uyama, H.; Suda, S.; Namekawa, S. Dehydration polymerization in aqueous medium catalyzed by lipase. *Chem. Lett.* 1997, *26*, 105–107.

156. Poojari, Y.; Palsule, A.S.; Cai, M.; Clarson, S.J.; Gross, R.A. Synthesis of organosiloxane copolymers using enzymatic polyesterification. *Eur. Polym. J.* 2008, *44*, 4139–4145.

157. Yang, Y.; Lu, W.; Cai, J.; Hou, Y.; Ouyang, S.; Xie, W.; Gross, R.A. Poly(oleicdiacid-co-glycerol): Comparison of polymer structure resulting from chemical and lipase catalysis.*Macromolecules* 2011, *44*, 1977–1985.

158. Korupp, C.; Weberskirch, R.; Muller, J.J.; Liese, A.; Hilterhaus, L. Scaleup of lipase-catalyzed polyester synthesis. *Org. Process. Res. Dev.* 2010, *14*, 1118–1124.

159. Fehling, E.; Bergander, K.; Klein, E.; Weber, N.; Vosmann, K. Thiol-functionalized copolymeric polyesters by lipase-catalyzed esterification and transesterification of 1,12-dodecanedioic acid and its diethyl ester, respectively, with 1-thioglycerol. *Biotechnol. Lett.*2010, *32*, 1463–1471.

160. Hu, J.; Gao, W.; Kulshrestha, A.S.; Gross, R.A. Sweet polyesters: Lipase-catalyzed condensation polymerization of alditols. *Macromolecules* 2006, *39*, 6789–6792.

161. Uyama, H.; Kobayashi, S. Enzymatic synthesis of polyesters via polycondensation. *Adv. Polym. Sci.* 2006, *194*, 133–158.

162. Brazwell, E.M.; Filos, D.Y.; Morrow, C.J. Biocatalytic synthesis of polymers. III. Formation of a high molecular weight polyester through limitation of hydrolysis by enzyme-bound water and through equilibrium control. *J. Polym. Sci. Pol. Chem.* 1995, *33*, 89–95.

163. Kobayashi, S. Recent developments in lipase-catalyzed synthesis of polyesters. *Macromol. Rapid Commun.* 2009, *30*, 237–266.

164. Azim, H.; Dekhterman, A.; Jiang, Z.; Gross, R.A. *Candida antarctica* lipase B-catalyzed synthesis of poly(butylene succinate): Shorter chain building blocks also work.*Biomacromolecules* 2006, *7*, 3093–3097.

165. Mesiano, A.J.; Beckman, E.J.; Russell, A.J. Biocatalytic synthesis of fluorinated polyesters.*Biotechnol. Prog.* 2000, *16*, 64–68.

166. Uyama, H.; Yaguchi, S.; Kobayashi, S. Lipase-catalyzed polycondensation of dicarboxylic acid-divinyl esters and glycols to aliphatic polyesters. *J. Polym. Sci. Pol. Chem.* 1999, *37*, 2737–2745.

167. Yao, D.; Li, G.; Kuila, T.; Li, P.; Kim, N.H.; Kim, S.I.; Lee, J.H. Lipase-catalyzed synthesis and characterization of biodegradable polyester containing l-malic acid unit in solvent system. *J. Appl. Polym. Sci.* 2011, *120*, 1114–1120.

168. Steunenberg, P.; Uiterweerd, M.; Sijm, M.; Scott, E.L.; Zuilhof, H.; Sanders, J.P.M.; Franssen, M.C.R. Enzyme-catalyzed polymerization of β-alanine esters, a sustainable route towards the formation of poly-β-alanine. *Curr. Org. Chem.* 2013, *17*, 682–690.

169. Kumar, D.; Bhalla, T.C. Microbial proteases in peptide synthesis: Approaches and applications. *Appl. Microbiol. Biotechnol.* 2005, *68*, 726–736.

170. Kuhn, D.; Durrschmidt, P.; Mansfeld, J.; Ulbrich-Hofmann, R. Boilysin and thermolysin in dipeptide synthesis: A comparative study. *Biotechnol. Appl. Biochem.* 2002, *36*, 71–76.

171. Bille, V.; Ripak, C.; Assche, I.; Forni, L.; Degelaen, J.; Searso, J. Semi-enzymic synthesis of somatostatin. In Proceedings of 21st European Peptide Symposium, Platja d'Aro, Spain, 2–8 September 1990.

172. Gu, Q.M.; Maslanka, W.W.; Cheng, H.N. Enzyme-catalyzed polyamides and their derivatives. In *Polymer Biocatalysis and Biomaterials II*; ACS Symposium Series 999; Cheng, H.N., Gross, R.A., Eds.; Oxford University Press: Washington, DC, USA, 2008; pp. 309–319.

173. Poulhès, F.; Mouysset, D.; Gil, G.; Bertrand, M.P.; Gastaldi, S. Speeding-up enzyme-catalyzed synthesis of polyamides using ω-amino-a-alkoxy-acetate as monomer. *Polymer* 2013, *54*, 3467–3471.

174. Ikeda, R.; Tanaka, H.; Uyama, H.; Kobayashi, S. Laccase-catalyzed polymerization of acrylamide. *Macromol. Rapid Commun.* 1998, *19*, 423–425.

175. Hollmann, F.; Gumulya, Y.; Tölle, C.; Liese, A.; Thum, O. Evaluation of the laccase from*Myceliophthora thermophila* as industrial biocatalyst for polymerization reactions.*Macromolecules* 2008, *41*, 8520–8524.

176. Emery, O.; Lalot, T.; Brigodiot, M.; Maréchal, E. Free-radical polymerization of acrylamide by horseradish peroxidase-mediated initiation. *J. Polym. Sci. Pol. Chem.* 1997, *35*, 3331–3333.

177. Mucientes, A.E.; Santiago, F.; Carrero, A.M.; Talavera, B. Superabsorbent hydrogels of poly(sodium acrylate) with crude and exfoliated vermiculites. *J. Polym. Eng.* 2013, *33*, 61–69.

178. Jones, N.A.; Atkins, E.D.T.; Hill, M.J.; Cooper, S.J.; Franco, L. Chain-folded lamellar crystals of aliphatic polyamides. comparisons between nylons 44, 64, 84, 104, and 124. *Macromolecules*1996, *29*, 6011–6018.

179. Ragupathy, L.; Ziener, U.; Dyllick-Brenzinger, R.; von Vacano, B.; Landfester, K. Enzyme-catalyzed polymerizations at higher temperatures: Synthetic methods to produce polyamides and new poly(amide-co-ester)s. *J. Mol. Catal. B Enzym.* 2012, *76*, 94–105.

180. Kalra, B.; Gross, R.A. Horseradish peroxidase mediated free radical polymerization of methyl methacrylate. *Biomacromolecules* 2000, *1*, 501–505.

181. Singh, A.; Ma, D.C.; Kaplan, D.L. Enzyme-mediated free radical polymerization of styrene.*Biomacromolecules* 2000, *1*, 592–596.

182. Sandoval, G.; Rivera, I.; Barrera-Rivera, K.A.; Martínez-Richa, A. Biopolymer synthesis catalyzed by tailored lipases. *Macromol. Symp.* 2010, *289*, 135–139.

183. Varma, I.K.; Albertsson, A.C.; Rajkhowa, R.; Srivastava, R.K. Enzyme catalyzed synthesis of polyesters. *Prog. Polym. Sci.* 2005, *30*, 949–981.

184. Kobayashi, S. Enzymatic polymerization: A new method of polymer synthesis. *J. Polym. Sci. Pol. Chem.* 1999, *37*, 3041–3056.

185. Kobayashi, S.; Uyama, H.; Ohmae, M. Enzymatic polymerization for precision polymer synthesis. *Bull. Chem. Soc. Jpn.* 2001, *74*, 613–635.

186. Uyama, H.; Kikuchi, H.; Takeya, K.; Kobayashi, S. Lipase-catalyzed ring-opening polymerization and copolymerization of 15-pentadecanolide. *Acta Polym.* 1996, *47*, 357–360.

187. Uyama, H.; Takeya, K.; Kobayashi, S. Enzymatic ring-opening polymerization of lactones to polyesters by lipase catalyst: Unusually high reactivity of macrolides. *Bull. Chem. Soc. Jpn.*1995, *68*, 56–61.

188. Uyama, H.; Takeya, K.; Hoshi, N.; Kobayashi, S. Lipase-catalyzed ring-opening polymerization of 12-dodecanolide. *Macromolecules* 1995, *28*, 7046–7050.

189. Bisht, K.S.; Henderson, L.A.; Gross, R.A. Enzyme-catalyzed ring-opening polymerization of ω-pentadecalactone. *Macromolecules* 1997, *30*, 2705–2711.

190. Hunsen, M.; Azim, A.; Mang, H.; Wallner, S.R.; Ronkvist, A.; Xie, W.C.; Gross, R.A. A cutinase with polyester synthesis activity. *Macromolecules* 2007, *40*, 148–150.

191. Nishida, H.; Yamashita, M.; Nagashima, M.; Endo, T.; Tokiwa, Y. Synthesis of metal-free poly(1,4-dioxan-2-one) by enzyme-catalyzed ring-opening polymerization. *J. Polym. Sci. Pol. Chem.* 2000, *38*, 1560–1567.

192. Jiang, Z.; Azim, H. Lipase-catalyzed copolymerization of ω-pentadecalactone with *p*-dioxanone and characterization of copolymer thermal and crystalline properties.*Biomacromolecules* 2007, *8*, 2262–2269.

193. Fickers, P.; Marty, A.; Nicaud, J.M. The lipases from *Yarrowia lipolytica*: Genetics, production, regulation, biochemical characterization and biotechnological applications. *Biotechnol. Adv.*2011, *29*, 632–644.

194. Uyama, H.; Kikuchi, H.; Takeya, K.; Hoshi, N.; Kobayashi, S. Immobilized lipase showing high catalytic activity toward enzymatic ring-opening polymerization of macrolides. *Chem. Lett.* 1996, *25*, 107–108.

195. Kobayashi, S.; Uyama, H. Precision enzymatic polymerization to polyesters with lipase catalysts. *Macromol. Symp.* 1999, *144*, 237–246.

196. Kobayashi, S.; Uyama, H.; Namekawa, S. *In vitro* biosynthesis of polyesters with isolated enzymes in aqueous systems and organic solvents. *Polym. Degrad. Stabil.* 1998, *59*, 195–201.

197. Kumar,A.;Kalra,B.;Dekhterman,A.;Gross,R.A.Efficient ring-opening polymerization and copolymerization of ε-caprolactone and ω-pentadecalactone catalyzed by *Candida antartica*lipase B. *Macromolecules* 2000, *33*, 6303–6309.

198. Kobayashi, S.; Uyama, H.; Namekawa, S.; Hayakawa, H. Enzymatic ring-opening polymerization and copolymerization of 8-octanolide by lipase catalyst. *Macromolecules* 1998,*31*, 5655–5659.

199. Namekawa, S.; Uyama, H.; Kobayashi, S. Lipase-catalyzed ring-opening polymerization of 16-hexadecanolide. *P. Jpn. Acad. B* 1998, *74*, 65–68.

200. Ebata, H.; Toshima, K.; Matsumura, S. Lipase-catalyzed transformation of poly(ε-caprolactone) into cyclic dicaprolactone. *Biomacromolecules* 2000, *1*, 511–514.

201. Sugihara, S.; Toshima, K.; Matsumura, S. New strategy for enzymatic synthesis of high-molecular-weight poly(butylene succinate) via cyclic oligomers. *Macromol. Rapid Commun.*2006, *27*, 203–207.

202. Kikuchi, H.; Uyama, H.; Kobayashi, S. Lipase-catalyzed ring-opening polymerization of substituted lactones. *Polym. J.* 2002, *34*, 835–893.

203. Küllmer, K.; Kikuchi, H.; Uyama, H.; Kobayashi, S. Lipase-catalyzed ring-opening polymerization of α-methyl-δ-valerolactone and α-methyl-ε-caprolacton. *Macromol. Rapid. Commun.* 1998, *19*, 127–130.

204. Cordova, A.; Iversen, T.; Martinelle, M. Lipase-catalysed formation of macrocycles by ring-opening polymerisation of ϵ-caprolactone. *Polymer* 1998, *39*, 6519–6524.

205. Namekawa, S.; Uyama, H.; Kobayashi, S. Lipase-catalyzed ring-opening and copolymerization of β-propiolactione. *Polym. J.* 1996, *28*, 730–731.

206. Suzuki, Y.; Taguchi, S.; Hisano, T.; Toshima, K.; Matsumura, S.; Doi, Y. Correlation between structure of the lactones and substrate specificity in enzyme-catalyzed polymerization for the synthesis of polyesters. *Biomacromolecules* 2003, *4*, 537–543.

207. Nagasawa, T.; Shimizu, H.; Yamada, H. The superiority of the third-generation catalyst,*Rhodococcus rhodochrous* J1 nitrile hydratase, for industrial production of acrylamide. *Appl. Microbiol. Biotechnol.* 1993, *40*, 189–195.

208. Thomas, S.M.; DiCosimo, R.; Nagarajan, V. Biocatalysis: Applications and potentials for the chemical industry. *Trends Biotechnol.* 2002, *20*, 238–242.

209. Chassin, C. A biotechnological process for the production of nicotinamide. *Chim. Oggi.* 1996,*14*, 9–12.

210. Nagasawa, T.; Matthew, C.D.; Mauger, J.; Yamada, H. Nitrile hydratase-catalyzed production of nicotinamide from 3-cyanopyridine in *Rhodococcus rhodochrous* J1. *Appl. Environ. Microbiol.*1988, *54*, 1766–1769.

211. Robins, K.T.; Nagasawa, T. Process for Preparing Amides. PCT Int. Appl. US 7,666,635 B2, 23 February 2010.

212. Heveling, J.; Armbruster, E.; Utiger, L.; Rhoner, M.; Dettwiler, H.R.; Chuck, R.J. Process for Preparing Nicotinamide. US Patent 5,719,045, 17 February 1998.

213. Straathof, A.J.J. Transformation of biomass into commodity chemicals using enzymes or cells. *Chem. Rev.* 2013.

214. Vuyyuru, K.R.; Strasser, P. Oxidation of biomass derived 5-hydroxymethylfurfural using heterogeneous and electrochemical catalysis. *Catal. Today* 2012, *195*, 144–154.

215. Casanova, O.; Iborra, S.; Corma, A. Biomass into chemicals: Aerobic oxidation of 5-hydroxymethyl-2-furfural into 2,5-furandicarboxylic acid with gold nanoparticle catalysts.*ChemSusChem* 2009, *2*, 1138–1144.

216. Koopman, F.; Wierckx, N.; de Winde, J.H.; Ruijssenaars, H.J. Efficient whole-cell biotransformation of 5-(hydroxymethyl)furfural into FDCA, 2,5-furandicarboxylic acid.*Bioresour. Technol.* 2010, *101*, 6291–6296.

217. Krystof, M.; Pérez-Sánchez, M.; Domínguez de María, P. Lipase-mediated selective oxidation of furfural and 5-hydroxymethylfurfural. *ChemSusChem* 2013, *6*, 826–830.

218. Van Deurzen, M.P.J.; van Rantwijk, F.; Sheldon, R.A. Chloroperoxidase-catalyzed oxidation of 5-hydroxymethylfurfural. *J. Carbohydr. Chem.* 1997, *16*, 299–309.

219. Panke, S.; Wubbolts, M. Advances in biocatalytic synthesis of pharmaceutical intermediates.*Curr. Opin. Chem. Biol.* 2005, *9*, 188–194.

220. Ooshima, H.; Mori, H.; Harano, Y. Synthesis of aspartame precursor by solid thermolysin in organic solvent. *Biotechnol. Lett.* 1985, *7*, 789–792.

221. Griengl, H.; Klempier, N.; Pochlauer, P.; Schmidt, M.; Shi, N.Y.; Zabelinskaja-Mackova, A.A. Enzyme catalysed formation of (*S*)-cyanohydrins derived from aldehydes and ketones in a biphasic solvent system. *Tetrahedron* 1998, *54*, 14477–14486.

222. Bieber, L.L. Carnitine. *Annu. Rev. Biochem.* 1988, *57*, 261–283.

223. Fritz, I.B. Action of carnitine on long chain fatty acid oxidation by liver. *Am. J. Physiol.* 1959,*197*, 297–304.

224. Meyer, H.P.; Kiener, A.; Imwinkelried, R.; Shaw, N. Biotransformations for fine chemicals production. *Chimia* 1997, *51*, 287–289.

225. Bommarius, A.S.; Schwarm, M.; Stingl, K.; Kottenhahn, M.; Huthmacher, K.; Drauz, K. Synthesis and use of enantiomerically pure *tert*-leucine. *Tetrahedron Asymmetry* 1995, *6*, 2851–2888.

226. Menzel, A.; Werner, H.; Altenbuchner, J.; Gröger, H. From enzymes to "designer bugs" in reductive amination: A new process for the synthesis of L-*tert*-leucine using a whole cell-catalyst. *Eng. Life Sci.* 2004, *4*, 573–576.

227. Krix, G.; Bommarius, A.S.; Drauz, K.; Kottenhahn, M.; Schwarm, M.; Kula, M.R. Enzymatic reduction of α-keto acids leading to L-amino acids, D- or L-hydroxy acids. *J. Biotechnol.* 1997,*53*, 29–39.

228. Furuhashi, K. Biological routes to optically active epoxides. In *Chirality in Industry*; Collins, A.N., Sheldrake, G.N., Crosby, J.C., Eds.; John Wiley & Sons: NJ, USA, 1992; pp. 167–186.

229. Li, K.; Frost, J. Synthesis of vanillin from glucose. *J. Am. Chem. Soc.* 1998, *120*, 10545–10546.

230. Aresta, M.; Quaranta, E.; Liberio, R.; Dileo, C.; Tommasi, I. Enzymatic synthesis of 4-OH-benzoic acid from phenol and CO_2: The first example of a biotechnological application of a carboxylase enzyme. *Tetrahedron* 1998, *54*, 8841–8846.

231. Baldessari, A. Lipases as catalysts in synthesis of fine chemicals. In *Lipases and Phospholipases. Methods in Molecular Biology*; Sandoval, G., Ed.; Springer-Humana Press: New York, NY, USA, 2012; Volume 861, pp. 445–448.

232. Pal, S.; Nair, V. Enzymatic synthesis of thymidine using bacterial whole cells and isolated purine nucleoside phosphorylase. *Biocatal. Biotransform.* 1997, *15*, 147–158.

233. Liese, A.; Villela-Filho, M. Production of fine chemicals using biocatalysis. *Curr. Opin. Biotechnol.* 1999, *10*, 595–603.

234. Rivera, I.; Mateos, J.C.; Marty, A.; Sandoval, G.; Duquesne, S. Lipase from Carica papaya latex presents high enantioselectivity toward the resolution of prodrug (*R,S*)-2-bromophenylacetic acid octyl ester. *Tetrahedron Lett.* 2013, *54*, 5523–5526

INDEX

N

O

P

R

S

T

U

V

W

X

Y